Applied Probability and Statistics (Continued)

CHAKRAVARTI, LAHA and ROY · Handbook of Methods of Applied Statistics, Vol. I
CHAKRAVARTI, LAHA and ROY · Handbook of Methods of Applied Statistics, Vol. II
CHERNOFF and MOSES · Elementary Decision Theory
CHIANG · Introduction to Stochastic Processes in Biostatistics
CLELLAND, deCANI, BROWN, BURSK, and MURRAY · Basic Statistics with Business Applications, *Second Edition*
COCHRAN · Sampling Techniques, *Second Edition*
COCHRAN and COX · Experimental Designs, *Second Edition*
COX · Planning of Experiments
COX and MILLER · The Theory of Stochastic Processes
DANIEL and WOOD · Fitting Equations to Data
DAVID · Order Statistics
DEMING · Sample Design in Business Research
DODGE and ROMIG · Sampling Inspection Tables, *Second Edition*
DRAPER and SMITH · Applied Regression Analysis
ELANDT-JOHNSON · Probability Models and Statistical Methods in Genetics
FLEISS · Statistical Methods for Rates and Proportions
GOLDBERGER · Econometric Theory
GUTTMAN, WILKS and HUNTER · Introductory Engineering Statistics, *Second Edition*
HAHN and SHAPIRO · Statistical Models in Engineering
HALD · Statistical Tables and Formulas
HALD · Statistical Theory with Engineering Applications
HOEL · Elementary Statistics, *Third Edition*
HOLLANDER and WOLFE · Nonparametric Statistical Methods
HUANG · Regression and Econometric Methods
JOHNSON and KOTZ · Distributions in Statistics
Discrete Distributions
Continuous Univariate Distributions-1
Continuous Univariate Distributions-2
Continuous Multivariate Distributions
JOHNSON and LEONE · Statistics and Experimental Design: In Engineering and the Physical Sciences, Volumes I and II
LANCASTER · The Chi Squared Distribution
LEWIS · Stochastic Point Processes
MILTON · Rank Order Probabilities: Two-Sample Normal Shift Alternatives
OTNES and ENOCHSON · Digital Time Series Analysis
RAO and MITRA · Generalized Inverse of Matrices and Its Applications
SARD and WEINTRAUB · A Book of Splines
SARHAN and GREENBERG · Contributions to Order Statistics
SEAL · Stochastic Theory of a Risk Business
SEARLE · Linear Models
THOMAS · An Introduction to Applied Probability and Random Processes

continued on back

Testing Statistical Hypotheses

A WILEY PUBLICATION IN MATHEMATICAL STATISTICS

Testing Statistical Hypotheses

E. L. LEHMANN
Professor of Statistics
University of California, Berkeley

New York · John Wiley & Sons, Inc.
London · Sydney

8 9 10

ISBN 0 471 52470 0

LIBRARY OF CONGRESS CATALOG CARD NUMBER: 59-11803

PRINTED IN THE UNITED STATES OF AMERICA

To Susanne

Preface

A mathematical theory of hypothesis testing in which tests are derived as solutions of clearly stated optimum problems was developed by Neyman and Pearson in the 1930's and since then has been considerably extended. The purpose of the present book is to give a systematic account of this theory and of the closely related theory of confidence sets, together with their principal applications. These include the standard one- and two-sample problems concerning normal, binomial, and Poisson distributions; some aspects of the analysis of variance and of regression analysis (linear hypothesis); certain multivariate and sequential problems. There is also an introduction to nonparametric tests, although here the theoretical approach has not yet been fully developed. One large area of methodology, the class of methods based on large-sample considerations, in particular χ^2 and likelihood ratio tests, essentially has been omitted because the approach and the mathematical tools used are so different that an adequate treatment would require a separate volume. The theory of these tests is only briefly indicated at the end of Chapter 7.

At present the theory of hypothesis testing is undergoing important changes in at least two directions. One of these stems from the realization that the standard formulation constitutes a serious oversimplification of the problem. The theory is therefore being re-examined from the point of view of Wald's statistical decision functions. Although these investigations throw new light on the classical theory, they essentially confirm its findings. I have retained the Neyman-Pearson formulation in the main part of this book but have included a discussion of the concepts of general decision theory in Chapter 1 to provide a basis for giving a broader justification of some of the results. It also serves as a background for the development of the theories of hypothesis testing and confidence sets.

Of much greater importance is the fact that many of the problems, which traditionally have been formulated in terms of hypothesis testing, are in reality multiple decision problems involving a choice between several decisions when the hypothesis is rejected. The development of suitable procedures for such problems is at present one of the most important tasks of statistics and is finding much attention in the current literature. However, since most of the work so far has been tentative, I have preferred to present the traditional tests even in cases in which the majority of the applications appear to call for a more elaborate procedure, adding only a warning regarding the limitations of this approach. Actually, it seems likely that the tests will remain useful because of their simplicity even when a more complete theory of multiple decision methods is available.

The natural mathematical framework for a systematic treatment of hypothesis testing is the theory of measure in abstract spaces. Since introductory courses in real variables or measure theory frequently present only Lebesgue measure, a brief orientation with regard to the abstract theory is given in Sections 1 and 2 of Chapter 2. Actually, much of the book can be read without knowledge of measure theory if the symbol $\int p(x)\,d\mu(x)$ is interpreted as meaning either $\int p(x)\,dx$ or $\Sigma p(x)$, and if the measure theoretic aspects of certain proofs together with all occurrences of the letters a.e. (almost everywhere) are ignored. With respect to statistics, no specific requirements are made, all statistical concepts being developed from the beginning. On the other hand, since readers will usually have had previous experience with statistical methods, applications of each method are indicated in general terms but concrete examples with data are not included. These are available in many of the standard textbooks.

The problems at the end of each chapter, many of them with outlines of solutions, provide exercises, further examples, and introductions to some additional topics. There is also given at the end of each chapter an annotated list of references regarding sources, both of ideas and of specific results. The notes are not intended to summarize the principal results of each paper cited but merely to indicate its significance for the chapter in question. In presenting these references I have not aimed for completeness but rather have tried to give a usable guide to the literature.

An outline of this book appeared in 1949 in the form of lecture notes taken by Colin Blyth during a summer course at the University of California. Since then, I have presented parts of the material in courses at Columbia, Princeton, and Stanford Universities and several times at the University of California. During these years I greatly benefited from comments of students and I regret that I cannot here thank them individually. At different stages of the writing I received many helpful suggestions from W. Gautschi, A. Høyland, and L. J. Savage, and particularly from Mrs. C. Striebel, whose critical reading of the next to final version of the manuscript resulted in many improvements. Also, I should like to mention gratefully the benefit I derived from many long discussions with Charles Stein.

It is a pleasure to acknowledge the generous support of this work by the Office of Naval Research; without it the book would probably not have been written. Finally, I should like to thank Mrs. J. Rubalcava, who typed and retyped the various drafts of the manuscript with unfailing patience, accuracy, and speed.

E. L. LEHMANN

Berkeley, Calif.
June, 1959

Contents

CHAPTER 1

The General Decision Problem

1. STATISTICAL INFERENCE AND STATISTICAL DECISIONS

The raw material of a statistical investigation is a set of observations; these are the values taken on by random variables X whose distribution P_θ is at least partly unknown. Of the parameter θ, which labels the distribution, it is assumed known only that it lies in a certain set Ω, the *parameter space*. *Statistical inference* is concerned with methods of using this observational material to obtain information concerning the distribution of X or the parameter θ with which it is labeled. To arrive at a more precise formulation of the problem we shall consider the purpose of the inference.

The need for statistical analysis stems from the fact that the distribution of X, and hence some aspect of the situation underlying the mathematical model, is not known. The consequence of such a lack of knowledge is uncertainty as to the best mode of behavior. To formalize this, suppose that a choice has to be made between a number of alternative actions. The observations, by providing information about the distribution from which they came, also provide guidance as to the best decision. The problem is to determine a rule which, for each set of values of the observations, specifies what decision should be taken. Mathematically such a rule is a function δ, which to each possible value x of the random variables assigns a decision $d = \delta(x)$, that is, a function whose domain is the set of values of X and whose range is the set of possible decisions.

In order to see how δ should be chosen, one must compare the consequences of using different rules. To this end suppose that the consequence of taking decision d when the distribution of X is P_θ is a *loss*, which can be expressed as a nonnegative real number $L(\theta, d)$. Then the long-term average loss that would result from the use of δ in a number of repetitions of the experiment is the expectation $E[L(\theta, \delta(X))]$ evaluated under the assumption that P_θ is the true distribution of X. This expectation, which depends on the decision rule δ and the distribution P_θ, is called the *risk function* of δ and will be denoted by $R(\theta, \delta)$. By basing the decision on the observations, the original problem of choosing a decision d with loss function $L(\theta, d)$ is thus replaced by that of choosing δ where the loss is now $R(\theta, \delta)$.*

The above discussion suggests that the aim of statistics is the selection of a decision function which minimizes the resulting risk. As will be seen later, this statement of aims is not sufficiently precise to be meaningful; its proper interpretation is in fact one of the basic problems of the theory.

2. SPECIFICATION OF A DECISION PROBLEM

The methods required for the solution of a specific statistical problem depend quite strongly on the three elements that define it: the class $\mathscr{P} = \{P_\theta, \theta \in \Omega\}$ to which the distribution of X is assumed to belong; the structure of the space D of possible decisions d; and the form of the loss function L. In order to obtain concrete results it is therefore necessary to make specific assumptions about these elements. On the other hand, if the theory is to be more than a collection of isolated results, the assumptions must be broad enough either to be of wide applicability or to define classes of problems for which a unified treatment is possible.

Consider first the specification of the class $\mathscr{P}$. Precise numerical assumptions concerning probabilities or probability distributions are usually not warranted. However, it is frequently possible to assume that certain events have equal probabilities and that certain others are statistically independent. Another type of assumption concerns the relative order of certain infinitesimal probabilities, for example the probability of occurrences in an interval of time or space as the length of the interval tends to zero. The following classes of distributions are derived on the basis of only such assumptions, and are therefore applicable in a great variety of situations.

The *binomial* distribution $b(p, n)$ with

$$(1) \quad P(X = x) = \binom{n}{x} p^x (1 - p)^{n-x}, \qquad x = 0, \cdots, n; \quad 0 \leqq p \leqq 1.$$

This is the distribution of the total number of successes in n independent trials when the probability of success for each trial is p.

* Sometimes, aspects of a decision rule other than the expectation of its loss are also taken into account.

The *Poisson* distribution $P(\tau)$ with

$$(2) \qquad P(X = x) = \frac{\tau^x}{x!} e^{-\tau}, \qquad x = 0, 1, \cdots; \quad 0 < \tau.$$

This is the distribution of the number of events occurring in a fixed interval of time or space if the probability of more than one occurrence in a very short interval is of smaller order of magnitude than that of a single occurrence, and if the numbers of events in nonoverlapping intervals are statistically independent. Under these assumptions, the process generating the events is called a *Poisson process.**

The *normal* distribution $N(\xi, \sigma^2)$ with probability density

$$(3) \quad p(x) = \frac{1}{\sqrt{2\pi}\,\sigma} \exp\left[-\frac{1}{2\sigma^2}(x - \xi)^2\right], \qquad -\infty < x, \xi < \infty; 0 < \sigma.$$

Under very general conditions, which are made precise by the central limit theorem, this is the approximate distribution of the sum of a large number of independent random variables when the relative contribution of each term to the sum is small.

We consider next the structure of the decision space D. The great variety of possibilities is indicated by the following examples.

Example 1. Let $X_1, \cdots, X_n$ be a *sample* from one of the distributions (1)–(3), that is, let the X's be distributed independently and identically according to one of these distributions. Let θ be p, τ, or the pair (ξ, σ) respectively, and let $\gamma = \gamma(\theta)$ be a real-valued function of θ.

(i) If one wishes to decide whether or not γ exceeds some specified value γ_0, the choice lies between the two decisions $d_0: \gamma > \gamma_0$ and $d_1: \gamma \leqq \gamma_0$. In specific applications these decisions might correspond to the acceptance or rejection of a lot of manufactured goods, of an experimental airplane as ready for flight testing, of a new treatment as an improvement over a standard one, etc. The loss function of course depends on the application to be made. Typically, the loss is 0 if the correct decision is chosen, while for an incorrect decision the losses $L(\gamma, d_0)$ and $L(\gamma, d_1)$ are increasing functions of $|\gamma - \gamma_0|$.

(ii) At the other end of the scale is the much more detailed problem of obtaining a numerical estimate of γ. Here a decision d of the statistician is a real number, the estimate of γ, and the losses might be $L(\gamma, d) = v(\gamma)w(|d - \gamma|)$ where w is a strictly increasing function of the error $|d - \gamma|$.

(iii) An intermediate case is the choice between the three alternatives $d_0: \gamma < \gamma_0$, $d_1: \gamma > \gamma_1$, $d_2: \gamma_0 \leqq \gamma \leqq \gamma_1$, for example accepting a new treatment, rejecting it, or recommending it for further study.

* Such processes are discussed in the books by Feller, *An Introduction to Probability Theory and Its Applications*, Vol. 1, New York, John Wiley & Sons, 2nd ed., 1957, and by Doob, *Stochastic Processes*, New York, John Wiley & Sons, 1953.

The distinction illustrated by this example is the basis for one of the principal classifications of statistical methods. Two-decision problems such as (i) are usually formulated in terms of *testing a hypothesis* which is to be accepted or rejected (see Chapter 3). It is the theory of this class of problems with which we shall be mainly concerned. The other principal branch of statistics is the theory of *point estimation* dealing with such problems as (ii). The investigation of *multiple-decision procedures* illustrated by (iii) has only begun in recent years.

Example 2. Suppose that the data consist of samples X_{ij}, $j = 1, \cdots, n_i$, from normal populations $N(\xi_i, \sigma^2)$, $i = 1, \cdots, s$.

(i) Consider first the case $s = 2$ and the question of whether or not there is a material difference between the two populations. This has the same structure as problem (iii) of the previous example. Here the choice lies between the three decisions d_0: $|\xi_2 - \xi_1| \leqq \Delta$, d_1: $\xi_2 > \xi_1 + \Delta$, d_2: $\xi_2 < \xi_1 - \Delta$ where Δ is preassigned. An analogous problem, involving $k + 1$ possible decisions, occurs in the general case of k populations. In this case one must choose between the decision that the k distributions do not differ materially, d_0: $\max |\xi_j - \xi_i| \leqq \Delta$, and the decisions d_k: $\max |\xi_j - \xi_i| > \Delta$ and ξ_k is the largest of the means.

(ii) A related problem is that of ranking the distributions in increasing order of their mean ξ.

(iii) Alternatively, a standard ξ_0 may be given and the problem is to decide which, if any, of the population means exceed that standard.

Example 3. Consider two distributions—to be specific, two Poisson distributions $P(\tau_1)$, $P(\tau_2)$—and suppose that τ_1 is known to be less that τ_2 but that otherwise the τ's are unknown. Let $Z_1, \cdots, Z_n$ be independently distributed, each according to either $P(\tau_1)$ or $P(\tau_2)$. Then each Z is to be classified as to which of the two distributions it comes from. Here the loss might be the number of Z's that are incorrectly classified, multiplied by a suitable function of τ_1 and τ_2. An example of the complexity that such problems can attain and the conceptual as well as mathematical difficulties that they may involve is provided by the efforts of anthropologists to classify the human population into a number of homogeneous races by studying the frequencies of the various blood groups and of other genetic characters.

All the problems considered so far could be termed *action problems*. It was assumed in all of them that if θ were known a unique correct decision would be available, that is, given any θ there exists a unique d for which $L(\theta, d) = 0$. However, not all statistical problems are so clear-cut. Frequently it is a question of providing a convenient summary of the data or indicating what information is available concerning the unknown parameter or distribution. This information will be used for guidance in various considerations but will not provide the sole basis for any specific decisions. In such cases the emphasis is on the inference rather than on the decision aspect of the problem, although formally it can still be considered a decision problem if the inferential statement itself is interpreted as the decision to be taken. An important class of such problems, estimation by interval,* is illustrated by the following example.

Example 4. Let $X = (X_1, \cdots, X_n)$ be a sample from $N(\xi, \sigma^2)$ and let a decision consist in selecting an interval $[\underline{L}, \bar{L}]$ and stating that it contains ξ. Suppose that decision procedures are restricted to intervals $[\underline{L}(X), \bar{L}(X)]$ whose expected length for all ξ and σ does not exceed $k\sigma$ where k is some preassigned constant. An appropriate loss function would be 0 if the decision is correct and would otherwise depend on the relative position of the interval to the true value of ξ. In this case there are many correct decisions corresponding to a given distribution $N(\xi, \sigma^2)$.

It remains to discuss the choice of loss function, and of the three elements defining the problem this is perhaps the most difficult to specify. Even in the simplest case, where all losses eventually reduce to financial ones, it can hardly be expected that one will be able to evaluate all the short- and long-term consequences of an action. Frequently it is possible to simplify the formulation by taking into account only certain aspects of the loss function. As an illustration consider Example 1(i) and let $L(\theta, d_0) = a$ for $\gamma(\theta) \leqq \gamma_0$ and $L(\theta, d_1) = b$ for $\gamma(\theta) > \gamma_0$. The risk function becomes

$$R(\theta, \delta) = \begin{cases} aP_\theta\{\delta(X) = d_0\} & \text{if } \gamma \leqq \gamma_0 \\ bP_\theta\{\delta(X) = d_1\} & \text{if } \gamma > \gamma_0, \end{cases} \tag{4}$$

and is seen to involve only the two probabilities of error with weights which can be adjusted according to the relative importance of these errors. Similarly, in Example 3 one may wish to restrict attention to the number of misclassifications.

Unfortunately, such a natural simplification is not always available, and in the absence of specific knowledge it becomes necessary to select the loss function in some conventional way, with mathematical simplicity usually an important consideration. In point estimation problems such as that considered in Example 1(ii), if one is interested in estimating a real-valued function $\gamma = \gamma(\theta)$ it is customary to take the square of the error, or somewhat more generally to put

$$L(\theta, d) = v(\theta)(d - \gamma)^2. \tag{5}$$

Besides being particularly simple mathematically, this can be considered as an approximation to the true loss function L provided that for each fixed θ, $L(\theta, d)$ is twice differentiable in d, that $L(\theta, \gamma(\theta)) = 0$ for all θ, and that the error is not large.

* For the more usual formulation in terms of confidence intervals, see Chapter 3, Section 5, and Chapter 5, Sections 4 and 5.

It is frequently found that, within one problem, quite different types of losses may occur, which are difficult to measure on a common scale. Consider once more Example 1(i) and suppose that γ_0 is the value of γ when a standard treatment is applied to a situation in medicine, agriculture, or industry. The problem is that of comparing some new process with unknown γ to the standard one. Turning down the new method when it is actually superior, or adopting it when it is not, clearly entails quite different consequences. In such cases it is sometimes convenient to treat the various components, say $L_1, L_2, \cdots, L_r$, separately. Suppose in particular that $r = 2$ and that L_1 represents the more serious possibility. One can then assign a bound to this risk component, that is, impose the condition

$$EL_1(\theta, \delta(X)) \leqq \alpha, \tag{6}$$

and subject to this condition minimize the other component of the risk. Example 4 provides an illustration of this procedure. The length of the interval $[\underline{L}, \bar{L}]$ (measured in σ-units) is one component of the loss function, the other being the loss that results if the interval does not cover the true ξ.

3. RANDOMIZATION; CHOICE OF EXPERIMENT

The description of the general decision problem given so far is still too narrow in certain respects. It has been assumed that for each possible value of the random variables a definite decision must be chosen. Instead, it is convenient to permit the selection of one out of a number of decisions according to stated probabilities, or more generally the selection of a decision according to a probability distribution defined over the decision space; which distribution depends of course on what x is observed. One way to describe such a randomized procedure is in terms of a nonrandomized procedure depending on X and a random variable Y whose values lie in the decision space and whose conditional distribution given x is independent of θ.

Although it may run counter to one's intuition that such extra randomization should have any value, there is no harm in permitting this greater freedom of choice. If the intuitive misgivings are correct it will turn out that the optimum procedures always are of the simple nonrandomized kind. Actually, the introduction of randomized procedures leads to an important mathematical simplification by enlarging the class of risk functions so that it becomes convex. In addition, there are problems in which some features of the risk function such as its maximum can be improved by using a randomized procedure.

Another assumption that tacitly has been made so far is that a definite experiment has already been decided upon so that it is known what observations will be taken. However, the statistical considerations involved in designing an experiment are no less important than those concerning its analysis. One question in particular that must be decided before an investigation is undertaken is how many observations should be taken so that the risk resulting from wrong decisions will not be excessive. Frequently it turns out that the required sample size depends on the unknown distribution and therefore cannot be determined in advance as a fixed number. Instead it is then specified as a function of the observations and the decision whether or not to continue experimentation is made *sequentially* at each stage of the experiment on the basis of the observations taken up to that point.

Example 5. On the basis of a sample $X_1, \cdots, X_n$ from a normal distribution $N(\xi, \sigma^2)$ one wishes to estimate ξ. Here the risk function of an estimate, for example its expected squared error, depends on σ. For large σ the sample contains only little information in the sense that two distributions $N(\xi_1, \sigma^2)$ and $N(\xi_2, \sigma^2)$ with fixed difference $\xi_2 - \xi_1$ become indistinguishable as $\sigma \to \infty$, with the result that the risk tends to infinity. Conversely, the risk approaches zero as $\sigma \to 0$ since then effectively the mean becomes known. Thus the number of observations needed to control the risk at a given level is unknown. However, as soon as some observations have been taken, it is possible to estimate σ^2 and hence to determine the additional number of observations required.

Example 6. In a sequence of trials with constant probability p of success, one wishes to decide whether $p \leq 1/2$ or $p > 1/2$. It will usually be possible to reach a decision at an early stage if p is close to 0 or 1 so that practically all observations are of one kind, while a larger sample will be needed for intermediate values of p. This difference may be partially balanced by the fact that for intermediate values a loss resulting from a wrong decision is presumably less serious than for the more extreme values.

Example 7. The possibility of determining the sample size sequentially is important not only because the distributions P_θ can be more or less informative but also because the same is true of the observations themselves. Consider, for example, observations from the uniform distribution* over the interval $(\theta - \frac{1}{2}, \theta + \frac{1}{2})$ and the problem of estimating θ. Here there is no difference in the amount of information provided by the different distributions P_θ. However, a sample $X_1, X_2, \cdots, X_n$ can practically pinpoint θ if $\max |X_j - X_i|$ is sufficiently close to 1, or it can give essentially no more information than a single observation if $\max |X_j - X_i|$ is close to 0. Again the required sample size should be determined sequentially.

Except in the simplest situations, the determination of the appropriate sample size is only one aspect of the design problem. In general, one

* This distribution is defined in Problem 1 at the end of the chapter.

must decide not only how many but also what kind of observations to take. Formally all these questions can be subsumed under the general decision problem described at the beginning of the section, by interpreting X as the set of all available variables, by introducing the decisions of whether or not to stop experimentation at the various stages, by specifying in case of continuance which type of variable to observe next, and by including the cost of observation in the loss function. However, in spite of this formal possibility, the determination of optimum designs in specific situations is typically of a higher order of difficulty than finding the optimum decision rule for a given experiment, and it has been carried out in only a few cases. Here, we shall be concerned primarily with the problem as it presents itself once the experiment has been set up, and only in a few special cases attempt a comparison of different designs.

4. OPTIMUM PROCEDURES

At the end of Section 1 the aim of statistical theory was stated to be the determination of a decision function δ which minimizes the risk function

$$R(\theta, \delta) = E_\theta[L(\theta, \delta(X))]. \tag{7}$$

Unfortunately, in general the minimizing δ depends on θ, which is unknown. Consider, for example, some particular decision d_0, and the decision procedure $\delta(x) \equiv d_0$ according to which decision d_0 is taken regardless of the outcome of the experiment. Suppose that d_0 is the correct decision for some θ_0 so that $L(\theta_0, d_0) = 0$. Then δ minimizes the risk at θ_0 since $R(\theta_0, \delta) = 0$, but presumably at the cost of a high risk for other values of θ.

In the absence of a decision function that minimizes the risk for all θ, the mathematical problem is still not defined since it is not clear what is meant by a best procedure. Although it does not seem possible to give a definition of optimality which will be appropriate in all situations, the following two methods of approach frequently are satisfactory.

The nonexistence of an optimum decision rule is a consequence of the possibility that a procedure devotes too much of its attention to a single parameter value at the cost of neglecting the various other values that might arise. This suggests the restriction to decision procedures which possess a certain degree of impartiality, and the possibility that within such a restricted class there may exist a procedure with uniformly smallest risk. Two conditions of this kind, invariance and unbiasedness, will be discussed in the next section.

Instead of restricting the class of procedures, one can approach the problem somewhat differently. Consider the risk functions corresponding to two different decision rules δ_1 and δ_2. If $R(\theta, \delta_1) < R(\theta, \delta_2)$ for all θ, then δ_1 is clearly preferable to δ_2 since its use will lead to a smaller risk no matter what the true value of θ is. However, the situation is not clear when the two risk functions intersect as in Figure 1. What is needed is a principle which in such cases establishes a preference of one of the two risk functions over the other, that is, which introduces an ordering into the set of all risk functions. A procedure will then be optimum if

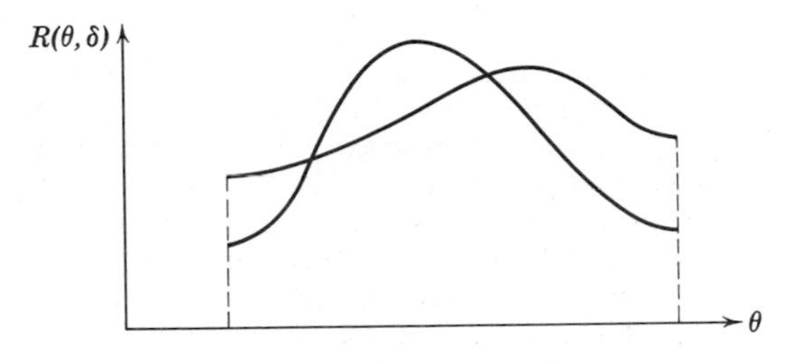

Figure 1.

its risk function is best according to this ordering. Some criteria that have been suggested for ordering risk functions will be discussed in Section 6.

A weakness of the theory of optimum procedures sketched above is its dependence on an extraneous restricting or ordering principle, and on knowledge concerning the loss function and the distributions of the observable random variables which in applications is frequently unavailable or unreliable. These difficulties, which may raise doubt concerning the value of an optimum theory resting on such shaky foundations, are in principle no different from those arising in any application of mathematics to reality. Mathematical formulations always involve simplification and approximation, so that solutions obtained through their use cannot be relied upon without additional checking. In the present case a check consists in an over-all evaluation of the performance of the procedure that the theory produces, and an investigation of its sensitivity to departure from the assumptions under which it was derived.

The difficulties can be overcome in part by considering the same problem with respect to a number of different formulations. If different optimality criteria lead to a common solution this will be the best procedure from several points of view and therefore will be more likely to be generally satisfactory. In the contrary case, the method indicates the strength and weaknesses of the various solutions and thereby possibly suggests a compromise procedure. Similarly, the sensitivity of a procedure to deviations from the assumptions under which it was derived can be tested, for example, by dropping one of the assumptions and comparing the procedure obtained from the wider model with the original one.

5. INVARIANCE AND UNBIASEDNESS*

A natural definition of impartiality suggests itself in situations which are symmetric with respect to the various parameter values of interest: *The procedure is then required to act symmetrically with respect to these values.*

Example 8. Suppose two treatments are to be compared and that each is applied n times. The resulting observations $X_{11}, \cdots, X_{1n}$ and $X_{21}, \cdots, X_{2n}$ are samples from $N(\xi_1, \sigma^2)$ and $N(\xi_2, \sigma^2)$ respectively. The three available decisions are $d_0: |\xi_2 - \xi_1| \leq \Delta, d_1: \xi_2 > \xi_1 + \Delta, d_2: \xi_2 < \xi_1 - \Delta$, and the loss is w_{ij} if decision d_j is taken when d_i would have been correct. If the treatments are to be compared solely in terms of the ξ's and no outside considerations are involved, the losses are symmetric with respect to the two treatments so that $w_{01} = w_{02}$, $w_{10} = w_{20}$, $w_{12} = w_{21}$. Suppose now that the labeling of the two treatments as 1 and 2 is reversed, and correspondingly also the labeling of the X's, the ξ's, and the decisions d_1 and d_2. This changes the meaning of the symbols but the formal decision problem, because of its symmetry, remains unaltered. It is then natural to require the corresponding symmetry from the procedure δ and ask that $\delta(x_{11}, \cdots, x_{1n}, x_{21}, \cdots, x_{2n}) = d_0, d_1$, or d_2 as $\delta(x_{21}, \cdots, x_{2n}, x_{11}, \cdots, x_{1n}) = d_0, d_2$, or d_1 respectively. If this condition were not satisfied the decision as to which population has the greater mean would depend on the presumably quite accidental and irrelevant labeling of the samples. Similar remarks apply to a number of further symmetries that are present in this problem.

Example 9. Consider a sample $X_1, \cdots, X_n$ from a distribution with density $\sigma^{-1}f[(x - \xi)/\sigma]$ and the problem of estimating the location parameter ξ, say the mean of the X's, when the loss is $(d - \xi)^2/\sigma^2$, the square of the error expressed in σ-units. Suppose that the observations are originally expressed in feet, and let $X_i' = aX_i$ with $a = 12$ be the corresponding observations in inches. In the transformed problem the density is $\sigma'^{-1}f[(x' - \xi')/\sigma']$ with $\xi' = a\xi$, $\sigma' = a\sigma$. Since $(d' - \xi')^2/\sigma'^2 = (d - \xi)^2/\sigma^2$ the problem is formally unchanged. The same estimation procedure that is used for the original observations is therefore appropriate after the transformation and leads to $\delta(aX_1, \cdots, aX_n)$ as an estimate of $\xi' = a\xi$, the parameter ξ expressed in inches. On reconverting the estimate into feet one finds that if the result is to be independent of the scale of measurements, δ must satisfy the condition of scale invariance

$$\delta(aX_1, \cdots, aX_n)/a = \delta(X_1, \cdots, X_n).$$

The general mathematical expression of symmetry is invariance under a suitable group of transformations. A group G of transformations g

* The concepts discussed here for general decision theory will be developed in more specialized form in later chapters. The present section may therefore be omitted at first reading.

of the sample space is said to leave a statistical decision problem invariant if it satisfies the following conditions.

(i) It leaves invariant the family of distributions $\mathscr{P} = \{P_\theta, \theta \in \Omega\}$, that is, for any possible distribution P_θ of X the distribution of gX, say $P_{\theta'}$, is also in $\mathscr{P}$. The resulting mapping $\theta' = \bar{g}\theta$ of Ω is assumed to be onto† Ω and 1 : 1.

(ii) To each $g \in G$, there exists a transformation $g^* = h(g)$ of the decision space D onto itself such that h is a homomorphism, that is, satisfies the relation $h(g_1 g_2) = h(g_1)h(g_2)$, and the loss function L is unchanged under the transformation so that

$$L(\bar{g}\theta, g^*d) = L(\theta, d).$$

Under these assumptions the transformed problem, in terms of $X' = gX$, $\theta' = \bar{g}\theta$, and $d' = g^*d$, is formally identical with the original problem in terms of X, θ, and d. Given a decision procedure δ for the latter, this is therefore still appropriate after the transformation. Interpreting the transformation as a change of coordinate system and hence of the names of the elements, one would, on observing x', select the decision which in the new system has the name $\delta(x')$ so that its old name is $g^{*-1}\delta(x')$. If the decision taken is to be independent of the particular coordinate system adopted, this should coincide with the original decision $\delta(x)$, that is, the procedure must satisfy the *invariance* condition

$$\delta(gx) = g^* \delta(x) \quad \text{for all} \quad x \in X, g \in G. \tag{8}$$

Invariance considerations are applicable only when a problem exhibits certain symmetries. An alternative impartiality restriction which is applicable to other types of problems is the following condition of unbiasedness. Suppose the problem is such that to each θ there exists a unique correct decision and that each decision is correct for some θ. Assume further that $L(\theta_1, d) = L(\theta_2, d)$ for all d whenever the same decision is correct for both θ_1 and θ_2. Then the loss $L(\theta, d')$ depends only on the actual decision taken, say d', and the correct decision d. The loss can thus be denoted by $L(d, d')$ and this function measures how far apart d and d' are. Under these assumptions a decision function δ is said to be unbiased if for all θ and d'

$$E_\theta L(d', \delta(X)) \geqq E_\theta L(d, \delta(X))$$

where the subscript θ indicates the distribution with respect to which the expectation is taken and where d is the decision that is correct for θ. Thus δ is unbiased if on the average $\delta(X)$ comes closer to the correct

† The term *onto* is used to indicate that $\bar{g}\Omega$ is not only contained in but actually equals Ω; that is, given any θ' in Ω there exists θ in Ω such that $\bar{g}\theta = \theta'$.

decision than to any wrong one. Extending this definition, δ is said to be *unbiased* for an arbitrary decision problem if for all θ and θ'

$$E_\theta L(\theta', \delta(X)) \geqq E_\theta L(\theta, \delta(X)). \tag{9}$$

Example 10. Suppose that in the problem of estimating a real-valued parameter θ by confidence intervals, as in Example 4, the loss is 0 or 1 as the interval $[\underline{L}, \bar{L}]$ does or does not cover the true θ. Then the set of intervals $[\underline{L}(X), \bar{L}(X)]$ is unbiased if the probability of covering the true value is greater than or equal to the probability of covering any false value.

Example 11. In a two-decision problem such as that of Example 1(i), let ω_0 and ω_1 be the sets of θ-values for which d_0 and d_1 are the correct decisions. Assume that the loss is 0 when the correct decision is taken, and otherwise is given by $L(\theta, d_0) = a$ for $\theta \in \omega_1$, and $L(\theta, d_1) = b$ for $\theta \in \omega_0$. Then

$$E_\theta L(\theta', \delta(X)) = \begin{cases} aP_\theta\{\delta(X) = d_0\} & \text{if} \quad \theta' \in \omega_1 \\ bP_\theta\{\delta(X) = d_1\} & \text{if} \quad \theta' \in \omega_0 \end{cases}$$

so that (9) reduces to

$$aP_\theta\{\delta(X) = d_0\} \geqq bP_\theta\{\delta(X) = d_1\} \quad \text{for} \quad \theta \in \omega_0$$

with the reverse inequality holding for $\theta \in \omega_1$. Since $P_\theta\{\delta(X) = d_0\} + P_\theta\{\delta(X) = d_1\} = 1$, the unbiasedness condition (9) becomes

$$\begin{aligned} P_\theta\{\delta(X) = d_1\} &\leqq \frac{a}{a+b} \quad \text{for} \quad \theta \in \omega_0 \\ P_\theta\{\delta(X) = d_1\} &\geqq \frac{a}{a+b} \quad \text{for} \quad \theta \in \omega_1. \end{aligned} \tag{10}$$

Example 12. In the problem of estimating a real-valued function $\gamma(\theta)$ with the square of the error as loss, the condition of unbiasedness becomes

$$E_\theta[\delta(X) - \gamma(\theta')]^2 \geqq E_\theta[\delta(X) - \gamma(\theta)]^2 \quad \text{for all} \quad \theta, \theta'.$$

On adding and subtracting $h(\theta) = E_\theta\, \delta(X)$ inside the brackets on both sides, this reduces to

$$[h(\theta) - \gamma(\theta')]^2 \geqq [h(\theta) - \gamma(\theta)]^2 \quad \text{for all} \quad \theta, \theta'.$$

If $h(\theta)$ is one of the possible values of the function γ, this condition holds if and only if

$$E_\theta\, \delta(X) = \gamma(\theta). \tag{11}$$

In the theory of point estimation, (11) is customarily taken as the definition of unbiasedness. Except under rather pathological conditions, it is both a necessary and sufficient condition for δ to satisfy (9). (See Problem 2.)

6. BAYES AND MINIMAX PROCEDURES

We now turn to a discussion of some preference orderings of decision procedures and their risk functions. One such ordering is obtained by assuming that in repeated experiments the parameter itself is a random variable Θ. If for the sake of simplicity one supposes that its distribution has a probability density $\rho(\theta)$, the over-all average loss resulting from the use of a decision procedure δ is

$$r(\rho, \delta) = \int E_\theta L(\theta, \delta(X))\rho(\theta)\, d\theta = \int R(\theta, \delta)\rho(\theta)\, d\theta \tag{12}$$

and the smaller $r(\rho, \delta)$, the better is δ. An optimum procedure is one that minimizes $r(\rho, \delta)$ and is called a *Bayes solution* of the given decision problem corresponding to the a priori density ρ. The resulting minimum of $r(\rho, \delta)$ is called the *Bayes risk* of ρ.

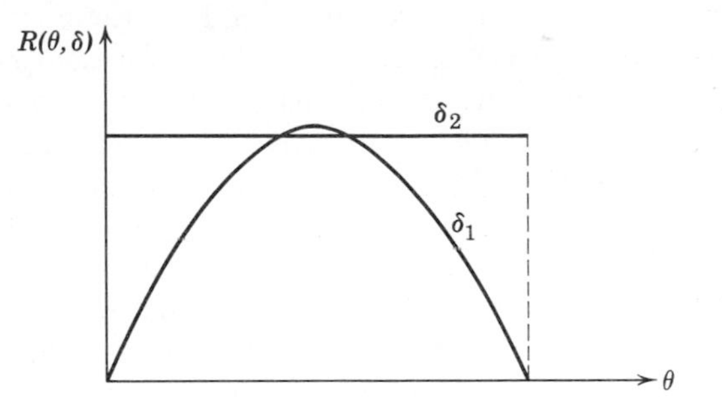

Figure 2.

Unfortunately, in order to apply this principle it is necessary to assume not only that θ is a random variable but also that its distribution is known. This assumption is usually not warranted in applications. Alternatively, the right-hand side of (12) can be considered as a weighted average of the risks; for $\rho(\theta) \equiv 1$ in particular, it is then the area under the risk curve. With this interpretation the choice of a weight function ρ expresses the importance the experimenter attaches to the various values of θ.

If no prior information regarding θ is available one might consider the maximum of the risk function its most important feature. Of two risk functions the one with the smaller maximum is then preferable, and the optimum procedures are those with the *minimax* property of minimizing the maximum risk. Since this maximum represents the worst (average) loss that can result from the use of a given procedure, a minimax solution is one that gives the greatest possible protection against large losses. That such a principle may sometimes be quite unreasonable is indicated in Figure 2, where under most circumstances one would prefer δ_1 to δ_2 although its risk function has the larger maximum.

Perhaps the most common situation is one intermediate to the two just described. On the one hand, past experience with the same or similar kind of experiment is available and provides an indication of what values of θ to expect; on the other, this information is neither sufficiently precise nor sufficiently reliable to warrant the assumptions that the Bayes approach requires. In such circumstances it seems desirable to make use of the available information without trusting it to such an extent that catastrophically high risks might result if it is inaccurate or misleading. To achieve this one can place a bound on the risk and restrict consideration to decision procedures δ for which

$$R(\theta, \delta) \leqq C \quad \text{for all} \quad \theta. \tag{13}$$

[Here the constant C will have to be larger than the maximum risk C_0 of the minimax procedure since otherwise there will exist no procedures satisfying (13).] Having thus assured that the risk can under no circumstances get out of hand, the experimenter can now safely exploit his knowledge of the situation, which may be based on theoretical considerations as well as on past experience; he can follow his hunches and guess at a distribution ρ for θ. This leads to the selection of a procedure δ (a *restricted Bayes solution*), which minimizes the average risk (12) for this a priori distribution subject to (13). The more certain one is of ρ, the larger one will select C, thereby running a greater risk in case of a poor guess but improving the risk if the guess is good.

Instead of specifying an ordering directly, one can postulate conditions that the ordering should satisfy. Various systems of such conditions have been investigated* and have generally led to the conclusion that the only orderings satisfying these systems are those which order the procedures according to their Bayes risk with respect to some a priori distribution of θ.

7. MAXIMUM LIKELIHOOD

Another approach, which is based on considerations somewhat different from those of the preceding sections, is the method of maximum likelihood. It has led to reasonable procedures in a great variety of problems, and is still playing a dominant role in the development of new tests and estimates. Suppose for a moment that X can taken on only a countable set of values $x_1, x_2, \cdots$, with $P_\theta(x) = P_\theta\,\{X = x\}$, and that one wishes to determine the correct value of θ, that is, the value that produced the

* See, for example, Savage, *The Foundations of Statistics*, New York, John Wiley & Sons, 1954, and Section 4.3 of Blackwell and Girshick, *Theory of Games and Statistical Decisions*, New York, John Wiley & Sons, 1954.

observed x. This suggests considering for each possible θ how probable the observed x would be if θ were the true value. The higher this probability, the more one is attracted to the explanation that the θ in question produced x, and the more likely the value of θ appears. Therefore, the expression $P_\theta(x)$ considered for fixed x as a function of θ has been called the *likelihood* of θ. To indicate the change in point of view, let it be denoted by $L_x(\theta)$. Suppose now that one is concerned with an action problem involving a countable number of decisions, and that it is formulated in terms of a gain function (instead of the usual loss function), which is 0 if the decision taken is incorrect and is $a(\theta) > 0$ if the decision taken is correct and θ is the true value. Then it seems natural to weight the likelihood $L_x(\theta)$ by the amount that can be gained if θ is true, to determine the value of θ that maximizes $a(\theta)L_x(\theta)$ and to select the decision that would be correct if this were the true value of θ.* Essentially the same remarks apply in the case in which $P_\theta(x)$ is a probability density rather than a discrete probability. The above motivation breaks down for the problem of estimating a continuous parameter since there is then no hope of determining the correct value of θ, but this can be considered as a limiting case.

In problems of point estimation, one usually assumes that $a(\theta)$ is independent of θ. This leads to estimating θ by the value that maximizes the likelihood $L_x(\theta)$, the *maximum likelihood estimate* of θ. Another case of interest is the class of two-decision problems illustrated by Example 1(i). Let ω_0 and ω_1 denote the sets of θ-values for which d_0 and d_1 are the correct decisions, and assume that $a(\theta) = a_0$ or a_1 as θ belongs to ω_0 or ω_1 respectively. Then decision d_0 or d_1 is taken as $a_1 \sup_{\theta\in\omega_1} L_x(\theta) <$ or $> a_0 \sup_{\theta\in\omega_0} L_x(\theta)$, that is, as

$$\frac{\sup\limits_{\theta\in\omega_0} L_x(\theta)}{\sup\limits_{\theta\in\omega_1} L_x(\theta)} > \text{ or } < \frac{a_1}{a_0}. \tag{14}$$

This is known as a *likelihood ratio procedure*.†

Although the maximum likelihood principle is not based on any clearly defined optimum considerations, it has been very successful in leading to satisfactory procedures in many specific problems. For wide classes of problems, maximum likelihood procedures have also been

* A variant of this approach has been suggested by Lindley, "Statistical inference," *J. Roy. Stat. Soc., Ser. B.*, Vol. XI (1953), pp. 30–76.

† This definition differs slightly from the usual one where in the denominator on the left-hand side of (14) the supremum is taken over the set $\omega_0 \cup \omega_1$. The two definitions agree whenever the left-hand side of (14) is $\leqq 1$, and the procedures therefore agree if $a_1 < a_0$.

shown to possess various asymptotic optimum properties as the sample size tends to infinity.* On the other hand, there exist examples for which the maximum likelihood procedure is worse than useless; where it is, in fact, so bad that one can do better without making any use of the observations (see Chapter 6, Problem 18).

8. COMPLETE CLASSES

None of the approaches described so far is reliable in the sense that the resulting procedure is necessarily satisfactory. There are problems in which a decision procedure δ_0 exists with uniformly minimum risk among all unbiased or invariant procedures, but where there exists a procedure δ_1 not possessing this particular impartiality property and preferable to δ_0. (Cf. Problems 14 and 16.) As was seen earlier, minimax procedures can also be quite undesirable, while the success of Bayes and restricted Bayes solutions depends on a priori information which is usually not very reliable if it is available at all. In fact, it seems that in the absence of reliable a priori information no principle leading to a unique solution can be entirely satisfactory.

This suggests the possibility, at least as a first step, of not insisting on a unique solution but asking only how far a decision problem can be reduced without loss of relevant information. It has already been seen that a decision procedure δ can sometimes be eliminated from consideration because there exists a procedure δ' *dominating* it in the sense that

$$
\begin{aligned}
&R(\theta, \delta') \leqq R(\theta, \delta) \quad \text{for all} \quad \theta \\
&R(\theta, \delta') < R(\theta, \delta) \quad \text{for some} \quad \theta.
\end{aligned}
\tag{15}
$$

In this case δ is said to be *inadmissible*; δ is called *admissible* if no such dominating δ' exists. A class $\mathscr{C}$ of decision procedures is said to be *complete* if for any δ not in $\mathscr{C}$ there exists δ' in $\mathscr{C}$ dominating it. A complete class is *minimal* if it does not contain a complete subclass. If a minimal complete class exists, as is typically the case, it consists exactly of the totality of admissible procedures.

It is convenient to define also the following variant of the complete class notion. A class $\mathscr{C}$ is said to be *essentially complete* if for any procedure δ there exists δ' in $\mathscr{C}$ such that $R(\theta, \delta') \leqq R(\theta, \delta)$ for all θ. Clearly, any complete class is also essentially complete. In fact, the two

* For some recent discussions see, for example, Wald, "Tests of statistical hypotheses concerning several parameters when the number of observations is large," *Trans. Am. Math. Soc.*, Vol. 54 (1943), pp. 426–482, and LeCam, "On some asymptotic properties of maximum likelihood estimates and related Bayes' estimates," *Univ. Calif. Publs. Statistics*, Vol. 1(1953), pp. 277–330.

definitions differ only in their treatment of equivalent decision rules, that is, decision rules with identical risk function. If δ belongs to the minimal complete class $\mathscr{C}$, any equivalent decision rule must also belong to $\mathscr{C}$. On the other hand, a minimal essentially complete class need contain only one member from such a set of equivalent procedures.

In a certain sense a minimal essentially complete class provides the maximum possible reduction of a decision problem. On the one hand, there is no reason to consider any of the procedures that have been weeded out. For each of them, there is included one in $\mathscr{C}$ that is as good or better. On the other hand, it is not possible to reduce the class further. Given any two procedures in $\mathscr{C}$, each of them is better in places than the other, so that without additional information it is not known which of the two is preferable.

The primary concern in statistics has been with the explicit determination of procedures, or classes of procedures, for various specific decision problems. Those studied most extensively have been estimation problems, and problems involving a choice between only two decisions (hypothesis testing) the theory of which constitutes the subject of the present volume. However, certain conclusions are possible without such specialization. In particular, two results concerning the structure of complete classes and minimax procedures have been proved to hold under very general assumptions:*

(i) The totality of Bayes solutions and limits of Bayes solutions constitute a complete class.

(ii) Minimax procedures are Bayes solutions with respect to a *least favorable* a priori distribution, that is, an a priori distribution that maximizes the associated Bayes risk, and the minimax risk equals this maximum Bayes risk. Somewhat more generally, if there exists no least favorable a priori distribution but only a sequence for which the Bayes risk tends to the maximum, the minimax procedures are limits of the associated sequence of Bayes solutions.

9. SUFFICIENT STATISTICS

A minimal complete class was seen in the preceding section to provide the maximum possible reduction of a decision problem without loss of information. Frequently it is possible to obtain a less extensive reduction of the data, which applies simultaneously to all problems relating to a given class $\mathscr{P} = \{P_\theta, \theta \in \Omega\}$ of distributions of the given random variable X. It consists essentially in discarding that part of the data which

* Precise statements and proofs of these results are given in the book by Wald, *Statistical Decision Functions*, New York, John Wiley & Sons, 1950.

contains no information regarding the unknown distribution P_θ, and which is therefore of no value for any decision problem concerning θ.

Example 13. Trials are performed with constant unknown probability p of success. If X_i is 1 or 0 as the ith trial is a success or failure, the sample $(X_1, \cdots, X_n)$ shows how many successes there were and in which trials they occurred. The second of these pieces of information contains no evidence as to the value of p. Once the total number of successes ΣX_i is known to be equal to t, each of the $\binom{n}{t}$ possible positions of these successes is equally likely regardless of p. It follows that knowing ΣX_i but neither the individual X_i nor p, one can, from a table of random numbers, construct a set of random variables $X_1', \cdots, X_n'$ whose joint distribution is the same as that of $X_1, \cdots, X_n$. Therefore, the information contained in the X_i is the same as that contained in ΣX_i and a table of random numbers.

Example 14. If $X_1, \cdots, X_n$ are independently normally distributed with zero mean and variance σ^2, the conditional distribution of the sample point over each of the spheres, $\Sigma X_i^2 =$ constant, is uniform irrespective of σ^2. One can therefore construct an equivalent sample $X_1', \cdots, X_n'$ from a knowledge of ΣX_i^2 and a mechanism that can produce a point randomly distributed over a sphere.

More generally, a statistic T is said to be *sufficient* for the family $\mathscr{P} = \{P_\theta, \theta \in \Omega\}$, or sufficient for θ if it is clear from the context what set Ω is being considered, if the conditional distribution of X given $T = t$ is independent of θ. As in the two examples it then follows under mild assumptions* that it is not necessary to utilize the original observations X. If one is permitted to observe only T instead of X, this does not restrict the class of available decision procedures. For any value t of T let X_t be a random variable possessing the conditional distribution of X given t. Such a variable can, at least theoretically, be constructed by means of a suitable random mechanism. If one then observes T to be t and X_t to be x', the random variable X' defined through this two-stage process has the same distribution as X. Thus, given any procedure based on X it is possible to construct an equivalent one based on X' which can be viewed as a randomized procedure based solely on T. Hence if randomization is permitted, and we shall assume throughout that this is the case, there is no loss of generality in restricting consideration to a sufficient statistic.

It is inconvenient to have to compute the conditional distribution of X given t in order to determine whether or not T is sufficient. A simple check is provided by the following *factorization criterion.*

Consider first the case that X is discrete and let $P_\theta(x) = P_\theta\{X = x\}$.

* These are connected with difficulties concerning the behavior of conditional probabilities. For a discussion of these difficulties see Chapter 2, Sections 3–5.

Then a necessary and sufficient condition for T to be sufficient for θ is that there exists a factorization

$$P_\theta(x) = g_\theta[T(x)]h(x), \tag{16}$$

where the first factor may depend on θ but depends on x only through $T(x)$ while the second factor is independent of θ.

Suppose that (16) holds and let $T(x) = t$. Then $P_\theta\{T = t\} = \Sigma P_\theta(x')$ summed over all points x' with $T(x') = t$, and the conditional probability

$$P_\theta\{X = x | T = t\} = P_\theta(x)/P_\theta\{T = t\} = h(x)/\Sigma h(x')$$

is independent of θ. Conversely, if this conditional distribution does not depend on θ and is equal to, say $k(x, t)$, then $P_\theta(x) = P_\theta\{T = t\}\, k(x, t)$ so that (16) holds.

Example 15. Let $X_1, \cdots, X_n$ be independently and identically distributed according to the Poisson distribution (2). Then

$$P_\tau(x_1, \cdots, x_n) = \frac{\tau^{\Sigma x_i} e^{-n\tau}}{\prod_{j=1}^{n} x_j!}$$

and it follows that ΣX_i is a sufficient statistic for τ.

In the case that the distribution of X is continuous and has probability density $p_\theta^X(x)$ let X and T be vector-valued, $X = (X_1, \cdots, X_n)$ and $T = (T_1, \cdots, T_r)$ say. Suppose that there exist functions $Y = (Y_1, \cdots, Y_{n-r})$ on the sample space such that the transformation

$$(x_1, \cdots, x_n) \leftrightarrow (T_1(x), \cdots, T_r(x), Y_1(x), \cdots, Y_{n-r}(x)) \tag{17}$$

is 1 : 1 on a suitable domain, and that the joint density of T and Y exists and is related to that of X by the usual formula*

$$p_\theta^X(x) = p_\theta^{T,Y}(T(x), Y(x)) \cdot |J|, \tag{18}$$

where J is the Jacobian of $(T_1, \cdots, T_r, Y_1, \cdots, Y_{n-r})$ with respect to $(x_1, \cdots, x_n)$. Thus in Example 14, $T = \sqrt{\Sigma X_i^2}$, $Y_1, \cdots, Y_{n-1}$ can be taken to be the polar coordinates of the sample point. From the joint density $p_\theta^{T,Y}(t, y)$ of T and Y, the conditional density of Y given $T = t$ is obtained as

$$p_\theta^{Y|t}(y) = \frac{p_\theta^{T,Y}(t, y)}{\int p_\theta^{T,Y}(t, y')\, dy'} \tag{19}$$

provided the denominator is different from zero.

* Regularity conditions for the validity of (18) are given by Tukey, "A smooth invertibility theorem," *Ann. Math. Stat.*, Vol. 29 (1958), pp. 581–584; see also Lehmann and Scheffé, "On the problem of similar regions," *Proc. Nat. Acad. Sci.*, Vol. 33 (1947), pp. 382–386.

Since in the conditional distribution given t only the Y's vary, T is sufficient for θ if the conditional distribution of Y given t is independent of θ. Suppose that T satisfies (19). Then analogously to the discrete case, a necessary and sufficient condition for T to be sufficient is a factorization of the density of the form

$$p_\theta^X(x) = g_\theta[T(x)]h(x). \tag{20}$$

(See Problem 19.) The following two examples illustrate the application of the criterion in this case. In both examples the existence of functions Y satisfying (17)–(19) will be assumed but not proved. As will be shown later (Chapter 2, Section 6), this assumption is actually not needed for the validity of the factorization criterion.

Example 16. Let $X_1, \cdots, X_n$ be independently distributed with normal probability density

$$p_{\xi,\sigma}(x) = (2\pi\sigma^2)^{-n/2} \exp\left(-\frac{1}{2\sigma^2}\Sigma x_i^2 + \frac{\xi}{\sigma^2}\Sigma x_i - \frac{n}{2\sigma^2}\xi^2\right).$$

Then the factorization criterion shows $(\Sigma X_i, \Sigma X_i^2)$ to be sufficient for (ξ, σ).

Example 17. Let $X_1, \cdots, X_n$ be independently distributed according to the rectangular distribution $R(0, \theta)$ over the interval $(0, \theta)$. Then $p_\theta(x) = \theta^{-n}u(\max x_i, \theta)$ where $u(a, b)$ is 1 or 0 as $a \leqq b$ or $a > b$, and hence $\max X_i$ is sufficient for θ.

An alternative criterion of sufficiency provides a direct connection between this concept and some of the basic notions of decision theory. As in the theory of Bayes solutions, consider the unknown parameter θ as a random variable Θ with an a priori distribution, and assume for simplicity that it has a density $\rho(\theta)$. Then if T is sufficient, the conditional distribution of Θ given $X = x$ depends only on $T(x)$. Conversely, if $\rho(\theta) \neq 0$ for all θ and if the conditional distribution of Θ given x depends only on $T(x)$, then T is sufficient for θ.

In fact, under the assumptions made, the joint density of X and Θ is $p_\theta(x)\rho(\theta)$. If T is sufficient it follows from (20) that the conditional density of Θ given x depends only on $T(x)$. Suppose, on the other hand, that for some a priori distribution for which $\rho(\theta) \neq 0$ for all θ the conditional distribution of Θ given x depends only on $T(x)$. Then

$$\frac{p_\theta(x)\rho(\theta)}{\int p_{\theta'}(x)\rho(\theta')\,d\theta'} = f_\theta[T(x)]$$

and by solving for $p_\theta(x)$ it is seen that T is sufficient.

Any Bayes solution depends only on the conditional distribution of Θ given x (see Problem 8) and hence on $T(x)$. Since typically Bayes solutions together with their limits form an essentially complete class, it follows that this is also true of the decision procedures based on T. The same conclusion had already been reached more directly at the beginning of the section.

By restricting attention to a sufficient statistic, one obtains a reduction of the data and it is then desirable to carry this reduction as far as possible. To illustrate the different possibilities, consider once more the binomial Example 13. If m is any integer less than n and $T_1 = \sum_{i=1}^m X_i$, $T_2 = \sum_{i=m+1}^n X_i$, then (T_1, T_2) constitutes a sufficient statistic since the conditional distribution of $X_1, \cdots, X_n$ given $T_1 = t_1$, $T_2 = t_2$ is independent of p. For the same reason, the full sample $(X_1, \cdots, X_n)$ itself is also a sufficient statistic. However, $T = \sum_{i=1}^n X_i$ provides a more thorough reduction than either of these and than various others that can be constructed. A sufficient statistic T is said to be *minimal sufficient* if the data cannot be reduced beyond T without losing sufficiency. For the binomial example in particular, $\sum_{i=1}^n X_i$ can be shown to be minimal (Problem 17). This illustrates the fact that in specific examples the sufficient statistic determined by inspection through the factorization criterion usually turns out to be minimal.*

10. PROBLEMS

Section 2

1. The following distributions arise on the basis of assumptions similar to those leading to (1)–(3).

(i) Independent trials with constant probability p of success are carried out until a preassigned number m of successes has been obtained. If the number of trials required is $X + m$, then X has the *negative binomial* distribution

$$P\{X = x\} = \binom{m + x - 1}{x} p^m (1 - p)^x, \qquad x = 0, 1, 2, \cdots.$$

(ii) In a sequence of random events, the number of events occurring in any time interval of length τ has the Poisson distribution $P(\lambda\tau)$, and the numbers of events in nonoverlapping time intervals are independent. Then the "waiting time" T, which elapses from the starting point, say $t = 0$, until the first event occurs, has the *exponential* probability density

$$p(t) = \lambda e^{-\lambda t}, \qquad t \geqq 0.$$

* Explicit procedures for constructing a minimal sufficient statistic (called necessary and sufficient by some writers) are given by Lehmann and Scheffé, "Completeness, similar regions and unbiased estimation," *Sankhyā*, Vol. 10 (1950), pp. 305–340, and by Bahadur, "Sufficiency and statistical decision functions," *Ann. Math. Stat.*, Vol. 25 (1954), pp. 423–462. See also Dynkin, "On sufficient and necessary statistics for families of probability distributions," *Doklady Akad. Nauk SSSR* (N. S.), Vol. 75 (1950), pp. 161–164 and *Uspehi Matem. Nauk* (N. S.), Vol. 6 (1951), No. 1, pp. 68–90.

(Let T_i, $i \geqq 2$, be the time elapsing from the occurrence of the $(i-1)$st event to that of the ith event. Then it is also true, although more difficult to prove,† that $T_1, T_2, \cdots$ are identically and independently distributed.)

(iii) A point X is selected "at random" in the interval (a, b), that is, the probability of X falling in any subinterval of (a, b) depends only on the length of the subinterval, not on its position. Then X has the *rectangular* or *uniform* distribution $R(a, b)$ with probability density

$$p(x) = 1/(b-a), \qquad a < x < b.$$

[(ii) If $t > 0$, then $T > t$ if and only if no event occurs in the time interval $(0, t)$.]

Section 5

2. *Unbiasedness in point estimation.* Suppose that the parameter space Ω is connected, that γ is a continuous real-valued function defined over Ω which is not constant in any open subset of Ω, and that the expectation $h(\theta) = E_\theta\delta(X)$ is a continuous function of θ for every estimate $\delta(X)$ of $\gamma(\theta)$. Then (11) is a necessary and sufficient condition for $\delta(X)$ to be unbiased when the loss function is the square of the error.

[Unbiasedness implies that $\gamma^2(\theta') - \gamma^2(\theta) \geqq 2h(\theta)[\gamma(\theta') - \gamma(\theta)]$ for all θ, θ'. If θ is neither a relative minimum or maximum of γ, it follows that there exist points θ' arbitrarily close to θ both such that $\gamma(\theta) + \gamma(\theta') \geqq$ and $\leqq 2h(\theta)$, and hence that $\gamma(\theta) = h(\theta)$. That this equality also holds for an extremum of γ follows by continuity since γ is not constant in any open set.]

3. *Median unbiasedness.* (i) A real number m is a median for the random variable Y if $P\{Y \geqq m\} \geqq 1/2$; $P\{Y \leqq m\} \geqq 1/2$. Then all real a_1, a_2 such that $m \leqq a_1 \leqq a_2$ or $m \geqq a_1 \geqq a_2$ satisfy $E|Y - a_1| \leqq E|Y - a_2|$.

(ii) For any estimate $\delta(X)$ of $\gamma(\theta)$, let $m^-(\theta)$ and $m^+(\theta)$ denote the infimum and supremum of the medians of $\delta(X)$, and suppose that they are continuous functions of θ. Let Ω be connected and let $\gamma(\theta)$ be continuous and not constant in any open subset of Ω. Then the estimate $\delta(X)$ of $\gamma(\theta)$ is unbiased with respect to the loss function $L(\theta, d) = |\gamma(\theta) - d|$ if and only if $\gamma(\theta)$ is a median of $\delta(X)$ for each θ. An estimate with this property is said to be *median-unbiased*.

4. *Nonexistence of unbiased procedures.* Consider a decision problem in which for each θ there exists a unique correct decision d, and suppose that

$$L(\theta, d') = h(\theta)V(d, d') \quad \text{for} \quad \theta \in \omega_d$$

where ω_d denotes the set of θ's for which d is correct. Then if the function h takes on at least two distinct values on each ω_d, the risk function of any unbiased procedure is identically zero, that is, typically no unbiased procedure exists. As an example, let $X_1, \cdots, X_n$ be independently distributed with density $(1/a)f((X-\xi)/a)$ and $\theta = (\xi, a)$. Then no estimate of ξ exists, which is unbiased with respect to the loss function $(\xi - d)^2/a^2$.

5. Let $\mathscr{C}$ be any class of procedures that is closed under the transformations of a group G in the sense that $\delta \in \mathscr{C}$ implies $g^*\delta g^{-1} \in \mathscr{C}$ for all $g \in G$. If there exists

† For a proof see Doob, *Stochastic Processes*, New York, John Wiley & Sons, 1953, p. 403.

a unique procedure δ_0 that uniformly minimizes the risk within the class $\mathscr{C}$, then δ_0 is invariant. If δ_0 is unique only up to sets of measure zero, then it is *almost invariant*, that is, for each g it satisfies the equation $\delta(gx) = g^*\delta(x)$ except on a set N_g of measure 0.

6. *Relation of unbiasedness and invariance.* (i) If δ_0 is the unique (up to sets of measure 0) unbiased procedure with uniformly minimum risk, it is almost invariant.

(ii) If $\bar{G}$ is transitive and G^* commutative, and if among all invariant (almost invariant) procedures there exists a procedure δ_0 with uniformly minimum risk, then it is unbiased.

[(i) This follows from the preceding problem and the fact that when δ is unbiased so is $g^*\delta g^{-1}$.

(ii) It is the defining property of transitivity that given θ, θ' there exists $\bar{g}$ such that $\theta' = \bar{g}\theta$. Hence for any θ, θ'

$$E_\theta L(\theta', \delta_0(X)) = E_\theta L(\bar{g}\theta, \delta_0(X)) = E_\theta L(\theta, g^{*-1}\delta_0(X)).$$

Since G^* is commutative $g^{*-1}\delta_0$ is invariant, so that

$$R(\theta, g^{*-1}\delta_0) \geqq R(\theta, \delta_0) = E_\theta L(\theta, \delta_0(X)).]$$

7. *Counterexample.* That conclusion (ii) of Problem 6 need not hold without the assumptions concerning G^* and $\bar{G}$ is shown by the problem of estimating the mean ξ of a normal distribution $N(\xi, \sigma^2)$ with loss function $(\xi - d)^2/\sigma^2$. This remains invariant under the groups G_1: $gx = x + b$, $-\infty < b < \infty$ and G_2: $gx = ax + b$, $0 < a < \infty$, $-\infty < b < \infty$. The best invariant estimate relative to both groups is X but there does not exist an estimate which is unbiased with respect to the given loss function.

Section 6

8. *Structure of Bayes solutions.* (i) Let Θ be an unobservable random quantity with probability density $\rho(\theta)$, and let the probability density of X be $p_\theta(x)$ when $\Theta = \theta$. Then δ is a Bayes solution of a given decision problem if for each x the decision $\delta(x)$ is chosen so as to minimize $\int L(\theta, \delta(x))\pi(\theta|x)\,d\theta$, where $\pi(\theta|x) = \rho(\theta)p_\theta(x)/\int\rho(\theta')p_{\theta'}(x)\,d\theta'$ is the conditional (a posteriori) probability density of Θ given x.

(ii) Let the problem be a two-decision problem with the losses as given in Example 11. Then the Bayes solution consists in choosing decision d_0 if

$$aP\{\Theta \in \omega_1|x\} < bP\{\Theta \in \omega_0|x\}$$

and decision d_1 if the reverse inequality holds. The choice of decision is immaterial in case of equality.

(iii) In case of point estimation of a real-valued function $g(\theta)$ with loss function $L(\theta, d) = (g(\theta) - d)^2$, the Bayes solution becomes $\delta(x) = E[g(\Theta)|x]$. When instead the loss function is $L(\theta, d) = |g(\theta) - d|$, the Bayes estimate $\delta(x)$ is any median of the conditional distribution of $g(\Theta)$ given x.

[(i) The Bayes risk $r(\rho, \delta)$ can be written as $\int[\int L(\theta, \delta(x))\pi(\theta|x)\,d\theta]p(x)\,dx$, where $p(x) = \int\rho(\theta')p_{\theta'}(x)\,d\theta'$.

(ii) The conditional expectation $\int L(\theta, d_0)\pi(\theta|x)\,d\theta$ reduces to $aP\{\Theta \in \omega_1|x\}$ and similarly for d_1.]

9. (i) As an example in which randomization reduces the maximum risk, suppose that a coin is known to be either standard (HT) or to have heads on both sides (HH). The nature of the coin is to be decided on the basis of a single toss, the loss being 1 for an incorrect decision and 0 for a correct one. Let the decision be HT when T is observed whereas in the contrary case the decision is made at random, with probability ρ for HT and $1 - \rho$ for HH. Then the maximum risk is minimized for $\rho = 1/3$.

(ii) A genetical setting in which such a problem might arise is that of a couple, of which the husband is either dominant homozygous (AA) or heterozygous (Aa) with respect to a certain characteristic and the wife is homozygous recessive (aa). Their child is heterozygous and it is of importance to determine to which genetical type the husband belongs. However, in such cases an a priori probability is usually available for the two possibilities. One is then dealing with a Bayes problem and randomization is no longer required. In fact, if the a priori probability is p that the husband is dominant, then the Bayes procedure classifies him as such if $p > 1/3$ and takes the contrary decision if $p < 1/3$.

10. *Unbiasedness and minimax.* Let $\Omega = \Omega_0 \cup \Omega_1$ where Ω_0, Ω_1 are mutually exclusive, and consider a two-decision problem with loss function $L(\theta, d_i) = a_i$ for $\theta \in \Omega_j$ $(j \neq i)$ and $L(\theta, d_i) = 0$ for $\theta \in \Omega_i$ $(i = 0, 1)$.

(i) Any minimax procedure is unbiased.

(ii) The converse of (i) holds provided $P_\theta(A)$ is a continuous function of θ for all A, and if the sets Ω_0 and Ω_1 have at least one common boundary point.

[(i) The condition of unbiasedness in this case is equivalent to $\sup R_\delta(\theta) \leqq a_0a_1/(a_0 + a_1)$. That this is satisfied by any minimax procedure is seen by comparison with the procedure $\delta(x) = d_0$ or $= d_1$ with probabilities $a_1/(a_0 + a_1)$ and $a_0/(a_0 + a_1)$ respectively.

(ii) If θ_0 is a common boundary point, continuity of the risk function implies that any unbiased procedure satisfies $R_\delta(\theta_0) = a_0a_1/(a_0 + a_1)$ and hence $\sup R_\delta(\theta) = a_0a_1/(a_0 + a_1)$.]

11. *Invariance and minimax.* Let a problem remain invariant relative to the groups G, $\bar{G}$, and G^* over the spaces $\mathscr{X}$, Ω, and D respectively. Then a randomized procedure Y_x is defined to be invariant if for all x and g the conditional distribution of Y_x given x is the same as that of $g^{*-1}Y_{gx}$.

(i) Consider a decision procedure which remains invariant under a finite group $G = \{g_1, \cdots, g_N\}$. If a minimax procedure exists, then there exists one that is invariant.

(ii) This conclusion does not necessarily hold for infinite groups as is shown by the following example. Let the parameter space Ω consist of all elements θ of the free group with two generators, that is, the totality of formal products $\pi_1 \cdots \pi_n$ $(n = 0, 1, 2, \cdots)$ where each π_i is one of the elements a, a^{-1}, b, b^{-1} and in which all products aa^{-1}, $a^{-1}a$, bb^{-1} and $b^{-1}b$ have been canceled. The empty product $(n = 0)$ is denoted by e. The sample point X is obtained by multiplying θ on the right by one of the 4 elements a, a^{-1}, b, b^{-1} with probability 1/4 each, and canceling if necessary, that is, if the random factor equals π_n^{-1}. The problem of estimating θ with $L(\theta, d)$ equal to 0 if $d = \theta$ and equal to 1 otherwise remains invariant under multiplication of X, θ, and d on the left by an arbitrary sequence $\pi_{-m} \cdots \pi_{-2}\pi_{-1}$ $(m = 0, 1, \cdots)$. The invariant procedure that minimizes the maximum risk has risk function $R(\theta, \delta) \equiv 3/4$. However, there exists a non-invariant procedure with maximum risk 1/4.

[(i) If Y_x is a (possibly randomized) minimax procedure, an invariant minimax procedure Y'_x is defined by $P(Y'_x = d) = \Sigma_{i=1}^N P(Y_{g_i x} = g_i^* d)/N$.

(ii) The better procedure consists in estimating θ to be $\pi_1 \cdots \pi_{k-1}$ when $\pi_1 \cdots \pi_k$ is observed ($k \geqq 1$), and to estimate θ to be a, a^{-1}, b, b^{-1} with probability 1/4 each in case the identity is observed. The estimate will be correct unless the last element of X was canceled, and hence will be correct with probability $\geqq 3/4$.]

Section 7

12. (i) Let X have probability density $p_\theta(x)$ with θ one of the values $\theta_1, \cdots, \theta_n$, and consider the problem of determining the correct value of θ, so that the choice lies between the n decisions $d_1 = \theta_1, \cdots, d_n = \theta_n$ with gain $a(\theta_i)$ if $d_i = \theta_i$ and 0 otherwise. Then the Bayes solution (which maximizes the average gain) when θ is a random variable taking on each of the n values with probability $1/n$ coincides with the maximum likelihood procedure.

(ii) Let X have probability density $p_\theta(x)$ with $0 \leqq \theta \leqq 1$. Then the maximum likelihood estimate is the mode (maximum value) of the a posteriori density of Θ given x when Θ is uniformly distributed over (0, 1).

13. (i) Let $X_1, \cdots, X_n$ be a sample from $N(\xi, \sigma^2)$ and consider the problem of deciding between ω_0: $\xi < 0$ and ω_1: $\xi \geqq 0$. If $\bar{x} = \Sigma x_i/n$ and $C = (a_1/a_0)^{2/n}$, the likelihood ratio procedure takes decision d_0 or d_1 as

$$\sqrt{n}\,\bar{x}/\sqrt{\Sigma(x_i - \bar{x})^2} < k \quad \text{or} \quad > k$$

where $k = -\sqrt{C-1}$ if $C > 1$ and $k = \sqrt{(1-C)/C}$ if $C < 1$.

(ii) For the problem of deciding between ω_0: $\sigma < \sigma_0$ and ω_1: $\sigma \geqq \sigma_0$, the likelihood ratio procedure takes decision d_0 or d_1 as

$$\Sigma(x_i - \bar{x})^2/n\sigma_0^2 < \text{or} > k$$

where k is the smaller root of the equation $Cx = e^{x-1}$ if $C > 1$ and the larger root of $x = Ce^{x-1}$ if $C < 1$, where C is defined as in (i).

Section 8

14. *Admissibility of unbiased procedures.* (i) Under the assumptions of Problem 10, if among the unbiased procedures there exists one with uniformly minimum risk, it is admissible.

(ii) That in general an unbiased procedure with uniformly minimum risk need not be admissible is seen by the following example. Let X have a Poisson distribution truncated at 0, so that $P_\theta\{X = x\} = \theta^x e^{-\theta}/[x!(1 - e^{-\theta})]$ for $x = 1, 2, \cdots$. For estimating $\gamma(\theta) = e^{-\theta}$ with loss function $L(\theta, d) = (d - \theta)^2$, there exists a unique unbiased estimate, and it is not admissible.

[(ii) The unique unbiased estimate $\delta_0(x) = (-1)^{x+1}$ is dominated by $\delta_1(x) = 0$ or 1 as x is even or odd.]

15. *Admissibility of invariant procedures.* If a decision problem remains invariant under a finite group, and if there exists a procedure δ_0 that uniformly minimizes the risk among all invariant procedures, then δ_0 is admissible.

[This follows from the identity $R(\theta, \delta) = R(\bar{g}\theta, g^*\delta g^{-1})$ and the hint given in Problem 11(i).]

16. (i) Let X take on the values $\theta - 1$ and $\theta + 1$ with probability 1/2 each. The problem of estimating θ with loss function $L(\theta, d) = \min(|\theta - d|, 1)$ remains invariant under the transformation $gX = X + c$, $\bar{g}\theta = \theta + c$, $g^*d = d + c$. Among invariant estimates, those taking on the values $X - 1$ and $X + 1$ with probabilities p and q (independent of X) uniformly minimize the risk.

(ii) That the conclusion of Problem 15 need not hold when G is infinite follows by comparing the best invariant estimates of (i) with the estimate $\delta_1(x)$ which is $X + 1$ when $X < 0$ and $X - 1$ when $X \geqq 0$.

Section 9

17. In n independent trials with constant probability p of success, let $X_i = 1$ or 0 as the ith trial is a success or not. Then $\Sigma_{i=1}^n X_i$ is minimal sufficient.

[Let $T = \Sigma X_i$ and suppose that $U = f(T)$ is sufficient and that $f(k_1) = \cdots = f(k_r) = u$. Then $P\{T = t | U = u\}$ depends on p.]

18. (i) Let $X_1, \cdots, X_n$ be a sample from the uniform distribution $R(0, \theta)$, $0 < \theta < \infty$, and let $T = \max(X_1, \cdots, X_n)$. Show that T is sufficient once by using the definition of sufficiency and once by using the factorization criterion and assuming the existence of statistics Y_i satisfying (17)–(19).

(ii) Let $X_1, \cdots, X_n$ be a sample from the exponential distribution with density $ae^{-a(x-b)}$ when $x \geqq b$ $(0 < a < \infty, -\infty < b < \infty)$. Use the factorization criterion to prove that $(\min(X_1, \cdots, X_n), \Sigma_{i=1}^n X_i)$ is sufficient for a, b assuming the existence of statistics Y_i satisfying (17)–(19).

19. A statistic T satisfying (17)–(19) is sufficient if and only if it satisfies (20).

11. REFERENCES

Some of the basic concepts of statistical theory were initiated during the first quarter of the 19th century by Laplace in his fundamental *Théorie Analytique des Probabilités* (1812), and by Gauss in his papers on the method of least squares. Loss and risk functions are mentioned in their discussions of the problem of point estimation, for which Gauss also introduced the condition of unbiasedness. A detailed account of this work and of its extensions during the 19th century is given in the book by Czuber (1891).

A period of intensive development of statistical methods began toward the end of the century with the work of Karl Pearson. In particular, two areas were explored in the researches of R. A. Fisher, J. Neyman, and many others: estimation and the testing of hypotheses. The work of Fisher can be found in his books (1925, 1935), and in the volume of his collected papers (1950). Many of Neyman's principal ideas are summarized in a published series of lectures (1938). General introductions to the modern methods of estimation and testing are given in the books by Wilks (1944); Cramér (1946); Kendall (1946); van der Waerden (1957); and the more theoretical work by Schmetterer (1956).

A formal unification of the theories of estimation and hypothesis testing, which also contains the possibility of many other specializations, was achieved by Wald, who gave a single comprehensive formulation in his general theory of decision procedures. A complete account of this theory, which is closely related to von Neumann's theory of games, is found in Wald's book (1950). A number of papers of Wald dealing with special aspects of the theory and many of his other contributions to statistics have been collected in one volume (1958). Two recent books dealing with the general theory are by Blackwell and Girshick (1954) and Savage (1954).

Blackwell, D., and M. A. Girschick
(1954) *Theory of Games and Statistical Decisions*, New York, John Wiley & Sons.

Brown, George
(1947) "On small sample estimation," *Ann. Math. Stat.*, Vol. 18, pp. 582–585.
[Definition of median unbiasedness.]

Cramér, Harald
(1946) *Mathematical Methods of Statistics*, Princeton University Press.

Czuber, E.
(1891) *Theorie der Beobachtungsfehler*, Leipzig, B. G. Teubner.

Edgeworth, F. Y.
(1908/09) "On the probable errors of frequency constants," *J. Roy. Stat. Soc.*, Vol. 71, pp. 381–397, 499–512, 651–678, Vol. 72, pp. 81–90.
[Following Laplace and Gauss, estimates are treated by the "genuine inverse method," that is, under the assumption of a uniform a priori distribution of the parameter. Since the estimates are determined so as to maximize the mode of the a posteriori distribution, they are also maximum likelihood estimates.]

Fisher, R. A.
(1922) "On the mathematical foundations of theoretical statistics, "*Phil. Trans. Roy. Soc., London, Ser. A*, Vol. 222, pp. 309–368.
[Development of a theory of point estimation on the basis of the maximum likelihood principle.]
(1920) "A mathematical examination of the methods of determining the accuracy of an observation by the mean error and by the mean square error," *Monthly Notices Roy. Astron. Soc.*, Vol. 80, pp. 758–770.
(1921) "On the mathematical foundation of theoretical statistics," *Phil. Trans. Roy. Soc., Ser. A*, Vol. 222, pp. 309–368.
(1925) "Theory of statistical estimation," *Proc. Cambridge Phil. Soc.*, Vol. 22, pp. 700–725.
[In these papers the concept of sufficiency is developed, principally in connection with the theory of point estimation. The factorization theorem is given in a form which is formally weaker but essentially equivalent to (20).]
(1925) *Statistical Methods for Research Workers*, Edinburgh, Oliver and Boyd, 1st ed., 1925; 11th ed., 1950.
(1935) *The Design of Experiments*, Edinburgh, Oliver and Boyd.
(1950) *Contributions to Mathematical Statistics*, New York, John Wiley & Sons.

Hodges, J. L., Jr., and E. L. Lehmann
(1952) "The use of previous experience in reaching statistical decisions, " *Ann. Math. Stat.*, Vol. 23, pp. 396–407.
[Theory of restricted Bayes solutions.]

Hotelling, H.
(1936) "Relations between two sets of variates," *Biometrika*, Vol. 28, pp. 321–377.
[One of the early papers making explicit use of invariance considerations.]

Hunt, G., and C. Stein
(1946) "Most stringent tests of statistical hypotheses."
[In this paper, which unfortunately has remained unpublished, a general theory of invariance is developed for hypothesis testing.]

Kendall, M. G.
(1946) *The Advanced Theory of Statistics*, Vol. 2, London, Charles Griffin and Co.

Kolmogorov, A.
(1942) "Sur l'estimation statistique des paramètres de la loi de Gauss," *Bull. Acad. Sci., URSS. Ser .Math.*, Vol. 6, pp. 3–32. (Russian-French summary.)
[Definition of sufficiency in terms of distributions for the parameters.]

Laplace, P. S.
(1812) *Théorie Analytique des Probabilités*, Paris.

LeCam, Lucien
(1953) "On some asymptotic properties of maximum likelihood estimates and related Bayes estimates," *Univ. Calif. Publs. Statistics*, Vol. 1, pp. 277–329, Univ. Calif. Press, Berkeley and Los Angeles.
[Rigorous and very general treatment of the large sample theory of maximum likelihood estimates, with a survey of the large previous literature on the subject.]

Lehmann, E. L.
(1947) "On families of admissible tests," *Ann. Math. Stat.*, Vol. 18, pp. 97–104.
[Introduction of the complete class concept in connection with a special class of testing problems.]
(1950) "Some principles of the theory of hypothesis testing," *Ann. Math. Stat.*, Vol. 21, pp. 1–26.
(1951) "A general concept of unbiasedness," *Ann. Math. Stat.*, Vol. 22, pp. 587–597.
[Definition (8); Problems 2, 3, 4, 6, 7, and 14.]

Neyman, J.
(1935) "Sur un teorema concernente le cosidette statistiche sufficienti," *Giorn. Ist. Ital. Att.*, Vol. 6, pp. 320–334.
[Obtains the factorization theorem in the form (20).]
(1938) "L'estimation statistique traitée comme un problème classique de probabilité," *Actualités sci. et ind.*, No. 739, pp. 25–57.
[Puts forth the point of view that statistics is primarily concerned with how to behave under uncertainty rather than with determining the values of unknown parameters, with inductive behavior rather than with inductive inference.]
(1938) *Lectures and Conferences on Mathematical Statistics and Probability*, Washington, Graduate School, U.S. Dept. Agriculture, 1st ed., 1938; 2nd ed., 1952.

Neyman, J., and E. S. Pearson
(1928) "On the use and interpretation of certain test criteria for purposes of statistical inference," *Biometrika*, Vol. 20A, pp. 175-240, 263–294.
[Proposes the likelihood ratio principle for obtaining reasonable tests, and applies it to a number of specific problems.]

(1933) "On the testing of statistical hypotheses in relation to probability a priori," *Proc. Cambridge Phil. Soc.*, Vol. 29, pp. 492–510.

[In connection with the problem of hypothesis testing, suggests assigning weights for the various possible wrong decisions and the use of the minimax principle.]

Peisakoff, Melvin

(1951) "Transformation of parameters," unpublished thesis, Princeton.

[Extends the Hunt-Stein theory of invariance to more general classes of decision problems; Problem 11(ii). The theory is generalized further by J. Kiefer, "Invariance, minimax sequential estimation, and continuous time processes," *Ann. Math. Stat.*, Vol. 28 (1957), pp. 573–601, and by H. Kudo, "On minimax invariant estimates of the transformations parameter," *Nat. Sci. Rept. Ochanomizu Univ., Tokyo*, Vol. 6 (1955), pp. 31–73.]

Pitman, E. J. G.

(1939) "Location and scale parameters," *Biometrika*, Vol. 30, pp. 391–421.

(1939) "Tests of hypotheses concerning location and scale parameters," *Biometrika*, Vol. 31, pp. 200–215.

[In these papers the restriction to invariant procedures is introduced for estimation and testing problems involving location and scale parameters.]

Savage, L. J.

(1954) *The Foundations of Statistics*, New York, John Wiley & Sons.

Schmetterer, L.

(1956) *Einfuehrung in die mathematische Statistik*, Wien, Springer-Verlag.

van der Waerden, B. L.

(1957) *Mathematische Statistik*, Berlin, Springer-Verlag.

Wald, Abraham

(1939) "Contributions to the theory of statistical estimation and testing hypotheses," *Ann. Math. Stat.*, Vol. 10, pp. 299–326.

[A general formulation of statistical problems containing estimation and testing problems as special cases. Discussion of Bayes and minimax procedures.]

(1947) "An essentially complete class of admissible decision functions," *Ann. Math. Stat.*, Vol. 18, pp. 549–555.

[Defines and characterizes complete classes of decision procedures for general decision problems. The ideas of this and the preceding paper were developed further in a series of papers culminating in Wald's book mentioned above.]

(1950) *Statistical Decision Functions*, New York, John Wiley & Sons.

(1958) *Selected Papers in Statistics and Probability by Abraham Wald*, Stanford University Press.

Wilks, S. S.

(1944) *Mathematical Statistics*, Princeton University Press.

Wolfwitz, J.

(1951) "On ϵ-complete classes of decision functions," *Ann. Math. Stat.*, Vol. 22, pp. 461–465.

(1953) "The method of maximum likelihood and the Wald theory of decision functions," *Indag. Math.*, Vol. 15, pp. 114–119.

CHAPTER 2

The Probability Background

1. PROBABILITY AND MEASURE

The mathematical framework for statistical decision theory is provided by the theory of probability which in turn has its foundations in the theory of measure and integration. The present and following sections serve to define some of the basic concepts of these theories, to establish some notation, and to state without proof some of the principal results. In the remainder of the chapter, certain special topics are treated in more detail.

Probability theory is concerned with situations which may result in different outcomes. The totality of these possible outcomes is represented abstractly by the totality of points in a space $\mathcal{X}$. Since the events to be studied are aggregates of such outcomes, they are represented by subsets of $\mathcal{X}$. The union of two sets A_1, A_2 will be denoted by $A_1 \cup A_2$, their intersection by $A_1 \cap A_2$, the complement of A by $\tilde{A} = \mathcal{X} - A$, and the empty set by 0. The probability $P(A)$ of an event A is a real number between 0 and 1; in particular

$$(1) \qquad P(0) = 0 \quad \text{and} \quad P(\mathcal{X}) = 1.$$

Probabilities have the property of *countable additivity*,

$$(2) \qquad P(\bigcup A_i) = \Sigma P(A_i) \quad \text{if } A_i \cap A_j = 0 \text{ for all } i \neq j.$$

Unfortunately it turns out that the set functions with which we shall be concerned usually cannot be defined in a reasonable manner for all subsets of $\mathcal{X}$ if they are to satisfy (2). It is, for example, not possible to give a reasonable definition of "area" for all subsets of a unit square in the plane.

The sets for which the probability function P will be defined are said to be "measurable." The domain of definition of P should include with any set A its complement $\tilde{A}$, and with any countable number of events their union. By (1), it should also include $\mathcal{X}$. A class of sets that contains

$\mathscr{X}$ and is closed under complementation and countable unions is a *σ-field.* Such a class is automatically also closed under countable intersections.

The starting point of any probabilistic considerations is therefore a space $\mathscr{X}$, representing the possible outcomes, and a σ-field $\mathscr{A}$ of subsets of $\mathscr{X}$, representing the events whose probability is to be defined. Such a couple $(\mathscr{X}, \mathscr{A})$ is called a *measurable space*, and the elements of $\mathscr{A}$ constitute the *measurable sets.* A countably additive nonnegative (not necessarily finite) set function μ defined over $\mathscr{A}$ and such that $\mu(0) = 0$ is called a *measure.* If it assigns the value 1 to $\mathscr{X}$ it is a *probability measure.* More generally, μ is *finite* if $\mu(\mathscr{X}) < \infty$ and *σ-finite* if there exist $A_1, A_2, \cdots$ in $\mathscr{A}$ (which may always be taken to be mutually exclusive) such that $\bigcup A_i = \mathscr{X}$ and $\mu(A_i) < \infty$ for $i = 1, 2, \cdots$. Important special cases are provided by the following examples.

Example 1. Let $\mathscr{X}$ be the n-dimensional Euclidean space E_n, and $\mathscr{A}$ the smallest σ-field containing all rectangles

$$R = \{(x_1, \cdots, x_n): a_i < x_i \leqq b_i, i = 1, \cdots, n\}.^*$$

The elements of $\mathscr{A}$ are called the *Borel sets* of E_n. Over $\mathscr{A}$ a unique measure μ can be defined, which to any rectangle R assigns as its measure the volume of R,

$$\mu(R) = \prod_{i=1}^{n} (b_i - a_i).$$

The measure μ can be *completed* by adjoining to $\mathscr{A}$ all subsets of sets of measure zero. The domain of μ is thereby enlarged to a σ-field $\mathscr{A}'$, the class of *Lebesgue measurable* sets. The term *Lebesgue measure* is used for μ both when it is defined over the Borel sets and over the Lebesgue measurable sets.

This example can be generalized to any nonnegative set function ν, which is defined and countably additive over the class of rectangles R. There exists then, as before, a unique measure μ over $(\mathscr{X}, \mathscr{A})$ that agrees with ν for all R. This measure can again be completed; however, the resulting σ-field depends on μ and need not agree with the σ-field $\mathscr{A}'$ obtained above.

Example 2. Suppose that $\mathscr{X}$ is countable, and let $\mathscr{A}$ be the class of all subsets of $\mathscr{X}$. For any set A, define $\mu(A)$ as the number of elements of A if that number is finite and otherwise as $+\infty$. This measure is sometimes called *counting measure.*

In applications, the probabilities over $(\mathscr{X}, \mathscr{A})$ refer to random experiments or observations, the possible outcomes of which are the points $x \in \mathscr{X}$. Let these observations be denoted by X, which may for example be real- or vector-valued, and let the probability of X falling in a set A

* If $\pi(x)$ is a statement concerning certain objects x, then $\{x: \pi(x)\}$ denotes the set of all those x for which $\pi(x)$ is true.

be $P\{X \in A\} = P(A)$. In this context, the probability $P(A)$ will sometimes be denoted by $P^X(A)$ and the probability measure P by P^X. We shall refer to X as a *random variable** over the space $(\mathscr{X}, \mathscr{A})$, and to the probability measure P or P^X as the *probability distribution* of X. Mathematically, a random variable is thus nothing but a carrier of its distribution. If $\pi(x)$ is any statement concerning the points x and if A is the set of points x for which $\pi(x)$ holds, we shall also write $P\{\pi(X)\}$ for the probability $P^X(A)$.

Let X be a real-valued random variable with probability distribution P^X defined over the Borel sets of the real line. Then the *cumulative distribution function* of X is defined as a point function F on the real line by $F(a) = P\{X \leqq a\}$ for all real a. The function F is nondecreasing and continuous on the right, and $F(-\infty) = 0$, $F(+\infty) = 1$. If F is any function with these properties, a measure can be defined over the intervals by $P\{a < X \leqq b\} = F(b) - F(a)$. It follows from the generalization of Example 1 that this measure uniquely determines a probability distribution over the Borel sets. Thus the probability distribution P^X and the cumulative distribution function F each uniquely determines the other. These remarks extend to probability distributions over an n-dimensional Euclidean space, where the cumulative distribution function is defined by

$$F(a_1, \cdots, a_n) = P\{X_1 \leqq a_1, \cdots, X_n \leqq a_n\}.$$

The distribution of X also determines that of any function of X. Let T be a function of the observations taking on values in some space $\mathscr{T}$. Such a function generates in $\mathscr{T}$ the σ-field $\mathscr{B}'$ of sets B whose inverse image

$$A = T^{-1}(B) = \{x\colon\ x \in \mathscr{X},\ T(x) \in B\}$$

is in $\mathscr{A}$. The values taken on by $T(X)$ are again the outcomes of a random experiment, so that $T = T(X)$ is a random variable over the space $(\mathscr{T}, \mathscr{B}')$. Since $X \in T^{-1}(B)$ if and only if $T(X) \in B$, the probability distribution of T over $(\mathscr{T}, \mathscr{B}')$ is given by

$$P^T(B) = P\{T \in B\} = P\{X \in T^{-1}(B)\} = P^X(T^{-1}(B)). \tag{3}$$

Frequently, there is given a σ-field $\mathscr{B}$ of sets in $\mathscr{T}$ such that the probability of the event $T \in B$ should be defined if and only if $B \in \mathscr{B}$. This requires that $T^{-1}(B) \in \mathscr{A}$ for all $B \in \mathscr{B}$, and the function (or transformation) T from $(\mathscr{X}, \mathscr{A})$ into† $(\mathscr{T}, \mathscr{B})$ is then said to be *measurable*. Another

* This differs from the definition given in most probability texts where a random variable is taken to be a function from an original space to a range space $(\mathscr{X}, \mathscr{A})$ and where in addition $\mathscr{X}$ is assumed to be the real line and $\mathscr{A}$ the class of Borel sets.

† The term *into* is used to indicate that the range $T(\mathscr{X})$ of T is in $\mathscr{T}$; if $T(\mathscr{X}) = \mathscr{T}$, the transformation is said to be from $\mathscr{X}$ *onto* $\mathscr{T}$.

implication is the sometimes convenient restriction of probability statements to the sets in $\mathscr{B}$ even though there may exist sets $B \notin \mathscr{B}$ for which $T^{-1}(B) \in \mathscr{A}$ and whose probability therefore could be defined.

In applications, there is given as the raw material of an investigation a set of observations constituting the totality of the available data. This is represented by a random variable X such that all other random variables that can be considered are functions of X. The space $(\mathscr{X}, \mathscr{A})$ over which X is defined is called the *sample space*, and any measurable transformation T from $(\mathscr{X}, \mathscr{A})$ into $(\mathscr{T}, \mathscr{B})$ is said to be a *statistic*. The distribution of T is then given by (3) applied to all $B \in \mathscr{B}$. With this definition, a statistic is specified by specifying both the function T and the σ-field $\mathscr{B}$. We shall, however, adopt the convention that when a function T takes on its values in a Euclidean space, unless otherwise stated the σ-field $\mathscr{B}$ of measurable sets will be taken to be the class of Borel sets. It then becomes unnecessary to mention it explicitly or to indicate it in the notation.

The distinction between statistics and random variables as defined here is slight. The term statistic is used to indicate that the quantity is a function of more basic observations; all statistics in a given problem are functions defined over the same sample space $(\mathscr{X}, \mathscr{A})$. On the other hand, any statistic T is a random variable since it has a distribution over $(\mathscr{T}, \mathscr{B})$, and it will be referred to as a random variable when its origin is irrelevant. Which term is used therefore depends on the point of view and to some extent is arbitrary.*

2. INTEGRATION

According to the convention of the preceding section, a real-valued function f defined over $(\mathscr{X}, \mathscr{A})$ is measurable if $f^{-1}(B) \in \mathscr{A}$ for every Borel set B on the real line. Such a function f is said to be *simple* if it takes on only a finite number of values. Let μ be a measure defined over $(\mathscr{X}, \mathscr{A})$, and let f be a simple function taking on the distinct values $a_1, \cdots, a_m$ on the sets $A_1, \cdots, A_m$, which are in $\mathscr{A}$ since f is measurable. If $\mu(A_i) < \infty$ when $a_i \neq 0$, the integral of f with respect to μ is defined by

$$\int f\,d\mu = \Sigma a_i \mu(A_i). \tag{4}$$

Given any nonnegative measurable function f, there exists a nondecreasing sequence of simple functions f_n converging to f. Then the

* The above definition of statistic is close to the definition of random variable customary in probability theory. However, the distinction made here corresponds more closely to the way the terms are used informally in most statistical writing.

integral of f is defined as

$$\int f\,d\mu = \lim_{n\to\infty} \int f_n\,d\mu, \tag{5}$$

which can be shown to be independent of the particular sequence of f_n's chosen. For any measurable function f its positive and negative parts

$$f^+(x) = \max\,[f(x), 0] \quad \text{and} \quad f^-(x) = \max\,[-f(x), 0] \tag{6}$$

are also measurable, and

$$f(x) = f^+(x) - f^-(x).$$

If the integrals of f^+ and f^- are both finite, then f is said to be *integrable*, and its integral is defined as

$$\int f\,d\mu = \int f^+\,d\mu - \int f^-\,d\mu.$$

If of the two integrals one is finite and one infinite, then the integral of f is defined to be the appropriate infinite value.

Example 3. Let $\mathscr{X}$ be the closed interval $[a, b]$, $\mathscr{A}$ be the class of Borel sets or of Lebesgue measurable sets in $\mathscr{X}$, and let μ be Lebesgue measure. Then the integral of f with respect to μ is written as $\int_a^b f(x)\,dx$, and is called the Lebesgue integral of f. This integral generalizes the Riemann integral in that it exists and agrees with the Riemann integral of f whenever the latter exists.

Example 4. Let $\mathscr{X}$ be countable and consist of the points $x_1, x_2, \cdots$; let $\mathscr{A}$ be the class of all subsets of $\mathscr{X}$, and let μ assign measure b_i to the point x_i. Then f is integrable provided $\Sigma f(x_i)b_i$ converges absolutely, and $\int f\,d\mu$ is given by this sum.

Let P^X be the probability distribution of a random variable X and let T be a real-valued statistic. If the function $T(x)$ is integrable, its *expectation* is defined by

$$E(T) = \int T(x)\,dP^X(x). \tag{7}$$

It will be seen from Lemma 2 in Section 3 below that the integration can be carried out alternatively in t-space with respect to the distribution of T defined by (3), so that also

$$E(T) = \int t\,dP^T(t). \tag{8}$$

The above definition of the integral permits the basic convergence theorem

Theorem 1. *Let f_n be a sequence of measurable functions and let $f_n(x) \to f(x)$ for all x. Then*

$$\int f_n \, d\mu \to \int f \, d\mu$$

if either one of the following conditions hold:

(i) (Lebesgue monotone convergence theorem)

the f_n's are nonnegative and the sequence is nondecreasing

or

(ii) (Lebesgue bounded convergence theorem)

there exists an integrable function g such that $|f_n(x)| \leqq g(x)$ for all n and x.

For any set $A \in \mathscr{A}$, let I_A be its *indicator function* defined by

$$I_A(x) = 1 \quad \text{or} \quad 0 \quad \text{as} \quad x \in A \quad \text{or} \quad x \in \tilde{A}, \tag{9}$$

and let

$$\int_A f \, d\mu = \int f I_A \, d\mu. \tag{10}$$

If μ is a measure and f a nonnegative measurable function over $(\mathscr{X}, \mathscr{A})$, then

$$\nu(A) = \int_A f \, d\mu \tag{11}$$

defines a new measure over $(\mathscr{X}, \mathscr{A})$. The fact that (11) holds for all $A \in \mathscr{A}$ is expressed by writing

$$d\nu = f \, d\mu \qquad \text{or} \qquad f = d\nu/d\mu. \tag{12}$$

Let μ and ν be two given σ-finite measures over $(\mathscr{X}, \mathscr{A})$. If there exists a function f satisfying (12), it is determined through this relation up to sets of measure zero, since

$$\int_A f \, d\mu = \int_A g \, d\mu \quad \text{for all} \quad A \in \mathscr{A}$$

implies that $f = g$ a.e. μ.* Such an f is called the *Radon-Nikodym derivative* of ν with respect to μ, and in the particular case that ν is a probability measure, the *probability density* of ν with respect to μ.

The question of existence of a function f satisfying (12) for given

* A statement that holds for all points x except possibly on a set of μ-measure zero is said to hold a.e. μ; or to hold $(\mathscr{A}, \mu)$ if it is desirable to indicate the σ-field over which μ is defined.

measures μ and ν is answered in terms of the following definition. A measure ν is *absolutely continuous* with respect to μ if

$$\mu(A) = 0 \quad \text{implies} \quad \nu(A) = 0.$$

Theorem 2. (*Radon-Nikodym.*) *If μ and ν are σ-finite measures over $(\mathscr{X}, \mathscr{A})$, then there exists a measurable function f satisfying* (12) *if and only if ν is absolutely continuous with respect to μ.*

The *direct* (or *Cartesian*) *product* $A \times B$ of two sets A and B is the set of all pairs (x, y) with $x \in A$, $y \in B$. Let $(\mathscr{X}, \mathscr{A})$ and $(\mathscr{Y}, \mathscr{B})$ be two measurable spaces, and let $\mathscr{A} \times \mathscr{B}$ be the smallest σ-field containing all sets $A \times B$ with $A \in \mathscr{A}$ and $B \in \mathscr{B}$. If μ and ν are two σ-finite measures over $(\mathscr{X}, \mathscr{A})$ and $(\mathscr{Y}, \mathscr{B})$ respectively, then there exists a unique measure $\lambda = \mu \times \nu$ over $(\mathscr{X} \times \mathscr{Y}, \mathscr{A} \times \mathscr{B})$, the *product* of μ and ν, such that for any $A \in \mathscr{A}$, $B \in \mathscr{B}$,

$$\lambda(A \times B) = \mu(A)\nu(B). \tag{13}$$

Example 5. Let $\mathscr{X}$, $\mathscr{Y}$ be Euclidean spaces of m and n dimensions, and let $\mathscr{A}$, $\mathscr{B}$ be the σ-fields of Borel sets in these spaces. Then $\mathscr{X} \times \mathscr{Y}$ is an $(m + n)$-dimensional Euclidean space, and $\mathscr{A} \times \mathscr{B}$ the class of its Borel sets.

Example 6. Let $Z = (X, Y)$ be a random variable defined over $(\mathscr{X} \times \mathscr{Y}, \mathscr{A} \times \mathscr{B})$ and suppose that the random variables X and Y have distributions P^X, P^Y over $(\mathscr{X}, \mathscr{A})$ and $(\mathscr{Y}, \mathscr{B})$. Then X and Y are said to be *independent* if the probability distribution P^Z of Z is the product $P^X \times P^Y$.

In terms of these concepts the reduction of a double integral to a repeated one is given by the following theorem.

Theorem 3. (*Fubini.*) *Let μ and ν be σ-finite measures over $(\mathscr{X}, \mathscr{A})$ and $(\mathscr{Y}, \mathscr{B})$ respectively, and let $\lambda = \mu \times \nu$. If $f(x, y)$ is integrable with respect to λ, then*

(i) *for almost all* (ν) *fixed y, the function $f(x, y)$ is integrable with respect to μ,*

(ii) *the function $\int f(x, y)\, d\mu(x)$ is integrable with respect to ν, and*

$$\int f(x, y)\, d\lambda(x, y) = \int \left[\int f(x, y)\, d\mu(x) \right] d\nu(y). \tag{14}$$

3. STATISTICS AND SUBFIELDS

According to the definition of Section 1, a statistic is a measurable transformation T from the sample space $(\mathscr{X}, \mathscr{A})$ into a measurable space $(\mathscr{T}, \mathscr{B})$. Such a transformation induces in the original sample space the subfield*

$$\mathscr{A}_0 = T^{-1}(\mathscr{B}) = \{T^{-1}(B) \colon B \in \mathscr{B}\}. \tag{15}$$

* We shall use this term in place of the more cumbersome "sub-σ-field."

Since the set $T^{-1}[T(A)]$ contains A but is not necessarily equal to A, the σ-field $\mathscr{A}_0$ need not coincide with $\mathscr{A}$ and hence can be a proper subfield of $\mathscr{A}$. On the other hand, suppose for a moment that $\mathscr{T} = T(\mathscr{X})$, that is, that the transformation T is onto rather than into $\mathscr{T}$. Then

$$T[T^{-1}(B)] = B \quad \text{for all} \quad B \in \mathscr{B}, \tag{16}$$

so that the relationship $A_0 = T^{-1}(B)$ establishes a 1 : 1 correspondence between the sets of $\mathscr{A}_0$ and $\mathscr{B}$, which is an isomorphism—that is, which preserves the set operations of intersection, union, and complementation. For most purposes it is therefore immaterial whether one works in the space $(\mathscr{X}, \mathscr{A}_0)$ or in $(\mathscr{T}, \mathscr{B})$. These generate two equivalent classes of events, and therefore of measurable functions, possible decision procedures, etc. If the transformation T is only into $\mathscr{T}$, the above 1 : 1 correspondence applies to the class $\mathscr{B}'$ of subsets of $\mathscr{T}' = T(\mathscr{X})$ which belong to $\mathscr{B}$, rather than to $\mathscr{B}$ itself. However, any set $B \in \mathscr{B}$ is equivalent to $B' = B \cap \mathscr{T}'$ in the sense that any measure over $(\mathscr{X}, \mathscr{A})$ assigns the same measure to B' as to B. Considered as classes of events, $\mathscr{A}_0$ and $\mathscr{B}$ therefore continue to be equivalent, with the only difference that $\mathscr{B}$ contains several (equivalent) representations of the same event.

As an example, let $\mathscr{X}$ be the real line and $\mathscr{A}$ the class of Borel sets, and let $T(x) = x^2$. Let $\mathscr{T}$ be either the positive real axis or the whole real axis and let $\mathscr{B}$ be the class of Borel subsets of $\mathscr{T}$. Then $\mathscr{A}_0$ is the class of Borel sets that are symmetric with respect to the origin. When considering, for example, real-valued measurable functions, one would, when working in $\mathscr{T}$-space, restrict attention to measurable functions of x^2. Instead, one could remain in the original space, where the restriction would be to the class of even measurable functions of x. The equivalence is clear. Which representation is more convenient depends on the situation.

That the correspondence between the sets $A_0 = T^{-1}(B) \in \mathscr{A}_0$ and $B \in \mathscr{B}$ establishes an analogous correspondence between measurable functions defined over $(\mathscr{X}, \mathscr{A}_0)$ and $(\mathscr{T}, \mathscr{B})$ is shown by the following lemma.

Lemma 1. *Let the statistic T from $(\mathscr{X}, \mathscr{A})$ into $(\mathscr{T}, \mathscr{B})$ induce the subfield $\mathscr{A}_0$. Then a real-valued $\mathscr{A}$-measurable function f is $\mathscr{A}_0$-measurable if and only if there exists a $\mathscr{B}$-measurable function g such that*

$$f(x) = g[T(x)]$$

for all x.

Proof. Suppose first that such a function g exists. Then the set

$$\{x\colon f(x) < r\} = T^{-1}(\{t\colon g(t) < r\})$$

is in $\mathscr{A}_0$, and f is $\mathscr{A}_0$-measurable. Conversely, if f is $\mathscr{A}_0$-measurable, then the sets

$$A_{in} = \left\{x\colon \frac{i}{2^n} < f(x) \leqq \frac{i+1}{2^n}\right\}, \qquad i = 0, \pm 1, \pm 2, \cdots$$

are (for fixed n) disjoint sets in $\mathscr{A}_0$ whose union is $\mathscr{X}$, and there exist $B_{in} \in \mathscr{B}$ such that $A_{in} = T^{-1}(B_{in})$. Let

$$B_{in}^* = B_{in} \cap \widetilde{\bigcup_{j \neq i} B_{jn}}.$$

Since A_{in} and A_{jn} are mutually exclusive for $i \neq j$, the set $T^{-1}(B_{in} \cap B_{jn})$ is empty and so is the set $T^{-1}(B_{in} \cap \widetilde{B_{in}^*})$. Hence, for fixed n, the sets B_{in}^* are disjoint, and still satisfy $A_{in} = T^{-1}(B_{in}^*)$. Defining

$$f_n(x) = \frac{i}{2^n} \quad \text{if} \quad x \in A_{in}, \qquad i = 0, \pm 1, \pm 2, \cdots,$$

one can write

$$f_n(x) = g_n[T(x)],$$

where

$$g_n(t) = \begin{cases} \dfrac{i}{2^n} & \text{for} \quad t \in B_{in}^*, \qquad i = 0, \pm 1, \pm 2, \cdots \\ 0 \text{ otherwise.} \end{cases}$$

Since the functions g_n are $\mathscr{B}$-measurable, the set B on which $g_n(t)$ converges to a finite limit is in $\mathscr{B}$. Let $R = T(\mathscr{X})$ be the range of T. Then for $t \in R$,

$$\lim g_n[T(x)] = \lim f_n(x) = f(x)$$

for all $x \in \mathscr{X}$ so that R is contained in B. Therefore, the function g defined by $g(t) = \lim g_n(t)$ for $t \in B$ and $g(t) = 0$ otherwise possesses the required properties.

The relationship between integrals of the functions f and g above is given by the following lemma.

Lemma 2. *Let T be a measurable transformation from $(\mathscr{X}, \mathscr{A})$ into $(\mathscr{T}, \mathscr{B})$, μ a σ-finite measure over $(\mathscr{X}, \mathscr{A})$, and g a real-valued measurable function of t. If μ^* is the measure defined over $(\mathscr{T}, \mathscr{B})$ by*

$$\mu^*(B) = \mu[T^{-1}(B)] \quad \text{for all} \quad B \in \mathscr{B}, \tag{17}$$

then for any $B \in \mathscr{B}$,

$$\int_{T^{-1}(B)} g[T(x)]\, d\mu(x) = \int_B g(t)\, d\mu^*(t) \tag{18}$$

in the sense that if either integral exists, so does the other and the two are equal.

Proof. Without loss of generality let B be the whole space $\mathscr{T}$. If g is the indicator of a set $B_0 \in \mathscr{B}$, the lemma holds since the left- and right-hand sides of (18) reduce respectively to $\mu[T^{-1}(B_0)]$ and $\mu^*(B_0)$ which are equal by the definition of μ^*. It follows that (18) holds successively for all simple functions, for all nonnegative measurable functions, and hence finally for all integrable functions.

4. CONDITIONAL EXPECTATION AND PROBABILITY

If two statistics induce the same subfield $\mathscr{A}_0$, they are equivalent in the sense of leading to equivalent classes of measurable events. This equivalence is particularly relevant to considerations of conditional probability. Thus if X is normally distributed with zero mean, the information carried by the statistics $|X|$, X^2, e^{-X^2}, etc., is the same. Given that $|X| = t$, $X^2 = t^2$, $e^{-X^2} = e^{-t^2}$, it follows that X is $\pm t$, and any reasonable definition of conditional probability will assign probability 1/2 to each of these values. The general definition of conditional probability to be given below will in fact involve essentially only $\mathscr{A}_0$ and not the range space $\mathscr{T}$ of T. However, when referred to $\mathscr{A}_0$ alone the concept loses much of its intuitive meaning, and the gap between the elementary definition and that of the general case becomes unnecessarily wide. For these reasons it is frequently more convenient to work with a particular representation of a statistic, involving a definite range space $(\mathscr{T}, \mathscr{B})$.

Let P be a probability measure over $(\mathscr{X}, \mathscr{A})$, T a statistic with range space $(\mathscr{T}, \mathscr{B})$, and $\mathscr{A}_0$ the subfield it induces. Consider a nonnegative function f which is integrable $(\mathscr{A}, P)$, that is, $\mathscr{A}$-measurable and P-integrable. Then $\int_A f\,dP$ is defined for all $A \in \mathscr{A}$ and therefore for all $A_0 \in \mathscr{A}_0$. It follows from the Radon-Nikodym theorem (Theorem 2) that there exists a function f_0, which is integrable $(\mathscr{A}_0, P)$ and such that

$$\int_{A_0} f\,dP = \int_{A_0} f_0\,dP \quad \text{for all} \quad A_0 \in \mathscr{A}_0, \tag{19}$$

and that f_0 is unique $(\mathscr{A}_0, P)$. By Lemma 1, f_0 depends on x only through $T(x)$. In the example of a normally distributed variable X with zero mean, and $T = X^2$, the function f_0 is determined by (19) holding for all sets A_0 that are symmetric with respect to the origin, so that $f_0(x) = \frac{1}{2}[f(x) + f(-x)]$.

The function f_0 defined through (19) is determined by two properties:

(i) Its average value over any set A_0 with respect to P is the same as that of f;

(ii) It depends on x only through $T(x)$ and hence is constant on the sets D_x over which T is constant.

Intuitively, what one attempts to do in order to construct such a function is to define $f_0(x)$ as the conditional P-average of f over the set D_x. One would thereby replace the single averaging process of integrating f represented by the left-hand side by a two-stage averaging process such as an iterated integral. Such a construction can actually be carried out when X is a discrete variable and in the regular case considered in Chapter 1, Section 9; $f_0(x)$ is then just the conditional expectation of $f(X)$ given $T(x)$. In general, it is not clear how to define this conditional expectation directly. Since it should, however, possess properties (i) and (ii), and since these through equation (19) determine f_0 uniquely $(\mathscr{A}_0, P)$, we shall take $f_0(x)$ of (19) as the general definition of the *conditional expectation* $E[f(X)|T(x)]$. Equivalently, if $f_0(x) = g[T(x)]$ one can write

$$E[f(X)|T = t] = g(t),$$

so that $E[f(X)t|]$ is a $\mathscr{B}$-measurable function defined up to equivalence $(\mathscr{B}, P^T)$. In the relationship of integrals given in Lemma 2, if $\mu = P^X$ then $\mu^* = P^T$, and it is seen that the function g can be defined directly in terms of f through

$$\text{(20)} \qquad \int_{T^{-1}(B)} f(x)\, dP^X(x) = \int_B g(t)\, dP^T(t) \quad \text{for all} \quad B \in \mathscr{B},$$

which is equivalent to (19).

So far, f has been assumed to be nonnegative. In the general case, the conditional expectation of f is defined as

$$E[f(X)|t] = E[f^+(X)|t] - E[f^-(X)|t].$$

Example 7. Let $X_1, \cdots, X_n$ be identically and independently distributed random variables with a continuous distribution function and let

$$T(x_1, \cdots, x_n) = (x^{(1)}, \cdots, x^{(n)})$$

where $x^{(1)} \leqq \cdots \leqq x^{(n)}$ denote the ordered x's. Without loss of generality one can restrict attention to the points with $x^{(1)} < \cdots < x^{(n)}$ since the probability of two coordinates being equal is 0. Then $\mathscr{X}$ is the set of all n-tuples with distinct coordinates, $\mathscr{T}$ the set of all ordered n-tuples, and $\mathscr{A}$ and $\mathscr{B}$ are the classes of Borel subsets of $\mathscr{X}$ and $\mathscr{T}$. Under T^{-1} the set consisting of the single point $a = (a_1, \cdots, a_n)$ is transformed into the set consisting of the $n!$ points $(a_{i_1}, \cdots, a_{i_n})$ that are obtained from a by permuting the coordinates in all possible ways. It follows that $\mathscr{A}_0$ is the class of all sets that are symmetric in the sense that if A_0 contains a point $x = (x_1, \cdots, x_n)$ then it also contains all points $(x_{i_1}, \cdots, x_{i_n})$.

For any integrable function f, let

$$f_0(x) = \frac{1}{n!} \Sigma f(x_{i_1} \cdots, x_{i_n})$$

where the summation extends over the $n!$ permutations of $(x_1, \cdots, x_n)$. Then f_0 is $\mathscr{A}_0$-measurable since it is symmetric in its n arguments. Also

$$\int_{A_0} f(x_1, \cdots, x_n)\, dP(x_1) \cdots dP(x_n) = \int_{A_0} f(x_{i_1}, \cdots, x_{i_n})\, dP(x_1) \cdots dP(x_n)$$

so that f_0 satisfies (19). It follows that $f_0(x)$ is the conditional expectation of $f(X)$ given $T(x)$.

The conditional expectation of $f(X)$ given the above statistic $T(x)$ can also be found without assuming the X's to be identically and independently distributed. Suppose that X has a density $h(x)$ with respect to a measure μ (such as Lebesgue measure), which is symmetric in the variables $x_1, \cdots, x_n$ in the sense that for any $A \in \mathscr{A}$ it assigns to the set $\{x\colon (x_{i_1} \cdots, x_{i_n}) \in A\}$ the same measure for all permutations $(i_1, \cdots, i_n)$. Let

$$f_0(x_1, \cdots, x_n) = \frac{\Sigma f(x_{i_1}, \cdots, x_{i_n})\, h\, (x_{i_1}, \cdots, x_{i_n})}{\Sigma h(x_{i_1}, \cdots, x_{i_n})}$$

where here and in the sums below the summation extends over the $n!$ permutations of $(x_1, \cdots, x_n)$. The function f_0 is symmetric in its n arguments and hence $\mathscr{A}_0$-measurable. For any symmetric set A_0, the integral

$$\int_{A_0} f_0(x_1, \cdots, x_n) h(x_{j_1} \cdots, x_{j_n})\, d\mu(x_1, \cdots, x_n)$$

has the same value for each permutation $(x_{j_1} \cdots, x_{j_n})$, and therefore

$$\int_{A_0} f_0(x_1, \cdots, x_n) h(x_1, \cdots, x_n)\, d\mu(x_1, \cdots, x_n)$$

$$= \int_{A_0} f_0(x_1, \cdots, x_n) \frac{1}{n!} \Sigma h(x_{i_1}, \cdots, x_{i_n})\, d\mu(x_1, \cdots, x_n)$$

$$= \int_{A_0} f(x_1, \cdots, x_n) h(x_1, \cdots, x_n)\, d\mu(x_1, \cdots, x_n).$$

It follows that $f_0(x) = E[f(X)|T(x)]$.

Equivalent to the statistic $T(x) = (x^{(1)} \cdots, x^{(n)})$, the set of *order statistics*, is $U(x) = (\Sigma x_i, \Sigma x_i^2, \cdots, \Sigma x_i^n)$. This is an immediate consequence of the fact, to be shown below, that if $T(x^0) = t^0$ and $U(x^0) = u^0$, then

$$T^{-1}(\{t^0\}) = U^{-1}(\{u^0\}) = S$$

where $\{t^0\}$ and $\{u^0\}$ denote the sets consisting of the single point t^0 and u^0 respectively, and where S consists of the totality of points $x = (x_1, \cdots, x_n)$ obtained by permuting the coordinates of $x^0 = (x_1^0, \cdots, x_n^0)$ in all possible ways.

That $T^{-1}(\{t^0\}) = S$ is obvious. To see the corresponding fact for U^{-1}, let

$$V(x) = (\sum_i x_i, \sum_{i<j} x_i x_j, \sum_{i<j<k} x_i x_j x_k, \cdots, x_1 x_2 \cdots x_n),$$

so that the components of $V(x)$ are the elementary symmetric functions $v_1 = \Sigma x_i, \cdots, v_n = x_1 \cdots x_n$ of the n arguments $x_1, \cdots, x_n$. Then

$$(x - x_1) \cdots (x - x_n) = x^n - v_1 x^{n-1} + v_2 x^{n-2} - \cdots + (-1)^n v_n.$$

Hence $V(x^0) = v^0 = (v_1^0, \cdots, v_n^0)$ implies that $V^{-1}(\{v^0\}) = S$. That then also $U^{-1}(\{u^0\}) = S$ follows from the $1:1$ correspondence between u and v established by the relations (known as Newton's identities),*

$$u_k - v_1 u_{k-1} + v_2 u_{k-2} - \cdots + (-1)^{k-1} v_{k-1} u_1 + (-1)^k k v_k = 0, \qquad 1 \leq k \leq n.$$

It is easily verified from the above definition that conditional expectation possesses most of the usual properties of expectation. It follows of course from the nonuniqueness of the definition that these properties can hold only $(\mathscr{B}, P^T)$. We state this formally in the following lemma.

Lemma 3. *If T is a statistic and the functions $f, g, \cdots$ are integrable $(\mathscr{A}, P)$, then a.e. $(\mathscr{B}, P^T)$*

(i) $E[af(X) + bg(X)|t] = aE[f(X)|t] + bE[g(X)|t]$;
(ii) $E[h(T)f(X)|t] = h(t)E[f(X)|t]$;
(iii) $a \leq f(x) \leq b$ $(\mathscr{A}, P)$ *implies* $a \leq E[f(X)|t] \leq b$;
(iv) $|f_n| \leq g, f_n(x) \to f(x)$ $(\mathscr{A}, P)$ *implies* $E[f_n(X)|t] \to E[f(X)|t]$.

A further useful result is obtained by specializing (20) to the case that B is the whole space $\mathscr{T}$. One then has

Lemma 4. *If $E|f(X)| < \infty$, and if $g(t) = E[f(X)|t]$, then*

$$Ef(X) = Eg(T), \tag{21}$$

that is, expectation can be obtained as the expected value of the conditional expectation.

Since $P\{X \in A\} = E[I_A(X)]$, where I_A denotes the indicator of the set A, it is natural to define the *conditional probability* of A given $T = t$ by

$$P(A|t) = E[I_A(X)|t]. \tag{22}$$

In view of (20) the defining equation for $P(A|t)$ can therefore be written as

$$\begin{aligned} P^X(A \cap T^{-1}(B)) &= \int_{A \cap T^{-1}(B)} dP^X(x) \\ &= \int_B P(A|t)\, dP^T(t) \quad \text{for all} \quad B \in \mathscr{B}. \end{aligned} \tag{23}$$

*For a proof of these relations see for example Dickson, *New First Course in the Theory of Equations*, New York, John Wiley & Sons, 1939, Chapter X.

It is an immediate consequence of Lemma 3 that subject to the appropriate null set* qualifications, $P(A|t)$ possesses the usual properties of probabilities, as summarized in the following lemma.

Lemma 5. *If T is a statistic with range space $(\mathcal{T}, \mathcal{B})$, and A, B, A_1, $A_2, \cdots$ are sets belonging to $\mathcal{A}$, then a.e. $(\mathcal{B}, P^T)$*

(i) $0 \leqq P(A|t) \leqq 1$;

(ii) *if the sets $A_1, A_2, \cdots$ are mutually exclusive,*

$$P(\bigcup A_i|t) = \Sigma P(A_i|t);$$

(iii) *$A \subset B$ implies $P(A|t) \leqq P(B|t)$.*

According to definition (22), the conditional probability $P(A|t)$ must be considered for fixed A as a $\mathcal{B}$-measurable function of t. This is in contrast to the elementary definition in which one takes t as fixed and considers $P(A|t)$ for varying A as a set function over $\mathcal{A}$. Lemma 5 suggests the possibility that the interpretation of $P(A|t)$ for fixed t as a probability distribution over $\mathcal{A}$ may be valid also in the general case. However, the equality $P(A_1 \cup A_2|t) = P(A_1|t) + P(A_2|t)$, for example, can break down on a null set that may vary with A_1 and A_2, and the union of all these null sets need no longer have measure zero.

For an important class of cases, this difficulty can be overcome through the nonuniqueness of the functions $P(A|t)$, which for each fixed A are determined only up to sets of measure zero in t. Since all determinations of these functions are equivalent, it is enough to find a specific determination for each A so that for each fixed t these determinations jointly constitute a probability distribution over $\mathcal{A}$. This possibility is illustrated by Example 7, in which the conditional probability distribution given $T(x) = t$ can be taken to assign probability $1/n!$ to each of the $n!$ points satisfying $T(x) = t$. The existence of such conditional distributions will be explored more generally in the next section.

5. CONDITIONAL PROBABILITY DISTRIBUTIONS†

We shall now investigate the existence of conditional probability distributions under the assumption, satisfied in most statistical applications, that $\mathcal{X}$ is a Borel set in a Euclidean space. We shall then say for short that $\mathcal{X}$ is Euclidean and assume that, unless otherwise stated, $\mathcal{A}$ is the class of Borel subsets of $\mathcal{X}$.

* This term is used as an alternative to the more cumbersome "set of measure zero."

† This section may be omitted at first reading. Its principal application is in the proof of Lemma 8(ii) in Section 7, which in turn is used only in the proof of Theorem 3 of Chapter 4.

Theorem 4. *If $\mathscr{X}$ is Euclidean, there exist determinations of the functions $P(A|t)$ such that for each t, $P(A|t)$ is a probability measure over $\mathscr{A}$.*

Proof. By setting equal to 0 the probability of any Borel set in the complement of $\mathscr{X}$, one can extend the given probability measure to the class of all Borel sets and can therefore assume without loss of generality that $\mathscr{X}$ is the full Euclidean space. For simplicity we shall give the proof only in the one-dimensional case. For each real x put $F(x, t) = P((-\infty, x]|t)$ for some version of this conditional probability function, and let $r_1, r_2, \cdots$ denote the set of all rational numbers in some order. Then $r_i < r_j$ implies that $F(r_i, t) \leq F(r_j, t)$ for all t except those in a null set N_{ij}, and hence that $F(x, t)$ is nondecreasing in x over the rationals for all t outside of the null set $N' = \bigcup N_{ij}$. Similarly, it follows from Lemma 3(iv) that for all t not in N'', as n tends to infinity $\lim F(r_i + 1/n, t) = F(r_i, t)$ for $i = 1, 2, \cdots$, $\lim F(n, t) = 1$, and $\lim F(-n, t) = 0$. Therefore, for all t outside of the null set $N' \cup N''$, $F(x, t)$ considered as a function of x is properly normalized, monotone, and continuous on the right over the rationals. For t not in $N' \cup N''$ let $F^*(x, t)$ be the unique function that is continuous on the right in x and agrees with $F(x, t)$ for all rational x. Then $F^*(x, t)$ is a cumulative distribution function and therefore determines a probability measure $P^*(A|t)$ over $\mathscr{A}$. We shall now show that $P^*(A|t)$ is a conditional probability of A given t, by showing that for each fixed A it is a $\mathscr{B}$-measurable function of t satisfying (23). This will be accomplished by proving that for each fixed $A \in \mathscr{A}$

$$P^*(A|t) = P(A|t) \qquad (\mathscr{B}, P^T).$$

By definition of P^* this is true whenever A is one of the sets $(-\infty, x]$ with x rational. It holds next when A is an interval $(a, b] = (-\infty, b] - (-\infty, a]$ with a, b rational, since P^* is a measure and P satisfies Lemma 5(ii). Therefore, the desired equation holds for the field $\mathscr{F}$ of all sets A which are finite unions of intervals $(a_i, b_i]$ with rational end points. Finally, the class of sets for which the equation holds is a monotone class (see Problem 1) and hence contains the smallest σ-field containing $\mathscr{F}$, which is $\mathscr{A}$. The measure $P^*(A|t)$ over $\mathscr{A}$ was defined above for all t in $N' \cup N''$. However, since neither the measurability of a function nor the values of its integrals are affected by its values on a null set, one can take arbitrary probability measures over $\mathscr{A}$ for t in $N' \cup N''$ and thereby complete the determination.

If X is a vector-valued random variable with probability distribution P^X and T is a statistic defined over $(\mathscr{X}, \mathscr{A})$, let $P^{X|t}$ denote any version of the family of conditional distributions $P(A|t)$ over $\mathscr{A}$ guaranteed by

Theorem 4. The connection with conditional expectation is given by the following theorem.

Theorem 5. *If X is a vector-valued random variable and $E|f(X)| < \infty$, then*

$$E[f(X)|t] = \int f(x)\, dP^{X|t}(x) \qquad (\mathscr{B}, P^T). \tag{24}$$

Proof. Equation (24) holds if f is the indicator of any set $A \in \mathscr{A}$. It then follows from Lemma 3 that it also holds for any simple function and hence for any integrable function.

The determination of the conditional expectation $E[f(X)|t]$ given by the right-hand side of (24) possesses for each t the usual properties of an expectation, (i), (iii), and (iv) of Lemma 3, which previously could be asserted only up to sets of measure zero depending on the functions $f, g, \cdots$ involved. Under the assumptions of Theorem 4 a similar strengthening is possible with respect to (ii) of Lemma 3, which can be shown to hold except possibly on a null set N not depending on the function h. It will be sufficient for the present purpose to prove this under the additional assumption that the range space of the statistic T is also Euclidean.*

Theorem 6. *If T is a statistic with Euclidean domain and range spaces $(\mathscr{X}, \mathscr{A})$ and $(\mathscr{T}, \mathscr{B})$, there exists a determination $P^{X|t}$ of the conditional probability distribution and a null set N such that the conditional expectation computed by*

$$E[f(X)|t] = \int f(x)\, dP^{X|t}(x)$$

satisfies for all $t \notin N$

$$E[h(T)f(X)|t] = h(t)E[f(X)|t]. \tag{25}$$

Proof. For the sake of simplicity and without essential loss of generality suppose that T is real-valued. Let $P^{X|t}(A)$ be a probability distribution over $\mathscr{A}$ for each t, the existence of which is guaranteed by Theorem 4. For $B \in \mathscr{B}$, the indicator function $I_B(t)$ is $\mathscr{B}$-measurable and

$$\int_{B'} I_B(t)\, dP^T(t) = P^T(B' \cap B) = P^X(T^{-1}B' \cap T^{-1}B).$$

Thus by (20)

$$I_B(t) = P^{X|t}(T^{-1}B) \qquad \text{a.e. } P^T.$$

* For a proof without this restriction see Section 26.2, Theorem A, of Loève, *Probability Theory*, New York, D. Van Nostrand Co., 1955.

Let B_n, $n = 1, 2, \cdots$, be the intervals of $\mathcal{T}$ with rational end points. Then there exists a P-null set $N = \bigcup N_n$ such that for $t \notin N$

$$I_{B_n}(t) = P^{X|t}(T^{-1}B_n)$$

for all n. For fixed $t \notin N$, the two set functions $P^{X|t}(T^{-1}B)$ and $I_B(t)$ are probability distributions over $\mathcal{B}$, the latter assigning probability 1 or 0 to a set as it does or does not contain the point t. Since these distributions agree over the rational intervals B_n, they agree for all $B \in \mathcal{B}$. In particular, for $t \notin N$, the set consisting of the single point t is in $\mathcal{B}$, and if

$$A^{(t)} = \{x\colon\ T(x) = t\}$$

it follows that for all $t \notin N$

$$P^{X|t}(A^{(t)}) = 1. \tag{26}$$

Thus

$$\int h[T(x)]f(x)\,dP^{X|t}(x) = \int_{A^{(t)}} h[T(x)]f(x)\,dP^{X|t}(x) = h(t)\int f(x)\,dP^{X|t}(x)$$

for $t \notin N$, as was to be proved.

It is a consequence of Theorem 6 that for all $t \notin N$, $E[h(T)|t] = h(t)$ and hence in particular $P(T \in B|t) = 1$ or 0 as $t \in B$ or $t \notin B$.

The conditional distributions $P^{X|t}$ still differ from those of the elementary case considered in Chapter 1, Section 9, in being defined over $(\mathcal{X}, \mathcal{A})$ rather than over the set $A^{(t)}$ and the σ-field $\mathcal{A}^{(t)}$ of its Borel subsets. However, (26) implies that for $t \notin N$

$$P^{X|t}(A) = P^{X|t}(A \cap A^{(t)}).$$

The calculations of conditional probabilities and expectations are therefore unchanged if for $t \notin N$, $P^{X|t}$ is replaced by the distribution $\bar{P}^{X|t}$, which is defined over $(A^{(t)}, \mathcal{A}^{(t)})$ and which assigns to any subset of $A^{(t)}$ the same probability as $P^{X|t}$.

Theorem 6 establishes for all $t \notin N$ the existence of conditional probability distributions $\bar{P}^{X|t}$, which are defined over $(A^{(t)}, \mathcal{A}^{(t)})$ and which by Lemma 4 satisfy

$$E[f(X)] = \int_{\mathcal{T}-N}\left[\int_{A^{(t)}} f(x)\,dP^{X|t}(x)\right] dP^T(t) \tag{27}$$

for all integrable functions f. Conversely, consider any family of distributions satisfying (27), and the experiment of observing first T, and if $T = t$, a random quantity with distribution $\bar{P}^{X|t}$. The result of this two-stage procedure is a point distributed over $(\mathcal{X}, \mathcal{A})$ with the same distribution as the original X. Thus $\bar{P}^{X|t}$ satisfies this "functional" definition of conditional probability.

If $(\mathscr{X}, \mathscr{A})$ is a product space $(\mathscr{T} \times \mathscr{Y}, \mathscr{B} \times \mathscr{C})$, then $A^{(t)}$ is the product of $\mathscr{Y}$ with the set consisting of the single point t. For $t \notin N$, the conditional distribution $\bar{P}^{X|t}$ then induces a distribution over $(\mathscr{Y}, \mathscr{C})$, which in analogy with the elementary case will be denoted by $P^{Y|t}$. In this case the definition can be extended to all of $\mathscr{T}$ by letting $P^{Y|t}$ assign probability 1 to a common specified point y_0 for all $t \in N$. With this definition, (27) becomes

$$Ef(T, Y) = \int_{\mathscr{T}} \left[\int_{\mathscr{Y}} f(t, y)\, dP^{Y|t}(y) \right] dP^T(t). \tag{28}$$

As an application, we shall prove the following lemma, which will be used in Section 7.

Lemma 6. *Let $(\mathscr{T}, \mathscr{B})$ and $(\mathscr{Y}, \mathscr{C})$ be Euclidean spaces, and let $P_0^{T,Y}$ be a distribution over the product space $(\mathscr{X}, \mathscr{A}) = (\mathscr{T} \times \mathscr{Y}, \mathscr{B} \times \mathscr{C})$. Suppose that another distribution P_1 over $(\mathscr{X}, \mathscr{A})$ is such that*

$$dP_1(t, y) = a(y)b(t)\, dP_0(t, y),$$

with $a(y) > 0$ for all y. Then under P_1 the marginal distribution of T and a version of the conditional distribution of Y given t are given by

$$dP_1^T(t) = b(t)\left[\int a(y)\, dP_0^{Y|t}(y) \right] dP_0^T(t)$$

and

$$dP_1^{Y|t}(y) = \frac{a(y)\, dP_0^{Y|t}(y)}{\displaystyle\int_{\mathscr{Y}} a(y')\, dP_0^{Y|t}(y')}.$$

Proof. The first statement of the lemma follows from the equation

$$\begin{aligned} P_1\{T \in B\} &= E_1[I_B(T)] = E_0[I_B(T)a(Y)b(T)] \\ &= \int_B b(t) \left[\int_{\mathscr{Y}} a(y)\, dP_0^{Y|t}(y) \right] dP_0^T(t). \end{aligned}$$

To check the second statement, one need only show that for any integrable f the expectation $E_1 f(Y, T)$ satisfies (28), which is immediate. The denominator of $dP_1^{Y|t}$ is positive since $a(y) > 0$ for all y.

6. CHARACTERIZATION OF SUFFICIENCY

We can now generalize the definition of sufficiency given in Chapter 1, Section 9. If $\mathscr{P} = \{P_\theta, \theta \in \Omega\}$ is any family of distributions defined over a common sample space $(\mathscr{X}, \mathscr{A})$, a statistic T is *sufficient* for $\mathscr{P}$ (or for θ)

if for each A in $\mathscr{A}$ there exists a determination of the conditional probability function $P_\theta(A|t)$ that is independent of θ. As an example suppose that $X_1, \cdots, X_n$ are identically and independently distributed with continuous distribution function F_θ, $\theta \in \Omega$. Then it follows from Example 7 that the set of order statistics $T(X) = (X^{(1)}, \cdots, X^{(n)})$ is sufficient for θ.

Theorem 7. *If $\mathscr{X}$ is Euclidean, and if the statistic T is sufficient for $\mathscr{P}$, then there exist determinations of the conditional probability distributions $P_\theta(A|t)$ which are independent of θ and such that for each fixed t, $P(A|t)$ is a probability measure over $\mathscr{A}$.*

Proof. This is seen from the proof of Theorem 4. By the definition of sufficiency one can, for each rational number r, take the functions $F(r, t)$ to be independent of θ, and the resulting conditional distributions will then also not depend on θ.

In Chapter 1 the definition of sufficiency was justified by showing that in a certain sense a sufficient statistic contains all the available information. In view of Theorem 7 the same justification applies quite generally when the sample space is Euclidean. With the help of a random mechanism one can then construct from a sufficient statistic T a random vector X' having the same distribution as the original sample vector X. Another generalization of the earlier result, not involving the restriction to a Euclidean sample space, is given in Problem 11.

The factorization criterion of sufficiency, derived in Chapter 1, can be extended to any *dominated* family of distributions, that is, any family $\mathscr{P} = \{P_\theta, \theta \in \Omega\}$ possessing probability densities p_θ with respect to some σ-finite measure μ over $(\mathscr{X}, \mathscr{A})$. The proof of this statement is based on the existence of a probability distribution $\lambda = \Sigma c_i P_{\theta_i}$ (Theorem 2 of the Appendix), which is *equivalent* to $\mathscr{P}$ in the sense that for any $A \in \mathscr{A}$

$$(29) \qquad \lambda(A) = 0 \quad \text{if and only if} \quad P_\theta(A) = 0 \quad \text{for all} \quad \theta \in \Omega.$$

Theorem 8. *Let $\mathscr{P} = \{P_\theta, \theta \in \Omega\}$ be a dominated family of probability distributions over $(\mathscr{X}, \mathscr{A})$ and let $\lambda = \Sigma c_i P_{\theta_i}$ satisfy* (29). *Then a statistic T with range space $(\mathscr{T}, \mathscr{B})$ is sufficient for $\mathscr{P}$ if and only if there exist nonnegative $\mathscr{B}$-measurable functions $g_\theta(t)$ such that*

$$(30) \qquad dP_\theta(x) = g_\theta[T(x)]\, d\lambda(x)$$

for all $\theta \in \Omega$.

Proof. Let $\mathscr{A}_0$ be the subfield induced by T and suppose that T is sufficient for θ. Then for all $\theta \in \Omega$, $A_0 \in \mathscr{A}_0$, and $A \in \mathscr{A}$

$$\int_{A_0} P(A|T(x))\, dP_\theta(x) = P_\theta(A \cap A_0);$$

and since $\lambda = \Sigma c_i P_{\theta_i}$

$$\int_{A_0} P(A|T(x))\, d\lambda(x) = \lambda(A \cap A_0),$$

so that $P(A|T(x))$ serves as conditional probability function also for λ. Let $g_\theta(T(x))$ be the Radon-Nikodym derivative $dP_\theta(x)/d\lambda(x)$ for $(\mathscr{A}_0, \lambda)$. To prove (30) it is necessary to show that $g_\theta(T(x))$ is also the derivative of P_θ for $(\mathscr{A}, \lambda)$. If A_0 is put equal to $\mathscr{X}$ in the first displayed equation, this follows from the relation

$$P_\theta(A) = \int P(A|T(x))\, dP_\theta(x) = \int E_\lambda[I_A(x)|T(x)]\, dP_\theta(x)$$

$$= \int E_\lambda[I_A(x)|T(x)]g_\theta(T(x))\, d\lambda(x) = \int E_\lambda[g_\theta(T(x))I_A(x)|T(x)]\, d\lambda(x)$$

$$= \int g_\theta(T(x))I_A(x)\, d\lambda(x) = \int_A g_\theta(T(x))\, d\lambda(x).$$

Here the second equality uses the fact, established at the beginning of the proof, that $P(A|T(x))$ is also the conditional probability for λ; the third equality holds since the function being integrated is $\mathscr{A}_0$-measurable and since $dP_\theta = g_\theta\, d\lambda$ for $(\mathscr{A}_0, \lambda)$; the fourth is an application of Lemma 3(ii); and the fifth employs the defining property of conditional expectation.

Suppose conversely that (30) holds. We shall then prove that the conditional probability function $P_\lambda(A|t)$ serves as a conditional probability function for all $P \in \mathscr{P}$. Let $g_\theta(T(x)) = dP_\theta(x)/d\lambda(x)$ on $\mathscr{A}$ and for fixed A and θ define a measure ν over $\mathscr{A}$ by the equation $d\nu = I_A\, dP_\theta$. Then over $\mathscr{A}_0$, $d\nu(x)/dP_\theta(x) = E_\theta[I_A(X)|T(x)]$, and therefore

$$d\nu(x)/d\lambda(x) = P_\theta[A|T(x)]g_\theta(T(x)) \text{ over } \mathscr{A}_0.$$

On the other hand, $d\nu(x)/d\lambda(x) = I_A(x)g_\theta(T(x))$ over $\mathscr{A}$, and hence

$$d\nu(x)/d\lambda(x) = E_\lambda[I_A(X)g_\theta(T(X))|T(x)] = P_\lambda[A|T(x)]g_\theta(T(x)) \text{ over } \mathscr{A}_0.$$

It follows that $P_\lambda(A|T(x))g_\theta(T(x)) = P_\theta(A|T(x))g_\theta(T(x))$ $(\mathscr{A}_0, \lambda)$ and hence $(\mathscr{A}_0, P_\theta)$. Since $g_\theta(T(x)) \neq 0$ $(\mathscr{A}_0, P_\theta)$ this shows that $P_\theta(A|T(x)) = P_\lambda(A|T(x))$ $(\mathscr{A}_0, P_\theta)$, and hence that $P_\lambda(A|T(x))$ is a determination of $P_\theta(A|T(x))$.

Instead of the above formulation, which explicitly involves the distribution λ, it is sometimes more convenient to state the result with respect to a given dominating measure μ.

Corollary 1. (*Factorization theorem.*) *If the distributions P_θ of $\mathscr{P}$ have probability densities $p_\theta = dP_\theta/d\mu$ with respect to a σ-finite measure μ,* *then T is sufficient for $\mathcal{P}$ if and only if there exist nonnegative $\mathcal{B}$-measurable functions g_θ on T and a nonnegative $\mathcal{A}$-measurable function h on $\mathcal{X}$ such that*

$$p_\theta(x) = g_\theta[(T(x)]h(x) \qquad (\mathcal{A}, \mu). \tag{31}$$

Proof. Let $\lambda = \Sigma c_i P_{\theta_i}$ satisfy (29). Then if T is sufficient, (31) follows from (30) with $h = d\lambda/d\mu$. Conversely, if (31) holds,

$$d\lambda(x) = \Sigma c_i g_{\theta_i}[T(x)]h(x)\, d\mu(x) = k[T(x)]h(x)\, d\mu(x)$$

and therefore $dP_\theta(x) = g_\theta^*(T(x))\, d\lambda(x)$, where $g_\theta^*(t) = g_\theta(t)/k(t)$ when $k(t) > 0$ and may be defined arbitrarily when $k(t) = 0$.

7. EXPONENTIAL FAMILIES

An important family of distributions which admits a reduction by means of sufficient statistics is the *exponential family*, defined by probability densities of the form

$$p_\theta(x) = C(\theta) \exp\left[\sum_{j=1}^{k} Q_j(\theta)T_j(x)\right] h(x) \tag{32}$$

with respect to a σ-finite measure μ over a Euclidean sample space $(\mathcal{X}, \mathcal{A})$. Particular cases are the distributions of a sample $X = (X_1, \cdots, X_n)$ from a binomial, Poisson, or normal distribution. In the binomial case, for example, the density (with respect to counting measure) is

$$\binom{n}{x} p^x(1-p)^{n-x} = (1-p)^n \exp\left[x \log\left(\frac{p}{1-p}\right)\right]\binom{n}{x}.$$

Example 8. If $Y_1, \cdots, Y_n$ are independently distributed, each with density (with respect to Lebesgue measure)

$$p_\sigma(y) = \frac{y^{[(f/2)-1]} \exp\left[-y/(2\sigma^2)\right]}{(2\sigma^2)^{f/2}\Gamma(f/2)}, \qquad y > 0, \tag{33}$$

then the joint distribution of the Y's constitutes an exponential family. For $\sigma = 1$, (33) is the density of the χ^2-distribution with f degrees of freedom; in particular for f an integer, this is the density of $\Sigma_{j=1}^f X_j^2$, where the X's are a sample from the normal distribution $N(0, 1)$.

Example 9. Consider n independent trials, each of them resulting in one of the s outcomes $E_1, \cdots, E_s$ with probabilities $p_1, \cdots, p_s$ respectively. If X_{ij} is 1 when the outcome of the ith trial is E_j and 0 otherwise, the joint distribution of the X's is

$$P\{X_{11} = x_{11}, \cdots, X_{ns} = x_{ns}\} = p_1^{\Sigma x_{i1}} p_2^{\Sigma x_{i2}} \cdots p_s^{\Sigma x_{is}},$$

where all $x_{ij} = 0$ or 1 and $\Sigma_j x_{ij} = 1$. This forms an exponential family with

$T_j(x) = \Sigma_{i=1}^n x_{ij}$ $(j = 1, \cdots, s-1)$. The joint distribution of the T's is the multinomial distribution

$$(34) \qquad P\{T_1 = t_1, \cdots, T_{s-1} = t_{s-1}\} = \frac{n!}{t_1! \cdots t_{s-1}!(n - t_1 - \cdots - t_{s-1})!} \cdot p_1^{t_1} \cdots p_{s-1}^{t_{s-1}} (1 - p_1 - \cdots - p_{s-1})^{n - t_1 - \cdots - t_{s-1}}.$$

If $X_1, \cdots, X_n$ is a sample from a distribution with density (32), the joint distribution of the X's constitutes an exponential family with the sufficient statistics $\sum_{i=1}^n T_j(X_i)$, $j = 1, \cdots, k$. Thus there exists a k-dimensional sufficient statistic for $(X_1, \cdots, X_n)$ regardless of the sample size. Suppose conversely that $X_1, \cdots, X_n$ is a sample from a distribution with some density $p_\theta(x)$ and that the set over which this density is positive is independent of θ. Then under regularity assumptions which make the concept of dimensionality meaningful, if there exists a k-dimensional sufficient statistic with $k < n$, the densities $p_\theta(x)$ constitute an exponential family.*

Employing a more natural parametrization and absorbing the factor $h(x)$ into μ, we shall write an exponential family in the form $dP_\theta(x) = p_\theta(x)\, d\mu(x)$ with

$$(35) \qquad p_\theta(x) = C(\theta) \exp\left[\sum_{j=1}^k \theta_j T_j(x)\right].$$

For suitable choice of the constant $C(\theta)$, the right-hand side of (35) is a probability density provided its integral is finite. The set Ω of parameter points $\theta = (\theta_1, \cdots, \theta_k)$ for which this is the case is the *natural parameter space* of the exponential family (35).

Optimum tests of certain hypotheses concerning any θ_j are obtained in Chapter 4. We shall now consider some properties of exponential families required for this purpose.

Lemma 7. *The natural parameter space of an exponential family is convex.*

Proof. Let $(\theta_1, \cdots, \theta_k)$ and $(\theta_1', \cdots, \theta_k')$ be two parameter points for which the integral of (35) is finite. Then by Hölder's inequality,

$$\int \exp\, [\Sigma[\alpha\theta_j + (1-\alpha)\theta_j']T_j(x)]\, d\mu(x) \leqq \left[\int \exp\, [\Sigma\theta_j T_j(x)]\, d\mu(x)\right]^\alpha \left[\int \exp\, [\Sigma\theta_j' T_j(x)]\, d\mu(x)\right]^{1-\alpha} < \infty$$

for any $0 < \alpha < 1$.

* For a proof and statement of the regularity conditions see Koopman, "On distributions admitting a sufficient statistic," *Trans. Am. Math. Soc.*, Vol. 39 (1936), pp. 399–409. The result is also discussed by Darmois, "Sur les lois de probabilité à estimation exhaustive," *Compt. Rend. Acad. Sci., Paris*, Vol. 260 (1935), pp. 1265–1266, and by Pitman, "Sufficient statistics and intrinsic accuracy," *Proc. Cambridge Phil. Soc.*, Vol. 32 (1936), pp. 567–579.

If the convex set Ω lies in a linear space of dimension $< k$, then (35) can be rewritten in a form involving fewer than k components of T. We shall therefore, without loss of generality, assume Ω to be k-dimensional.

It follows from the factorization theorem that $T(x) = (T_1(x), \cdots, T_k(x))$ is sufficient for $\mathscr{P} = \{P_\theta, \theta \in \Omega\}$.

Lemma 8. *Let X be distributed according to the exponential family*

$$dP^X_{\theta,\vartheta}(x) = C(\theta, \vartheta) \exp\left[\sum_{i=1}^{r} \theta_i U_i(x) + \sum_{j=1}^{s} \vartheta_j T_j(x)\right] d\mu(x).$$

Then there exist measures λ_θ and probability measures ν_t over s and r dimensional Euclidean space respectively such that

(i) *the distribution of $T = (T_1, \cdots, T_s)$ is an exponential family of the form*

$$dP^T_{\theta,\vartheta}(t) = C(\theta, \vartheta) \exp\left(\sum_{j=1}^{s} \vartheta_j t_j\right) d\lambda_\theta(t), \tag{36}$$

(ii) *the conditional distribution of $U = (U_1, \cdots, U_r)$ given $T = t$ is an exponential family of the form*

$$dP^{U|t}_\theta(u) = C_t(\theta) \exp\left(\sum_{i=1}^{r} \theta_i u_i\right) d\nu_t(u), \tag{37}$$

and hence in particular is independent of ϑ.

Proof. Let (θ^0, ϑ^0) be a point of the natural parameter space, and let $\mu^* = P^X_{\theta^0,\vartheta^0}$. Then

$$dP^X_{\theta,\vartheta}(x) = \frac{C(\theta, \vartheta)}{C(\theta^0, \vartheta^0)} \exp\left[\sum_{i=1}^{r} (\theta_i - \theta_i^0) U_i(x) + \sum_{j=1}^{s} (\vartheta_j - \vartheta_j^0) T_j(x)\right] d\mu^*(x),$$

and the result follows from Lemma 6, with

$$d\lambda_\theta(t) = \exp\left(\Sigma \vartheta_i^0 t_i\right)\left[\int \exp\left[\sum_{i=1}^{r} (\theta_i - \theta_i^0) u_i\right] dP^{U|t}_{\theta^0,\vartheta^0}(u)\right] dP^T_{\theta^0,\vartheta^0}(t)$$

and

$$d\nu_t(u) = dP^{U|t}_{\theta^0,\vartheta^0}(u).$$

Theorem 9. *Let ϕ be any bounded measurable function on $(\mathscr{X}, \mathscr{A})$. Then*

(i) *the integral*

$$\int \phi(x) \exp\left[\sum_{j=1}^{k} \theta_j T_j(x)\right] d\mu(x) \tag{38}$$

considered as a function of the complex variables $\theta_j = \xi_j + i\eta_j$ ($j = 1, \cdots, k$) is an analytic function in each of these variables in the region

R of parameter points for which $(\xi_1, \cdots, \xi_k)$ *is an interior point of the natural parameter space* Ω;

(ii) *the derivatives of all orders with respect to the* θ's *of the integral* (38) *can be computed under the integral sign.*

Proof. If $|\phi| \leqq M$, then

$$|\phi(x) \exp [\Sigma\theta_j T_j(x)]| \leqq M \exp [\Sigma\xi_j T_j(x)]$$

so that the integral (38) exists and is finite for all points $(\xi_1, \cdots, \xi_k)$ of Ω. Let $(\xi_1^0, \cdots, \xi_k^0)$ be any fixed point in the interior of Ω, and consider one of the variables in question, say θ_1. Breaking up the factor

$$\phi(x) \exp [(\xi_2^0 + i\eta_2^0)T_2(x) + \cdots + (\xi_k^0 + i\eta_k^0)T_k(x)]$$

into its real and complex part and each of these into its positive and negative part, and absorbing this factor in each of the four terms thus obtained into the measure μ, one sees that as a function of θ_1 the integral (38) can be written as

$$\int \exp [\theta_1 T_1(x)]\, d\mu_1(x) - \int \exp [\theta_1 T_1(x)]\, d\mu_2(x)$$
$$+ i \int \exp [\theta_1 T_1(x)]\, d\mu_3(x) - i \int \exp [\theta_1 T_1(x)]\, d\mu_4(x).$$

It is therefore sufficient to prove the result for integrals of the form

$$\psi(\theta_1) = \int \exp [\theta_1 T_1(x)]\, d\mu(x).$$

Since $(\xi_1^0, \cdots, \xi_k^0)$ is in the interior of Ω, there exists $\delta > 0$ such that $\psi(\theta_1)$ exists and is finite for all θ_1 with $|\xi_1 - \xi_1^0| \leqq \delta$. Consider the difference quotient

$$\frac{\psi(\theta_1) - \psi(\theta_1^0)}{\theta_1 - \theta_1^0} = \int \frac{\exp [\theta_1 T_1(x)] - \exp [\theta_1^0 T_1(x)]}{\theta_1 - \theta_1^0}\, d\mu(x).$$

The integrand can be written as

$$\exp [\theta_1^0 T_1(x)] \left[\frac{\exp [(\theta_1 - \theta_1^0)T_1(x)] - 1}{\theta_1 - \theta_1^0}\right].$$

Applying to the second factor the inequality

$$\left|\frac{\exp (az) - 1}{z}\right| \leqq \frac{\exp (\delta|a|)}{\delta} \quad \text{for } |z| \leqq \delta,$$

the integrand is seen to be bounded above in absolute value by

$$\frac{1}{\delta}|\exp (\theta_1^0 T_1 + \delta|T_1|)| \leqq \frac{1}{\delta}|\exp [(\theta_1^0 + \delta)T_1] + \exp [(\theta_1^0 - \delta)T_1]|$$

for $|\theta_1 - \theta_1^0| \leqq \delta$. Since the right-hand side is integrable, it follows from the Lebesgue bounded convergence theorem [Theorem 1(ii)] that for any sequence of points $\theta_1^{(n)}$ tending to θ_1^0, the difference quotient of ψ tends to

$$\int T_1(x) \exp[\theta_1^0 T_1(x)]\, d\mu(x).$$

This completes the proof of (i), and proves (ii) for the first derivative. The proof for the higher derivatives is by induction and is completely analogous.

8. PROBLEMS

Section 1

1. *Monotone class.* A class $\mathcal{F}$ of subsets of a space is a *field* if it contains the whole space, is closed under complementation and under finite unions; a class $\mathcal{M}$ is *monotone* if the union and intersection of every increasing and decreasing sequence of sets of $\mathcal{M}$ is again in $\mathcal{M}$. The smallest monotone class $\mathcal{M}_0$ containing a given field $\mathcal{F}$ coincides with the smallest σ-field $\mathcal{A}$ containing $\mathcal{F}$.

[One proves first that $\mathcal{M}_0$ is a field. To show, for example, that $A \cap B \in \mathcal{M}_0$ when A and B are in $\mathcal{M}_0$, consider for a fixed set $A \in \mathcal{F}$, the class $\mathcal{M}_A$ of all B in $\mathcal{M}_0$ for which $A \cap B \in \mathcal{M}_0$. Then $\mathcal{M}_A$ is a monotone class containing $\mathcal{F}$, and hence $\mathcal{M}_A = \mathcal{M}_0$. Thus $A \cap B \in \mathcal{M}_A$ for all B. The argument can now be repeated with a fixed set $B \in \mathcal{M}_0$ and the class $\mathcal{M}_B$ of sets A in $\mathcal{M}_0$ for which $A \cap B \in \mathcal{M}_0$. Since $\mathcal{M}_0$ is a field and monotone, it is a σ-field containing $\mathcal{F}$ and hence contains $\mathcal{A}$. But any σ-field is a monotone class so that also $\mathcal{M}_0$ is contained in $\mathcal{A}$.]

Section 2

2. *Radon-Nikodym derivatives.* (i) If λ and μ are σ-finite measures over $(\mathcal{X}, \mathcal{A})$ and μ is absolutely continuous with respect to λ, then

$$\int f\, d\mu = \int f \frac{d\mu}{d\lambda}\, d\lambda.$$

for any μ-integrable function f.

(ii) If λ, μ, and ν are σ-finite measures over $(\mathcal{X}, \mathcal{A})$ such that ν is absolutely continuous with respect to μ and μ with respect to λ, then

$$\frac{d\nu}{d\lambda} = \frac{d\nu}{d\mu} \cdot \frac{d\mu}{d\lambda} \quad \text{a.e. } \lambda.$$

(iii) If μ and ν are σ-finite measures, which are *equivalent* in the sense that each is absolutely continuous with respect to the other, then

$$\frac{d\nu}{d\mu} = \left(\frac{d\mu}{d\nu}\right)^{-1} \quad \text{a.e. } \mu, \nu.$$

(iv) If μ_k, $k = 1, 2, \cdots$, and μ are finite measures over $(\mathscr{X}, \mathscr{A})$ such that $\sum_{k=1}^{\infty} \mu_k(A) = \mu(A)$ for all $A \in \mathscr{A}$, and if the μ_k are absolutely continuous with respect to a σ-finite measure λ, then μ is absolutely continuous with respect to λ, and

$$\frac{d\sum_{k=1}^{n} \mu_k}{d\lambda} = \sum_{k=1}^{n} \frac{d\mu_k}{d\lambda}, \qquad \lim_{n\to\infty} \frac{d\sum_{k=1}^{n} \mu_k}{d\lambda} = \frac{d\mu}{d\lambda} \qquad \text{a.e. } \lambda.$$

[(i) The equation in question holds when f is the indicator of a set, hence when f is simple, and therefore for all integrable f.
(ii) Apply (i) with $f = d\nu/d\mu$.]

Section 3

3. Let $(\mathscr{X}, \mathscr{A})$ be a measurable space, and $\mathscr{A}_0$ a σ-field contained in $\mathscr{A}$. Suppose that for any function T, the σ-field $\mathscr{B}$ is taken as the totality of sets B such that $T^{-1}(B) \in \mathscr{A}$. Then it is not necessarily true that there exists a function T such that $T^{-1}(\mathscr{B}) = \mathscr{A}_0$.

[An example is furnished by any $\mathscr{A}_0$ such that for all x the set consisting of the single point x is in $\mathscr{A}_0$.]

Section 4

4. (i) Let $\mathscr{P}$ be any family of distributions of $X = (X_1, \cdots, X_n)$ such that

$$P\{(X_i, X_{i+1}, \cdots, X_n, X_1, \cdots, X_{i-1}) \in A\} = P\{(X_1, \cdots, X_n) \in A\}$$

for all Borel sets A and all $i = 1, \cdots, n$. For any sample point $(x_1, \cdots, x_n)$ define $(y_1, \cdots, y_n) = (x_i, x_{i+1}, \cdots, x_n, x_1, \cdots, x_{i-1})$ where $x_i = x^{(1)} = \min(x_1, \cdots, x_n)$. Then the conditional expectation of $f(X)$ given $Y = y$ is

$$f_0(y_1, \cdots, y_n) = \frac{1}{n}[f(y_1, \cdots, y_n) + f(y_2, \cdots, y_n, y_1) + \cdots + (f(y_n, y_1, \cdots, y_{n-1})].$$

(ii) Let $G = \{g_1, \cdots, g_r\}$ be any group of permutations of the coordinates $x_1, \cdots, x_n$ of a point x in n-space, and denote by gx the point obtained by applying g to the coordinates of x. Let $\mathscr{P}$ be any family of distributions P of $X = (X_1, \cdots, X_n)$ such that

$$P\{gX \in A\} = P\{X \in A\} \quad \text{for all} \quad g \in G. \tag{39}$$

For any point x let $t = T(x)$ be any rule that selects a unique point from the r points $g_k x$, $k = 1, \cdots, r$ (for example the smallest first coordinate if this defines it uniquely, otherwise also the smallest second coordinate, etc.). Then

$$E[f(X)|t] = \frac{1}{r}\sum_{k=1}^{r} f(g_k t).$$

(iii) Suppose that in (ii) the distributions P do not satisfy the invariance condition (39) but are given by

$$dP(x) = h(x)\, d\mu(x)$$

where μ is invariant in the sense that $\mu\{x: gx \in A\} = \mu(A)$. Then

$$E[f(X)|t] = \frac{\sum_{k=1}^{r} f(g_k t)h(g_k t)}{\sum_{k=1}^{r} h(g_k t)}.$$

Section 5

5. Prove Theorem 4 for the case of an n-dimensional sample space.

[The condition that the cumulative distribution function be nondecreasing is replaced by $P\{x_1 < X_1 \leqq x_1', \cdots, x_n < X_n \leqq x_n'\} \geqq 0$; the condition that it be continuous on the right can be stated as $\lim_{m\to\infty} F(x_1 + 1/m, \cdots, x_n + 1/m) = F(x_1, \cdots, x_n)$.]

6. Let $\mathscr{X} = \mathscr{Y} \times \mathscr{T}$ and suppose that P_0, P_1 are two probability distributions given by

$$dP_0(y, t) = f(y)g(t)\, d\mu(y)\, d\nu(t)$$

$$dP_1(y, t) = h(y, t)\, d\mu(y)\, d\nu(t)$$

where $h(y, t)/f(y)g(t) < \infty$. Then under P_1 the probability density of Y with respect to μ is

$$p_1^Y(y) = f(y)E_0\left[\frac{h(y, T)}{f(y)g(T)}\middle| Y = y\right].$$

$$\left[p_1^Y(y) = \int_{\mathscr{T}} h(y, t)\, d\nu(t) = f(y)\int_{\mathscr{T}} \frac{h(y, t)}{f(y)g(t)} g(t)\, d\nu(t).\right]$$

Section 6

7. *Symmetric distributions.* (i) Let $\mathscr{P}$ be any family of distributions of $X = (X_1, \cdots, X_n)$ which are symmetric in the sense that

$$P\{(X_{i_1}, \cdots, X_{i_n}) \in A\} = P\{(X_1, \cdots, X_n) \in A\}$$

for all Borel sets A and all permutations $(i_1, \cdots, i_n)$ of $(1, \cdots, n)$. Then the statistic T of Example 7 is sufficient for $\mathscr{P}$, and the formula given in the first part of the example for the conditional expectation $E[f(X)|T(x)]$ is valid.

(ii) The statistic Y of Problem 4 is sufficient.

(iii) Let $X_1, \cdots, X_n$ be identically and independently distributed according to a continuous distribution $P \in \mathscr{P}$, and suppose that the distributions of $\mathscr{P}$ are symmetric with respect to the origin. Let $V_i = |X_i|$ and $W_i = V^{(i)}$. Then $(W_1, \cdots, W_n)$ is sufficient for $\mathscr{P}$.

8. *Sufficiency of likelihood ratios.* Let P_0, P_1 be two distributions with densities p_0, p_1. Then $T(x) = p_1(x)/p_0(x)$ is sufficient for $\mathscr{P} = \{P_0, P_1\}$.

[This follows from the factorization criterion by writing $p_1 = T \cdot p_0$, $p_0 = 1 \cdot p_0$.]

9. *Pairwise sufficiency.* A statistic T is pairwise sufficient for $\mathscr{P}$ if it is sufficient for every pair of distributions in $\mathscr{P}$.

(i) If $\mathscr{P}$ is countable and T is pairwise sufficient for $\mathscr{P}$, then T is sufficient for $\mathscr{P}$.

(ii) If $\mathscr{P}$ is a dominated family and T is pairwise sufficient for $\mathscr{P}$, then T is sufficient for $\mathscr{P}$.

[(i) Let $\mathscr{P} = \{P_0, P_1, \cdots\}$ and let $\mathscr{A}_0$ be the sufficient subfield induced by T. Let $\lambda = \Sigma c_i P_i$ ($c_i > 0$) be equivalent to $\mathscr{P}$. For each $j = 1, 2, \cdots$ the probability measure λ_j that is proportional to $(c_0/n)P_0 + c_j P_j$ is equivalent to $\{P_0, P_j\}$. Thus by pairwise sufficiency, the derivative $f_j = dP_0/[(c_0/n)\,dP_0 + c_j\,dP_j)]$ is $\mathscr{A}_0$-measurable. Let $S_j = \{x: f_j(x) = 0\}$ and $S = \bigcup_{j=1}^n S_j$. Then $S \in \mathscr{A}_0$, $P_0(S) = 0$, and on $\mathscr{X} - S$ the derivative $dP_0/d\Sigma_{j=1}^n c_j P_j$ equals $\Sigma_{j=1}^n 1/f_j$ which is $\mathscr{A}_0$-measurable. It then follows from Problem 2 that

$$\frac{dP_0}{d\lambda} = \frac{dP_0}{d\sum_{j=1}^{n} c_j P_j} \cdot \frac{d\sum_{j=1}^{n} c_j P_j}{d\lambda}$$

is also $\mathscr{A}_0$-measurable.

(ii) Let $\lambda = \Sigma_{j=1}^{\infty} c_j P_{\theta_j}$ be equivalent to $\mathscr{P}$. Then pairwise sufficiency of T implies for any θ_0 that $dP_{\theta_0}/(dP_{\theta_0} + d\lambda)$ and hence $dP_{\theta_0}/d\lambda$ is a measurable function of T.]

10. If a statistic T is sufficient for $\mathscr{P}$, then for every function f which is $(\mathscr{A}, P_\theta)$-integrable for all $\theta \in \Omega$ there exists a determination of the conditional expectation function $E_\theta[f(X)|t]$ that is independent of θ.

[If $\mathscr{X}$ is Euclidean, this follows from Theorems 5 and 7. In general, if f is nonnegative there exists a nondecreasing sequence of simple nonnegative functions f_n tending to f. Since the conditional expectation of a simple function can be taken to be independent of θ by Lemma 3(ii), the desired result follows from Lemma 3(iv).]

11. For a decision problem with a finite number of decisions, the class of procedures depending on a sufficient statistic T only is essentially complete.*

[For Euclidean sample spaces this follows from Theorem 4 without any restriction on the decision space. For the present case, let a decision procedure be given by $\delta(x) = (\delta^{(1)}(x), \cdots, \delta^{(m)}(x))$ where $\delta^{(i)}(x)$ is the probability with which decision d_i is taken when x is observed. If T is sufficient and $\eta^{(i)}(t) = E[\delta^{(i)}(X)|t]$, the procedures δ and η have identical risk functions.]

Section 7

12. Let X_i ($i = 1, \cdots, s$) be independently distributed with Poisson distribution $P(\lambda_i)$, and let $T_0 = \Sigma X_j$, $T_i = X_i$, $\lambda = \Sigma\lambda_j$. Then T_0 has the Poisson distribution $P(\lambda)$, and the conditional distribution of $T_1, \cdots, T_{s-1}$ given $T_0 = t_0$ is the multinomial distribution (34) with $n = t_0$ and $p_i = \lambda_i/\lambda$.

[Direct computation.]

13. *Life testing*. Let $X_1, \cdots, X_n$ be independently distributed with exponential density $(2\theta)^{-1}e^{-x/2\theta}$ for $x \geqq 0$, and let the ordered X's be denoted by $Y_1 \leqq Y_2 \leqq \cdots \leqq Y_n$. It is assumed that Y_1 becomes available first, then Y_2, etc., and that observation is continued until Y_r has been observed. This might arise, for example, in life testing where each X measures the length of life of, say, an

* For a more general result see Bahadur, "A characterization of sufficiency," *Ann. Math. Stat.*, Vol. 26 (1955), pp. 286–293, and Elfving, "Sufficiency and completeness," *Ann. Acad. Sci. Fennicae, Ser. A*, No. 135, 1952.

electron tube, and n tubes are being tested simultaneously. Another application is to the disintegration of radioactive material, where n is the number of atoms, and observation is continued until r α-particles have been emitted.

(i) The joint distribution of $Y_1, \cdots, Y_r$ is an exponential family with density

$$\frac{1}{(2\theta)^r}\frac{n!}{(n-r)!}\exp\left[-\frac{\sum_{i=1}^{r} y_i + (n-r)y_r}{2\theta}\right], \qquad 0 \leqq y_1 \leqq \cdots \leqq y_r.$$

(ii) The distribution of $[\Sigma_{i=1}^r Y_i + (n-r)Y_r]/\theta$ is χ^2 with $2r$ degrees of freedom.

(iii) Let $Y_1, Y_2, \cdots$ denote the time required until the first, second, etc., event occurs in a Poisson process with parameter $1/2\theta'$ (see Chapter 1, Problem 1). Then $Z_1 = Y_1/\theta'$, $Z_2 = (Y_2 - Y_1)/\theta'$, $Z_3 = (Y_3 - Y_2)/\theta', \cdots$ are independently distributed as χ^2 with 2 degrees of freedom, and the joint density of $Y_1, \cdots, Y_r$ is an exponential family with density

$$\frac{1}{(2\theta')^r}\exp(-y_r/2\theta'), \qquad 0 \leqq y_1 \leqq \cdots \leqq y_r.$$

The distribution of Y_r/θ' is again χ^2 with $2r$ degrees of freedom.

(iv) The same model arises in the application to life testing if the number n of tubes is held constant by replacing each burned-out tube by a new one, and if Y_1 denotes the time at which the first tube burns out, Y_2 the time at which the second tube burns out, etc., measured from some fixed time.

[(ii) The random variables $Z_i = (n - i + 1)(Y_i - Y_{i-1})/\theta$ $(i = 1, \cdots, r)$ are independently distributed as χ^2 with 2 degrees of freedom, and $[\Sigma_{i=1}^r Y_i + (n-r)Y_r]/\theta = \Sigma_{i=1}^r Z_i$.]

14. The expectations and covariances of the statistics T_j in the exponential family (35) are given by

$$E[T_j(X)] = -\partial \log C(\theta)/\partial\theta_j \qquad (j = 1, \cdots, k)$$

$$E[T_i(X)T_j(X)] - [ET_i(X)ET_j(X)] = -\partial^2 \log C(\theta)/\partial\theta_i\partial\theta_j \qquad (i, j = 1, \cdots, k).$$

15. Let Ω be the natural parameter space of the exponential family (35), and for any fixed $t_{r+1}, \cdots, t_k$ $(r < k)$ let $\Omega'_{\theta_1,\cdots,\theta_r}$ be the natural parameter space of the family of conditional distributions given $T_{r+1} = t_{r+1}, \cdots, T_k = t_k$.

(i) Then $\Omega'_{\theta_1,\cdots,\theta_r}$ contains the projection $\Omega_{\theta_1,\cdots,\theta_r}$ of Ω onto $\theta_1, \cdots, \theta_r$.

(ii) An example in which $\Omega_{\theta_1,\cdots,\theta_r}$ is a proper subset of $\Omega'_{\theta_1,\cdots,\theta_r}$ is the family of densities

$$p_{\theta_1\theta_2}(x, y) = C(\theta_1, \theta_2)\exp(\theta_1 x + \theta_2 y - xy), \qquad x, y > 0.$$

9. REFERENCES

The theory of measure and integration in abstract spaces is treated in a number of books, among them: Halmos (1950); Loève (1955); Saks (1937).

The basic definitions and properties of conditional probability and expectation were given by Kolmogorov (1933). A more detailed account containing many additional results may be found in the books by Doob (1953), and by Loève (1955).

Detailed references to these books and some more specific references for Sections 3, 6, and 7 are given below.

Bahadur, R. R.

(1954) "Sufficiency and statistical decision functions," *Ann. Math. Stat.*, Vol. 25, pp. 423–462.

[A detailed abstract treatment of sufficient statistics, including the factorization theorem, the structure theorem for minimal sufficient statistics, and a discussion of sufficiency for the case of sequential experiments.]

(1955) "Statistics and subfields," *Ann. Math. Stat.*, Vol. 26, pp. 490–497.

Bahadur, R. R., and E. L. Lehmann

(1955) "Two comments on 'sufficiency and statistical decision functions,'" *Ann. Math. Stat.*, Vol. 26, pp. 139–142.

[Problem 3].

Doob, J. L.

(1953) *Stochastic Processes*, New York, John Wiley & Sons.

Epstein, B., and M. Sobel

(1954) "Some theorems relevant to life testing from an exponential distribution," *Ann. Math. Stat.*, Vol. 25, pp. 373–381.

[Problem 13.]

Halmos, P. R.

(1950) *Measure Theory*, New York, D. Van Nostrand Co.

Halmos, Paul R., and L. J. Savage

(1949) "Application of the Radon-Nikodym theorem to the theory of sufficient statistics," *Ann. Math. Stat.*, Vol. 20, pp. 225–241.

[First abstract treatment of sufficient statistics; the factorization theorem. Problem 9.]

Kolmogorov, A.

(1933) Grundbegriffe der Wahrscheinlichkeitsrechnung, Berlin, J. Springer.

Loève, M.

(1955) *Probability Theory*, New York, D. Van Nostrand Co.

Saks, S.

(1937) *Theory of the Integral*, New York, G. E. Stechert and Co.

CHAPTER 3

Uniformly Most Powerful Tests

1. STATING THE PROBLEM

We now begin the study of the statistical problem whose theory has been explored most thoroughly, the problem of hypothesis testing. As the term suggests, one wishes to decide whether or not some hypothesis that has been formulated is correct. The choice here lies between only two decisions: accepting or rejecting the hypothesis. A decision procedure for such a problem is called a *test* of the hypothesis in question.

The decision is to be based on the value of a certain random variable X, the distribution P_θ of which is known to belong to a class $\mathscr{P} = \{P_\theta, \theta \in \Omega\}$. We shall assume that if θ were known one would also know whether or not the hypothesis is true. The distributions of $\mathscr{P}$ can then be classified into those for which the hypothesis is true and those for which it is false. The resulting two mutually exclusive classes are denoted by H and K and the corresponding subsets of Ω by Ω_H and Ω_K respectively, so that $H \cup K = \mathscr{P}$ and $\Omega_H \cup \Omega_K = \Omega$. Mathematically, the hypothesis is equivalent to the statement that P_θ is an element of H. It is therefore convenient to identify the hypothesis with this statement and to use the letter H also to denote the hypothesis. Analogously we call the distributions in K the alternatives to H, so that K is the *class of alternatives*.

Let the decisions of accepting or rejecting H be denoted by d_0 and d_1 respectively. A nonrandomized test procedure assigns to each possible value x of X one of these two decisions and thereby divides the sample space into two complementary regions S_0 and S_1. If X falls into S_0 the hypothesis is accepted, otherwise it is rejected. The set S_0 is called the region of acceptance, and the set S_1 the region of rejection or *critical* region.

When performing a test one may arrive at the correct decision, or one may commit one of two errors: rejecting the hypothesis when it is true (error of the first kind) or accepting it when it is false (error of the second kind). The consequences of these are often quite different. For example, if one tests for the presence of some disease, incorrectly deciding on the necessity of treatment may cause the patient discomfort and financial loss. On the other hand, failure to diagnose the presence of the ailment may lead to his death.

It is desirable to carry out the test in a manner which keeps the probabilities of the two types of error to a minimum. Unfortunately, when the number of observations is given, both probabilities cannot be controlled simultaneously. It is customary therefore to assign a bound to the probability of incorrectly rejecting H when it is true, and to attempt to minimize the other probability subject to this condition. Thus one selects a number α between 0 and 1, called the *level of significance*, and imposes the condition that

$$P_\theta\{\delta(X) = d_1\} = P_\theta\{X \in S_1\} \leqq \alpha \quad \text{for all} \quad \theta \in \Omega_H. \tag{1}$$

Subject to this condition, it is desired to minimize $P_\theta\{\delta(X) = d_0\}$ for θ in Ω_K or, equivalently, to maximize

$$P_\theta\{\delta(X) = d_1\} = P_\theta\{X \in S_1\} \quad \text{for all} \quad \theta \in \Omega_K. \tag{2}$$

Although usually (2) implies that

$$\sup_{\Omega_H} P_\theta\{X \in S_1\} = \alpha, \tag{3}$$

it is convenient to introduce a term for the left-hand side of (3): it is called the *size* of the test or critical region S_1. Condition (1) therefore restricts consideration to tests whose size does not exceed the given level of significance. The probability of rejection (2) evaluated for a given θ in Ω_K is called the *power* of the test against the alternative θ. Considered as a function of θ for all $\theta \subset \Omega$, the probability (2) is called the *power function* of the test and is denoted by $\beta(\theta)$.

The choice of a level of significance α will usually be somewhat arbitrary since in most situations there is no precise limit to the probability of an error of the first kind that can be tolerated. It has become customary to choose for α one of a number of standard values such as .005, .01, or .05. There is some convenience in such standardization since it permits a reduction in certain tables needed for carrying out various tests. Otherwise there appears to be no particular reason for selecting these values. In fact, when choosing a level of significance one should also consider the power that the test will achieve against various alternatives. If the power is too low one may wish to use much higher values of α than the customary ones, for example, .1 or .2.*

* A rule of thumb for choosing α in relation to the power of the test is suggested by Lehmann, "Significance level and power," *Ann. Math. Stat.*, Vol. 29 (1958), pp. 1167–1176.

Another consideration that frequently enters into the specification of a significance level is the attitude toward the hypothesis before the experiment is performed. If one firmly believes the hypothesis to be true, extremely convincing evidence will be required before one is willing to give up this belief, and the significance level will accordingly be set very low. (A low significance level results in the hypothesis being rejected only for a set of values of the observations whose total probability under the hypothesis is small, so that such values would be most unlikely to occur if H were true.)

In applications, there is usually available a nested family of rejection regions, corresponding to different significance levels. It is then good practice to determine not only whether the hypothesis is accepted or rejected at the given significance level, but also to determine the smallest significance level $\hat{\alpha} = \hat{\alpha}(x)$, the *critical level*, at which the hypothesis would be rejected for the given observation. This number gives an idea of how strongly the data contradict (or support) the hypothesis, and enables others to reach a verdict based on the significance level of their choice. (Cf. Problem 7 and Chapter 4, Problem 2.)

Let us next consider the structure of a randomized test. For any value x such a test chooses among the two decisions, rejection or acceptance, with certain probabilities that depend on x and will be denoted by $\phi(x)$ and $1 - \phi(x)$ respectively. If the value of X is x, a random experiment is performed with two possible outcomes R and $\bar{R}$ the probabilities of which are $\phi(x)$ and $1 - \phi(x)$. If in this experiment R occurs, the hypothesis is rejected, otherwise it is accepted. A randomized test is therefore completely characterized by a function ϕ, the *critical function*, with $0 \leq \phi(x) \leq 1$ for all x. If ϕ takes on only the values 1 and 0, one is back in the case of a nonrandomized test. The set of points x for which $\phi(x) = 1$ is then just the region of rejection, so that in a nonrandomized test ϕ is simply the indicator function of the critical region.

If the distribution of X is P_θ, and the critical function ϕ is used, the probability of rejection is

$$E_\theta \phi(X) = \int \phi(x) \, dP_\theta(x),$$

the conditional probability $\phi(x)$ of rejection given x, integrated with respect to the probability distribution of X. The problem is to select ϕ so as to maximize the power

$$\beta_\phi(\theta) = E_\theta \phi(X) \quad \text{for all} \quad \theta \in \Omega_K \tag{4}$$

subject to the condition

$$E_\theta \phi(X) \leq \alpha \quad \text{for all} \quad \theta \in \Omega_H. \tag{5}$$

The same difficulty now arises that presented itself in the general discussion of Chapter 1. Typically, the test that maximizes the power against a particular alternative in K depends on this alternative, so that some additional principle has to be introduced to define what is meant by an optimum test. There is one important exception: if K contains only one distribution, that is, if one is concerned with a single alternative, the problem is completely specified by (4) and (5). It then reduces to the mathematical problem of maximizing an integral subject to certain side conditions. The theory of this problem, and its statistical applications, constitutes the principal subject of the present chapter. In special cases it may of course turn out that the same test maximizes the power for all alternatives in K even when there is more than one. Examples of such *uniformly most powerful* (UMP) tests will be given in Sections 3 and 7.

In the above formulation the problem can be considered as a special case of the general decision problem with two types of losses. Corresponding to the two kinds of error one can introduce the two component loss functions,

$$\begin{cases} L_1(\theta, d_1) = 1 \quad \text{or} \quad 0 \quad \text{as} \quad \theta \in \Omega_H \quad \text{or} \quad \theta \in \Omega_K \\ L_1(\theta, d_0) = 0 \quad \text{for all} \quad \theta \end{cases}$$

and

$$\begin{cases} L_2(\theta, d_0) = 0 \quad \text{or} \quad 1 \quad \text{as} \quad \theta \in \Omega_H \quad \text{or} \quad \theta \in \Omega_K \\ L_2(\theta, d_1) = 0 \quad \text{for all} \quad \theta. \end{cases}$$

With this definition the minimization of $EL_2(\theta, \delta(X))$ subject to the restriction $EL_1(\theta, \delta(X)) \leqq \alpha$ is exactly equivalent to the problem of hypothesis testing as given above.

The formal loss functions L_1 and L_2 clearly do not represent in general the true losses. The loss resulting from an incorrect acceptance of the hypothesis, for example, will not be the same for all alternatives. The more the alternative differs from the hypothesis the more serious are the consequences of such an error. As was discussed earlier, we have purposely foregone the more detailed approach implied by this criticism. Rather than working with a loss function which in practice one does not know, it seems preferable to base the theory on the simpler and intuitively appealing notion of error. It will be seen later that at least some of the results can be justified also in the more elaborate formulation.

2. THE NEYMAN-PEARSON FUNDAMENTAL LEMMA

A class of distributions is called *simple* if it contains only a single distribution and otherwise is said to be *composite*. The problem of hypothesis testing is completely specified by (4) and (5) if K is simple. Its solution is easiest and can be given explicitly when the same is true of H. Let the distribution under a simple hypothesis H and alternative K be P_0 and P_1, and suppose for a moment that these distributions are discrete with $P_i\{X = x\} = P_i(x)$ for $i = 0, 1$. If at first one restricts attention to nonrandomized tests, the optimum test is defined as the critical region S satisfying

$$\sum_{x \in S} P_0(x) \leqq \alpha \tag{6}$$

and

$$\sum_{x \in S} P_1(x) = \text{maximum}.$$

It is easy to see which points should be included in S. To each point are attached two values, its probability under P_0 and under P_1. The selected points are to have a total value not exceeding α on the one scale, and as large as possible on the other. This is a situation that occurs in many contexts. A buyer with a limited budget who wants to get "the most for his money" will rate the items according to their *value per dollar*. In order to travel a given distance in the shortest possible time, one must choose the speediest mode of transportation, that is, the one that yields the largest number of *miles per hour*. Analogously in the present problem the most valuable points x are those with the highest value of

$$r(x) = P_1(x)/P_0(x).$$

The points are therefore rated according to the value of this ratio and selected for S in this order, as many as one can afford under restriction (6). Formally this means that S is the set of all points x for which $r(x) > c$, where c is determined by the condition

$$P_0\{X \in S\} = \sum_{x:r(x)>c} P_0(x) = \alpha.$$

Here a difficulty is seen to arise. It may happen that when a certain point is included, the value α has not yet been reached but that it would be exceeded if the next point were also included. The exact value α can then either not be achieved at all, or it can be attained only by passing over the next desirable point and in its place taking one further down the list. The difficulty can be overcome by permitting randomization. This makes it possible to split the next point, including only a portion of it, and thereby to obtain the exact value α without breaking the order of preference that has been established for the various sample points. These considerations are formalized in the following theorem, the *fundamental lemma of Neyman and Pearson.*

Theorem 1. *Let P_0 and P_1 be probability distributions possessing densities p_0 and p_1 respectively with respect to a measure μ.**

(i) Existence. *For testing H: p_0 against the alternative K: p_1 there exists a test ϕ and a constant k such that*

$$E_0\,\phi(X) = \alpha \tag{7}$$

and

$$\phi(x) = \begin{cases} 1 & \text{when } p_1(x) > kp_0(x) \\ 0 & \text{when } p_1(x) < kp_0(x). \end{cases} \tag{8}$$

(ii) Sufficient condition for a most powerful test. *If a test satisfies* (7) *and* (8) *for some k, then it is most powerful for testing p_0 against p_1 at level α.*

(iii) Necessary condition for a most powerful test. *If ϕ is most powerful at level α for testing p_0 against p_1, then for some k it satisfies* (8) *a.e. μ. It also satisfies* (7) *unless there exists a test of size $<\alpha$ and with power* 1.

Proof. For $\alpha = 0$ and $\alpha = 1$ the theorem is easily seen to be true provided the value $k = +\infty$ is admitted in (8) and $0\cdot\infty$ is interpreted as 0. Throughout the proof we shall therefore assume $0 < \alpha < 1$.

(i) Let $\alpha(c) = P_0\{p_1(X) > cp_0(X)\}$. Since the probability is computed under P_0, the inequality need be considered only for the set where $p_0(x) > 0$, so that $\alpha(c)$ is the probability that the random variable $p_1(X)/p_0(X)$ exceeds c. Thus $1 - \alpha(c)$ is a cumulative distribution function, and $\alpha(c)$ is nonincreasing and continuous on the right, $\alpha(c - 0) - \alpha(c) = P_0\{p_1(X)/p_0(X) = c\}$, $\alpha(-\infty) = 1$, and $\alpha(\infty) = 0$. Given any $0 < \alpha < 1$, let c_0 be such that $\alpha(c_0) \leqq \alpha \leqq \alpha(c_0 - 0)$ and consider the test ϕ defined by

$$\phi(x) = \begin{cases} 1 & \text{when } p_1(x) > c_0p_0(x) \\ \dfrac{\alpha - \alpha(c_0)}{\alpha(c_0 - 0) - \alpha(c_0)} & \text{when } p_1(x) = c_0p_0(x) \\ 0 & \text{when } p_1(x) < c_0p_0(x). \end{cases}$$

Here the middle expression is meaningful unless $\alpha(c_0) = \alpha(c_0 - 0)$; since then $P_0\{p_1(X) = c_0p_0(X)\} = 0$, ϕ is defined a.e. The size of ϕ is

$$E_0\,\phi(X) = P_0\left\{\frac{p_1(X)}{p_0(X)} > c_0\right\} + \frac{\alpha - \alpha(c_0)}{\alpha(c_0 - 0) - \alpha(c_0)}\,P_0\left\{\frac{p_1(X)}{p_0(X)} = c_0\right\} = \alpha,$$

so that c_0 can be taken as the k of the theorem.

It is of interest to note that c_0 is essentially unique. The only exception

* There is no loss of generality in this assumption since one can take $\mu = P_0 + P_1$.

is the case that an interval of c's exists for which $\alpha(c) = \alpha$. If (c', c'') is such an interval, and

$$C = \left\{x: p_0(x) > 0 \quad \text{and} \quad c' < \frac{p_1(x)}{p_0(x)} < c''\right\},$$

then $P_0(C) = \alpha(c') - \alpha(c'' - 0) = 0$; and this implies $\mu(C) = 0, P_1(C) = 0$. Thus the sets corresponding to two different values of c differ only in a set of points which has probability 0 under both distributions, that is, points that could be excluded from the sample space.

(ii) Suppose that ϕ is a test satisfying (7) and (8) and that ϕ^* is any other test with $E_0\,\phi^*(X) \leqq \alpha$. Denote by S^+ and S^- the sets in the sample space where $\phi(x) - \phi^*(x) > 0$ and < 0 respectively. If x is in S^+, $\phi(x)$ must be > 0 and $p_1(x) \geqq kp_0(x)$. In the same way $p_1(x) \leqq kp_0(x)$ for all x in S^-, and hence

$$\int (\phi - \phi^*)(p_1 - kp_0)\,d\mu = \int_{S^+ \cup S^-} (\phi - \phi^*)(p_1 - kp_0)\,d\mu \geqq 0.$$

The difference in power between ϕ and ϕ^* therefore satisfies

$$\int (\phi - \phi^*)p_1\,d\mu \geqq k \int (\phi - \phi^*)p_0\,d\mu \geqq 0,$$

as was to be proved.

(iii) Let ϕ^* be most powerful at level α for testing p_0 against p_1, and let ϕ satisfy (7) and (8). Let S be the intersection of the set $S^+ \cup S^-$, on which ϕ and ϕ^* differ, with the set $\{x: p_1(x) \neq kp_0(x)\}$ and suppose that $\mu(S) > 0$. Since $(\phi - \phi^*)(p_1 - kp_0)$ is positive on S, it follows that

$$\int_{S^+ \cup S^-} (\phi - \phi^*)(p_1 - kp_0)\,d\mu = \int_S (\phi - \phi^*)(p_1 - kp_0)\,d\mu > 0$$

and hence that ϕ is more powerful against p_1 than ϕ^*. This is a contradiction, and therefore $\mu(S) = 0$, as was to be proved.

If ϕ^* were of size $< \alpha$ and power < 1, it would be possible to include in the rejection region additional points or portions of points and thereby to increase the power until either the power is 1 or the size is α. Thus either $E_0\,\phi^*(X) = \alpha$ or $E_1\,\phi^*(X) = 1$.

The proof of part (iii) shows that the most powerful test is uniquely determined by (7) and (8) except on the set on which $p_1(x) = kp_0(x)$. On this set, ϕ can be defined arbitrarily provided the resulting test has size α. Actually, we have shown that it is always possible to define ϕ to be constant over this boundary set. In the trivial case that there exists a test of power 1, the constant k of (8) is 0, and one will accept H for all points for which $p_1(x) = kp_0(x)$ even though the test may then have size $< \alpha$.

It follows from these remarks that the most powerful test is determined uniquely (up to sets of measure zero) by (7) and (8) whenever the set on which $p_1(x) = kp_0(x)$ has μ-measure zero. This unique test is then clearly nonrandomized. More generally, it is seen that randomization is not required except possibly on the boundary set where it may be necessary to randomize in order to get the size equal to α. In practice one will frequently prefer to adopt a different value for the level of significance which does not require randomization. In the case that there exists a test of power 1, (7) and (8) will determine a most powerful test but it may not be unique in that there may exist a test also most powerful and satisfying (7) and (8) for some $\alpha' < \alpha$.

Corollary 1. *Let β denote the power of the most powerful level α test* $(0 < \alpha < 1)$ *for testing P_0 against P_1. Then $\alpha < \beta$ unless $P_0 = P_1$.*

Proof. Since the level α test given by $\phi(x) \equiv \alpha$ has power α, it is seen that $\alpha \leqq \beta$. If $\alpha = \beta < 1$, the test $\phi(x) \equiv \alpha$ is most powerful and by Theorem 1(iii) must satisfy (8). Then $p_0(x) = p_1(x)$ a.e. μ, and hence $P_0 = P_1$.

An alternative method for proving the results of this section is based on the following geometric representation of the problem of testing a simple hypothesis against a simple alternative. Let N be the set of all points (α, β) for which there exists a test ϕ such that

$$\alpha = E_0\,\phi(X), \qquad \beta = E_1\,\phi(X).$$

This set is convex, contains the points (0, 0) and (1, 1), and is symmetric with respect to the point $(\frac{1}{2}, \frac{1}{2})$ in the sense that with any point (α, β) it also contains the point $(1 - \alpha, 1 - \beta)$. In addition, the set N is closed. [This follows from the weak compactness theorem for critical functions, Theorem 3 of the Appendix; the argument is the same as that in the proof of Theorem 5(i).]

For each value $0 < \alpha_0 < 1$, the level α_0 tests are represented by the points whose abscissa is $\leqq \alpha_0$. The most powerful of these tests (whose existence follows from the fact that N is closed) corresponds to the point on the upper boundary of N with abscissa α_0. This is the only point corresponding to a most powerful level α_0 test unless there exists a point $(\alpha, 1)$ in N with $\alpha < \alpha_0$ (Figure 1*b*).

As an example of this geometric approach, consider the following alternative proof of Corollary 1. Suppose that for some $0 < \alpha_0 < 1$ the power of the most powerful level α_0 test is α_0. Then it follows from the convexity of N that $(\alpha, \beta) \in N$ implies $\beta \leqq \alpha$, and hence from the symmetry of N that N consists exactly of the line segment connecting the points (0, 0) and (1, 1). This means that $\int \phi p_0\, d\mu = \int \phi p_1\, d\mu$ for all ϕ

and hence that $p_0 = p_1$ (a.e. μ), as was to be proved. A proof of Theorem 1 along these lines is given in a more general setting in the proof of Theorem 5.

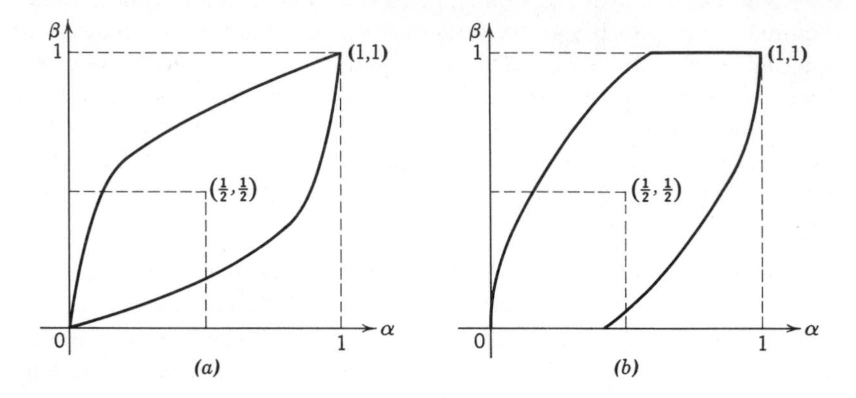

Figure 1.

3. DISTRIBUTIONS WITH MONOTONE LIKELIHOOD RATIO

The case that both the hypothesis and the class of alternatives are simple is mainly of theoretical interest since problems arising in applications typically involve a parametric family of distributions depending on one or more continuous parameters. In the simplest situation of this kind the distributions depend on a single real-valued parameter θ, and the hypothesis is one-sided, say $H: \theta \leqq \theta_0$. In general, the most powerful test of H against an alternative $\theta_1 > \theta_0$ depends on θ_1 and is then not UMP. However, a UMP test does exist if an additional assumption is satisfied. The real-parameter family of densities $p_\theta(x)$ is said to have *monotone likelihood ratio* if there exists a real-valued function $T(x)$ such that for any $\theta < \theta'$ the distributions P_θ and $P_{\theta'}$ are distinct, and the ratio $p_{\theta'}(x)/p_\theta(x)$ is a nondecreasing function of $T(x)$.

Theorem 2. *Let θ be a real parameter, and let the random variable X have probability density $p_\theta(x)$ with monotone likelihood ratio in $T(x)$.*

(i) *For testing $H: \theta \leqq \theta_0$ against $K: \theta > \theta_0$, there exists a UMP test, which is given by*

$$\phi(x) = \begin{cases} 1 & \text{when} \quad T(x) > C \\ \gamma & \text{when} \quad T(x) = C \\ 0 & \text{when} \quad T(x) < C \end{cases} \tag{9}$$

where C and γ are determined by

$$E_{\theta_0} \phi(X) = \alpha. \tag{10}$$

(ii) *The power function*

$$\beta(\theta) = E_\theta \phi(X)$$

of this test is strictly increasing for all points θ for which $\beta(\theta) < 1$.

(iii) *For all θ', the test determined by* (9) *and* (10) *is UMP for testing $H': \theta \leqq \theta'$ against $K': \theta > \theta'$ at level $\alpha' = \beta(\theta')$.*

(iv) *For any $\theta < \theta_0$ the test minimizes $\beta(\theta)$ (the probability of an error of the first kind) among all tests satisfying* (10).

Proof. (i) and (ii). Consider first the hypothesis $H_0: \theta = \theta_0$ and some simple alternative $\theta_1 > \theta_0$. Applying the fundamental lemma, one finds that the most powerful test rejects when

$$p_{\theta_1}(x)/p_{\theta_0}(x) > C$$

or equivalently when

$$T(x) > C.^*$$

It follows from Theorem 1(i) that there exist C and γ such that (9) and (10) hold. By Theorem 1(ii), the resulting test is also most powerful for testing $P_{\theta'}$ against $P_{\theta''}$ at level $\alpha' = \beta(\theta')$ provided $\theta' < \theta''$. Part (ii) of the present theorem now follows from Corollary 1. Since $\beta(\theta)$ is therefore nondecreasing the test satisfies

$$E_\theta \phi(X) \leqq \alpha \quad \text{for} \quad \theta \leqq \theta_0. \tag{11}$$

The class of tests satisfying (11) is contained in the class satisfying $E_{\theta_0} \phi(X) \leqq \alpha$. Since the given test maximizes $\beta(\theta_1)$ within this wider class, it also maximizes $\beta(\theta_1)$ subject to (11); since it is independent of the particular alternative $\theta_1 > \theta_0$ chosen, it is UMP against K.

(iii) is proved by an analogous argument.

(iv) follows from the fact that the test which minimizes the power for testing a simple hypothesis against a simple alternative is obtained by applying the fundamental lemma (Theorem 1) with all inequalities reversed.

By interchanging inequalities throughout, one obtains in an obvious manner the solution of the dual problem, $H: \theta \geqq \theta_0$, $K: \theta < \theta_0$.

A few examples of families with monotone likelihood ratio, and hence of UMP one-sided tests, will be given below. However, the main applications of Theorem 2 will come later, when such families appear as the set of conditional distributions given a sufficient statistic (Chapters 4 and 5) and as distributions of a maximal invariant (Chapters 6 and 7).

* Here and in similar derivations C is used as a generic notation.

Example 1. From a lot containing N items of a manufactured product, a sample of size n is selected at random, and each item in the sample is inspected. If the total number of defective items in the lot is D, the number X of defectives found in the sample has the *hypergeometric* distribution

$$P\{X = x\} = P_D(x) = \frac{\binom{D}{x}\binom{N-D}{n-x}}{\binom{N}{n}}, \qquad x = 0, 1, \cdots, D.$$

Interpreting $P_D(x)$ as a density with respect to the measure μ that assigns to any set on the real line as measure the number of integers 0, 1, 2, $\cdots$ that it contains, and noting that

$$\frac{P_{D+1}(x)}{P_D(x)} = \frac{D+1}{N-D} \frac{N-D-n+x}{D+1-x},$$

it is seen that the distributions satisfy the assumption of monotone likelihood ratios with $T(x) = x$. Therefore there exists a UMP test for testing the hypothesis H: $D \leq D_0$ against K: $D > D_0$, which rejects H when X is too large, and an analogous test for testing H': $D \geq D_0$.

An important class of families of distributions that satisfy the assumptions of Theorem 2 are the one-parameter exponential families.

Corollary 2. *Let θ be a real parameter, and let X have probability density (with respect to some measure μ)*

$$p_\theta(x) = C(\theta)\, e^{Q(\theta)T(x)} h(x) \tag{12}$$

where Q is strictly monotone. Then there exists a UMP test ϕ for testing H: $\theta \leq \theta_0$ against K: $\theta > \theta_0$. If Q is increasing,

$$\phi(x) = 1, \gamma, 0 \quad \text{as} \quad T(x) >, =, < C$$

where C and γ are determined by $E_{\theta_0}\phi(X) = \alpha$. If Q is decreasing, the inequalities are reversed.

As in Example 1, we shall denote the right-hand side of (12) by $P_\theta(x)$ instead of $p_\theta(x)$ when it is a probability, that is, when X is discrete and μ is counting measure.

Example 2. The binomial distributions $b(p, n)$ with

$$P_p(x) = \binom{n}{x} p^x (1-p)^{n-x}$$

satisfy (12) with $T(x) = x$, $\theta = p$, $Q(p) = \log[p/(1-p)]$. The problem of testing H: $p \geq p_0$ arises, for instance, in the situation of Example 1 if one supposes that the production process is in statistical control, so that the various items constitute independent trials with constant probability p of being defective. The number of defectives X in a sample of size n is then a sufficient statistic for the distribution of the variables $X_i (i = 1, \cdots, n)$, where X_i is 1 or 0 as the ith item drawn is defective or not, and X is distributed as $b(p, n)$. There exists therefore a UMP test of H, which rejects H when X is too small.

An alternative sampling plan which is sometimes used in binomial situations is *inverse binomial sampling*. Here the experiment is continued until a specified number m of successes—for example, cures effected by some new medical treatment—have been obtained. If Y_i denotes the number of trials after the $(i-1)$st success up to but not including the ith success, the probability that $Y_i = y$ is pq^y for $y = 0, 1, \cdots$, so that the joint distribution of $Y_1, \cdots, Y_m$ is

$$P_p(y_1, \cdots, y_m) = p^m q^{\Sigma y_i}, \qquad y_k = 0, 1, \cdots; \qquad k = 1, \cdots, m.$$

This is an exponential family with $T(y) = \Sigma y_i$ and $Q(p) = \log(1-p)$. Since $Q(p)$ is a decreasing function of p, the UMP test of $H: p \leqq p_0$ rejects H when T is too small. This is what one would expect since the realization of m successes in only a few more than m trials indicates a high value of p. The test statistic T, which is the number of trials required in excess of m to get m successes, has the negative binomial distribution [Chapter 1, Problem 1(i)]

$$P(t) = \binom{m+t-1}{m-1} p^m q^t, \qquad t = 0, 1, \cdots.$$

Example 3. If $X_1, \cdots, X_n$ are independent Poisson variables with $E(X_i) = \lambda$, their joint distribution is

$$P_\lambda(x_1, \cdots, x_n) = \frac{\lambda^{x_1 + \cdots + x_n}}{x_1! \cdots x_n!} e^{-n\lambda}.$$

This constitutes an exponential family with $T(x) = \Sigma x_i$, and $Q(\lambda) = \log \lambda$. One-sided hypotheses concerning λ might arise if λ is a bacterial density and the X's are a number of bacterial counts, or if the X's denote the number of α-particles produced in equal time intervals by a radioactive substance, etc. The UMP test of the hypothesis $\lambda \leqq \lambda_0$ rejects when ΣX_i is too large. Here the test statistic ΣX_i has itself a Poisson distribution with parameter $n\lambda$.

Instead of observing the radioactive material for given time periods or counting the number of bacteria in given areas of a slide, one can adopt an inverse sampling method. The experiment is then continued, or the area over which the bacteria are counted is enlarged, until a count of m has been obtained. The observations consist of the times $T_1, \cdots, T_m$ that it takes for the first occurrence, from the first to the second, etc. If one is dealing with a Poisson process and the number of occurrences in a time or space interval τ has the distribution

$$P(x) = \frac{(\lambda\tau)^x}{x!} e^{-\lambda\tau}, \qquad x = 0, 1, \cdots,$$

then the observed times are independently distributed, each with the exponential probability density $\lambda e^{-\lambda t}$ for $t \geqq 0$ [Problem 1(ii) of Chapter 1]. The joint densities

$$p_\lambda(t_1, \cdots, t_m) = \lambda^m \exp\left(-\lambda \sum_{i=1}^{m} t_i\right), \qquad t_1, \cdots, t_m \geqq 0,$$

form an exponential family with $T(t_1, \cdots, t_m) = \Sigma t_i$ and $Q(\lambda) = -\lambda$. The UMP test of $H: \lambda \leqq \lambda_0$ rejects when $T = \Sigma T_i$ is too small. Since $2\lambda T_i$ has density $\frac{1}{2}e^{-u/2}$ for $u \geqq 0$, which is the density of a χ^2-distribution with 2 degrees of freedom, $2\lambda T$ has a χ^2-distribution with $2m$ degrees of freedom. The boundary of the rejection region can therefore be determined from a table of χ^2.

The formulation of the problem of hypothesis testing given at the beginning of the chapter takes account of the losses resulting from wrong decisions only in terms of the two types of error. To obtain a more detailed description of the problem of testing $H: \theta \leqq \theta_0$ against the alternatives $\theta > \theta_0$, one can consider it as a decision problem with the decisions d_0 and d_1 of accepting and rejecting H and a loss function $L(\theta, d_i) = L_i(\theta)$. Typically, $L_0(\theta)$ will be 0 for $\theta \leqq \theta_0$ and strictly increasing for $\theta \geqq \theta_0$, and $L_1(\theta)$ will be strictly decreasing for $\theta \leqq \theta_0$ and equal to 0 for $\theta \geqq \theta_0$. The difference then satisfies

$$L_1(\theta) - L_0(\theta) \gtrless 0 \quad \text{as} \quad \theta \lessgtr \theta_0. \tag{13}$$

Theorem 3. (i) *Under the assumptions of Theorem* 2, *the family of tests given by* (9) *and* (10) *with* $0 \leqq \alpha \leqq 1$ *is essentially complete provided the loss function satisfies* (13).

(ii) *This family is also minimal essentially complete if the set of points* x *for which* $p_\theta(x) > 0$ *is independent of* θ.

Proof. (i) The risk function of any test ϕ is

$$\begin{aligned} R(\theta, \phi) &= \int p_\theta(x) \{\phi(x)L_1(\theta) + [1 - \phi(x)]L_0(\theta)\} \, d\mu(x) \\ &= \int p_\theta(x) \{L_0(\theta) + [L_1(\theta) - L_0(\theta)]\phi(x)\} \, d\mu(x), \end{aligned}$$

and hence the difference of two risk functions is

$$R(\theta, \phi') - R(\theta, \phi) = [L_1(\theta) - L_0(\theta)] \int (\phi' - \phi) p_\theta \, d\mu.$$

This is $\leqq 0$ for all θ if

$$\beta_{\phi'}(\theta) - \beta_\phi(\theta) = \int (\phi' - \phi) p_\theta \, d\mu \gtreqqless 0 \quad \text{for} \quad \theta \gtreqqless \theta_0.$$

Given any test ϕ, let $E_{\theta_0} \phi(X) = \alpha$. It follows from Theorem 2(i) that there exists a UMP level α test ϕ' for testing $\theta = \theta_0$ against $\theta > \theta_0$, which satisfies (9) and (10). By Theorem 2(iv), ϕ' also minimizes the power for $\theta < \theta_0$. Thus the two risk functions satisfy $R(\theta, \phi') \leqq R(\theta, \phi)$ for all θ, as was to be proved.

(ii) Let ϕ_α and $\phi_{\alpha'}$ be of sizes $\alpha < \alpha'$ and UMP for testing θ_0 against $\theta > \theta_0$. Then $\beta_{\phi_\alpha}(\theta) < \beta_{\phi_{\alpha'}}(\theta)$ for all $\theta > \theta_0$ unless $\beta_{\phi_\alpha}(\theta) = 1$. By considering the problem of testing $\theta = \theta_0$ against $\theta < \theta_0$ it is seen analogously that this inequality also holds for all $\theta < \theta_0$ unless $\beta_{\phi_{\alpha'}}(\theta) = 0$. Since the exceptional possibilities are excluded by the assumptions, it follows that $R(\theta, \phi') \lessgtr R(\theta, \phi)$ as $\theta \gtrless \theta_0$. Hence each of the two risk functions is better than the other for some values of θ.

The class of tests previously derived as UMP at the various significance levels α is now seen to constitute an essentially complete class for a much more general decision problem, in which the loss function is only required to satisfy certain broad qualitative conditions. From this point of view, the formulation involving the specification of a level of significance can be considered as a simple way of selecting a particular procedure from an essentially complete family.

The property of monotone likelihood ratio defines a very strong ordering of a family of distributions. For later use, we consider also the following somewhat weaker definition. A family of cumulative distribution functions F_θ on the real line is said to be *stochastically increasing* (and the same term is applied to random variables possessing these distributions) if the distributions are distinct and if $\theta < \theta'$ implies $F_\theta(x) \geqq F_{\theta'}(x)$ for all x. If then X and X' have distributions F_θ and F'_θ respectively, it follows that $P\{X > x\} \leqq P\{X' > x\}$ for all x so that X' tends to have larger values than X. In this case the variable X' is said to be *stochastically larger* than X. This relationship is made more precise by the following characterization of the stochastic ordering of two distributions.

Lemma 1. *Let F_0 and F_1 be two cumulative distribution functions on the real line. Then $F_1(x) \leqq F_0(x)$ for all x if and only if there exist two nondecreasing functions f_0 and f_1, and a random variable V, such that (a) $f_0(v) \leqq f_1(v)$ for all v, and (b) the distributions of $f_0(V)$ and $f_1(V)$ are F_0 and F_1 respectively.*

Proof. Suppose first that the required f_0, f_1, and V exist. Then

$$F_1(x) = P\{f_1(V) \leqq x\} \leqq P\{f_0(V) \leqq x\} = F_0(x)$$

for all x. Conversely, suppose that $F_1(x) \leqq F_0(x)$ for all x, and let $f_i(y) = \inf\{x\colon F_i(x - 0) \leqq y \leqq F_i(x)\}$, $i = 0, 1$. These functions are nondecreasing and for $f_i = f$, $F_i = F$ satisfy

$$f[F(x)] \leqq x \qquad \text{and} \qquad F[f(y)] \geqq y \quad \text{for all} \quad x \text{ and } y.$$

It follows that $y \leqq F(x_0)$ implies $f(y) \leqq f[F(x_0)] \leqq x_0$ and that conversely $f(y) \leqq x_0$ implies $F[f(y)] \leqq F(x_0)$ and hence $y \leqq F(x_0)$, so that the two inequalities $f(y) \leqq x_0$ and $y \leqq F(x_0)$ are equivalent. Let V be uniformly distributed on $(0, 1)$. Then $P\{f_i(V) \leqq x\} = P\{V \leqq F_i(x)\} = F_i(x)$. Since $F_1(x) \leqq F_0(x)$ for all x implies $f_0(y) \leqq f_1(y)$ for all y, this completes the proof.

One of the simplest examples of a stochastically ordered family is a location parameter family, that is, a family satisfying

$$F_\theta(x) = F(x - \theta).$$

To see that this is stochastically increasing, let X be a random variable with distribution $F(x)$. Then $\theta < \theta'$ implies

$$F(x-\theta) = P\{X \leqq x - \theta\} \geqq P\{X \leqq x - \theta'\} = F(x - \theta'),$$

as was to be shown.

Another example is furnished by families with monotone likelihood ratio. This is seen from the following lemma, which establishes some basic properties of these families.

Lemma 2.* *Let $p_\theta(x)$ be a family of densities on the real line with monotone likelihood ratio in x.*

(i) *If ψ is a nondecreasing function of x, then $E_\theta\psi(X)$ is a nondecreasing function of θ; if $X_1, \cdots, X_n$ are independently distributed with density p_θ and ψ' is a function of $x_1, \cdots, x_n$ which is nondecreasing in each of its arguments, then $E_\theta\psi'(X_1, \cdots, X_n)$ is a nondecreasing function of θ.*

(ii) *For any $\theta < \theta'$, the cumulative distribution functions of X under θ and θ' satisfy*

$$F_{\theta'}(x) \leqq F_\theta(x) \quad \text{for all} \quad x.$$

(iii) *Let ψ be a function with a single change of sign. More specifically, suppose there exists a value x_0 such that $\psi(x) \leqq 0$ for $x < x_0$ and $\psi(x) \geqq 0$ for $x \geqq x_0$. Then there exists θ_0 such that $E_\theta\psi(X) \leqq 0$ for $\theta < \theta_0$ and $E_\theta\psi(X) \geqq 0$ for $\theta > \theta_0$, unless $E_\theta\psi(X)$ is either positive for all θ or negative for all θ.*

Proof. (i) Let $\theta < \theta'$ and let A and B be the sets for which $p_{\theta'}(x) < p_\theta(x)$ and $p_{\theta'}(x) > p_\theta(x)$ respectively. If $a = \sup_A \psi(x)$ and $b = \inf_B \psi(x)$, then $b - a \geqq 0$ and

$$\int \psi(p_{\theta'} - p_\theta)\,d\mu \geqq a\int_A (p_{\theta'} - p_\theta)\,d\mu + b\int_B (p_{\theta'} - p_\theta)\,d\mu$$

$$= (b-a)\int_B (p_{\theta'} - p_\theta)\,d\mu \geqq 0,$$

which proves the first assertion. The result for general n follows by induction.

(ii) This follows from (i) by letting $\psi(x) = 1$ for $x > x_0$ and $\psi(x) = 0$ otherwise.

(iii) We shall show first that for any $\theta' < \theta''$, $E_{\theta'}\psi(X) > 0$ implies $E_{\theta''}\psi(X) \geqq 0$. If $p_{\theta''}(x_0)/p_{\theta'}(x_0) = \infty$, then $p_{\theta'}(x) = 0$ for $x \geqq x_0$ and hence $E_{\theta'}\psi(X) \leqq 0$. Suppose therefore that $p_{\theta''}(x_0)/p_{\theta'}(x_0) = c < \infty$.

* This is a special case of a theorem of Karlin relating the number of changes of sign of $E_\theta\psi(X)$ to those of $\psi(x)$ when the densities p_θ are of Pólya type. See Karlin, "Pólya type distributions II," *Ann. Math. Stat.*, Vol. 28 (1957), pp. 281–308.

Then $\psi(x) \geqq 0$ on the set $S = \{x: p_{\theta'}(x) = 0 \text{ and } p_{\theta''}(x) > 0\}$, and

$$E_{\theta''}\psi(X) \geqq \int_{\tilde{S}} \psi \frac{p_{\theta''}}{p_{\theta'}} p_{\theta'}\, d\mu$$

$$\geqq \int_{-\infty}^{x_0 -} c\psi p_{\theta'}\, d\mu + \int_{x_0}^{\infty} c\psi p_{\theta'}\, d\mu = cE_{\theta'}\, \psi(X) \geqq 0.$$

The result now follows by letting $\theta_0 = \inf\{\theta: E_\theta \psi(X) > 0\}$.

Part (ii) of the lemma shows that any family of distributions with monotone likelihood ratio in x is stochastically increasing. That the converse does not hold is shown for example by the Cauchy densities

$$\frac{1}{\pi} \frac{1}{1 + (x - \theta)^2}.$$

The family is stochastically increasing since θ is a location parameter; however, the likelihood ratio is not monotone. Conditions under which a location parameter family possesses monotone likelihood ratio are given in Chapter 8, Example 1.

4. COMPARISON OF EXPERIMENTS*

Suppose that different experiments are available for testing a simple hypothesis H against a simple alternative K. One experiment results in a random variable X, which has probability densities f and g under H and K respectively; the other one leads to the observation of X' with densities f' and g'. Let $\beta(\alpha)$ and $\beta'(\alpha)$ denote the power of the most powerful level α test based on X and X'. In general, the relationship between $\beta(\alpha)$ and $\beta'(\alpha)$ will depend on α. However, if $\beta'(\alpha) \leqq \beta(\alpha)$ for all α, then X or the experiment (f, g) is said to be *more informative* than X'. As an example, suppose that the family of densities $p_\theta(x)$ is the exponential family (12) and that $f = f' = p_{\theta_0}$, $g = p_{\theta_2}$, $g' = p_{\theta_1}$, where $\theta_0 < \theta_1 < \theta_2$. Then (f, g) is more informative than (f', g') by Theorem 2.

A simple sufficient condition† for X to be more informative than X' is the existence of a function $h(x, u)$ and a random quantity U, independent of X and having a known distribution, such that the density of $Y = h(X, U)$ is f' or g' as that of X is f or g. This follows, as in the theory of sufficient statistics, from the fact that one can then construct from X (with the help of U) a variable Y, which is equivalent to X'. One can also argue

* This section constitutes a digression and may be omitted.

† For a proof that this condition is also necessary see Blackwell, "Comparison of experiments," *Proc. Second Berkeley Symposium on Mathematical Statistics and Probability*, Berkeley, Univ. Calif. Press, 1951.

more specifically that if $\phi(x')$ is the most powerful level α test for testing f' against g' and if $\psi(x) = E\phi[h(x, U)]$, then $E\psi(X) = E\phi(X')$ both under H and K. The test $\psi(x)$ is therefore a level α test with power $\beta'(\alpha)$, and hence $\beta(\alpha) \geq \beta'(\alpha)$.

When such a transformation h exists, the experiment (f, g) is said to be *sufficient* for (f', g'). If then $X_1, \cdots, X_n$ and $X'_1, \cdots, X'_n$ are samples from X and X' respectively, the first of these samples is more informative than the second one. It is also more informative than $(Z_1, \cdots, Z_n)$ where each Z_i is either X_i or X'_i with certain probabilities.

Example 4. Two characteristics A and B, which each member of a population may or may not possess, are to be tested for independence. The probabilities $p = P(A)$ and $\pi = P(B)$, that is, the proportions of individuals possessing properties A and B, are assumed to be known. This might be the case, for example, if the characteristics have previously been studied separately but not in conjunction. The probabilities of the four possible combinations AB, $A\tilde{B}$, $\tilde{A}B$, and $\tilde{A}\tilde{B}$ under the hypothesis of independence and under the alternative that $P(AB)$ has a specified value ρ are

	Under H:		Under K:	
	B	$\tilde{B}$	B	$\tilde{B}$
A	$p\pi$	$p(1-\pi)$	ρ	$p-\rho$
$\tilde{A}$	$(1-p)\pi$	$(1-p)(1-\pi)$	$\pi-\rho$	$1-p-\pi+\rho$

The experimental material is to consist of a sample of size s. This can be selected, for example, at random from those members of the population possessing property A. One then observes for each member of the sample whether or not it possesses property B, and hence is dealing with a sample from a binomial distribution with probabilities

$$H\colon P(B|A) = \pi \qquad \text{and} \qquad K\colon P(B|A) = \rho/p.$$

Alternatively, one can draw the sample from one of the other categories B, $\tilde{B}$, or $\tilde{A}$, obtaining in each case a sample from a binomial distribution with probabilities given by the following table.

Population Sampled	Probability	H	K
A	$P(B\|A)$	π	ρ/p
B	$P(A\|B)$	p	ρ/π
$\tilde{B}$	$P(A\|\tilde{B})$	p	$(p-\rho)/(1-\pi)$
$\tilde{A}$	$P(B\|\tilde{A})$	π	$(\pi-\rho)/(1-p)$

Without loss of generality let the categories A, $\tilde{A}$, B, and $\tilde{B}$ be labeled so that $p \leq \pi \leq 1/2$. We shall now show that of the four experiments, which consist in observing an individual from one of the four categories, the first one (sampling from A) is most informative and in fact is sufficient for each of the others.

To compare A with B, let X and X' be 1 or 0 and let the probability of their being equal to 1 be given by the first and second row of the table respectively. Let U be uniformly distributed on (0, 1) and independent of X, and let $Y = h(X, U) = 1$ when $X = 1$ and $U \leq p/\pi$, and $Y = 0$ otherwise. Then $P\{Y = 1\}$ is p under H and ρ/π under K so that Y has the same distribution as X'. This proves that X is sufficient for X', and hence is the more informative of the two. For the comparison of A with $\tilde{B}$ define Y to be 1 when $X = 0$ and $U \leq p/(1 - \pi)$, and to be 0 otherwise. Then the probability that $Y = 1$ coincides with the third row of the table. Finally, the probability that $Y = 1$ is given by the last row of the table if one defines Y to be equal to 1 when $X = 1$ and $U \leq (\pi - p)/(1 - p)$ and when $X = 0$ and $U > (1 - \pi - p)/(1 - p)$.

It follows from the general remarks preceding the example that if the experimental material is to consist of s individuals these should be drawn from category A, that is, the rarest of the four categories, in preference to any of the others. This is preferable also to drawing the s from the population at large, since the latter procedure is equivalent to drawing each of them from either A or $\tilde{A}$ with probabilities p and $1 - p$ respectively.

The comparison between these various experiments is independent not only of α but also of ρ. Furthermore, if a sample is taken from A, there exists by Corollary 2 a UMP test of H against the one-sided alternatives of positive dependence, $P(B|A) > \pi$ and hence $\rho > p\pi$, according to which the probabilities of AB and $\tilde{A}\tilde{B}$ are larger, those of $A\tilde{B}$ and $\tilde{A}B$ smaller than under the assumption of independence. This test therefore provides the best power that can be obtained for the hypothesis of independence on the basis of a sample of size s.

Example 5. In a Poisson process the number of events occurring in a time interval of length v has the Poisson distribution $P(\lambda v)$. The problem of testing λ_0 against λ_1 for these distributions arises also for spatial distributions of particles where one is concerned with the number of particles in a region of volume v. To see that the experiment is the more informative the longer the interval v, let $v < w$ and denote by X and Y the number of occurrences in the intervals $(t, t + v)$ and $(t + v, t + w)$. Then X and Y are independent Poisson variables, and $Z = X + Y$ is a sufficient statistic for λ. Thus any test based on X can be duplicated by one based on Z, and Z is more informative than X. That it is in fact strictly more informative in an obvious sense is seen from the fact that the unique most powerful test for testing λ_0 against λ_1 depends on $X + Y$ and therefore cannot be duplicated from X alone.

Sometimes it is not possible to count the number of occurrences but only to determine whether or not at least one event has taken place. In the dilution method in bacteriology, for example, a bacterial culture is diluted in a certain volume of water, from which a number of samples of fixed size are taken and tested for the presence or absence of bacteria. In general, one observes then for each of n intervals whether an event occurred. The result is a binomial variable with probability of success (at least one occurrence)

$$p = 1 - e^{-\lambda v}.$$

Since a very large or small interval leads to nearly certain success or failure, one might suspect that for testing λ_0 against λ_1 intermediate values of v would be more informative than extreme ones. However, it turns out that the experiments $(\lambda_0 v, \lambda_1 v)$ and $(\lambda_0 w, \lambda_1 w)$ are not comparable for any values of v and w.* (See Problem 15.)

5. CONFIDENCE BOUNDS

The theory of UMP one-sided tests can be applied to the problem of obtaining a lower or upper bound for a real-valued parameter θ. The problem of setting a lower bound arises, for example, when θ is the breaking strength of a new alloy; that of setting an upper bound when θ is the toxicity of a drug or the probability of an undesirable event. The discussion of lower and upper bounds is completely parallel, and it is therefore enough to consider the case of a lower bound, say $\underline{\theta}$.

Since $\underline{\theta} = \underline{\theta}(X)$ will be a function of the observations, it cannot be required to fall below θ with certainty but only with specified high probability. One selects a number $1 - \alpha$, the *confidence level*, and restricts attention to bounds $\underline{\theta}$ satisfying

$$P_\theta\{\underline{\theta}(X) \leqq \theta\} \geqq 1 - \alpha \quad \text{for all} \quad \theta. \tag{14}$$

The function $\underline{\theta}$ is called a lower *confidence bound* for θ at confidence level $1 - \alpha$; the infimum of the left-hand side of (14), which in practice will be equal to $1 - \alpha$, is called the *confidence coefficient* of $\underline{\theta}$.

Subject to (14), $\underline{\theta}$ should underestimate θ by as little as possible. One can ask, for example, that the probability of $\underline{\theta}$ falling below any $\theta' < \theta$ should be a minimum. A function $\underline{\theta}$ for which

$$P_\theta\{\underline{\theta}(X) \leqq \theta'\} = \text{minimum} \tag{15}$$

for all $\theta' < \theta$ subject to (14) is a uniformly *most accurate* lower confidence bound for θ at confidence level $1 - \alpha$.

Let $L(\theta, \underline{\theta})$ be a measure of the loss resulting from underestimating θ, so that for each fixed θ the function $L(\theta, \underline{\theta})$ is defined and nonnegative for $\underline{\theta} < \theta$, and is nonincreasing in its second argument. One would then wish to minimize

$$E_\theta L(\theta, \underline{\theta}) \tag{16}$$

subject to (14). It can be shown that a uniformly most accurate lower confidence bound $\underline{\theta}$ minimizes (16) subject to (14) for every such loss function L. (See Problem 17.)

The derivation of uniformly most accurate confidence bounds is facilitated by introducing the following more general concept, which will

* For a discussion of how to select v in this and similar situations see Hodges, "The choice of inspection stringency in acceptance sampling by attributes," *Univ. Calif. Publ. Statistics*, Vol. 1 (1949), pp. 1–14.

be considered in more detail in Chapter 5. A family of subsets $S(x)$ of the parameter space Ω is said to constitute a family of *confidence sets* at confidence level $1 - \alpha$ if

$$P_\theta\{\theta \in S(X)\} \geqq 1 - \alpha \quad \text{for all} \quad \theta \in \Omega, \tag{17}$$

that is, if the random set $S(X)$ covers the true parameter point with probability $\geqq 1 - \alpha$. A lower confidence bound corresponds to the special case that $S(x)$ is a one-sided interval

$$S(x) = \{\theta : \underline{\theta}(x) \leqq \theta < \infty\}.$$

Theorem 4. (i) *For each $\theta_0 \in \Omega$ let $A(\theta_0)$ be the acceptance region of a level α test for testing $H(\theta_0)$: $\theta = \theta_0$, and for each sample point x let $S(x)$ denote the set of parameter values*

$$S(x) = \{\theta : x \in A(\theta), \theta \in \Omega\}.$$

Then $S(x)$ is a family of confidence sets for θ at confidence level $1 - \alpha$.

(ii) *If $A(\theta_0)$ is UMP for testing $H(\theta_0)$ at level α against the alternatives $K(\theta_0)$, then $S(X)$ minimizes the probability*

$$P_\theta\{\theta' \in S(X)\} \quad \text{for all} \quad \theta \in K(\theta')$$

among all level $1 - \alpha$ families of confidence sets for θ.

Proof. (i) By definition of $S(x)$,

$$\theta \in S(x) \quad \text{if and only if} \quad x \in A(\theta), \tag{18}$$

and hence

$$P_\theta\{\theta \in S(X)\} = P_\theta\{X \in A(\theta)\} \geqq 1 - \alpha.$$

(ii) If $S^*(x)$ is any other family of confidence sets at level $1 - \alpha$, and if $A^*(\theta) = \{x : \theta \in S^*(x)\}$, then

$$P_\theta\{X \in A^*(\theta)\} = P_\theta\{\theta \in S^*(X)\} \geqq 1 - \alpha,$$

so that $A^*(\theta_0)$ is the acceptance region of a level α test of $H(\theta_0)$. It follows from the assumed property of $A(\theta_0)$ that for any $\theta \in K(\theta_0)$

$$P_\theta\{X \in A^*(\theta_0)\} \geqq P_\theta\{X \in A(\theta_0)\}$$

and hence that

$$P_\theta\{\theta_0 \in S^*(X)\} \geqq P_\theta\{\theta_0 \in S(X)\}$$

as was to be proved.

The equivalence (18) shows the structure of the confidence sets $S(x)$ as the totality of parameter values θ for which the hypothesis $H(\theta)$ is accepted when x is observed. A confidence set can therefore be viewed as a combined statement regarding the tests of the various hypotheses

$H(\theta)$, which exhibits the values for which the hypothesis is accepted ($\theta \in S(x)$) and those for which it is rejected ($\theta \in \widetilde{S(x)}$).

Corollary 3. *Let the family of densities $p_\theta(x)$, $\theta \in \Omega$ have monotone likelihood ratio in $T(x)$ and suppose that the cumulative distribution function $F_\theta(t)$ of $T = T(X)$ is a continuous function of t for each fixed θ.*

(i) *There exists a uniformly most accurate confidence bound $\underline{\theta}$ for θ at each confidence level $1 - \alpha$.*

(ii) *If x denotes the observed values of X and $t = T(x)$, and if the equation*

$$F_\theta(t) = 1 - \alpha \tag{19}$$

has a solution $\theta = \hat{\theta}$ in Ω, then this solution is unique and $\underline{\theta}(x) = \hat{\theta}$.

Proof. (i) There exists for each θ_0 a constant $C(\theta_0)$ such that

$$P_{\theta_0}\{T > C(\theta_0)\} = \alpha$$

and by Theorem 2, $T > C(\theta_0)$ is a UMP level α rejection region for testing $\theta = \theta_0$ against $\theta > \theta_0$. By Corollary 1, the power of this test against any alternative $\theta_1 > \theta_0$ exceeds α, and hence $C(\theta_0) < C(\theta_1)$ so that the function C is strictly increasing. Let $A(\theta_0)$ denote the acceptance region $T \leqq C(\theta_0)$ and let $S(x)$ be defined by (18). It follows from the monotonicity of the function C that $S(x)$ consists of those values $\theta \in \Omega$ which satisfy $\underline{\theta} \leqq \theta$ where

$$\underline{\theta} = \inf\{\theta: T(x) \leqq C(\theta)\}.$$

By Theorem 4, the sets $\{\theta: \underline{\theta}(x) \leqq \theta\}$, restricted to possible values of the parameter, thus constitute a family of confidence sets at level $1 - \alpha$, which minimize $P\{\underline{\theta} \leq \theta'\}$ for all $\theta \in K(\theta')$, that is, for all $\theta > \theta'$. This shows $\underline{\theta}$ to be a uniformly most accurate confidence bound for θ.

(ii) It follows from Corollary 1 that $F_\theta(t)$ is a strictly decreasing function of θ at any point t for which $0 < F_\theta(t) < 1$, and hence that (19) can have at most one solution. Suppose now that t is the observed value of T and that the equation $F_\theta(t) = 1 - \alpha$ has the solution $\hat{\theta} \in \Omega$. Then $F_{\hat{\theta}}(t) = 1 - \alpha$ and by definition of the function C, $C(\hat{\theta}) = t$. The inequality $t \leqq C(\theta)$ is then equivalent to $C(\hat{\theta}) \leqq C(\theta)$ and hence to $\hat{\theta} \leqq \theta$. It follows that $\underline{\theta} = \hat{\theta}$, as was to be proved.

Under the same assumptions, the corresponding upper confidence bound with confidence coefficient $1 - \alpha$ is the solution $\bar{\theta}$ of the equation $P_\theta\{T \geqq t\} = 1 - \alpha$ or equivalently of $F_\theta(t) = \alpha$.

Example 6. To determine an upper bound for the degree of radioactivity λ of a radioactive substance, the substance is observed until a count of m has been obtained on a Geiger counter. The joint probability density of the times $T_i (i = 1, \cdots, m)$ elapsing between the $(i - 1)$st count and the ith one is

$$p(t_1, \cdots, t_m) = \lambda^m e^{-\lambda \Sigma t_i}, \qquad t_1, \cdots, t_m \geqq 0.$$

If $T = \Sigma T_i$ denotes the total time of observation, $2\lambda T$ has a χ^2-distribution with $2m$ degrees of freedom and, as was shown in Example 3, the acceptance region of the most powerful test of $H(\lambda_0)$: $\lambda = \lambda_0$ against $\lambda < \lambda_0$ is $2\lambda_0 T \leqq C$ where C is determined by the equation

$$\int_0^C \chi_{2m}^2 = 1 - \alpha.$$

The set $S(t_1, \cdots, t_m)$ defined by (18) is then the set of values λ such that $\lambda \leqq C/2T$ and it follows from Theorem 4 that $\bar{\lambda} = C/2T$ is a uniformly most accurate upper confidence bound for λ. This result can also be obtained through Corollary 3.

If the variables X or T are discrete, Corollary 3 cannot be applied directly since the distribution functions $F_\theta(t)$ are not continuous, and for most values θ_0 the optimum tests of H: $\theta = \theta_0$ are randomized. However, any randomized test based on X has the following representation as a nonrandomized test depending on X and an independent variable U distributed uniformly over (0, 1). Given a critical function ϕ, consider the rejection region

$$R = \{(x, u): u \leqq \phi(x)\}.$$

Then

$$P\{(X, U) \in R\} = P\{U \leqq \phi(X)\} = E\phi(X),$$

whatever the distribution of X, so that R has the same power function as ϕ and the two tests are equivalent. The pair of variables (X, U) has a particularly simple representation when X is integer-valued. In this case the statistic

$$T = X + U$$

is equivalent to the pair (X, U) since with probability 1

$$X = [T], \qquad U = T - [T],$$

where $[T]$ denotes the largest integer $\leqq T$. The distribution of T is continuous, and confidence bounds can be based on this statistic.

Example 7. An upper bound is required for a binomial probability p—for example, the probability that a batch of polio vaccine manufactured according to a certain procedure contains any live virus. Let $X_1, \cdots, X_n$ denote the outcomes of n trials, X_i being 1 or 0 with probabilities p and q respectively, and let $X = \Sigma X_i$. Then $T = X + U$ has probability density

$$\binom{n}{[t]} p^{[t]} q^{n-[t]}, \qquad 0 \leqq t < n + 1.$$

This satisfies the conditions of Corollary 3, and the upper confidence bound $\bar{p}$ is therefore the solution, if it exists, of the equation

$$P_p\{T < t\} = \alpha,$$

where t is the observed value of T. A solution does exist for all values $\alpha \leqq t \leqq n + \alpha$. For $n + \alpha < t$, the hypothesis $H(p_0)$: $p = p_0$ is accepted against the alternatives $p < p_0$ for all values of p_0 and hence $\bar{p} = 1$. For $t < \alpha$, $H(p_0)$ is rejected for all values of p_0 and the confidence set $S(t)$ is therefore empty. Consider instead the sets $S^*(t)$ which are equal to $S(t)$ for $t \geqq \alpha$ and which for $t < \alpha$ consist of the single point $p = 0$. They are also confidence sets at level $1 - \alpha$ since for all p,

$$P_p\{p \in S^*(T)\} \geqq P_p\{p \in S(T)\} = 1 - \alpha.$$

On the other hand, $P_p\{p' \in S^*(T)\} = P_p\{p' \in S(T)\}$ for all $p' > 0$ and hence

$$P_p\{p' \in S^*(T)\} = P_p\{p' \in S(T)\} \quad \text{for all} \quad p' > p.$$

Thus the family of sets $S^*(t)$ minimizes the probability of covering p' for all $p' > p$ at confidence level $1 - \alpha$. The associated confidence bound $\bar{p}^*(t) = \bar{p}(t)$ for $t \geqq \alpha$ and $\bar{p}^*(t) = 0$ for $t < \alpha$ is therefore a uniformly most accurate upper confidence bound for p at level $1 - \alpha$.

In practice, so as to avoid randomization and obtain a bound not dependent on the extraneous variable U, one usually replaces T by $X + 1 = [T] + 1$. Since $\bar{p}^*(t)$ is a nondecreasing function of t, the resulting upper confidence bound $\bar{p}^*([t] + 1)$ is then somewhat larger than necessary; as a compensation it also gives a correspondingly higher probability of not falling below the true p.

Let $\underline{\theta}$ and $\bar{\theta}$ be lower and upper bounds for θ with confidence coefficients $1 - \alpha_1$ and $1 - \alpha_2$, and suppose that $\underline{\theta}(x) < \bar{\theta}(x)$ for all x. This will be the case under the assumptions of Corollary 3 if $\alpha_1 + \alpha_2 < 1$. The intervals $(\underline{\theta}, \bar{\theta})$ are then *confidence intervals* for θ with confidence coefficient $1 - \alpha_1 - \alpha_2$; that is, they contain the true parameter value with probability $1 - \alpha_1 - \alpha_2$, since

$$P_\theta\{\underline{\theta} \leqq \theta \leqq \bar{\theta}\} = 1 - \alpha_1 - \alpha_2 \quad \text{for all} \quad \theta.$$

If $\underline{\theta}$ and $\bar{\theta}$ are uniformly most accurate, they minimize $E_\theta L_1(\theta, \underline{\theta})$ and $E_\theta L_2(\theta, \bar{\theta})$ at their respective levels for any function L_1 that is nonincreasing in $\underline{\theta}$ for $\underline{\theta} < \theta$ and 0 for $\underline{\theta} \geqq \theta$ and any L_2 that is nondecreasing in $\bar{\theta}$ for $\bar{\theta} > \theta$ and 0 for $\bar{\theta} \leqq \theta$. Letting

$$L(\theta;\ \underline{\theta}, \bar{\theta}) = L_1(\theta, \underline{\theta}) + L_2(\theta, \bar{\theta}),$$

the intervals $(\underline{\theta}, \bar{\theta})$ therefore minimize $E_\theta L(\theta;\ \underline{\theta}, \bar{\theta})$ subject to

$$P_\theta\{\underline{\theta} > \theta\} \leqq \alpha_1, \qquad P_\theta\{\bar{\theta} < \theta\} \leqq \alpha_2.$$

An example of such a loss function is

$$L(\theta;\ \underline{\theta}, \bar{\theta}) = \begin{cases} \bar{\theta} - \underline{\theta} & \text{if} \quad \underline{\theta} \leqq \theta \leqq \bar{\theta} \\ \bar{\theta} - \theta & \text{if} \quad \theta < \underline{\theta} \\ \theta - \underline{\theta} & \text{if} \quad \bar{\theta} < \theta, \end{cases}$$

which provides a natural measure of the accuracy of the intervals. The actual length $\bar{\theta} - \underline{\theta}$ is not as meaningful in this context since there is no merit in short intervals that are far away from the true θ.

An important limiting case corresponds to the levels $\alpha_1 = \alpha_2 = \frac{1}{2}$. Under the assumptions of Corollary 3 and if the region of positive density is independent of θ so that tests of power 1 are impossible when $\alpha < 1$, the upper and lower confidence bounds $\bar{\theta}$ and $\underline{\theta}$ coincide in this case. The common bound satisfies

$$P_\theta\{\underline{\theta} \leqq \theta\} = P_\theta\{\underline{\theta} \geqq \theta\} = \tfrac{1}{2},$$

and the estimate $\underline{\theta}$ of θ is therefore as likely to underestimate as to overestimate the true value. An estimate with this property is said to be *median unbiased.* (For the relation of this to other concepts of unbiasedness, see Chapter 1, Problem 3.) It follows from the above result for arbitrary α_1 and α_2 that among all median unbiased estimates, $\underline{\theta}$ minimizes $EL(\theta, \underline{\theta})$ for any loss function which for fixed θ has a minimum of 0 at $\underline{\theta} = \theta$ and is nondecreasing as $\underline{\theta}$ moves away from θ in either direction. By taking in particular $L(\theta, \underline{\theta}) = 0$ when $|\theta - \underline{\theta}| \leqq \Delta$ and $=1$ otherwise, it is seen that among all median unbiased estimates, $\underline{\theta}$ minimizes the probability of differing from θ by more than any given amount; more generally it maximizes the probability

$$P_\theta\{-\Delta_1 \leqq \theta - \underline{\theta} \leqq \Delta_2\}$$

for any $\Delta_1, \Delta_2 \geqq 0$.

6. A GENERALIZATION OF THE FUNDAMENTAL LEMMA

The following is a useful extension of Theorem 1 to the case of more than one side condition.

Theorem 5. *Let* $f_1, \cdots, f_{m+1}$ *be real-valued functions defined on a Euclidean space* $\mathscr{X}$ *and integrable* μ, *and suppose that for given constants* $c_1, \cdots, c_m$ *there exists a critical function* ϕ *satisfying*

$$\int \phi f_i \, d\mu = c_i, \qquad i = 1, \cdots, m. \tag{20}$$

Let $\mathscr{C}$ *be the class of critical functions* ϕ *for which* (20) *holds.*

(i) *Among all members of* $\mathscr{C}$ *there exists one that maximizes*

$$\int \phi f_{m+1} \, d\mu.$$

(ii) *A sufficient condition for a member of $\mathscr{C}$ to maximize*

$$\int \phi f_{m+1}\, d\mu$$

is the existence of constants $k_1, \cdots, k_m$ *such that*

$$(21)\qquad \begin{aligned} \phi(x) &= 1 \quad \text{when} \quad f_{m+1}(x) > \sum_{i=1}^{m} k_i f_i(x) \\ \phi(x) &= 0 \quad \text{when} \quad f_{m+1}(x) < \sum_{i=1}^{m} k_i f_i(x). \end{aligned}$$

(iii) *If a member of $\mathscr{C}$ satisfies* (21) *with* $k_1, \cdots, k_m \geqq 0$, *then it maximizes*

$$\int \phi f_{m+1}\, d\mu$$

among all critical functions satisfying

$$(22)\qquad \int \phi f_i\, d\mu \leqq c_i, \qquad i = 1, \cdots, m.$$

(iv) *The set M of points in m-dimensional space whose coordinates are*

$$\left(\int \phi f_1\, d\mu, \cdots, \int \phi f_m\, d\mu\right)$$

for some critical function ϕ is convex and closed. If $(c_1, \cdots, c_m)$ *is an inner point* of M, then there exist constants* $k_1, \cdots, k_m$ *and a test ϕ satisfying* (20) *and* (21), *and a necessary condition for a member of $\mathscr{C}$ to maximize*

$$\int \phi f_{m+1}\, d\mu$$

is that (21) *holds a.e. μ.*

Here the term "inner point of M" in statement (iv) can be interpreted as meaning a point interior to M relative to m-space or relative to the smallest linear space (of dimension $\leqq m$) containing M. The theorem is correct with both interpretations but is stronger with respect to the latter, for which it will be proved.

We also note that exactly analogous results hold for the minimization of $\int \phi f_{m+1}\, d\mu$.

Proof. (i) Let $\{\phi_n\}$ be a sequence of functions in $\mathscr{C}$ such that $\int \phi_n f_{m+1}\, d\mu$ tends to $\sup_\phi \int \phi f_{m+1}\, d\mu$. By the weak compactness theorem for critical

* A discussion of the problem when this assumption is not satisfied is given by Dantzig and Wald, "On the fundamental lemma of Neyman and Pearson," *Ann. Math. Stat.*, Vol. 22 (1951), pp. 87–93.

functions (Theorem 3 of the Appendix), there exists a subsequence $\{\phi_{n_i}\}$ and a critical function ϕ such that

$$\int \phi_{n_i} f_k \, d\mu \to \int \phi f_k \, d\mu \quad \text{for} \quad k = 1, \cdots, m+1.$$

It follows that ϕ is in $\mathscr{C}$ and maximizes the integral with respect to $f_{m+1}\,d\mu$ within $\mathscr{C}$.

(ii) and (iii) are proved exactly as was part (ii) of Theorem 1.

(iv) That M is closed follows again from the weak compactness theorem and its convexity is a consequence of the fact that if ϕ_1 and ϕ_2 are critical

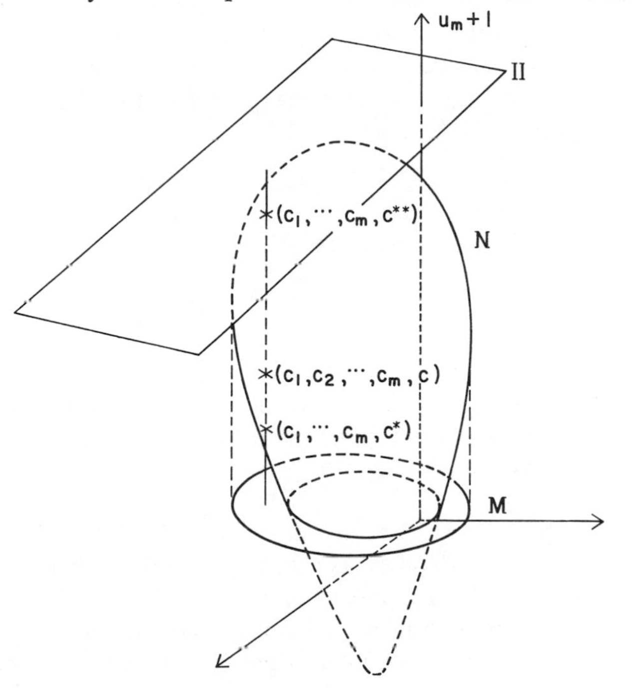

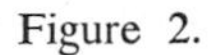
Figure 2.

functions, so is $\alpha\phi_1 + (1-\alpha)\phi_2$ for any $0 \leqq \alpha \leqq 1$. If N is the totality of points in $(m+1)$-dimensional space with coordinates

$$\left(\int \phi f_1 \, d\mu, \cdots, \int \phi f_{m+1} \, d\mu\right),$$

where ϕ ranges over the class of all critical functions, then N is convex and closed by the same argument. Denote the coordinates of a general point in M and N by $(u_1, \cdots, u_m)$ and $(u_1, \cdots, u_{m+1})$ respectively. The points of N, the first m coordinates of which are $c_1, \cdots, c_m$, form a closed interval $[c^*, c^{**}]$.

Assume first that $c^* < c^{**}$. Since $(c_1, \cdots, c_m, c^{**})$ is a boundary point of N, there exists a hyperplane Π through it such that every point of N lies below or on Π. Let the equation of Π be

$$\sum_{i=1}^{m+1} k_i u_i = \sum_{i=1}^{m} k_i c_i + k_{m+1} c^{**}.$$

Since $(c_1, \cdots, c_m)$ is an inner point of M, the coefficient $k_{m+1} \neq 0$. To see this, let $c^* < c < c^{**}$, so that $(c_1, \cdots, c_m, c)$ is an inner point of N. Then there exists a sphere with this point as center lying entirely in N and hence below Π. It follows that the point $(c_1, \cdots, c_m, c)$ does not lie on Π and hence that $k_{m+1} \neq 0$. We may therefore take $k_{m+1} = -1$ and see that for any point of N

$$u_{m+1} - \sum_{i=1}^{m} k_i u_i \leqq c_{m+1}^{**} - \sum_{i=1}^{m} k_i c_i.$$

That is, all critical functions ϕ satisfy

$$\int \phi \left(f_{m+1} - \sum_{i=1}^{m} k_i f_i \right) d\mu \leqq \int \phi^{**} \left(f_{m+1} - \sum_{i=1}^{m} k_i f_i \right) d\mu,$$

where ϕ^{**} is the test giving rise to the point $(c_1, \cdots, c_m, c^{**})$. Thus ϕ^{**} is the critical function that maximizes the left-hand side of this inequality. Since the integral in question is maximized by putting ϕ equal to 1 when the integrand is positive and equal to 0 when it is negative, ϕ^{**} satisfies (21) a.e. μ.

If $c^* = c^{**}$, let $(c_1', \cdots, c_m')$ be any point of M other than $(c_1, \cdots, c_m)$. We shall show now that there exists exactly one real number c' such that $(c_1', \cdots, c_m', c')$ is in N. Suppose to the contrary that $(c_1' \cdots, c_m', \underline{c}')$ and $(c_1', \cdots, c_m', \bar{c}')$ are both in N, and consider any point $(c_1'', \cdots, c_m'', c'')$ of N such that $(c_1, \cdots, c_m)$ is an interior point of the line segment joining $(c_1', \cdots, c_m')$ and $(c_1'', \cdots, c_m'')$. Such a point exists since $(c_1, \cdots, c_m)$ is an inner point of M. Then the convex set spanned by the three points $(c_1', \cdots, c_m' \underline{c}')$, $(c_1', \cdots, c_m', \bar{c}')$, and $(c_1'', \cdots, c_m'', c'')$ is contained in N and contains points $(c_1, \cdots, c_m, \underline{c})$ and $(c_1, \cdots, c_m, \bar{c})$ with $\underline{c} < \bar{c}$, which is a contradiction. Since N is convex, contains the origin, and has at most one point on any vertical line $u_1 = c_1', \cdots, u_m = c_m'$, it is contained in a hyperplane, which passes through the origin and is not parallel to the u_{m+1}-axis. It follows that

$$\int \phi f_{m+1} \, d\mu = \sum_{i=1}^{m} k_i \int \phi f_i \, d\mu$$

for all ϕ. This arises of course only in the trivial case that

$$f_{m+1} = \sum_{i=1}^{m} k_i f_i, \qquad \text{a.e. } \mu,$$

and (21) is satisfied vacuously.

Corollary 4. *Let $p_1, \cdots, p_m, p_{m+1}$ be probability densities with respect to a measure μ, and let $0 < \alpha < 1$. Then there exists a test ϕ such that $E_i\,\phi(X) = \alpha$ $(i = 1, \cdots, m)$ and $E_{m+1}\,\phi(X) > \alpha$, unless $p_{m+1} = \sum_{i=1}^{m} k_i p_i$, a.e. μ.*

Proof. The proof will be by induction over m. For $m = 1$ the result reduces to Corollary 1. Assume now that it has been proved for any set of m distributions, and consider the case of $m + 1$ densities $p_1, \cdots, p_{m+1}$. If $p_1, \cdots, p_m$ are linearly dependent, the number of p_i can be reduced and the result follows from the induction hypothesis. Assume therefore that $p_1, \cdots, p_m$ are linearly independent. Then for each $j = 1, \cdots, m$ there exist by the induction hypothesis tests ϕ_j and ϕ_j' such that $E_i\,\phi_j(X) = E_i\,\phi_j'(X) = \alpha$ for all $i = 1, \cdots, j - 1, j + 1, \cdots, m$ and $E_j\,\phi_j(X) < \alpha < E_j\,\phi_j'(X)$. It follows that the point of m-space for which all m coordinates are equal to α is an inner point of M, so that Theorem 5(iv) is applicable. The test $\phi(x) \equiv \alpha$ is such that $E_i\,\phi(X) = \alpha$ for $i = 1, \cdots, m$. If among all tests satisfying the side conditions this one is most powerful, it has to satisfy (21). Since $0 < \alpha < 1$, this implies

$$p_{m+1} = \sum_{i=1}^{m} k_i p_i, \qquad \text{a.e. } \mu,$$

as was to be proved.

The most useful parts of Theorems 1 and 5 are the parts (ii), which give sufficient conditions for a critical function to maximize an integral subject to certain side conditions. These results can be derived very easily as follows by the method of undetermined multipliers.

Lemma 3. *Let $F_1, \cdots, F_{m+1}$ be real-valued functions defined over a space U, and consider the problem of maximizing $F_{m+1}(u)$ subject to $F_i(u) = c_i$ $(i = 1, \cdots, m)$. A sufficient condition for a point u^0 satisfying the side conditions to be a solution of the given problem is that among all points of U it maximizes*

$$F_{m+1}(u) - \sum_{i=1}^{m} k_i F_i(u)$$

for some $k_1, \cdots, k_m$.

When applying the lemma one usually carries out the maximization for arbitrary k's, and then determines the constants so as to satisfy the side conditions.

Proof. If u is any point satisfying the side conditions, then

$$F_{m+1}(u) - \sum_{i=1}^{m} k_i F_i(u) \leqq F_{m+1}(u^0) - \sum_{i=1}^{m} k_i F_i(u^0),$$

and hence $F_{m+1}(u) \leqq F_{m+1}(u^0)$.

As an application consider the problem treated in Theorem 5. Let U be the space of critical functions ϕ, and let $F_i(\phi) = \int \phi f_i \, d\mu$. Then a sufficient condition for ϕ to maximize $F_{m+1}(\phi)$, subject to $F_i(\phi) = c_i$, is that it maximizes $F_{m+1}(\phi) - \Sigma k_i F_i(\phi) = \int (f_{m+1} - \Sigma k_i f_i)\phi \, d\mu$. This is achieved by setting $\phi(x) = 1$ or 0 as $f_{m+1}(x) >$ or $< \Sigma k_i f_i(x)$.

7. TWO-SIDED HYPOTHESES

UMP tests exist not only for one-sided but also for certain two-sided hypotheses of the form

$$H: \theta \leqq \theta_1 \quad \text{or} \quad \theta \geqq \theta_2 \qquad (\theta_1 < \theta_2). \tag{23}$$

Such testing problems occur when one wishes to determine whether given specifications have been met concerning the proportion of an ingredient in a drug or some other compound, or whether a measuring instrument, for example a scale, is properly balanced. One then sets up the hypothesis that θ does not lie within the required limits so that an error of the first kind consists in declaring θ to be satisfactory when in fact it is not. In practice, the decision to accept H will typically be accompanied by a statement of whether θ is believed to be $\leqq \theta_1$ or $\geqq \theta_2$. The implications of H are, however, frequently sufficiently important so that acceptance will in any case be followed by a more detailed investigation. If a manufacturer tests each precision instrument before releasing it and the test indicates an instrument to be out of balance, further work will be done to get it properly adjusted. If in a scientific investigation the inequalities $\theta \leqq \theta_1$ and $\theta \geqq \theta_2$ contradict some assumptions that have been formulated, a more complex theory may be needed and further experimentation will be required. In such situations there may be only two basic choices, to act as if $\theta_1 < \theta < \theta_2$ or to carry out some further investigation, and the formulation of the problem as that of testing the hypothesis H may be appropriate. In the present section the existence of a UMP test of H will be proved for exponential families.

Theorem 6. (i) *For testing the hypothesis* $H: \theta \leqq \theta_1$ or $\theta \geqq \theta_2$

$(\theta_1 < \theta_2)$ against the alternatives K: $\theta_1 < \theta < \theta_2$ in a one-parameter exponential family there exists a UMP test given by

$$(24) \qquad \phi(x) = \begin{cases} 1 & \text{when } C_1 < T(x) < C_2 \quad (C_1 < C_2) \\ \gamma_i & \text{when } T(x) = C_i, \quad i = 1, 2 \\ 0 & \text{when } T(x) < C_1 \quad \text{or} \quad > C_2, \end{cases}$$

where the C's and γ's are determined by

$$(25) \qquad E_{\theta_1}\phi(X) = E_{\theta_2}\phi(X) = \alpha.$$

(ii) *This test minimizes $E_\theta\phi(X)$ subject to* (25) for all $\theta < \theta_1$ *and* $> \theta_2$.

(iii) *For* $0 < \alpha < 1$ *the power function of this test has a maximum at a point θ_0 between θ_1 and θ_2 and decreases strictly as θ tends away from θ_0 in either direction, unless there exist two values t_1, t_2 such that $P_\theta\{T(X) = t_1\} + P_\theta\{T(X) = t_2\} = 1$ for all θ.*

Proof. (i) One can restrict attention to the sufficient statistic $T = T(X)$, the distribution of which by Lemma 8 of Chapter 2 is

$$dP_\theta(t) = C(\theta)\, e^{Q(\theta)t}\, d\nu(t),$$

where $Q(\theta)$ is assumed to be strictly increasing. Let $\theta_1 < \theta' < \theta_2$, and consider first the problem of maximizing $E_{\theta'}\psi(T)$ subject to (25) with $\phi(x) = \psi[T(x)]$. If M denotes the set of all points $(E_{\theta_1}\psi(T), E_{\theta_2}\psi(T))$ as ψ ranges over the totality of critical functions, then the point (α, α) is an inner point of M. This follows from the fact that by Corollary 1 the set M contains points (α, u_1) and (α, u_2) with $u_1 < \alpha < u_2$ and that it contains all points (u, u) with $0 < u < 1$. Hence by part (iv) of Theorem 5 there exist constants k_1, k_2 and a test $\psi_0(t)$ such that $\phi_0(x) = \psi_0[T(x)]$ satisfies (25) and that $\psi_0(t) = 1$ when

$$k_1 C(\theta_1)e^{Q(\theta_1)t} + k_2 C(\theta_2)e^{Q(\theta_2)t} < C(\theta')e^{Q(\theta')t}$$

and therefore when

$$a_1 e^{b_1 t} + a_2 e^{b_2 t} < 1 \qquad (b_1 < 0 < b_2),$$

and $\psi_0(t) = 0$ when the left-hand side is > 1. Here not both a's can be $\leqq 0$ since then the test would always reject. If one of the a's is $\leqq 0$ and the other one is > 0, then the left-hand side is strictly monotone, and the test is of the one-sided type considered in Corollary 2, which has a strictly monotone power function and hence cannot satisfy (25). Since therefore both a's are positive, the test satisfies (24); by Theorem 5(iii) it also maximizes $E_{\theta'}\psi(T)$ subject to the weaker restriction $E_{\theta_i}\psi(T) \leqq \alpha$ $(i = 1, 2)$. To complete the proof that this test is UMP for testing H, it is necessary to show that it satisfies $E_\theta\psi(T) \leqq \alpha$ for $\theta \leqq \theta_1$ and $\theta \geqq \theta_2$. This follows from (ii) by comparison with the test $\psi(t) \equiv \alpha$.

(ii) Let $\theta' < \theta_1$ and apply Theorem 5(iv) to minimize $E_{\theta'}\,\phi(X)$ subject to (25). Dividing through by $e^{Q(\theta_1)t}$, the desired test is seen to have a rejection region of the form

$$a_1 e^{b_1 t} + a_2 e^{b_2 t} < 1 \qquad (b_1 < 0 < b_2).$$

Thus it coincides with the test $\psi_0(t)$ obtained in (i). By Theorem 5(iv), the first and third conditions of (24) are also necessary, and the optimum test is therefore unique provided $P\{T = C_i\} = 0$.

(iii) Without loss of generality let $Q(\theta) = \theta$. It follows from (i) and the continuity of $\beta(\theta) = E_\theta\,\phi(X)$ that either $\beta(\theta)$ satisfies (iii) or there exist three points $\theta' < \theta'' < \theta'''$ such that $\beta(\theta') = \beta(\theta'') = \beta(\theta''')$. If this common value is c, then $0 < c < 1$ since $\beta(\theta') = 0$ (or 1) implies $\phi(t) = 0$ (or 1) a.e. ν and this is excluded by (25). As is seen by the proof of (i), the test maximizes $E_\theta\,\phi(X)$ subject to $E_{\theta'}\,\phi(X) = E_{\theta'''}\,\phi(X) = c$ for all $\theta' < \theta < \theta'''$, and the possibility $E_{\theta''}\,\phi(X) = c$ is therefore excluded by Corollary 4 unless $p_{\theta''} = k_1 p_{\theta'} + k_2 p_{\theta'''}$ a.e. ν. By the assumptions made in (iii) this would imply the existence of three points t_1, t_2, t_3 such that

$$1 = k_1 \frac{p_{\theta'}(t_i)}{p_{\theta''}(t_i)} + k_2 \frac{p_{\theta'''}(t_i)}{p_{\theta''}(t_i)}, \qquad i = 1, 2, 3,$$

which is impossible since $[k_1 p_{\theta'}(t) + k_2 p_{\theta'''}(t)]/p_{\theta''}(t)$ is convex.

In order to determine the C's and γ's, one will in practice start with some trial values C_1^*, γ_1^*, find C_2^*, γ_2^* such that $\beta^*(\theta_1) = \alpha$, and compute $\beta^*(\theta_2)$, which will usually be either too large or too small. For the selection of the next trial values it is then helpful to note that if $\beta^*(\theta_2) < \alpha$, the correct acceptance region is to the right of the one chosen, that is, it satisfies either $C_1 > C_1^*$ or $C_1 = C_1^*$ and $\gamma_1 < \gamma_1^*$, and that the converse holds if $\beta^*(\theta_2) > \alpha$. This is a consequence of Lemma 2 applied to $T(x)$. Any test ϕ^* satisfying (24) and $\beta^*(\theta_1) = \alpha$ must be either to the right or the left of the test ϕ satisfying (24) and (25). As ϕ is to the left or right of ϕ^*, the function $\psi(t) = \phi^*(t) - \phi(t)$ is monotone increasing or decreasing and from the lemma $\beta^*(\theta_2) \geqq \alpha$ or $\leqq \alpha$.

Although a UMP test exists for testing that $\theta \leqq \theta_1$ or $\geqq \theta_2$ in an exponential family, the same is not true for the dual hypothesis H: $\theta_1 \leqq \theta \leqq \theta_2$ or for testing $\theta = \theta_0$ (Problem 26). There do, however, exist UMP unbiased tests of these hypotheses, as will be shown in Chapter 4.

8. LEAST FAVORABLE DISTRIBUTIONS

It is a consequence of Theorem 1 that there always exists a most powerful test for testing a simple hypothesis against a simple alternative.

More generally, consider the case of a Euclidean sample space, probability densities f_θ, $\theta \in \omega$, and g with respect to a measure μ, and the problem of testing $H: f_\theta$, $\theta \in \omega$, against the simple alternative K: g. The existence of a most powerful level α test then follows from the weak compactness theorem for critical functions (Theorem 3 of the Appendix) as in Theorem 5(i).

Theorem 1 also provides an explicit construction for the most powerful test in the case of a simple hypothesis. We shall now extend this theorem to composite hypotheses in the direction of Theorem 5 by the method of undetermined multipliers. However, in the process of extension the result becomes much less explicit. Essentially it leaves open the determination of the multipliers, which now take the form of an arbitrary distribution. In specific problems this usually still involves considerable difficulty.

From another point of view the method of attack, as throughout the theory of hypothesis testing, is to reduce the composite hypothesis to a simple one. This is achieved by considering weighted averages of the distributions of H. The composite hypothesis H is replaced by the simple hypothesis H_λ that the probability density of X is given by

$$h_\lambda(x) = \int_\omega f_\theta(x)\, d\lambda(\theta),$$

where λ is a probability distribution over ω. The problem of finding a suitable λ is frequently made easier by the following consideration. Since H provides no information concerning θ and since H_λ is to be equivalent to H for the purpose of testing against g, knowledge of the distribution λ should provide as little help for this task as possible. To make this precise suppose that θ is known to have a distribution λ. Then the maximum power β_λ that can be attained against g is that of the most powerful test ϕ_λ for testing H_λ against g. The distribution λ is said to be *least favorable* (at level α) if for all λ' the inequality $\beta_\lambda \leqq \beta_{\lambda'}$ holds.

Theorem 7. *Let a σ-field be defined over ω such that the densities $f_\theta(x)$ are jointly measurable in θ and x. Suppose that over this σ-field there exists a probability distribution λ such that the most powerful level α test ϕ_λ for testing H_λ against g is of size $\leqq \alpha$ also with respect to the original hypothesis H.*

(i) *The test ϕ_λ is most powerful for testing H against g.*

(ii) *If ϕ_λ is the unique most powerful level α test for testing H_λ against g, it is also the unique most powerful test of H against g.*

(iii) *The distribution λ is least favorable.*

Proof. We note first that h_λ is again a density with respect to μ since by Fubini's theorem (Theorem 3 of Chapter 2)

$$\int h_\lambda(x)\,d\mu(x) = \int_\omega d\lambda(\theta) \int f_\theta(x)\,d\mu(x) = \int_\omega d\lambda(\theta) = 1.$$

Suppose that ϕ_λ is a level α test for testing H and let ϕ^* be any other level α test. Then since $E_\theta\,\phi^*(X) \leqq \alpha$ for all $\theta \in \omega$, we have

$$\int \phi^*(x) h_\lambda(x)\,d\mu(x) = \int_\omega E_\theta\,\phi^*(X)\,d\lambda(\theta) \leqq \alpha.$$

Therefore ϕ^* is a level α test also for testing H_λ and its power cannot exceed that of ϕ_λ. This proves (i) and (ii). If λ' is any distribution, it follows further that ϕ_λ is a level α test also for testing $H_{\lambda'}$, and hence that its power against g cannot exceed that of the most powerful test which by definition is $\beta_{\lambda'}$.

The conditions of this theorem can be given a somewhat different form by noting that ϕ_λ can satisfy $\int_\omega E_\theta\,\phi_\lambda(X)\,d\lambda(\theta) = \alpha$ and $E_\theta\,\phi_\lambda(X) \leqq \alpha$ for all θ only if the set of θ's with $E_\theta\,\phi_\lambda(X) = \alpha$ has λ-measure one.

Corollary 5. *Suppose that λ is a probability distribution over ω and that ω' is a subset of ω with $\lambda(\omega') = 1$. Let ϕ_λ be a test such that*

$$\phi_\lambda(x) = \begin{cases} 1 & \text{if} \quad g(x) > k \int f_\theta(x)\,d\lambda(\theta) \\ 0 & \text{if} \quad g(x) < k \int f_\theta(x)\,d\lambda(\theta). \end{cases} \tag{26}$$

Then ϕ_λ is a most powerful level α test for testing H against g provided

$$E_{\theta'}\,\phi_\lambda(X) = \sup_{\theta\in\omega} E_\theta\,\phi_\lambda(X) = \alpha \quad \text{for all} \quad \theta' \in \omega'. \tag{27}$$

Theorems 2 and 6 constitute two simple applications of Theorem 7. The set ω' over which the least favorable distribution λ is concentrated consists of the single point θ_0 in the first of these examples and of the two points θ_1 and θ_2 in the second. This is what one might expect since in both cases these are the distributions of H that appear to be "closest" to K. Another example in which the least favorable distribution is concentrated at a single point is the following.

Example 8. The quality of items produced by a manufacturing process is measured by a characteristic X such as the tensile strength of a piece of material, or the length of life or brightness of a light bulb. For an item to be satisfactory

X must exceed a given constant u, and one wishes to test the hypothesis H: $p \geqq p_0$ where

$$p = P\{X \leqq u\}$$

is the probability of an item being defective. Let $X_1, \cdots, X_n$ be the measurements of n sample items, so that the X's are independently distributed with common distribution about which no knowledge is assumed. Any distribution on the real line can be characterized by the probability p together with the conditional probability distributions P_- and P_+ of X given $X \leqq u$ and $X > u$ respectively. If the distributions P_- and P_+ have probability densities p_- and p_+, for example with respect to $\mu = P_- + P_+$, then the joint density of $X_1, \cdots, X_n$ at a sample point $x_1, \cdots, x_n$ satisfying

$$x_{i_1}, \cdots, x_{i_m} \leqq u < x_{j_1}, \cdots, x_{j_{n-m}}$$

is

$$p^m(1-p)^{n-m}p_-(x_{i_1}) \cdots p_-(x_{i_m})p_+(x_{j_1}) \cdots p_+(x_{j_{n-m}}).$$

Consider now a fixed alternative to H, say (p_1, P_-, P_+), with $p_1 < p_0$. One would then expect the least favorable distribution λ over H to assign probability 1 to the distribution (p_0, P_-, P_+) since this appears to be closest to the selected alternative. With this choice of λ, the test (26) becomes

$$\phi_\lambda(x) = 1 \text{ or } 0 \quad \text{as} \quad \left(\frac{p_1}{p_0}\right)^m \left(\frac{q_1}{q_0}\right)^{n-m} > \text{ or } < C,$$

and hence as $m <$ or $> C$. The test therefore rejects when the number M of defectives is sufficiently small or more precisely when $M < C$ and with probability γ when $M = C$ where

$$P\{M < C\} + \gamma P\{M = C\} = \alpha \quad \text{for} \quad p = p_0. \tag{28}$$

The distribution of M is the binomial distribution $b(p, n)$, and does not depend on P_+ and P_-. As a consequence, the power function of the test depends only on p and is a decreasing function of p, so that under H it takes on its maximum for $p = p_0$. This proves λ to be least favorable and ϕ_λ to be most powerful. Since the test is independent of the particular alternative chosen, it is UMP.

Expressed in terms of the variables $Z_i = X_i - u$, the test statistic M is the number of variables $\leqq 0$, and the test is the so-called *sign test* (cf. Chapter 4, Section 7). It is an example of a *nonparametric* test since it is derived without assuming a given functional form for the distribution of the X's such as the normal, rectangular, or Poisson, in which only certain parameters are unknown.

The above argument applies, with only the obvious modifications, to the case that an item is satisfactory if X lies within certain limits: $u < X < v$. This occurs, for example, if X is the length of a metal part or the proportion of an ingredient in a chemical compound, for which certain tolerances have been specified. More generally the argument applies also to the situation in which X is vector-valued. Suppose that an item is satisfactory only when X lies in a certain set S, for example if all the dimensions of a metal part or the proportions of several ingredients lie within specified limits. The probability of a defective is then

$$p = P\{X \in \bar{S}\},$$

and P_- and P_+ denote the conditional distributions of X given $X \in S$ and

$X \in \tilde{S}$ respectively. As before there exists a UMP test of $H: p \geqq p_0$, and it rejects H when the number M of defectives is sufficiently small, with the boundary of the test being determined by (28).

A distribution λ satisfying the conditions of Theorem 7 exists in most of the usual statistical problems, and in particular under the following assumptions.* Let the sample space be Euclidean, let ω be a Borel set in s-dimensional Euclidean space, and suppose that $f_\theta(x)$ is a continuous function of θ for almost all x. Then given any g there exists a distribution λ satisfying the conditions of Theorem 7 provided

$$\lim_{n \to \infty} \int_S f_{\theta_n}(x)\, d\mu(x) = 0$$

for every bounded set S in the sample space and for every sequence of vectors θ_n whose distance from the origin tends to infinity.

From this it follows, as did Corollaries 1 and 4 from Theorems 1 and 5, that if the above conditions hold and if $0 < \alpha < 1$, there exists a test of power $\beta > \alpha$ for testing $H: f_\theta,\ \theta \in \omega$, against g unless $g = \int f_\theta\, d\lambda(\theta)$ for some λ. An example of the latter possibility is obtained by letting f_θ and g be the normal densities $N(\theta, \sigma_0^2)$ and $N(0, \sigma_1^2)$ respectively with $\sigma_0^2 < \sigma_1^2$. (See p. 97.)

9. TESTING THE MEAN AND VARIANCE OF A NORMAL DISTRIBUTION

Because of their wide applicability, the problems of testing the mean ξ and variance σ^2 of a normal distribution are of particular importance. Here and in similar problems later the parameter not being tested is assumed to be unknown but will not be shown explicitly in a statement of the hypothesis. We will write, for example, $\sigma \leqq \sigma_0$ instead of the more complete statement $\sigma \leqq \sigma_0, -\infty < \xi < \infty$. The standard (likelihood ratio) tests of the two hypotheses $\sigma \leqq \sigma_0$ and $\xi \leqq \xi_0$ are given by the rejection regions

$$\Sigma(x_i - \bar{x})^2 \geqq C \tag{29}$$

and

$$\frac{\sqrt{n}\,(\bar{x} - \xi_0)}{\sqrt{\dfrac{1}{n-1}\Sigma(x_i - \bar{x})^2}} \geqq C. \tag{30}$$

* See Lehmann, "On the existence of least favorable distributions," *Ann. Math. Stat.*, Vol. 23 (1952), pp. 408–416.

The corresponding tests for the hypotheses $\sigma \geqq \sigma_0$ and $\xi \geqq \xi_0$ are obtained from the rejection regions (29) and (30) by reversing the inequalities. As will be shown in later chapters, these four tests are UMP both within the class of unbiased and the class of invariant tests. However, at the usual significance levels only the first of them is actually UMP.

Let $X_1, \cdots, X_n$ be a sample from $N(\xi, \sigma^2)$ and consider first the hypotheses H_1: $\sigma \geqq \sigma_0$ and H_2: $\sigma \leqq \sigma_0$, and a simple alternative K: $\xi = \xi_1$, $\sigma = \sigma_1$. It seems reasonable to suppose that the least favorable distribution λ in the (ξ, σ)-plane is concentrated on the line $\sigma = \sigma_0$. Since $Y = \Sigma X_i/n = \bar{X}$ and $U = \Sigma(X_i - \bar{X})^2$ are sufficient statistics for the parameters (ξ, σ), attention can be restricted to these variables. Their joint density under H_λ is

$$C_0 u^{\frac{1}{2}(n-3)} \exp\left(-\frac{u}{2\sigma_0^2}\right) \int \exp\left[-\frac{n}{2\sigma_0^2}(y - \xi)^2\right] d\lambda(\xi)$$

while under K it is

$$C_1 u^{\frac{1}{2}(n-3)} \exp\left(-\frac{u}{2\sigma_1^2}\right) \exp\left[-\frac{n}{2\sigma_1^2}(y - \xi_1)^2\right].$$

The choice of λ is seen to affect only the distribution of Y. A least favorable λ should therefore have the property that the density of Y under H_λ,

$$\int \frac{\sqrt{n}}{\sqrt{2\pi\sigma_0^2}} \exp\left[-\frac{n}{2\sigma_0^2}(y - \xi)^2\right] d\lambda(\xi),$$

comes as close as possible to the alternative density,

$$\frac{\sqrt{n}}{\sqrt{2\pi\sigma_1^2}} \exp\left[-\frac{n}{2\sigma_1^2}(y - \xi_1)^2\right].$$

At this point one must distinguish between H_1 and H_2. In the first case $\sigma_1 < \sigma_0$. By suitable choice of λ the mean of Y can be made equal to ξ_1, but the variance will if anything be increased over its initial value σ_0^2. This suggests that the least favorable distribution assigns probability 1 to the point $\xi = \xi_1$ since in this way the distribution of Y is normal both under H and K with the same mean in both cases and the smallest possible difference between the variances. The situation is somewhat different for H_2 for which $\sigma_0 < \sigma_1$. If the least favorable distribution λ has a density, say λ', the density of Y under H_λ becomes

$$\int_{-\infty}^{\infty} \frac{\sqrt{n}}{\sqrt{2\pi}\sigma_0} \exp\left[-\frac{n}{2\sigma_0^2}(y - \xi)^2\right] \lambda'(\xi)\, d\xi.$$

This is the probability density of the sum of two independent random variables, one distributed as $N(0, \sigma_0^2/n)$ and the other with density $\lambda'(\xi)$. If λ is taken to be $N(\xi_1, (\sigma_1^2 - \sigma_0^2)/n)$, the distribution of Y under H_λ becomes $N(\xi_1, \sigma_1^2/n)$, the same as under K.

We now apply Corollary 5 with the distributions λ suggested above. For H_1 it is more convenient to work with the original variables than with Y and U. Substitution in (26) gives $\phi(x) = 1$ when

$$\frac{(2\pi\sigma_1^2)^{-n/2} \exp\left[-\frac{1}{2\sigma_1^2}\Sigma(x_i - \xi_1)^2\right]}{(2\pi\sigma_0^2)^{-n/2} \exp\left[-\frac{1}{2\sigma_0^2}\Sigma(x_i - \xi_1)^2\right]} > C,$$

that is, when

$$\Sigma(x_i - \xi_1)^2 \leqq C. \tag{31}$$

To justify the choice of λ, one must show that

$$P\{\Sigma(X_i - \xi_1)^2 \leqq C | \xi, \sigma\}$$

takes on its maximum over the half plane $\sigma \geqq \sigma_0$ at the point $\xi = \xi_1$, $\sigma = \sigma_0$. For any fixed σ, the above is the probability of the sample point falling in a sphere of fixed radius, computed under the assumption that the X's are independently distributed as $N(\xi, \sigma^2)$. This probability is maximized when the center of the sphere coincides with that of the distribution, that is, when $\xi = \xi_1$. The probability then becomes

$$P\left\{\Sigma\left(\frac{X_i - \xi_1}{\sigma}\right)^2 \leqq \frac{C}{\sigma^2} \middle| \xi_1, \sigma\right\} = P\left\{\Sigma V_i^2 \leqq \frac{C}{\sigma^2}\right\}$$

where $V_1, \cdots, V_n$ are independently distributed as $N(0, 1)$. This is a decreasing function of σ and therefore takes on its maximum when $\sigma = \sigma_0$.

In the case of H_2 application of Corollary 5 to the sufficient statistics (Y, U) gives $\phi(y, u) = 1$ when

$$\frac{C_1 u^{\frac{1}{2}(n-3)} \exp\left(-\frac{u}{2\sigma_1^2}\right) \exp\left[-\frac{n}{2\sigma_1^2}(y - \xi_1)^2\right]}{C_0 u^{\frac{1}{2}(n-3)} \exp\left(-\frac{u}{2\sigma_0^2}\right) \int \exp\left[-\frac{n}{2\sigma_0^2}(y - \xi)^2\right] \lambda'(\xi)\, d\xi}$$
$$= K \exp\left[-\frac{u}{2}\left(\frac{1}{\sigma_1^2} - \frac{1}{\sigma_0^2}\right)\right] \geqq C,$$

that is, when

$$u = \Sigma(x_i - \bar{x})^2 \geqq C. \tag{32}$$

Since the distribution of $\Sigma(X_i - \bar{X})^2/\sigma^2$ does not depend on ξ or σ, the probability $P\{\Sigma(X_i - \bar{X})^2 \geqq C|\xi, \sigma\}$ is independent of ξ and increases with σ, so that the conditions of Corollary 5 are satisfied. The test (32), being independent of ξ_1 and σ_1, is UMP for testing $\sigma \leqq \sigma_0$ against $\sigma > \sigma_0$. It is also seen to coincide with the likelihood ratio test (29). On the other hand, the most powerful test (31) for testing $\sigma \geqq \sigma_0$ against $\sigma < \sigma_0$ does depend on the value ξ_1 of ξ under the alternative.

It was tacitly assumed so far that $n > 1$. If $n = 1$, the argument applies without change with respect to H_1, leading to (31) with $n = 1$. However, in the discussion of H_2 the statistic U now drops out, and Y coincides with the single observation X. Using the same λ as before one sees that X has the same distribution under H_λ as under K, and the test ϕ_λ therefore becomes $\phi_\lambda(x) \equiv \alpha$. This satisfies the conditions of Corollary 5 and is therefore the most powerful test for the given problem. It follows that a single observation is of no value for testing the hypothesis H_2 as seems intuitively obvious, but that it could be used to test H_1 if the class of alternatives were sufficiently restricted.

The corresponding derivation for the hypothesis $\xi \leqq \xi_0$ is less straightforward. It turns out* that Student's test given by (30) is most powerful if the level of significance α is $\geqq 1/2$, regardless of the alternative $\xi_1 > \xi_0, \sigma_1$. This test is therefore UMP for $\alpha \geqq 1/2$. On the other hand, when $\alpha < 1/2$ the most powerful test of H rejects when $\Sigma(x_i - a)^2 \leqq b$, where the constants a and b depend on the alternative (ξ_1, σ_1) and on α. Thus for the significance levels that are of interest, a UMP test of H does not exist. No new problem arises for the hypothesis $\xi \geqq \xi_0$ since this reduces to the case just considered through the transformation $Y_i = \xi_0 - (X_i - \xi_0)$.

10. SEQUENTIAL PROBABILITY RATIO TESTS

According to the Neyman-Pearson fundamental lemma, the best procedure for testing the simple hypothesis H that the probability density of X is p_0 against the simple alternative that it is p_1 accepts or rejects H as

$$\frac{p_{1n}}{p_{0n}} = \frac{p_1(x_1) \cdots p_1(x_n)}{p_0(x_1) \cdots p_0(x_n)}$$

is less or greater than a suitable constant C. However, further improvement is possible if the sample size is not fixed in advance but is permitted to depend on the observations. The best procedure, in a certain sense, is then the following *sequential probability ratio test*. Let $A_0 < A_1$ be

* See Lehmann and Stein, "Most powerful tests of composite hypotheses. I. Normal distributions," *Ann. Math. Stat.*, Vol. 19 (1948), pp. 495–516.

two given constants and suppose that observation is continued as long as the probability ratio p_{1n}/p_{0n} satisfies the inequality

$$A_0 < \frac{p_{1n}}{p_{0n}} < A_1. \tag{33}$$

The hypothesis H is accepted or rejected at the first violation of (33) as $p_{1n}/p_{0n} \leqq A_0$ or $\geqq A_1$.

The usual measures of the performance of such a procedure are the probabilities, say α_0 and α_1, of rejecting H when $p = p_0$ and of accepting it when $p = p_1$ and the expected number of observations $E_i(N)$ when $p = p_i$ $(i = 0, 1)$.

Theorem 8. *Among all tests (sequential or not) for which*

$$P_0\,(\text{rejecting } H) \leqq \alpha_0, \qquad P_1\,(\text{accepting } H) \leqq \alpha_1$$

and for which $E_0(N)$ and $E_1(N)$ are finite, the sequential probability ratio test with error probabilities α_0 and α_1 minimizes both $E_0(N)$ and $E_1(N)$.

In particular, the sequential probability ratio test therefore requires on the average fewer observations than the fixed sample size test which controls the errors at the same levels. The proof of this result will be deferred to Section 12. In this and the following sections some of the basic properties of sequential probability ratio tests will be sketched.

Because of the difficulty of determining exactly the boundaries A_0 and A_1 for which α_0 and α_1 take on preassigned values, the following inequalities are useful. Let R_n be the part of n-space defined by the inequalities

$$A_0 < \frac{p_{1k}}{p_{0k}} < A_1 \quad \text{for} \quad k = 1, \cdots, n-1 \qquad \text{and} \qquad A_1 \leqq \frac{p_{1n}}{p_{0n}}.$$

This is the set of points $(x_1, \cdots, x_n)$ for which the procedure stops with $N = n$ observations and rejects H. Then

$$\alpha_0 = \sum_{n=1}^{\infty} \int_{R_n} p_{0n} \leqq \frac{1}{A_1} \sum_{n=1}^{\infty} \int_{R_n} p_{1n} = \frac{1 - \alpha_1}{A_1}.$$

Similarly, if S_n denotes the part of n-space in which $N = n$ and H is accepted, one has

$$1 - \alpha_0 = \sum_{n=1}^{\infty} \int_{S_n} p_{0n} \geqq \frac{\alpha_1}{A_0}.$$

Here it has been tacitly assumed that

$$\sum_{n=1}^{\infty} P_i\{N = n\} = \sum_{n=1}^{\infty} \int_{R_n \cup S_n} p_{in} = 1 \quad \text{for} \quad i = 0, 1,$$

that is, that the probability is 0 of the procedure continuing indefinitely. For a proof of this fact see Problems 34 and 35. The inequalities

$$A_0 \geqq \frac{\alpha_1}{1-\alpha_0}, \qquad A_1 \leqq \frac{1-\alpha_1}{\alpha_0} \tag{34}$$

suggest the possibility of approximating the boundaries A_0 and A_1 that would yield the desired α_0 and α_1 by

$$A_0' = \frac{\alpha_1}{1-\alpha_0}, \qquad A_1' = \frac{1-\alpha_1}{\alpha_0}.$$

By (34) the error probabilities of the approximate procedure then satisfy

$$\frac{\alpha_1'}{1-\alpha_0'} \leqq A_0' = \frac{\alpha_1}{1-\alpha_0} \quad \text{and} \quad \frac{1-\alpha_1'}{\alpha_0'} \geqq A_1' = \frac{1-\alpha_1}{\alpha_0}$$

and hence

$$\alpha_0' \leqq \frac{\alpha_0}{1-\alpha_1} \quad \text{and} \quad \alpha_1' \leqq \frac{\alpha_1}{1-\alpha_0}.$$

If typically α_0 and α_1 are of the order .01 to .1, the amount by which α_i' can exceed α_i $(i = 1, 0)$ is negligible so that the probabilities of the two kinds of error are very nearly bounded above by the specified α_0 and α_1. This conclusion is strengthened by the fact that $\alpha_0' + \alpha_1' \leqq \alpha_0 + \alpha_1$, as is seen by adding the inequalities $\alpha_1'(1-\alpha_0) \leqq \alpha_1(1-\alpha_0')$ and $\alpha_0'(1-\alpha_1') \leqq \alpha_0(1-\alpha_1')$.

The only serious risk in using the approximate boundaries A_0', A_1' is therefore that α_0' and α_1' are much smaller than required, which would lead to an excessive number of observations. There is some reason to hope that this effect is also moderate. For let

$$z_i = \log\,[p_1(x_i)/p_0(x_i)]. \tag{35}$$

Then (33) becomes

$$\log A_0 < \sum_{i=1}^{n} z_i < \log A_1,$$

and when H is rejected the z's satisfy

$$z_1 + \cdots + z_{n-1} < \log A_1 \leqq z_1 + \cdots + z_n.$$

The approximation consists in replacing $z_1 + \cdots + z_n$ by $\log A_1$. The error will usually be moderate since after $n-1$ observations Σz_i is still $< A_1$ and the excess has therefore had no possibility to accumulate, but is due to a single observation. An analogous argument applies to the other boundary.

Example 9. Consider a sequence of binomial trials with constant probability p of success, and the problem of testing $p = p_0$ against $p = p_1 (p_0 < p_1)$. Then

$$\frac{p_{1n}}{p_{0n}} = \frac{p_1^{\Sigma x_i}(1 - p_1)^{n - \Sigma x_i}}{p_0^{\Sigma x_i}(1 - p_0)^{n - \Sigma x_i}} = \left(\frac{p_1 q_0}{p_0 q_1}\right)^{\Sigma x_i}\left(\frac{q_1}{q_0}\right)^n.$$

In the case that $\log (p_1 p_0^{-1})/\log (q_0 q_1^{-1})$ is rational, exact formulas have been obtained† for the error probabilities and expected sample size which make it possible to compute the effects involved in the approximation of A_0, A_1 by A_0', A_1'. As an illustration,‡ suppose that $p_0 = .05$, $p_1 = .17$, $\alpha_0 = .05$, $\alpha_1 = .10$. It then turns out that $\alpha_0' = .031$, $\alpha_1' = .099$, and that the expectations of the sample size for the approximate procedure are $E_0'(N) = 31.4$, $E_1'(N) = 30.0$. There is an alternate plan, determined by trial and error, with $\alpha_0^* = .046$, $\alpha_1^* = .097$, $E_0^*(N) = 30.5$, $E_1^*(N) = 26.1$. On the other hand, the fixed sample size procedure with error probabilities .05 and .10 requires 57 observations.

In order to be specific, we assumed in the definition of a sequential probability ratio test that observation continues only as long as the probability ratio is strictly between A_0 and A_1. The discussion applies equally well to the rule of continuing as long as $A_0 < p_{1n}/p_{0n} < A_1$, coming to the indicated conclusion the first time that $p_{1n}/p_{0n} < A_0$ or $> A_1$, and deciding on the boundaries according to any fixed probabilities. The term *sequential probability ratio test* is applied also to this more general procedure. If the probability ratio $p_1(X)/p_0(X)$ has a continuous distribution, all these procedures are equivalent. However, in case of discrete probability ratios the possibility of randomization on the boundary is necessary to achieve preassigned error probabilities. If randomization is permitted also between taking at least one observation or reaching a decision without taking any observations, it can be shown that actually any preassigned error probabilities can be achieved.§

11. POWER AND EXPECTED SAMPLE SIZE OF SEQUENTIAL PROBABILITY RATIO TESTS

The preceding section is somewhat misleading in that it discusses the problem in a setting, that of testing a simple hypothesis against a simple alternative, which is interesting mainly because of its implications for the more realistic situation of a continuous parameter family of distributions.

† Girshick, "Contributions to the theory of sequential analysis, II, III," *Ann. Math. Stat.*, Vol. 17 (1946), pp. 282–298, and Polya, "Exact formulas in the sequential analysis of attributes," *Univ. Calif. Publs. Mathematics*, New Series, Vol. 1 (1948), pp. 229–240.

‡ Taken from Robinson, "A note on exact sequential analysis," *Univ. Calif. Publs. Mathematics*, New Series, Vol. 1 (1948), pp. 241–246.

§ This result is contained in an as yet unpublished paper by Stein, "Existence of sequential probability ratio tests." See also the abstract by Wijsman, "On the existence of Wald's sequential test," *Ann. Math. Stat.*, Vol. 29 (1958), pp. 938–939.

Unfortunately, the property of being uniformly most powerful, which the fixed sample size probability ratio test possesses for families with monotone likelihood ratio (Theorem 2), does not extend to the sequential case. More specifically, consider the sequential probability ratio test for testing $H: \theta_0$ against $K: \theta_1$, and let its power function be $\beta(\theta) = P_\theta$ (rejecting H). Then if θ_2 is some other alternative, the sequential probability ratio test for testing θ_0 against θ_2 with error probabilities α_0 and α_1 does not in general coincide with the original test, which therefore does not minimize $E_{\theta_2}(N)$. It seems in fact likely that from an over-all point of view the sequential probability ratio test is not the best sequential procedure in the continuous parameter case, although it is usually better than the best competitive test with fixed sample size.

When the probability density depends on a real parameter θ and one is testing the hypothesis $\theta \leq \theta_0$, one is usually not concerned with the power of the test against alternatives θ close to θ_0, but would like to be able to control the probability of detecting alternatives sufficiently far away. The test should therefore satisfy

$$\begin{aligned} \beta(\theta) &\leq \alpha \quad \text{for} \quad \theta \leq \theta_0 \\ \beta(\theta) &\geq \beta \quad \text{for} \quad \theta \geq \theta_1 \end{aligned} \qquad (\theta_0 < \theta_1), \tag{36}$$

which it will do in particular if

$$\beta(\theta_0) = \alpha, \qquad \beta(\theta_1) = \beta,$$

and if $\beta(\theta)$ is a nondccrcasing function of θ. The sequential probability ratio test for testing θ_0 against θ_1 with error probabilities $\alpha_0 = \alpha$, $\alpha_1 = 1 - \beta$ thus is a solution of the stated problem provided its power function is nondecreasing.

Lemma 4. *Let $X_1, X_2, \cdots$ be independently distributed with probability density $p_\theta(x)$, and suppose that the densities $p_\theta(x)$ have monotone likelihood ratio in $T(x)$. Then any sequential probability ratio test for testing θ_0 against θ_1 $(\theta_0 < \theta_1)$ has a nondecreasing power function.*

Proof. Let $Z_i = \log [p_{\theta_1}(X_i)/p_{\theta_0}(X_i)] = h(T_i)$, where h is nondecreasing, and let $\theta < \theta'$. By Lemma 2, the cumulative distribution function $F_\theta(t)$ of T_i satisfies $F_{\theta'}(t) \leq F_\theta(t)$ for all t, and by Lemma 1 there exists therefore a random variable V_i and functions f and f' such that $f(v) \leq f'(v)$ for all v and that the distributions of $f(V_i)$ and $f'(V_i)$ are F_θ and $F_{\theta'}$ respectively. The sequential test under consideration has the following graphical representation in the $(n, \sum_{i=1}^n h(t_i))$ plane. Observation is continued as long as the sample points fall inside the band formed by the parallel straight lines

$$\sum_{i=1}^{n} h(t_i) = \log A_j, \qquad j = 0, 1.$$

The hypothesis is rejected if the path formed by the points $(1, h(t_1))$, $(2, h(t_1) + h(t_2)), \cdots, (N, h(t_1) + \cdots + h(t_N))$ leaves the band through the upper boundary. The probability of this event is therefore the probability of rejection, for θ when each T_i is replaced by $f(V_i)$ and for θ' when T_i is replaced by $f'(V_i)$. Since $f(V_i) \leqq f'(V_i)$ for all i, the path generated by the $f'(V_i)$ leaves the band through the upper boundary whenever this is true for the path generated by the $f(V_i)$. Hence $\beta(\theta) \leqq \beta(\theta')$, as was to be proved.

In the case of monotone likelihood ratios, the sequential probability ratio test with error probabilities $\alpha_0 = \alpha$, $\alpha_1 = 1 - \beta$ therefore satisfies (36). It follows from the optimum property stated in Section 10 that among all tests satisfying (36) the sequential probability ratio test minimizes the expected sample size for $\theta = \theta_0$ and $\theta = \theta_1$. However, one is now concerned with $E_\theta(N)$ for all values of θ. Typically, the function $E_\theta(N)$ has a maximum at a point between θ_0 and θ_1, and decreases as θ moves away from this point in either direction. It frequently turns out that the maximum is $< n_0$, the smallest fixed sample size for which there exists a test satisfying (36). On the other hand, this is not always the case. Thus, in Example 9 for $p_0 = .4$, $p_1 = .6$, $\alpha_0 = \alpha_1 = .005$ for example, the fixed sample size n_0 is 160, and $E_p(N)$, while below this for most values of p, equals 170 for $p = 1/2$. The important problem of determining the test that minimizes $\sup E_\theta(N)$ subject to (36) is still unsolved.

An exact evaluation of the power function $\beta(\theta)$ and the expected sample size $E_\theta(N)$ of a sequential probability ratio test is in general extremely difficult. However, a simple approximation is available provided the equation

$$E_\theta \{[p_{\theta_1}(X)/p_{\theta_0}(X)]^h\} = 1 \tag{37}$$

has a nonzero solution $h = h(\theta)$, as is the case under mild assumptions. (See Problem 38.) Then

$$p_\theta^*(x) = \left[\frac{p_{\theta_1}(x)}{p_{\theta_0}(x)}\right]^h p_\theta(x)$$

is again a probability density. Suppose now that $h > 0$—the other case can be treated similarly—and consider the sequential probability ratio test with boundaries A_0^h, A_1^h for testing p_θ against p_θ^*. With this procedure observation is continued as long as

$$A_0^h < \frac{p_\theta^*(x_1)}{p_\theta(x_1)} \cdots \frac{p_\theta^*(x_n)}{p_\theta(x_n)} < A_1^h.$$

If α_0^* and $1 - \alpha_1^*$ denote the probability of rejection when p_θ and p_θ^* are the true densities, it is seen from (34) that the boundaries are given approximately by

$$A_0^h \sim \frac{\alpha_1^*}{1 - \alpha_0^*}, \qquad A_1^h \sim \frac{1 - \alpha_1^*}{\alpha_0^*}.$$

However, the test under consideration is exactly the same as the sequential probability ratio test with error probabilities $\alpha_0 = \alpha$, $\alpha_1 = 1 - \beta$ for testing θ_0 against θ_1. Hence α_0^* and $\beta(\theta)$, the probability of rejection for the two tests when p_θ is the true density, must be equal. Solving for α_0^* from the above two approximate equations one therefore finds

$$\beta(\theta) \sim \frac{1 - A_0^h}{A_1^h - A_0^h}. \tag{38}$$

An approximation for $E_\theta(N)$ can be based on *Wald's equation*

$$E_\theta(Z_1 + \cdots + Z_N) = E_\theta(N)E_\theta(Z), \tag{39}$$

which is valid whenever the Z's are identically and independently distributed and the procedure is such that the expected sample size $E_\theta(N)$ is finite. For a proof of this equation see Problem 37. If the Z's are defined by (35) and the procedure is a sequential probability ratio test, $Z_1 + \cdots + Z_N$ can be approximated as before by $\log A_1$ and $\log A_0$ when H is rejected and accepted respectively, so that from (39) one obtains

$$E_\theta(N) \sim \frac{\beta(\theta) \log A_1 + [1 - \beta(\theta)] \log A_0}{E_\theta(Z)} \tag{40}$$

provided $E_\theta(Z) \neq 0$.

Example 10. In the binomial problem of Example 9, equation (37) becomes

$$p\left(\frac{p_1}{p_0}\right)^h + q\left(\frac{q_1}{q_0}\right)^h = 1. \tag{41}$$

Since the left-hand side is a convex function of h which is 1 for $h = 0$, it is seen that the equation has a unique nonzero solution except when $p = \log(q_0/q_1)/\log(p_1q_0/p_0q_1)$, in which case the left-hand side has its minimum at $h = 0$. Equations (38) and (41) provide a parametric representation of the approximate power function, which can now be computed by giving different values to h and obtaining the associated values p and β from (38) and (41). (For $h = 0$,

β can be obtained by continuity.) The following is a comparison of the approximate with the exact values of $\beta(p)$ and $E_p(N)$ in the numerical case considered in Example 9, with $p_0 = .05$, $p = .099$, $p_1 = .17$:*

$\beta(p_0)$	$\beta(p)$	$\beta(p_1)$	$E_{p_0}(N)$	$E_p(N)$	$E_{p_1}(N)$	
.05	.44	.90	30	39	25	Approx.
.031	.409	.901	31.4	46.8	30.0	Exact

12. OPTIMUM PROPERTY OF SEQUENTIAL PROBABILITY RATIO TESTS†

The main part of the proof of Theorem 8 is contained in the solution of the following auxiliary problem. For testing the hypothesis H that p_0 is the true probability density of X against the alternative that it is p_1, let the losses resulting from false rejection and acceptance of H be w_0 and w_1, and let the cost of each observation be c. The risk (expected loss plus expected cost) of a sequential procedure is then

$$\alpha_i w_i + cE_i(N)$$

when p_i is the true density, where

$$\alpha_0 = P_0\,(\text{rejecting } H), \qquad \alpha_1 = P_1\,(\text{accepting } H)$$

are the two probabilities of error. If one supposes that the subscript i of the probability density is itself a random variable, which takes on the values 0 and 1 with probability π and $1 - \pi$ respectively, the total average risk of a procedure δ is

$$r(\pi, \delta) = \pi[\alpha_0 w_0 + cE_0(N)] + (1 - \pi)[\alpha_1 w_1 + cE_1(N)]. \tag{42}$$

We shall now determine the Bayes procedure for this problem, that is, the procedure that minimizes (42). Here the interpretation of (42) as a Bayes risk is helpful for an understanding of the proof and gives the auxiliary problem independent interest. However, from the point of view of Theorem 8, the introduction of the w's, c, and π is only a mathematical device, and the problem is simply that of minimizing the formal expression (42).

The Bayes solutions involve two numbers $\pi' \leqq \pi''$ which are uniquely determined by w_0, w_1, and c through equations (44) and (45) below, and which are independent of π. It will be sufficient to restrict attention to the case that $0 < \pi' < \pi'' < 1$ and to a priori probabilities π satisfying $\pi' \leqq \pi \leqq \pi''$.

* Taken from Robinson, *loc. cit.*, where a number of further examples are given.

† This section treats a special topic to which no reference is made in the remainder of the book.

Lemma 5. *Let π', π'' satisfy the equations* (44). *If $0 < \pi' < \pi'' < 1$, then for all $\pi' \leqq \pi \leqq \pi''$ the Bayes risk* (42) *is minimized by any sequential probability ratio test with boundaries*

$$A_0 = \frac{\pi}{1-\pi} \cdot \frac{1-\pi''}{\pi''}, \qquad A_1 = \frac{\pi}{1-\pi} \cdot \frac{1-\pi'}{\pi'}. \tag{43}$$

Proof. (1) We begin by investigating whether at least one observation should be taken, in which case the resulting risk will be at least c, or whether it is better to come to a decision immediately. Let δ_0 denote the procedure that rejects H without taking any observations, and δ_1 the corresponding procedure that accepts H, so that

$$r(\pi, \delta_0) = \pi w_0 \qquad \text{and} \qquad r(\pi, \delta_1) = (1-\pi)w_1.$$

Let

$$\rho(\pi) = \inf_{\delta \in \mathscr{C}} r(\pi, \delta)$$

where $\mathscr{C}$ is the class of all procedures requiring at least one observation. Then for any $0 < \lambda < 1$ and any π_0, π_1,

$$\begin{aligned}\rho[\lambda\pi_0 + (1-\lambda)\pi_1] &= \inf_{\delta \in \mathscr{C}} [\lambda r(\pi_0, \delta) + (1-\lambda) r(\pi_1, \delta)] \\ &\geqq \lambda\rho(\pi_0) + (1-\lambda)\rho(\pi_1).\end{aligned}$$

Hence ρ is concave, and since it is bounded below by zero it is continuous in the interval (0, 1).* If

$$\rho\left(\frac{w_1}{w_0 + w_1}\right) < \frac{w_0 w_1}{w_0 + w_1},$$

define π' and π'' by

$$r(\pi', \delta_0) = \rho(\pi') \qquad \text{and} \qquad r(\pi'', \delta_1) = \rho(\pi''). \tag{44}$$

(See Figure 3.) Otherwise let

$$\pi' = \pi'' = \frac{w_1}{w_0 + w_1}. \tag{45}$$

In the case $0 < \pi' < \pi'' < 1$ with which we are concerned, δ_0 minimizes (42) if and only if $\pi \leqq \pi'$, and δ_1 minimizes (42) if and only if $\pi \geqq \pi''$. This establishes the following uniquely as an optimum first step for $\pi \neq \pi', \pi''$: If $\pi < \pi'$ or $> \pi''$, no observation is taken and H is rejected or accepted respectively; if $\pi' < \pi < \pi''$ the variable X_1 is observed.

(2) The proof is now completed by induction. Suppose that $\pi' < \pi < \pi''$ and that n observations have been taken with outcomes $X_1 = x_1, \cdots, X_n = x_n$, and that one is faced with the alternatives of not taking another

* See, for example, section 3.18 of Hardy, Littlewood, Pólya, *Inequalities*, Cambridge Univ. Press, 1934.

observation and rejecting or accepting H with losses w_0, w_1 for possible wrong decisions, or of going on to observe X_{n+1}. The situation is very similar to the one analyzed in part (1). An unlimited supply of observations $X_{n+1}, X_{n+2}, \cdots$ is available. The fact that one has already incurred the expense of nc units does not affect the problem, since once this loss has been sustained no future action can retrieve it. The procedure is therefore as before: No further observation is taken if the probability of H

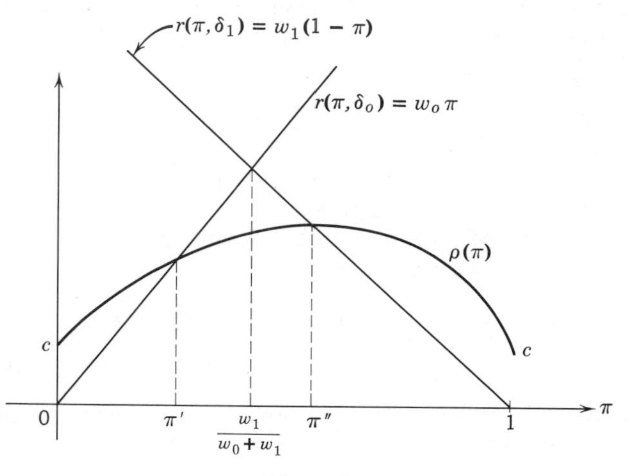

Figure 3.

being true is $< \pi'$ or $> \pi''$, whereas X_{n+1} is observed if this probability is strictly between π' and π''.

One aspect of the situation has changed as a result of observing $x_1, \cdots, x_n$. The probability of H being true is no longer π but has become

$$\pi(x_1, \cdots, x_n) = \frac{\pi p_{0n}}{\pi p_{0n} + (1 - \pi)p_{1n}},$$

the conditional (a posteriori) probability of H given $X_1 = x_1, \cdots, X_n = x_n$. A complete procedure therefore consists in continuing as long as

$$\pi' < \pi(x_1, \cdots, x_n) < \pi''$$

or equivalently as long as

$$A_0 = \frac{\pi}{1 - \pi} \cdot \frac{1 - \pi''}{\pi''} < \frac{p_{1n}}{p_{0n}} < \frac{\pi}{1 - \pi} \cdot \frac{1 - \pi'}{\pi'} = A_1.$$

H is accepted if, at the first violation of these inequalities, p_{1n}/p_{0n} is $< A_0$ and rejected if it is $> A_1$.

(3) In part (1) of this proof the first step of the procedure was uniquely determined as δ_0 for $\pi < \pi'$, as δ_1 for $\pi > \pi''$, and as taking at least one observation when $\pi' < \pi < \pi''$. For $\pi = \pi'$, the procedure δ_0 still minimizes (42) but it is no longer unique, that is, there also exists a procedure $\delta \in \mathscr{C}$ for which $r(\pi', \delta) = \rho(\pi')$. In order to belong to $\mathscr{C}$, such a procedure must require at least one observation. Once X_1 has been observed, it follows from part (2) that the best procedure in $\mathscr{C}$ is obtained by continuing observation as long as $\pi' < \pi(x_1, \cdots x_n) < \pi''$. At the first step it is therefore immaterial whether on the boundary experimentation is continued or the indicated decision is taken. The same is then true at the subsequent steps. This establishes in particular that for $\pi' \leqq \pi \leqq \pi''$ the procedure of taking a first observation and then following the sequential probability ratio test with boundaries (43) is Bayes.

The required connection between the auxiliary problem and the original one is established by the following lemma.

Lemma 6. *Given any $0 < \pi_0' < \pi_0'' < 1$, there exist numbers $0 < w < 1$, $0 < c$ such that the Bayes solution of the auxiliary problem defined by $w_0 = 1 - w$, $w_1 = w$, c, and an a priori probability π satisfying $\pi_0' < \pi < \pi_0''$ is a sequential probability ratio test with boundaries*

$$A_0 = \frac{\pi}{1-\pi} \cdot \frac{1-\pi_0''}{\pi_0''}, \qquad A_1 = \frac{\pi}{1-\pi} \cdot \frac{1-\pi_0'}{\pi_0'} .$$

*Proof.** (1) By Lemma 5, the quantities π' and π'' are functions of w and c, and it is therefore sufficient to find w and c such that $\pi'(w, c) = \pi_0'$, $\pi''(w, c) = \pi_0''$. For fixed w, let $\pi'(c) = \pi'(w, c)$ and $\pi''(c) = \pi''(w, c)$. If c_0 is the smallest value of c such that $\pi'(c_0) = \pi''(c_0)$, then for $0 < c < c_0$ the quantities $\pi'(c)$ and $\pi''(c)$ are determined by the equations

$$(1 - w)\pi' = \rho(\pi', c), \qquad (1 - \pi'')w = \rho(\pi'', c),$$

where $\rho(\pi, c)$ stands for the quantity previously denoted by $\rho(\pi)$. The function $\rho(\pi', c)$ considered as a function of c for fixed π' has the following properties. (i) It is continuous. This follows as before from its being concave. (ii) It is strictly increasing, since for any $\delta \in \mathscr{C}$ the risk $r(\delta, \pi')$ increases strictly with c and since the minimum risk $\rho(\pi', c)$ is taken on by a procedure $\delta \in \mathscr{C}$. (iii) As c tends to zero, so do $\rho(\pi', c)$ and $\rho(\pi'', c)$. This follows from the fact that for n sufficiently large there exists a test of fixed sample size n for which the two error probabilities are arbitrarily small.

* This proof was communicated to me by L. LeCam.

These properties of the function ρ imply that for $0 < c < c_0$ the functions π' and π'' are also continuous, strictly increasing and decreasing respectively, and that $\pi'(c) \to 0$, $\pi''(c) \to 1$ as $c \to 0$. On the other hand, as $c \to c_0$, $\pi''(c) - \pi'(c) \to 0$ so that both quantities tend to the solution $\pi' = \pi'' = w$ of the equation $\pi'(1 - w) = (1 - \pi')w$. It follows from these properties that for fixed w

$$\lambda(c) = \frac{\pi'(c)}{1 - \pi'(c)} \cdot \frac{1 - \pi''(c)}{\pi''(c)}$$

is a continuous, strictly increasing function of c, which increases from 0 to 1 as c varies from 0 to $c_0 = c_0(w)$.

(2) Let

$$\lambda(w, c) = \frac{\pi'(w, c)}{1 - \pi'(w, c)} \cdot \frac{1 - \pi''(w, c)}{\pi''(w, c)}, \qquad \gamma(w, c) = \frac{\pi''(w, c)}{1 - \pi''(w, c)} \cdot$$

Instead of working with the variables π' and π'', it is equivalent and more convenient to work with λ and γ, and to prove the existence of w, c such that

$$\lambda(w, c) = \frac{\pi_0'}{1 - \pi_0'} \cdot \frac{1 - \pi_0''}{\pi_0''} = \lambda_0, \qquad \gamma(w, c) = \frac{\pi_0''}{1 - \pi_0''} = \gamma_0.$$

For any w, there exists by part (1) a unique cost $c = c(w)$ such that $\lambda(w, c) = \lambda_0$. It will be shown below that $\gamma(w) = \gamma[w, c(w)]$ is a 1:1 mapping of the interval $0 < w < 1$ onto $0 < \gamma < \infty$, and hence that there exists a unique value w such that $\gamma(w) = \gamma_0$. This will complete the proof of the lemma.

(3) For the auxiliary problem defined by w, $c = c(w)$, and $\pi = \pi'[w, c(w)]$ there exists by Lemma 5 a Bayes solution δ' which is a sequential probability ratio test with boundaries

$$A_0' = \frac{\pi'[w, c(w)]}{1 - \pi'[w, c(w)]} \cdot \frac{1 - \pi''[w, c(w)]}{\pi''[w, c(w)]} = \lambda[w, c(w)] = \lambda_0, \qquad A_1' = 1.$$

Let δ'' be the corresponding solution of the problem defined by w, $c = c(w)$, and $\pi = \pi''[w, c(w)]$, so that its boundaries are

$$A_0'' = 1, \qquad A_1'' = \frac{\pi''[w, c(w)]}{1 - \pi''[w, c(w)]} \cdot \frac{1 - \pi'[w, c(w)]}{\pi'[w, c(w)]} = \frac{1}{\lambda_0} \cdot$$

Then the error probabilities and the expectations of the sample size α_0', α_1', $E_0'(N)$, $E_1'(N)$ of δ' and α_0'', α_1'', $E_0''(N)$, $E_1''(N)$ of δ'' depend on w and c only through λ_0 and not through γ, so that for fixed λ_0 they are fixed numbers. The Bayes risks for $\pi = \pi'[w, c(w)]$ and $\pi = \pi''[w, c(w)]$ are given by

$$\rho(\pi') = r(\pi', \delta') \qquad \text{and} \qquad \rho(\pi'') = r(\pi'', \delta'')$$

and it follows from (44) that

$$r(\pi', \delta_0) = r(\pi', \delta') \qquad \text{and} \qquad r(\pi'', \delta_1) = r(\pi'', \delta'').$$

These equations can be written more explicitly as

$$\pi'(1 - w) = \pi'[\alpha_0'(1 - w) + cE_0'(N)] + (1 - \pi')[\alpha_1' w + cE_1'(N)]$$

and

$$(1 - \pi'')w = \pi''[\alpha_0''(1 - w) + cE_0''(N)] + (1 - \pi'')[\alpha_1'' w + cE_1''(N)].$$

If one substitutes $\lambda_0\gamma$ for $\pi'/(1 - \pi')$ and γ for $\pi''/(1 - \pi'')$ and eliminates c, this reduces to a single equation connecting γ and w:

$$\{\lambda_0\gamma(1 - \alpha_0') - w[\lambda_0\gamma(1 - \alpha_0') + \alpha_1']\}\{\gamma E_0''(N) + E_1''(N)\}$$
$$= \{-\gamma\alpha_0'' + w[(1 - \alpha_1'') + \gamma\alpha_0'']\}\{\lambda_0\gamma E_0'(N) + E_1'(N)\}.$$

This is linear in w and for any $\gamma > 0$ has a solution $0 < w < 1$. As a function of γ it is quadratic, and the coefficients of the constant and quadratic terms have opposite signs provided $0 < w < 1$. In this case there exists therefore a unique positive solution γ, which establishes the required 1:1 relation between γ and w.

To complete the proof of Theorem 8, consider now any sequential probability ratio test with $A_0 < 1 < A_1$, and any constant $0 < \pi < 1$. Let

$$\pi' = \frac{\pi}{A_1(1 - \pi) + \pi}, \qquad \pi'' = \frac{\pi}{A_0(1 - \pi) + \pi}.$$

These values satisfy (43) and $0 < \pi' < \pi < \pi'' < 1$, and by Lemma 6 there exist therefore constants $0 < w < 1$ and $c > 0$ such that the given test is a Bayes solution for the auxiliary problem with an a priori probability π of p_0 being the true density, with losses $w_0 = 1 - w$ and $w_1 = w$, and cost c. Let the error probabilities and expectations of the sample size be α_0, α_1, $E_0(N)$, $E_1(N)$ for the given test, and consider any competitive procedure δ^*, with error probabilities $\alpha_i^* \leqq \alpha_i$ and expectations of sample size $E_i^*(N) < \infty$ $(i = 0, 1)$. Since the given test minimizes the Bayes risk, it satisfies

$$\pi[(1 - w)\alpha_0 + cE_0(N)] + (1 - \pi)[w\alpha_1 + cE_1(N)]$$
$$\leqq \pi[(1 - w)\alpha_0^* + cE_0^*(N)] + (1 - \pi)[w\alpha_1^* + cE_1^*(N)]$$

and hence

$$\pi E_0(N) + (1 - \pi)E_1(N) \leqq \pi E_0^*(N) + (1 - \pi)E_1^*(N).$$

The validity of this inequality for all $0 < \pi < 1$ implies

$$E_0(N) \leqq E_0^*(N) \qquad \text{and} \qquad E_1(N) \leqq E_1^*(N),$$

as was to be proved.

13. PROBLEMS

Section 2

1. *UMP test for $R(0, \theta)$.* Let $X = (X_1, \cdots, X_n)$ be a sample from the uniform distribution on $(0, \theta)$.

(i) For testing $H\colon\ \theta \leqq \theta_0$ against $K\colon\ \theta > \theta_0$ any test is UMP at level α for which $E_{\theta_0}\,\phi(X) = \alpha$, $E_\theta\,\phi(X) \leqq \alpha$ for $\theta \leqq \theta_0$, and $\phi(x) = 1$ when $\max(x_1, \cdots, x_n) > \theta_0$.

(ii) For testing $H\colon\ \theta = \theta_0$ against $K\colon\ \theta \neq \theta_0$ a unique UMP test exists, and is given by $\phi(x) = 1$ when $\max(x_1, \cdots, x_n) > \theta_0$ or $\max(x_1, \cdots, x_n) \leqq \theta_0 \sqrt[n]{\alpha}$, and $\phi(x) = 0$ otherwise.

[(ii) Determine the UMP tests for testing $\theta = \theta_0$ against $\theta < \theta_0$ and combine this result with that of part (i).]

2. *UMP test for exponential densities.* Let $X_1, \cdots, X_n$ be a sample from the distribution with exponential density $ae^{-a(x-b)}$, $x \geqq b$.

(i) Determine the UMP test for testing $H\colon\ b = b_0$ against $K\colon\ b \neq b_0$ when a is assumed known.

(ii) Determine the UMP test for testing $H\colon\ a = a_0,\ b = b_0$ against the alternatives $a > a_0$, $b < b_0$. Explain the (very unusual) existence in this case of a UMP test in a two-parameter problem.

[(i) The variables $Y_i = e^{-aX_i}$ are a sample from the uniform distribution on $(0, e^{-ab})$.]

3. If the sample space $\mathscr{X}$ is Euclidean and P_0, P_1 have densities with respect to Lebesgue measure, there exists a nonrandomized most powerful test for testing P_0 against P_1 at every significance level α.†

[This is a consequence of Theorem 1 and the following lemma.‡ Let $f \geqq 0$ and $\int_A f(x)\,dx = a$. Given any $0 \leqq b \leqq a$, there exists a subset B of A such that $\int_B f(x)\,dx = b$.]

† For more general results concerning the possibility of dispensing with randomized procedures, see Dvoretzky, Wald, and Wolfowitz, "Elimination of randomization in certain statistical decision procedures and zero-sum two-person games," *Ann. Math. Stat.*, Vol. 22 (1951), pp. 1–21.

‡ For a proof of this lemma see Halmos, *Measure Theory*, New York, D. Van Nostrand Co., 1950, p. 174. The lemma is a special case of a theorem of Liapounoff, "Sur les fonctions-vecteurs complétement additives," *Bull. Acad. Sci., URSS*, Vol. 4 (1940), pp. 465–478.

4. *Fully informative statistics.* A statistic T is *fully informative* if for every decision problem the decision procedures based only on T form an essentially complete class. If $\mathcal{P}$ is dominated and T is fully informative, then T is sufficient.

[Consider any pair of distributions $P_0, P_1 \in \mathcal{P}$ with densities p_0, p_1, and let $g_i = p_i/(p_0 + p_1)$. Suppose that T is fully informative, and let $\mathcal{A}_0$ be the subfield induced by T. Then $\mathcal{A}_0$ contains the subfield induced by (g_0, g_1) since it contains every rejection region which is unique most powerful for testing P_0 against P_1 (or P_1 against P_0) at some level α. Therefore, T is sufficient for every pair of distributions (P_0, P_1), and hence by Problem 9 of Chapter 2 it is sufficient for $\mathcal{P}$.]

Section 3

5. Let X be the number of successes in n independent trials with probability p of success, and let $\phi(x)$ be the UMP test (9) for testing $p \leqq p_0$ against $p > p_0$ at level of significance α.

(i) For $n = 6$, $p_0 = .25$ and the levels $\alpha = .05, .1, .2$ determine C and γ, and find the power of the test against $p_1 = .3, .4, .5, .6, .7$.

(ii) If $p_0 = .2$ and $\alpha = .05$, and it is desired to have power $\beta \geqq .9$ against $p_1 = .4$, determine the necessary sample size (a) by using tables of the binomial distribution, (b) by using the normal approximation.*

(iii) Use the normal approximation to determine the sample size required when $\alpha = .05$, $\beta = .9$, $p_0 = .01$, $p_1 = .02$.

6. (i) A necessary and sufficient condition for densities $p_\theta(x)$ to have monotone likelihood ratio in x, if the mixed second derivative $\partial^2 \log p_\theta(x)/\partial\theta\, \partial x$ exists, is that this derivative be $\geqq 0$ for all θ and x.

(ii) An equivalent condition is that

$$p_\theta(x)\frac{\partial^2 p_\theta(x)}{\partial\theta\,\partial x} \geqq \frac{\partial p_\theta(x)}{\partial\theta}\frac{\partial p_\theta(x)}{\partial x} \quad \text{for all } \theta \text{ and } x.$$

7. Let the probability density p_θ of X have monotone likelihood ratio in $T(x)$, and consider the problem of testing H: $\theta \leqq \theta_0$ against $\theta > \theta_0$. If the distribution of T is continuous, the critical level $\hat{\alpha}$ is given by $\hat{\alpha} = P_{\theta_0}\{T \geqq t\}$ where t is the observed value of T. This holds also without the assumption of continuity if for randomized tests $\hat{\alpha}$ is defined as the smallest significance level at which the hypothesis is rejected with probability 1.

8. Let $X_1, \cdots, X_n$ be independently distributed with density $(2\theta)^{-1}e^{-x/2\theta}$, $x \geqq 0$ and let $Y_1 \leqq \cdots \leqq Y_n$ be the ordered X's. Assume that Y_1 becomes available first, then Y_2, etc., and that observation is continued until Y_r has been observed. On the basis of $Y_1, \cdots, Y_r$ it is desired to test H: $\theta > \theta_0 = 1000$ at level $\alpha = .05$ against $\theta < \theta_0$.

(i) Determine the rejection region when $r = 4$, and find the power of the test against $\theta_1 = 500$.

(ii) Find the value of r required to get power $\beta \geqq .95$ against this alternative.

[In Problem 13, Chapter 2, the distribution of $[\sum_{i=1}^r Y_i + (n - r)Y_r]/\theta$ was found to be χ^2 with $2r$ degrees of freedom.]

* For a discussion of another convenient method applying to this and many related problems, see Mosteller and Tukey, "The uses and usefulness of binomial probability paper," *J. Am. Stat. Assoc.*, Vol. 44 (1949), pp. 174–212.

9. When a Poisson process is observed for a time interval of length τ, the number X of events occurring has the Poisson distribution $P(\lambda\tau)$. Under an alternative scheme, the process is observed until r events have occurred, and the time T of observation is then a random variable such that $2\lambda T$ has a χ^2-distribution with $2r$ degrees of freedom. For testing H: $\lambda \leqq \lambda_0$ at level α one can, under either design, obtain a specified power β against an alternative λ_1 by choosing τ and r sufficiently large.

(i) The ratio of the time of observation required for this purpose under the first design to the expected time required under the second is $\lambda\tau/r$.

(ii) Determine for which values of λ each of the two designs is preferable when $\lambda_0 = 1$, $\lambda_1 = 2$, $\alpha = .05$, $\beta = .9$.

10. *Extension of Lemma* 2. Let P_0 and P_1 be two distributions with densities p_0, p_1 such that $p_1(x)/p_0(x)$ is a nondecreasing function of a real-valued statistic $T(x)$.

(i) If T has probability density p_i' when the original distribution is P_i, then $p_1'(t)/p_0'(t)$ is nondecreasing in t.

(ii) $E_0\psi(T) \leqq E_1\psi(T)$ for any nondecreasing function ψ.

(iii) If $p_1(x)/p_0(x)$ is a strictly increasing function of $t = T(x)$, so is $p_1'(t)/p_0'(t)$ and $E_0\psi(T) < E_1\psi(T)$ unless $\psi[T(x)]$ is constant a.e. $(P_0 + P_1)$ or $E_0\psi(T) = E_1\psi(T) = \pm\infty$.

(iv) For any distinct distributions with densities p_0, p_1,

$$-\infty \leqq E_0 \log [p_1(X)/p_0(X)] < E_1 \log [p_1(X)/p_0(X)] \leqq \infty.$$

[(i) Without loss of generality suppose that $p_1(x)/p_0(x) = T(x)$. Then for any integrable ϕ,

$$\int \phi(t)p_1'(t)\, d\nu(t) = \int \phi[T(x)]T(x)p_0(x)\, d\mu(x) = \int \phi(t)tp_0'(t)\, d\nu(t),$$

and hence $p_1'(t)/p_0'(t) = t$ a.e.

(iv) The possibility $E_0 \log [p_1(X)/p_0(X)] = \infty$ is excluded since by the convexity of the function log,

$$E_0 \log [p_1(X)/p_0(X)] \leqq \log E_0[p_1(X)/p_0(X)] = 0.$$

Similarly for E_1. The strict inequality now follows from (iii) with $T(x) = p_1(x)/p_0(x)$.]

11. If F_0, F_1 are two cumulative distribution functions on the real line such that $F_1(x) \leqq F_0(x)$ for all x, then $E_0\psi(X) \leqq E_1\psi(X)$ for any nondecreasing function ψ.

Section 4

12. If the experiment (f, g) is more informative than (f', g'), then (g, f) is more informative than (g', f').

13. *Conditions for comparability*. (i) Let X and X' be two random variables taking on the values 1 and 0, and suppose that $P\{X = 1\} = p_0$, $P\{X' = 1\} = p_0'$ or that $P\{X = 1\} = p_1$, $P\{X' = 1\} = p_1'$. Without loss of generality let $p_0 < p_0'$, $p_0 < p_1$, $p_0' < p_1'$. (This can be achieved by exchanging X with X' and by exchanging the values 0 and 1 of one or both of the variables.) Then X is more informative than X' if and only if $(1 - p_1)(1 - p_0') \leqq (1 - p_0)(1 - p_1')$.

(ii) Let U_0, U_1 be independently uniformly distributed over (0, 1) and let $Y = 1$ if $X = 1$ and $U_1 \leqq \gamma_1$ and if $X = 0$ and $U_0 \leqq \gamma_0$ and $Y = 0$ otherwise. Under the assumptions of (i) there exist $0 \leqq \gamma_0, \gamma_1 \leqq 1$ such that $P\{Y = 1\} = p_i'$ when $P\{X = 1\} = p_i$ $(i = 0, 1)$ provided $(1 - p_1)(1 - p_0') \leqq (1 - p_0)(1 - p_1')$. This inequality, which is therefore sufficient for a sample $X_1, \cdots, X_n$ from X to be more informative than a sample $X_1', \cdots, X_n'$ from X', is also necessary. Similarly, the condition $p_0'p_1 \leqq p_0p_1'$ is necessary and sufficient for a sample from X' to be more informative than one from X.

[(i) The power $\beta(\alpha)$ of the most powerful level α test of p_0 against p_1 based on X is $\alpha p_1/p_0$ if $\alpha \leqq p_0$ and $p_1 + q_1q_0^{-1}(\alpha - p_0)$ if $p_0 \leqq \alpha$. One obtains the desired result by comparing the graphs of $\beta(\alpha)$ and $\beta'(\alpha)$.

(ii) The last part of (ii) follows from a comparison of the power $\beta_n(\alpha)$ and $\beta_n'(\alpha)$ of the most powerful level α tests based on ΣX_i and $\Sigma X_i'$ for α close to 1. The dual condition is obtained from Problem 12.]

14. For the 2×2 table described in Example 4, and under the assumption $p \leqq \pi \leqq 1/2$ made there, a sample from $\tilde{B}$ is more informative than one from $\tilde{A}$. On the other hand, samples from B and $\tilde{B}$ are not comparable.

[A necessary and sufficient condition for comparability is given in the preceding problem.]

15. In the experiment discussed in Example 5, n binomial trials with probability of success $p = 1 - e^{-\lambda v}$ are performed for the purpose of testing $\lambda = \lambda_0$ against $\lambda = \lambda_1$. Experiments corresponding to two different values of v are not comparable.

Section 5

16. (i) For $n = 5$, 10 and $1 - \alpha = .95$, graph the upper confidence limits $\bar{p}$ and $\bar{p}^*$ of Example 7 as functions of $t = x + u$.

(ii) For the same values of n and $\alpha_1 = \alpha_2 = .05$, graph the lower and upper confidence limits $\underline{p}$ and $\bar{p}$.

17. *Confidence bounds with minimum risk.* Let $L(\theta, \underline{\theta})$ be nonnegative and nonincreasing in its second argument for $\underline{\theta} < \theta$, and equal to 0 for $\underline{\theta} \geqq \theta$. If $\underline{\theta}$ and $\underline{\theta}^*$ are two lower confidence bounds for θ such that

$$P_\theta\{\underline{\theta} \leqq \theta'\} \leqq P_\theta\{\underline{\theta}^* \leqq \theta'\} \quad \text{for all} \quad \theta' \leqq \theta,$$

then

$$E_\theta L(\theta, \underline{\theta}) \leqq E_\theta L(\theta, \underline{\theta}^*).$$

[Define two cumulative distribution functions F and F^* by $F(u) = P_\theta\{\underline{\theta} \leqq u\}/P_\theta\{\underline{\theta}^* \leqq \theta\}$; $F^*(u) = P_\theta\{\underline{\theta}^* \leqq u\}/P_\theta\{\underline{\theta}^* \leqq \theta\}$ for $u < \theta$ and $F(u) = F^*(u) = 1$ for $u \geqq \theta$. Then $F(u) \leqq F^*(u)$ for all u, and it follows from Problem 11 that

$$E_\theta[L(\theta, \underline{\theta})] \leqq P_\theta\{\underline{\theta}^* \leqq \theta\} \int L(\theta, u)\, dF(u)$$

$$\leqq P_\theta\{\underline{\theta}^* \leqq \theta\} \int L(\theta, u)\, dF^*(u) = E_\theta[L(\theta, \underline{\theta}^*)].]$$

Section 6

18. If $\beta(\theta)$ denotes the power function of the UMP test of Corollary 2, and if the function Q of (12) is differentiable, then $\beta'(\theta) > 0$ for all θ for which $Q'(\theta) > 0$.

[To show that $\beta'(\theta_0) > 0$, consider the problem of maximizing, subject to $E_{\theta_0}\phi(X) = \alpha$, the derivative $\beta'(\theta_0)$ or equivalently the quantity $E_{\theta_0}[T(X)\phi(X)]$.]

19. *Optimum selection procedures.* On each member of a population n measurements $(X_1, \cdots, X_n) = X$ are taken, for example the scores of n aptitude tests which are administered to judge the qualifications of candidates for a certain training program. A future measurement Y such as the score in a final test at the end of the program is of interest but unavailable. The joint distribution of X and Y is assumed known.

(i) One wishes to select a given proportion α of the candidates in such a way as to maximize the expectation of Y for the selected group. This is achieved by selecting the candidates for which $E(Y|x) \geqq C$, where C is determined by the condition that the probability of a member being selected is α. When $E(Y|x) = C$, it may be necessary to randomize in order to get the exact value α.

(ii) If instead the problem is to maximize the probability with which in the selected population Y is greater than or equal to some preassigned score y_0, one selects the candidates for which the conditional probability $P\{Y \geqq y_0|x\}$ is sufficiently large.

[(i) Let $\phi(x)$ denote the probability with which a candidate with measurements x is to be selected. Then the problem is that of maximizing

$$\int\left[\int yp^{Y|x}(y)\phi(x)\,dy\right]p^X(x)\,dx$$

subject to

$$\int \phi(x)p^X(x)\,dx = \alpha.]$$

20. The following example shows that Corollary 4 does not extend to a countably infinite family of distributions. Let p_n be the uniform probability density on $[0, 1 + 1/n]$ and p_0 the uniform density on $(0, 1)$.

(i) Then p_0 is linearly independent of $(p_1, p_2, \cdots)$, that is, there do not exist constants $c_1, c_2, \cdots$ such that $p_0 = \Sigma c_n p_n$.

(ii) There does not exist a test ϕ such that $\int\phi p_n = \alpha$ for $n = 1, 2, \cdots$ but $\int \phi p_0 > \alpha$.

21. Let $F_1, \cdots, F_{m+1}$ be real-valued functions defined over a space U. A sufficient condition for u_0 to maximize F_{m+1} subject to $F_i(u) \leqq c_i$ $(i = 1, \cdots, m)$ is that it maximizes $F_{m+1}(u) - \Sigma k_i F_i(u)$ for some constants $k_i \geqq 0$ and that $F_i(u_0) = c_i$ for those values i for which $k_i > 0$.

Section 7

22. For a random variable X with binomial distribution $b(p, n)$ determine the constants C_i, γ_i $(i = 1, 2)$ in the UMP test (24) for testing H: $p \leqq .2$ or $\geqq .7$ when $\alpha = .1$ and $n = 15$. Find the power of the test against the alternative $p = .4$.

23. *Pólya type.* A family of distributions with probability densities $p_\theta(x)$ which are continuous in the real variables θ and x is said to be of *Pólya type* if for all $x_1 < \cdots < x_n$ and $\theta_1 < \cdots < \theta_n$

$$\Delta_n = \begin{vmatrix} p_{\theta_1}(x_1) \cdots p_{\theta_1}(x_n) \\ \\ p_{\theta_n}(x_1) \cdots p_{\theta_n}(x_n) \end{vmatrix} \geqq 0 \quad \text{for all} \quad n = 1, 2, \cdots, \tag{46}$$

and *strictly of Pólya type* if strict inequality holds in (46). For $n = 1$ the condition states that $p_\theta(x) \geqq 0$, for $n = 2$ that $p_\theta(x)$ has monotone likelihood ratio. The exponential families (12) with $T(x) = x$ and $Q(\theta) = \theta$ are strictly of Pólya type.

[That the determinant $|e^{\theta_i x_j}|$, $i, j = 1, \cdots, n$, is positive can be proved by induction. Divide the ith column by $e^{\theta_1 x_i}$, $i = 1, \cdots, n$; subtract in the resulting determinant the $(n-1)$st column from the nth, the $(n-2)$nd from the $(n-1)$st, $\cdots$, the 1st from the 2nd; and expand the determinant obtained in this way by the first row. Then Δ_n is seen to have the same sign as

$$\Delta'_n = |e^{\eta_i x_j} - e^{\eta_i x_{j-1}}|, \qquad i, j = 2, \cdots, n,$$

where $\eta_i = \theta_i - \theta_1$. If this determinant is expanded by the first column one obtains a sum of the form

$$a_2(e^{\eta_2 x_2} - e^{\eta_2 x_1}) + \cdots + a_n(e^{\eta_n x_2} - e^{\eta_n x_1}) = h(x_2) - h(x_1) = (x_2 - x_1)h'(y_2),$$

where $x_1 \leqq y_2 \leqq x_2$. Rewriting $h'(y_2)$ as a determinant of which all columns but the first coincide with those of Δ'_n and proceeding in the same manner with the other columns, one reduces the determinant to $|e^{\eta_i y_j}|$, $i, j = 2, \cdots, n$, which is positive by the induction hypothesis.]

24. *Pólya type* 3. Let θ and x be real-valued and suppose that the probability densities $p_\theta(x)$ are such that $p_{\theta'}(x)/p_\theta(x)$ is strictly increasing in x for $\theta < \theta'$. Then the following two conditions are equivalent: (a) For $\theta_1 < \theta_2 < \theta_3$ and $k_1, k_2, k_3 > 0$, let

$$g(x) = k_1 p_{\theta_1}(x) - k_2 p_{\theta_2}(x) + k_3 p_{\theta_3}(x).$$

If $g(x_1) = g(x_3) = 0$, then the function g is positive outside the interval (x_1, x_3) and negative inside. (b) The determinant Δ_3 given by (46) is positive for all $\theta_1 < \theta_2 < \theta_3$, $x_1 < x_2 < x_3$. (It follows from (a) that the equation $g(x) = 0$ has at most two solutions.)

[That (b) implies (a) can be seen for $x_1 < x_2 < x_3$ by considering the determinant

$$\begin{vmatrix} g(x_1) & g(x_2) & g(x_3) \\ p_{\theta_2}(x_1) & p_{\theta_2}(x_2) & p_{\theta_2}(x_3) \\ p_{\theta_3}(x_1) & p_{\theta_3}(x_2) & p_{\theta_3}(x_3) \end{vmatrix}.$$

Suppose conversely that (a) holds. Monotonicity of the likelihood ratios implies that the rank of Δ_3 is at least two, so that there exist constants k_1, k_2, k_3 such that $g(x_1) = g(x_3) = 0$. That the k's are positive follows again from the monotonicity of the likelihood ratios.]

25. *Extension of Theorem* 6. The conclusions of Theorem 6 remain valid if the densities p_θ of a sufficient statistic T, which without loss of generality will

be taken to be X, satisfy the following conditions: (a) $p_\theta(x)$ is continuous in x for each θ; (b) $p_{\theta'}(x)/p_\theta(x)$ is strictly increasing in x for $\theta < \theta'$; and the determinant Δ_3 defined by (46) is positive for all $\theta_1 < \theta_2 < \theta_3$ and $x_1 < x_2 < x_3$.

[The two properties of exponential families that are used in the proof of Theorem 6 are continuity in x and (a) of the preceding problem.]

26. For testing the hypothesis H': $\theta_1 \leqq \theta \leqq \theta_2 (\theta_1 \leqq \theta_2)$ against the alternatives $\theta < \theta_1$ or $\theta > \theta_2$, or the hypothesis $\theta = \theta_0$ against the alternatives $\theta \neq \theta_0$, in an exponential family or more generally in a family of distributions satisfying the assumptions of Problem 25, a UMP test does not exist.

[This follows from a consideration of the UMP tests for the one-sided hypotheses H_1: $\theta \geqq \theta_1$ and H_2: $\theta \leqq \theta_2$.]

Section 8

27. Let the variables $X_i (i = 1, \cdots, s)$ be independently distributed with Poisson distribution $P(\lambda_i)$. For testing the hypothesis H: $\Sigma\lambda_j \leqq a$ (for example, that the combined radioactivity of a number of pieces of radioactive material does not exceed a), there exists a UMP test, which rejects when $\Sigma X_j > C$.

[If the joint distribution of the X's is factored into the marginal distribution of ΣX_j (Poisson with mean $\Sigma\lambda_j$) times the conditional distribution of the variables $Y_i = X_i/\Sigma X_j$ given ΣX_j (multinomial with probabilities $p_i = \lambda_i/\Sigma\lambda_j$), the argument is analogous to that given in Example 8.]

28. *Confidence bounds for a median.* Let $X_1, \cdots, X_n$ be a sample from a continuous cumulative distribution function F. Let ξ be the unique median of F if it exists or more generally let $\xi = \inf\{\xi': F(\xi') = 1/2\}$.

(i) If the ordered X's are $X^{(1)} < \cdots < X^{(n)}$ a uniformly most accurate lower confidence bound for ξ is $\underline{\xi} = X^{(k)}$ with probability ρ, $\underline{\xi} = X^{(k+1)}$ with probability $(1 - \rho)$ where k and ρ are determined by

$$\rho \sum_{j=k}^{n} \binom{n}{j} \frac{1}{2^n} + (1 - \rho) \sum_{j=k+1}^{n} \binom{n}{j} \frac{1}{2^n} = 1 - \alpha.$$

(ii) This bound has confidence coefficient $1 - \alpha$ for any median of F.

(iii) Determine most accurate lower confidence bounds for the $100p$-percentile ξ of F defined by $\xi = \inf\{\xi': F(\xi') = p\}$.

[For fixed ξ_0 the problem of testing H: $\xi = \xi_0$ against K: $\xi > \xi_0$ is equivalent to testing H': $p = 1/2$ against K': $p < 1/2$.]

29. *A counterexample.* Typically, as α varies the most powerful level α tests for testing a hypothesis H against a simple alternative are nested in the sense that the associated rejection regions, say R_α, satisfy $R_\alpha \subset R_{\alpha'}$ for any $\alpha < \alpha'$. This relation always holds when H is simple, but the following example shows that it need not be satisfied for composite H.

Let X take on the values 1, 2, 3, 4 with probabilities under distributions P_0, P_1, Q:

	1	2	3	4
P_0	$\frac{2}{13}$	$\frac{4}{13}$	$\frac{3}{13}$	$\frac{4}{13}$
P_1	$\frac{4}{13}$	$\frac{2}{13}$	$\frac{1}{13}$	$\frac{6}{13}$
Q	$\frac{4}{13}$	$\frac{3}{13}$	$\frac{2}{13}$	$\frac{4}{13}$

Then the most powerful test for testing the hypothesis that the distribution of X is P_0 or P_1 against the alternative that it is Q rejects at level $\alpha = \frac{5}{13}$ when $X = 1$ or 3, and at level $\alpha = \frac{6}{13}$ when $X = 1$ or 2.

30. Let X and Y be the number of successes in two sets of n binomial trials with probabilities p_1 and p_2 of success.

(i) The most powerful test of the hypothesis H: $p_2 \leqq p_1$ against an alternative (p_1', p_2') with $p_1' < p_2'$ and $p_1' + p_2' = 1$ at level $\alpha < \frac{1}{2}$ rejects when $Y - X > C$ and with probability γ when $Y - X = C$.

(ii) This test is not UMP against the alternatives $p_1 < p_2$.

[(i) Take the distribution λ assigning probability 1 to the point $p_1 = p_2 = 1/2$ as an a priori distribution over H. The most powerful test against (p_1', p_2') is then the one proposed above. To see that λ is least favorable, consider the probability of rejection $\beta(p_1, p_2)$ for $p_1 = p_2 = p$. By symmetry this is given by

$$2\beta(p, p) = P\{|Y - X| > C\} + \gamma P\{|Y - X| = C\}.$$

Let X_i be 1 or 0 as the ith trial in the first series is a success or failure, and let Y_i be defined analogously with respect to the second series. Then $Y - X = \sum_{i=1}^n (Y_i - X_i)$, and the fact that $2\beta(p, p)$ attains its maximum for $p = 1/2$ can be proved by induction over n.

(ii) Since $\beta(p, p) < \alpha$ for $p \neq 1/2$, the power $\beta(p_1, p_2)$ is $< \alpha$ for alternatives $p_1 < p_2$ sufficiently close to the line $p_1 = p_2$. That the test is not UMP now follows from a comparison with $\phi(x, y) \equiv \alpha$.]

31. *Sufficient statistics with nuisance parameters.* (i) A statistic T is said to be sufficient for θ in the presence of a nuisance parameter η if the parameter space is the direct product of the set of possible θ- and η-values, and if the following two conditions hold: (a) the conditional distribution given $T = t$ depends only on η; (b) the marginal distribution of T depends only on θ. If these conditions are satisfied, there exists a UMP test for testing the composite hypothesis H: $\theta = \theta_0$ against the composite class of alternatives $\theta = \theta_1$, which depends only on T.

(ii) Part (i) provides an alternative proof that the test of Example 8 is UMP.

[Let $\psi_0(t)$ be the most powerful level α test for testing θ_0 against θ_1 that depends only on t, let $\phi(x)$ be any level α test, and let $\psi(t) = E_{\eta_1}[\phi(X)|t]$. Since $E_{\theta_i}\psi(T) = E_{\theta_i,\eta_1}\phi(X)$, it follows that ψ is a level α test of H and its power, and therefore the power of ϕ does not exceed the power of ψ_0.]

Section 9

32. Let $X_1 \cdots, X_m$ and $Y_1, \cdots, Y_n$ be independent samples from $N(\xi, 1)$ and $N(\eta, 1)$, and consider the hypothesis H: $\eta \leqq \xi$ against K: $\eta > \xi$. There exists a UMP test, and it rejects the hypothesis when $\bar{Y} - \bar{X}$ is too large.

[If $\xi_1 < \eta_1$ is a particular alternative, the distribution assigning probability 1 to the point $\eta = \xi = (m\xi_1 + n\eta_1)/(m + n)$ is least favorable.]

33. Let $X_1, \cdots, X_m$; $Y_1, \cdots, Y_n$ be independently, normally distributed with means ξ and η, and variances σ^2 and τ^2 respectively, and consider the hypothesis H: $\tau \leqq \sigma$ against K: $\sigma < \tau$.

(i) If ξ and η are known there exists a UMP test given by the rejection region $\Sigma(Y_j - \eta)^2/\Sigma(X_i - \xi)^2 \geqq C$.

(ii) No UMP test exists when ξ and η are unknown.

Section 10

34. *Distribution of sequential sample size.* Let X_i $(i = 1, 2, \cdots)$ be identically and independently distributed and let Z_i be defined by (35). If N is the number of observations required by the sequential probability ratio test (33) where $A_0 < 1 < A_1$, and if the true distribution of X is such that $P\{Z = 0\} < 1$, then there exists $0 < \delta < 1$ and $C > 0$ such that $P\{N \geqq n\} \leqq C\delta^n$.

[Let $c = \log A_1 - \log A_0$ and suppose first that $P\{|Z| \leqq c\} = p < 1$. The event $N \geqq n$ implies that

$$\log A_0 < z_1, z_1 + z_2, \cdots, z_1 + \cdots + z_{n-1} < \log A_1,$$

and hence that $|z_1|, |z_2|, \cdots, |z_{n-1}|$ are all $\leqq c$. Therefore $P\{N \geqq n\} \leqq p^{n-1} = p^{-1} \cdot p^n$. If $P\{|Z| \leqq c\} = 1$, there exists r such that $P\{|Z_1 + \cdots + Z_r| \leqq c\} = p < 1$, and this implies that $P\{N \geqq rm\} \leqq p^{m-1}$ and hence that

$$P\{N \geqq n\} \leqq p^{[n/r]-1} \leqq p^{n/r-2} = p^{-2}(p^{1/r})^n.]$$

35. *Moments of sequential sample size.* Under the assumptions of the preceding problem (i) $E(N) < \infty$ and (ii) $E(e^{tN}) < \infty$ for some $t > 0$ so that $E(N^k) < \infty$ for all $k = 1, 2, \cdots$. Also quite generally (iii) $P\{Z = 0\} < 1$ if the true distribution of the X's is either P_0 or P_1 and if $P_0 \neq P_1$.

[(i)

$$E(N) = \sum_{n=1}^{\infty} nP\{N = n\} = \sum_{n=1}^{\infty} P\{N \geqq n\} \leqq C \sum_{n=1}^{\infty} \delta^n < \infty.$$

(ii)

$$E(e^{tN}) \leqq \Sigma e^{tn} P\{N \geqq n\} \leqq C\Sigma(\delta e^t)^n < \infty$$

provided $e^t < \delta^{-1}$.]

Section 11

36. *Power of binomial sequential probability ratio test.* In the sequential probability ratio test of Examples 9 and 10 for testing that a binomial probability is p_0 against the alternative that it is p_1, let $p_1 = q_0$ and suppose that $\log A_0/\log(q_0p_0^{-1}) = -a$ and $\log A_1/\log(q_0p_0^{-1}) = b$ where a and b are positive integers.

(i) Then the inequalities (34) become equalities and the approximations (38) and (40) become exact formulas.

(ii) The power function of the test is

$$\beta(p) = \frac{q^a p^b - p^{a+b}}{q^{a+b} - p^{a+b}} \quad \text{for} \quad p \neq 1/2$$

$$\beta(1/2) = a/(a + b) \quad \text{(by continuity)}.$$

(iii) The stopping rule is the same as that imposed by chance on two gamblers with capitals a and b who play a sequence of games for unit stakes with probabilities p and q of winning each game, and who continue playing until one of them has exhausted his capital.*

* For an alternative derivation of the formula for $\beta(p)$ in this setting see, for example, Chapter 14, Section 2, of Feller, *An Introduction to Probability Theory and Its Applications*, Vol. I, New York, John Wiley & Sons, 2nd ed., 1957.

[The test continues as long as $-a < 2\Sigma x_i - n < b$, and (i) and (ii) follow from the fact that the middle term of this inequality is 0 for $n = 0$ and with each observation either increases or decreases by 1.]

37. *Wald's equation.* If $Z_1, Z_2, \cdots$ are identically and independently distributed with $E|Z| < \infty$, and if the number of observations is decided according to a sequential rule with $E(N) < \infty$, then

$$E(Z_1 + \cdots + Z_N) = E(N)E(Z). \tag{47}$$

[The left-hand side equals

$$\sum_{n=1}^{\infty}\left[P\{N = n\}\sum_{i=1}^{n} E(Z_i|N = n)\right] = \sum_{i=1}^{\infty}\sum_{n=i}^{\infty} P\{N = n\}E(Z_i|N = n)$$
$$= \sum_{i=1}^{\infty} P\{N \geqq i\}E(Z_i|N \geqq i).$$

Since the event $N \geqq i$ depends only on $Z_1, \cdots, Z_{i-1}$, it is independent of Z_i; also $\Sigma_{i=1}^{\infty} P\{N \geqq i\} = E(N)$, and this establishes the desired equation. To justify the rearranging of the infinite series, replace Z_i by $|Z_i|$ throughout. This shows that

$$E|Z_1 + \cdots + Z_N| \leqq E(|Z_1| + \cdots + |Z_N|) = E|Z| \cdot E(N) < \infty,$$

which proves the required absolute convergence.]

38. (i) Let Z be a random variable such that (a) $E(Z) \neq 0$, (b) $\psi(h) = E(e^{hZ})$ exists for all real h, (c) $P\{e^Z < 1 - \delta\}$ and $P\{e^Z > 1 + \delta\}$ are positive for some $\delta > 0$. Then there exists one and only one solution $h \neq 0$ of the equation $\psi(h) = 1$.

(ii) This provides sufficient conditions for the existence of a nonzero solution of (37).

[(i) The function ψ is convex since $\psi''(h) = E(Z^2e^{hZ}) > 0$; also $\psi(h) \to \infty$ as $h \to \pm\infty$. Therefore ψ has a minimum h_0, at which $\psi'(h_0) = E(Ze^{h_0Z}) = 0$ so that by (a) $h_0 \neq 0$. Since $\psi(0) = 1$, there exists a unique $h_1 \neq 0$ for which $\psi(h_1) = 1$.

(ii) With Z defined by (35), (37) can be written as $E(e^{hZ}) = 1$.]

39. The following example shows that the power of a test can sometimes be increased by selecting a random rather than a fixed sample size even when the randomization does not depend on the observations.* Let $X_1, \cdots, X_n$ be independently distributed as $N(\theta, 1)$ and consider the problem of testing H: $\theta = 0$ against K: $\theta = \theta_1 > 0$.

(i) The power of the most powerful test as a function of the sample size n is not necessarily concave.

(ii) In particular for $\alpha = .005$, $\theta_1 = \frac{1}{2}$, better power is obtained by taking 2 or 16 observations with probability 1/2 each than by taking a fixed sample of 9 observations.

(iii) The power can be increased further if the test is permitted to have different significance levels α_1 and α_2 for the two sample sizes and it is required only that

* This and related examples were discussed by Kruskal in a seminar held at Columbia University in 1954. Recently, a more detailed investigation of the phenomenon has been undertaken by Cohen, "On mixed single sample experiments," *Ann. Math. Stat.*, Vol. 29 (1958), pp. 947–971.

the expected significance level be equal to $\alpha = .005$. Examples are: (a) with probability 1/2 take $n_1 = 2$ observations and perform the test of significance at level $\alpha_1 = .001$ or take $n_2 = 16$ observations and perform the test at level $\alpha_2 = .009$; (b) with probability 1/2 take $n_1 = 0$ or $n_2 = 18$ observations and let the respective significance levels be $\alpha_1 = 0$, $\alpha_2 = .01$.

14. REFERENCES

The method of hypothesis testing developed gradually, with early instances frequently being rather vague statements of the significance or nonsignificance of a set of observations. Isolated applications [the earliest one is perhaps due to Laplace (1773)] are found throughout the 19th century, for example in the writings of Gavarret (1840), Lexis (1875, 1877), and Edgeworth (1885). A systematic use of hypothesis testing began with the work of Karl Pearson, particularly his χ^2 paper of 1900.

The first authors to recognize that the rational choice of a test must involve consideration not only of the hypothesis but also of the alternatives against which it is being tested were Neyman and Pearson (1928). They introduced the distinction between errors of the first and second kind, and thereby motivated their proposal of the likelihood ratio criterion as a general method of test construction. These considerations were carried to their logical conclusion by Neyman and Pearson in their paper of 1933, in which they developed the theory of UMP tests.*

The earliest example of confidence intervals appears to occur in the work of Laplace (1812) who points out how a (approximate) probability statement concerning the difference between an observed frequency and a binomial probability p can be inverted to obtain an associated interval for p. Other examples can be found in the work of Gauss (1816), Fourier (1826), and Lexis (1875). However, in all these cases, although the statements made are formally correct, the authors appear to consider the parameter as the variable which with the stated probability falls in the fixed confidence interval. The proper interpretation seems to have been pointed out for the first time by E. B. Wilson (1927). About the same time two examples of exact confidence statements were given by Working and Hotelling (1929) and Hotelling (1931).

A general method for obtaining exact confidence bounds for a real-valued parameter in a continuous distribution was proposed by Fisher (1930), who however later disavowed this interpretation of his work.

* A different approach to hypothesis testing, based on prior probabilities, has been developed by Jeffreys, *Theory of Probability*, Oxford, Clarendon Press, 2nd ed., 1948. Some aspects of the relation between the two theories are discussed by Lindley, "A statistical paradox," *Biometrika*, Vol. 44 (1957), pp. 187–192, and Bartlett, "A comment on D. V. Lindley's statistical paradox," *Biometrika*, Vol. 44 (1957), pp. 523–534.

[For a bibliography concerning Fisher's concept of fiducial probability, in terms of which his theory is formulated, see Tukey (1957).] At about the same time,* a completely general theory of confidence statements was developed by Neyman and shown by him to be intimately related to the theory of hypothesis testing. A detailed account of this work, which underlies the treatment given here, was published by Neyman in his papers of 1937 and 1938.

Arrow, K. J., D. Blackwell, and M. A. Girshick

(1949) "Bayes and minimax solutions of sequential decision problems," *Econometrica*, Vol. 17, pp. 213–244.

Birnbaum, Z. W., and D. G. Chapman

(1950) "On optimum selections from multinormal populations," *Ann. Math. Stat.*, Vol. 21, pp. 443–447.

[Problem 19.]

Blackwell, David

(1951) "Comparison of experiments," *Proc. Second Berkeley Symposium on Mathematical Statistics and Probability*, Berkeley, Univ. Calif. Press, pp. 93–102.

(1953) "Equivalent comparisons of experiments," *Ann. Math. Stat.*, Vol. 24, pp. 265–272.

[Theory, Example 4, and problems of Section 4.]

Chernoff, Herman, and Henry Scheffé

(1952) "A generalization of the Neyman-Pearson fundamental lemma," *Ann. Math. Stat.*, Vol. 23, pp. 213–225.

Dantzig, George B., and A. Wald

(1951) "On the fundamental lemma of Neyman and Pearson," *Ann. Math. Stat.*, Vol. 22, pp. 87–93.

[Gives necessary conditions, including those of Theorem 5, for a critical function which maximizes an integral subject to a number of integral side conditions, to satisfy (21).]

Dvoretzky, A., J. Kiefer, and J. Wolfowitz

(1953) "Sequential decision problems for processes with continuous time parameter. Testing hypotheses," *Ann. Math. Stat.*, Vol. 24, pp. 254–264.

[Extends the optimum property of the sequential probability ratio test to stochastic processes that are observed continuously.]

Edgeworth, F. Y.

(1885) "Methods of statistics," Jubilee volume of the Stat. Soc., London, E. Stanford.

Epstein, Benjamin, and Milton Sobel

(1953) "Life testing," *J. Am. Stat. Assoc.*, Vol. 48, pp. 486–502.

[Problem 8.]

Fisher, R. A.

(1930) "Inverse probability," *Proc. Cambridge Phil. Soc.*, Vol. 26, pp. 528–535.

Fourier, J. B. J.

(1826) *Recherches statistiques sur la ville de Paris et le département de la Seine*, Vol. 3.

* Cf. Neyman, "Fiducial argument and the theory of confidence intervals," *Biometrika*, Vol. 32 (1941), pp. 128–150.

Fraser, D. A. S.

(1953) "Non-parametric theory: Scale and location parameters," *Canad. J. Math.*, Vol. 6, pp. 46–68.

[Example 8.]

(1956) "Sufficient statistics with nuisance parameters," *Ann. Math. Stat.*, Vol. 27, pp. 838–842.

[Problem 31.]

Gauss, C. F.

(1816) "Bestimmung der Genauigkeit der Beobachtungen," *Z. Astronomie und verwandte Wissenschaften*, Vol. 1. (Reprinted in Gauss' collected works, Vol. 4, pp. 109–119.)

Gavarret, J.

(1840) *Principes généraux de statistique médicale*, Paris.

Grenander, Ulf

(1950) "Stochastic processes and statistical inference," *Arkiv för Matematik*, Vol. 1, pp. 195–277.

[Application of the fundamental lemma to problems in stochastic processes.]

Hotelling, Harold

(1931) "The generalization of Student's ratio," *Ann. Math. Stat.*, Vol. 2, pp. 360–378.

Karlin, Samuel

(1955) "Decision theory for Pólya type distributions. Case of two actions, I.," *Proc. Third Berkeley Symposium on Mathematical Statistics and Probability*, Vol. 1, Berkeley, Univ. Calif. Press, pp. 115–129.

(1957) "Pólya type distributions, II.," *Ann. Math. Stat.*, Vol. 28, pp. 281–308.

[Properties of Pólya type distributions including Problems 23–25.]

Karlin, Samuel, and Herman Rubin

(1956) "The theory of decision procedures for distributions with monotone likelihood ratio," *Ann. Math. Stat.*, Vol. 27, pp. 272–299.

[General theory of families with monotone likelihood ratio, including Theorem 3.]

Laplace, P. S.

(1773) "Mémoire sur l'inclinaison moyenne des orbites des comètes," *Mem. acad. roy. sci. Paris*, Vol. VII (1776), pp. 503–524.

(1812) *Théorie Analytique des Probabilités*, Paris.

(The 3rd edition of 1820 is reprinted as Vol. 7 of Laplace's collected works.)

Lehmann, E. L.

(1955) "Ordered families of distributions," *Ann. Math. Stat.*, Vol. 26, pp. 399–419.

[Lemmas 1, 2, and 4.]

Lehmann, E. L., and C. Stein

(1948) "Most powerful tests of composite hypotheses," *Ann. Math. Stat.*, Vol. 19, pp. 495–516.

[Theorem 7 and applications.]

Lexis, W.

(1875) *Einleitung in die Theorie der Bevölkerungsstatistik*, Strassburg.

(1877) *Zur Theorie der Massenerscheinungen in der menschlichen Gesellschaft*, Freiburg.

Neyman, J.

(1937) "Outline of a theory of statistical estimation based on the classical theory of probability," *Phil. Trans. Roy. Soc.*, Vol. 236, pp. 333–380.

[Develops the theory of optimum confidence sets so that it reduces to the determination of optimum tests of associated classes of hypotheses.]

(1938) "L'estimation statistique traitée comme un problème classique de probabilité," *Actualités sci. et ind.*, No. 739, pp. 25–57.
(1952) *Lectures and Conferences on Mathematical Statistics*, Washington, Graduate School, U. S. Dept. Agriculture, 2nd ed. pp. 43–66.
[An account of various approaches to the problem of hypothesis testing.]

Neyman, J., and E. S. Pearson
(1928) "On the use and interpretation of certain test criteria," *Biometrika*, Vol. 20A, pp. 175–240, 263–294.
(1933) "On the problem of the most efficient tests of statistical hypotheses," *Phil. Trans. Roy. Soc., Ser. A.*, Vol. 231, pp. 289–337.
[The basic paper on the theory of hypothesis testing. Formulates the problem in terms of the two kinds of error, and develops a body of theory including the fundamental lemma. Applications including Problem 1.]
(1936) "Contributions to the theory of testing statistical hypotheses. I. Unbiased critical regions of type A and type A_1," *Stat. Res. Mem.*, Vol. 1, pp. 1–37.
[Generalization of the fundamental lemma to more than one side condition.]
(1936) "Sufficient statistics and uniformly most powerful tests of statistical hypotheses," *Stat. Res. Mem.* Vol. 1, pp. 113–137.
[Problem 2(ii).]

Pearson, Karl
(1900) "On the criterion that a given system of deviations from the probable in the case of a correlated system of variables is such that it can be reasonably supposed to have arisen from random sampling," *Phil. Mag.*, Ser. 5, Vol. 50, pp. 157–172.

Stein, C. M.
(1946) "A note on cumulative sums, "*Ann. Math. Stat.*, Vol. 17, pp. 489–499.
[Problems 34 and 35.]
(1951) "A property of some tests of composite hypotheses," *Ann. Math. Stat.*, Vol. 22, pp. 475–476.
[Problem 29.]

Thompson, W. R.
(1936) "On confidence ranges for the median and other expectation distributions for populations of unknown distribution form," *Ann. Math. Stat.*, Vol. 7, pp. 122–128.
[Problem 28.]

Tukey, John W.
(1957) "Some examples with fiducial relevance," *Ann. Math. Stat.*, Vol. 28, pp. 687–695.

Wald, Abraham
(1947) *Sequential Analysis*, New York, John Wiley & Sons.
[Theory and application of sequential probability ratio tests.]

Wald, A., and J. Wolfowitz
(1948) "Optimum character of the sequential probability ratio test," *Ann. Math. Stat.*, Vol. 19, pp. 326–339.
(1950) "Bayes solutions of sequential decision problems," *Ann. Math. Stat.*, Vol. 21, pp. 82–99.
[These papers prove the optimum properties of the sequential probability ratio test given in Section 12. In this connection see also the paper by Arrow, Blackwell, and Girshick.]

Wilson, E. B.
(1927) "Probable inference, the law of succession, and statistical inference," *J. Am. Stat. Assoc.*, Vol. 22, pp. 209–212.

Wolfowitz, J.
(1947) "The efficiency of sequential estimates and Wald's equation for sequential processes," *Ann. Math. Stat.*, Vol. 18, pp. 215–230.
[The proof of Wald's equation (45) given in Problem 37.]

Working, Holbrook, and Harold Hotelling
(1929) "Applications of the theory of error to the interpretation of trends," *J. Am. Stat. Assoc.*, Suppl., Vol. 24, pp. 73–85.

CHAPTER 4

Unbiasedness: Theory and First Applications

1. UNBIASEDNESS FOR HYPOTHESIS TESTING

A simple condition that one may wish to impose on tests of the hypothesis $H: \theta \in \Omega_H$ against the composite class of alternatives $K: \theta \in \Omega_K$ is that for no alternative in K the probability of rejection should be less than the size of the test. Unless this condition is satisfied there will exist alternatives under which acceptance of the hypothesis is more likely than in some cases in which the hypothesis is true. A test ϕ for which the above condition holds, that is, for which the power function $\beta_\phi(\theta) = E_\theta \phi(X)$ satisfies

$$\begin{aligned} &\beta_\phi(\theta) \leqq \alpha \quad \text{if} \quad \theta \in \Omega_H \\ &\beta_\phi(\theta) \geqq \alpha \quad \text{if} \quad \theta \in \Omega_K \end{aligned} \tag{1}$$

is said to be *unbiased.* For an appropriate loss function this was seen in Chapter 1 to be a particular case of the general definition of unbiasedness given there. Whenever a UMP test exists, it is unbiased since its power cannot fall below that of the test $\phi(x) \equiv \alpha$.

For a large class of problems for which a UMP test does not exist, there does exist a UMP unbiased test. This is the case in particular for certain hypotheses of the form $\theta \leqq \theta_0$ or $\theta = \theta_0$, where the distribution of the random observables depends on other parameters besides θ.

When $\beta_\phi(\theta)$ is a continuous function of θ, unbiasedness implies

$$\beta_\phi(\theta) = \alpha \quad \text{for all} \quad \theta \text{ in } \omega \tag{2}$$

where ω is the common boundary of Ω_H and Ω_K, that is, the set of all points θ that are points or limit points of both Ω_H and Ω_K. Tests satisfying this condition are said to be *similar on the boundary* (of H and K). Since it is more convenient to work with (2) than with (1), the following lemma plays an important role in the determination of UMP unbiased tests.

Lemma 1. *If the distributions P_θ are such that the power function of every test is continuous, and if ϕ_0 is UMP among all tests satisfying* (2) *and is a level α test of H, then ϕ_0 is UMP unbiased.*

Proof. The class of tests satisfying (2) contains the class of unbiased tests, and hence ϕ_0 is uniformly at least as powerful as any unbiased test. On the other hand, ϕ_0 is unbiased since it is uniformly at least as powerful as $\phi(x) \equiv \alpha$.

2. ONE-PARAMETER EXPONENTIAL FAMILIES

Let θ be a real parameter, and $X = (X_1, \cdots, X_n)$ a random vector with probability density (with respect to some measure μ)

$$p_\theta(x) = C(\theta)e^{\theta T(x)}h(x).$$

It was seen in Chapter 3 that a UMP test exists when the hypothesis H and the class K of alternatives are given by (i) $H\colon \theta \leqq \theta_0$, $K\colon \theta > \theta_0$ (Corollary 2) and (ii) $H\colon \theta \leqq \theta_1$ or $\theta \geqq \theta_2$ $(\theta_1 < \theta_2)$, $K\colon \theta_1 < \theta < \theta_2$ (Theorem 6) but not for (iii) $H\colon \theta_1 \leqq \theta \leqq \theta_2$, $K\colon \theta < \theta_1$ or $\theta > \theta_2$. We shall now show that in case (iii) there does exist a UMP unbiased test given by

$$\phi(x) = \begin{cases} 1 & \text{when} \quad T(x) < C_1 \text{ or } > C_2 \\ \gamma_i & \text{when} \quad T(x) = C_i, \qquad i = 1, 2 \\ 0 & \text{when} \quad C_1 < T(x) < C_2, \end{cases} \tag{3}$$

where the C's and γ's are determined by

$$E_{\theta_1}\phi(X) = E_{\theta_2}\phi(X) = \alpha. \tag{4}$$

The power function $E_\theta\,\phi(X)$ is continuous by Theorem 9 of Chapter 2, so that Lemma 1 is applicable. The set ω consists of the two points θ_1 and θ_2, and we therefore consider first the problem of maximizing $E_{\theta'}\,\phi(X)$ for some θ' outside the interval $[\theta_1, \theta_2]$, subject to (4). If this problem is restated in terms of $1 - \phi(x)$, it follows from part (ii) of Theorem 6, Chapter 3, that its solution is given by (3) and (4). This test is therefore UMP among those satisfying (4), and hence UMP unbiased by Lemma 1. It further follows from part (iii) of the theorem that the power function of the test has a minimum at a point between θ_1 and θ_2, and is strictly increasing as θ tends away from this minimum in either direction.

A closely related problem is that of testing $H\colon \theta = \theta_0$ against the alternatives $\theta \neq \theta_0$. For this there also exists a UMP unbiased test given by (3), but the constants are now determined by

$$E_{\theta_0}[\phi(X)] = \alpha \tag{5}$$

and

$$E_{\theta_0}[T(X)\phi(X)] = E_{\theta_0}[T(X)]\alpha. \tag{6}$$

To see this, let θ' be any particular alternative and restrict attention to the sufficient statistic T, the distribution of which by Chapter 2, Lemma 8, is of the form

$$dP_\theta(t) = C(\theta)e^{\theta t}\, d\nu(t).$$

Unbiasedness of a test $\psi(t)$ implies (5) with $\phi(x) = \psi[T(x)]$; also that the power function $\beta(\theta) = E_\theta[\psi(T)]$ must have a minimum at $\theta = \theta_0$. By Theorem 9 of Chapter 2 the function $\beta(\theta)$ is differentiable, and the derivative can be computed by differentiating $E_\theta\psi(T)$ under the expectation sign, so that for all tests $\psi(t)$

$$\beta'(\theta) = E_\theta[T\psi(T)] + \frac{C'(\theta)}{C(\theta)}\, E_\theta[\psi(T)].$$

For $\psi(t) \equiv \alpha$, this equation becomes

$$0 = E_\theta(T) + \frac{C'(\theta)}{C(\theta)}\cdot$$

Substituting this in the expression for $\beta'(\theta)$ gives

$$\beta'(\theta) = E_\theta[T\psi(T)] - E_\theta(T)E_\theta[\psi(T)],$$

and hence unbiasedness implies (6) in addition to (5).

Let M be the set of points $(E_{\theta_0}[\psi(T)], E_{\theta_0}[T\psi(T)])$ as ψ ranges over the totality of critical functions. Then M is convex and contains all points $(u, uE_{\theta_0}(T))$ with $0 < u < 1$. It also contains points (α, u_2) with $u_2 > \alpha E_{\theta_0}(T)$. This follows from the fact that there exist tests with $E_{\theta_0}[\psi(T)] = \alpha$ and $\beta'(\theta_0) > 0$ (see Problem 18 of Chapter 3). Since similarly M contains points (α, u_1) with $u_1 < \alpha E_{\theta_0}(T)$, the point $(\alpha, \alpha E_{\theta_0}(T))$ is an inner point of M. Therefore, by Theorem 5(iv) of Chapter 3 there exist constants k_1, k_2 and a test $\psi(t)$ satisfying (5) and (6) with $\phi(x) = \psi[T(x)]$, such that $\psi(t) = 1$ when

$$C(\theta_0)(k_1 + k_2 t)e^{\theta_0 t} < C(\theta')\, e^{\theta' t}$$

and therefore, when

$$a_1 + a_2 t < e^{bt}.$$

This region is either one-sided or the outside of an interval. By Theorem 2(ii) of Chapter 3 a one-sided test has a strictly monotone power function and therefore cannot satisfy (6). Thus $\psi(t)$ is 1 when $t < C_1$ or $> C_2$, and the most powerful test subject to (5) and (6) is given by (3). This test is unbiased, as is seen by comparing it with $\phi(x) \equiv \alpha$. It is then also UMP unbiased since the class of tests satisfying (5) and (6) includes the class of unbiased tests.

A simplification of this test is possible if for $\theta = \theta_0$ the distribution of T is symmetric about some point a, that is, if $P_{\theta_0}\{T < a - u\} = P_{\theta_0}\{T > a + u\}$ for all real u. Any test which is symmetric about a

and satisfies (5) must also satisfy (6) since $E_{\theta_0}[T\psi(T)] = E_{\theta_0}[(T - a)\psi(T)] + aE_{\theta_0}\psi(T) = a\alpha = E_{\theta_0}(T)\alpha$. The C's and γ's are therefore determined by

$$P_{\theta_0}\{T < C_1\} + \gamma_1 P_{\theta_0}\{T = C_1\} = \alpha/2$$

$$C_2 = 2a - C_1, \qquad \gamma_2 = \gamma_1.$$

The above tests of the hypotheses $\theta_1 \leqq \theta \leqq \theta_2$ and $\theta = \theta_0$ are *strictly unbiased* in the sense that the power is $>\alpha$ for all alternatives θ. For the first of these tests, given by (3) and (4), strict unbiasedness is an immediate consequence of Theorem 6(iii) of Chapter 3. This states in fact that the power of the test has a minimum at a point θ_0 between θ_1 and θ_2 and increases strictly as θ tends away from θ_0 in either direction. The second of the tests, determined by (3), (5), and (6), has a continuous power function with a minimum of α at $\theta = \theta_0$. Thus there exist $\theta_1 < \theta_0 < \theta_2$ such that $\beta(\theta_1) = \beta(\theta_2) = c$ where $\alpha \leqq c < 1$. The test therefore coincides with the UMP unbiased level c test of the hypothesis $\theta_1 \leqq \theta \leqq \theta_2$, and the power increases strictly as θ moves away from θ_0 in either direction. This proves the desired result.

Example 1. Let X be the number of successes in n binomial trials with probability p of success. A theory to be tested assigns to p the value p_0 so that one wishes to test the hypothesis H: $p = p_0$. When rejecting H one will usually wish to state also whether p appears to be less or greater than p_0. If, however, the conclusion that $p \neq p_0$ in any case requires further investigation, the preliminary decision is essentially between the two possibilities that the data do or do not contradict the hypothesis $p = p_0$. The formulation of the problem as one of hypothesis testing may then be appropriate.

The UMP unbiased test of H is given by (3) with $T(X) = X$. Condition (5) becomes

$$\sum_{x=C_1+1}^{C_2-1} \binom{n}{x} p_0^x q_0^{n-x} + \sum_{i=1}^{2} (1 - \gamma_i) \binom{n}{C_i} p_0^{C_i} q_0^{n-C_i} = 1 - \alpha,$$

and the left-hand side of this can be obtained from tables of the individual probabilities and cumulative distribution function of X. Condition (6), with the help of the identity

$$x\binom{n}{x} p_0^x q_0^{n-x} = np_0 \binom{n-1}{x-1} p_0^{x-1} q_0^{(n-1)-(x-1)}$$

reduces to

$$\sum_{x=C_1+1}^{C_2-1} \binom{n-1}{x-1} p_0^{x-1} q_0^{(n-1)-(x-1)} + \sum_{i=1}^{2} (1 - \gamma_i) \binom{n-1}{C_i - 1} p_0^{C_i-1} q_0^{(n-1)-(C_i-1)} = 1 - \alpha,$$

the left-hand side of which can be computed from the binomial tables.

As n increases, the distribution of $(X - np_0)/\sqrt{np_0q_0}$ tends to the normal distribution $N(0, 1)$. For sample sizes which are not too small, and values of p_0 which are not too close to 0 or 1, the distribution of X is therefore approximately symmetric with respect to the origin. In this case, the much simpler "equal tails" test, for which the C's and γ's are determined by

$$\sum_{x=0}^{C_1-1} \binom{n}{x} p_0^x q_0^{n-x} + \gamma_1 \binom{n}{C_1} p_0^{C_1} q_0^{n-C_1}$$
$$= \gamma_2 \binom{n}{C_2} p_0^{C_2} q_0^{n-C_2} + \sum_{x=C_2+1}^{n} \binom{n}{x} p_0^x q_0^{n-x} = \frac{\alpha}{2},$$

is approximately unbiased, and constitutes a reasonable approximation to the unbiased test. Of course, when n is sufficiently large, the constants can be determined directly from the normal distribution.

Example 2. Let $X = (X_1, \cdots, X_n)$ be a sample from a normal distribution with mean 0 and variance σ^2, so that the density of the X's is

$$\left(\frac{1}{\sqrt{2\pi}\sigma}\right)^n \exp\left(-\frac{1}{2\sigma^2}\sum x_i^2\right) \cdot$$

Then $T(x) = \Sigma x_i^2$ is sufficient for σ^2, and has probability density $(1/\sigma^2)f_n(y/\sigma^2)$, where

$$f_n(y) = \frac{1}{2^{n/2}\Gamma(n/2)} y^{(n/2)-1} e^{-(y/2)}, \qquad y > 0$$

is the density of a χ^2-distribution with n degrees of freedom. For varying σ, these distributions form an exponential family, which arises also in problems of life testing (see Problem 13 of Chapter 2), and concerning normally distributed variables with unknown mean and variance (Section 3 of Chapter 5). The acceptance region of the UMP unbiased test of the hypothesis H: $\sigma = \sigma_0$ is

$$C_1 \leqq \Sigma x_i^2/\sigma_0^2 \leqq C_2$$

with

$$\int_{C_1}^{C_2} f_n(y)\, dy = 1 - \alpha$$

and

$$\int_{C_1}^{C_2} y f_n(y)\, dy = (1 - \alpha)E_{\sigma_0}(\Sigma X_i^2)/\sigma_0^2 = n(1 - \alpha).$$

For the determination of the constants from tables of the χ^2-distribution, it is convenient to use the identity

$$y f_n(y) = n f_{n+2}(y),$$

to rewrite the second condition as

$$\int_{C_1}^{C_2} f_{n+2}(y)\, dy = 1 - \alpha.$$

Alternatively, one can integrate $\int_{C_1}^{C_2} y f_n(y)\,dy$ by parts to reduce the second condition to

$$C_1^{n/2}e^{-C_1/2} = C_2^{n/2}e^{-C_2/2}.$$

Actually, unless n is very small or σ_0 very close to 0 or ∞, the equal tails test given by

$$\int_0^{C_1} f_n(y)\,dy = \int_{C_2}^{\infty} f_n(y)\,dy = \frac{\alpha}{2}$$

is a good approximation to the unbiased test. This follows from the fact that T, suitably normalized, tends to be normally and hence symmetrically distributed for large n.

3. SIMILARITY AND COMPLETENESS

In many important testing problems, the hypothesis concerns a single real-valued parameter, but the distribution of the observable random variables depends in addition on certain nuisance parameters. For a large class of such problems a UMP unbiased test exists and can be found through the method indicated by Lemma 1. This requires the characterization of the tests ϕ, which satisfy

$$E_\theta\,\phi(X) = \alpha$$

for all distributions of X belonging to a given family $\mathscr{P}^X = \{P_\theta, \theta \in \omega\}$. Such tests are called *similar* with respect to $\mathscr{P}^X$ or ω since if ϕ is nonrandomized with critical region S, the latter is "similar to the sample space" $\mathscr{X}$ in that both the probability $P_\theta\{X \in S\}$ and $P_\theta\{X \in \mathscr{X}\}$ are independent of $\theta \in \omega$.

Let T be a sufficient statistic for $\mathscr{P}^X$, and let $\mathscr{P}^T$ denote the family $\{P_\theta^T, \theta \in \omega\}$ of distributions of T as θ ranges over ω. Then any test satisfying

$$E[\phi(X)|t] = \alpha \qquad \text{a.e. } \mathscr{P}^T\dagger \tag{7}$$

is similar with respect to $\mathscr{P}^X$ since then

$$E_\theta[\phi(X)] = E_\theta\{E[\phi(X)|T]\} = \alpha \quad \text{for all} \quad \theta \in \omega.$$

A test satisfying (7) is said to have *Neyman structure* with respect to T. It is characterized by the fact that the conditional probability of rejection is α on each of the surfaces $T = t$. Since the distribution on each such surface is independent of θ for $\theta \in \omega$, condition (7) essentially reduces the problem to that of testing a simple hypothesis for each value of t.

† A statement is said to hold a.e. $\mathscr{P}$ if it holds except on a set N with $P(N) = 0$ for all $P \in \mathscr{P}$.

It is frequently easy to obtain a most powerful test among those having Neyman structure, by solving the optimum problem on each surface separately. The resulting test is then most powerful among all similar tests provided every similar test has Neyman structure. A condition for this to be the case can be given in terms of the following definition.

A family $\mathscr{P}$ of probability distributions P is *complete* if

$$E_P[f(X)] = 0 \quad \text{for all} \quad P \in \mathscr{P} \tag{8}$$

implies

$$f(x) = 0 \qquad \text{a.e. } \mathscr{P}. \tag{9}$$

In applications, $\mathscr{P}$ will be the family of distributions of a sufficient statistic.

Example 3. Consider n independent trials with probability p of success, and let X_i be 1 or 0 as the ith trial is a success or failure. Then $T = X_1 + \cdots + X_n$ is a sufficient statistic for p, and the family of its possible distributions is $\mathscr{P} = \{b(p, n), 0 \leqq p \leqq 1\}$. For this family (8) implies that

$$\sum_{t=0}^{n} f(t)\binom{n}{t}\rho^t = 0 \quad \text{for all} \quad 0 < \rho < \infty$$

where $\rho = p/(1-p)$. The left-hand side is a polynomial in ρ, all the coefficients of which must be zero. Hence $f(t) = 0$ for $t = 0, \cdots, n$ and the binomial family of distributions of T is complete.

Example 4. Let $X_1, \cdots, X_n$ be a sample from the uniform distribution $R(0, \theta)$, $0 < \theta < \infty$. Then $T = \max(X_1, \cdots, X_n)$ is a sufficient statistic for θ, and (8) becomes

$$\int f(t)\, dP_\theta^T(t) = n\theta^{-n}\int_0^\theta f(t)\cdot t^{n-1}\, dt = 0 \quad \text{for all} \quad \theta.$$

Let $f(t) = f^+(t) - f^-(t)$ where f^+ and f^- denote the positive and negative parts of f respectively. Then

$$\nu^+(A) = \int_A f^+(t)t^{n-1}\, dt \qquad \text{and} \qquad \nu^-(A) = \int_A f^-(t)t^{n-1}\, dt$$

are two measures over the Borel sets on $(0, \infty)$, which agree for all intervals and hence for all A. This implies $f^+(t) = f^-(t)$ except possibly on a set of Lebesgue measure zero, and hence $f(t) = 0$ a.e. $\mathscr{P}^T$.

Example 5. Let $X_1, \cdots, X_m$; $Y_1 \cdots, Y_n$ be independently normally distributed as $N(\xi, \sigma^2)$ and $N(\xi, \tau^2)$ respectively. Then the joint density of the variables is

$$C(\xi, \sigma, \tau)\exp\left(-\frac{1}{2\sigma^2}\Sigma x_i^2 + \frac{\xi}{\sigma^2}\Sigma x_i - \frac{1}{2\tau^2}\Sigma y_j^2 + \frac{\xi}{\tau^2}\Sigma y_j\right).$$

The statistic

$$T = (\Sigma X_i, \Sigma X_i^2, \Sigma Y_j, \Sigma Y_j^2)$$

is sufficient; it is, however, not complete, since $E(\Sigma Y_j/n - \Sigma X_i/m)$ is identically zero. If the Y's are instead distributed with a mean $E(Y) = \eta$ which varies independently of ξ, the set of possible values of the parameters $\theta_1 = -1/2\sigma^2$, $\theta_2 = \xi/\sigma^2$, $\theta_3 = -1/2\tau^2$, $\theta_4 = \eta/\tau^2$ contains a four-dimensional rectangle, and it follows from Theorem 1 below that $\mathscr{P}^T$ is complete.

Completeness of a large class of families of distributions including that of Example 3 is covered by the following theorem.

Theorem 1. *Let X be a random vector with probability distribution*

$$dP_\theta(x) = C(\theta) \exp\left[\sum_{j=1}^{k} \theta_j T_j(x)\right] d\mu(x)$$

and let $\mathscr{P}^T$ be the family of distributions of $T = (T_1(X), \cdots, T_k(X))$ as θ ranges over the set ω. Then $\mathscr{P}^T$ is complete provided ω contains a k-dimensional rectangle.

Proof. By making a translation of the parameter space one can assume without loss of generality that ω contains the rectangle

$$I = \{(\theta_1, \cdots, \theta_k): -a \leqq \theta_j \leqq a, j = 1, \cdots, k\}.$$

Let $f(t) = f^+(t) - f^-(t)$ be such that

$$E_\theta f(T) = 0 \quad \text{for all} \quad \theta \in \omega.$$

Then for all $\theta \in I$, if ν denotes the measure induced in T-space by the measure μ,

$$\int e^{\Sigma\theta_j t_j} f^+(t)\, d\nu(t) = \int e^{\Sigma\theta_j t_j} f^-(t)\, d\nu(t)$$

and hence in particular

$$\int f^+(t)\, d\nu(t) = \int f^-(t)\, d\nu(t).$$

Dividing f by a constant, one can take the common value of these two integrals to be 1, so that

$$dP^+(t) = f^+(t)\, d\nu(t) \qquad \text{and} \qquad dP^-(t) = f^-(t)\, d\nu(t)$$

are probability measures, and

$$\int e^{\Sigma\theta_j t_j}\, dP^+(t) = \int e^{\Sigma\theta_j t_j}\, dP^-(t)$$

for all θ in I. Changing the point of view, consider these integrals now as functions of the complex variables $\theta_j = \xi_j + i\eta_j$, $j = 1, \cdots, k$. For any fixed $\theta_1, \cdots, \theta_{j-1}, \theta_{j+1}, \cdots, \theta_k$, with real parts strictly between $-a$ and $+a$, they are by Theorem 9 of Chapter 2 analytic functions of θ_j in the strip R_j: $-a < \xi_j < a$, $-\infty < \eta_j < \infty$ of the complex plane.

For $\theta_2, \cdots, \theta_k$ fixed, real, and between $-a$ and a, equality of the integrals holds on the line segment $\{(\xi_1, \eta_1): -a < \xi_1 < a, \eta_1 = 0\}$ and can therefore be extended to the strip R_1, in which the integrals are analytic. By induction the equality can be extended to the complex region $\{(\theta_1, \cdots, \theta_k): (\xi_j, \eta_j) \in R_j \text{ for } j = 1, \cdots, k\}$. It follows in particular that for all real $(\eta_1, \cdots, \eta_k)$

$$\int e^{i\Sigma\eta_j t_j}\, dP^+(t) = \int e^{i\Sigma\eta_j t_j}\, dP^-(t).$$

These integrals are the characteristic functions of the distributions P^+ and P^- respectively, and by the uniqueness theorem for characteristic functions,* the two distributions P^+ and P^- coincide. From the definition of these distributions it then follows that $f^+(t) = f^-(t)$, a.e. ν, and hence that $f(t) = 0$ a.e. $\mathscr{P}^T$, as was to be proved.

Example 6. Let $X_1, \cdots, X_N$ be independently and identically distributed with cumulative distribution function $F \in \mathscr{F}$, where $\mathscr{F}$ is the family of all continuous distributions. Then the set of order statistics $T(X) = (X^{(1)}, \cdots, X^{(N)})$ was shown to be sufficient for $\mathscr{F}$ in Chapter 2, Section 6. We shall now prove it to be complete. Since $T'(X) = (\Sigma X_i, \Sigma X_i^2, \cdots, \Sigma X_i^N)$ is equivalent to $T(X)$ in the sense that both induce the same subfield of the sample space, $T'(X)$ is also sufficient and is complete if and only if $T(X)$ is complete. To prove the completeness of $T'(X)$ and thereby that of $T(X)$, consider the family of densities

$$f(x) = C(\theta_1, \cdots, \theta_N) \exp(-x^{2N} + \theta_1 x + \cdots + \theta_N x^N)$$

where C is a normalizing constant. These densities are defined for all values of the θ's since the integral of the exponential is finite, and being continuous they belong to $\mathscr{F}$. The density of a sample of size N is

$$C^N \exp(-\Sigma x_j^{2N} + \theta_1 \Sigma x_j + \cdots + \theta_N \Sigma x_j^N)$$

and these densities constitute an exponential family $\mathscr{F}_0$. By Theorem 1, $T'(X)$ is complete for $\mathscr{F}_0$, and hence also for $\mathscr{F}$, as was to be proved. (For an alternative proof, see Problems 12, 13.)

The same method of proof establishes also the following more general result. Let X_{ij}, $j = 1, \cdots, N_i$; $i = 1, \cdots, c$, be independently distributed with continuous distributions F_i, and let $X_i^{(1)} < \cdots < X_i^{(N_i)}$ denote the N_i observations $X_{i1}, \cdots, X_{iN_i}$ arranged in increasing order. Then the set of order statistics

$$(X_1^{(1)}, \cdots, X_1^{(N_1)}; \cdots; X_c^{(1)}, \cdots, X_c^{(N_c)})$$

is sufficient and complete for the family of distributions obtained by letting $F_1, \cdots, F_c$ range over all distributions of $\mathscr{F}$. Here completeness is proved by considering the subfamily $\mathscr{F}_0$ of $\mathscr{F}$ in which the distributions F_i have densities of the form

$$f_i(x) = C_i(\theta_{i1}, \cdots, \theta_{iN_i}) \exp(-x^{2N_i} + \theta_{i1} x + \cdots + \theta_{iN_i} x^{N_i}).$$

* See, for example, section 10.6 of Cramér: *Mathematical Methods of Statistics*, Princeton Univ. Press, Princeton, 1946.

For the present purpose the slightly weaker property of bounded completeness is appropriate, a family $\mathscr{P}$ of probability distributions being *boundedly complete* if for all bounded functions f, (8) implies (9). If $\mathscr{P}$ is complete it is a fortiori boundedly complete.

Theorem 2. *Let X be a random variable with distribution $P \in \mathscr{P}$ and let T be a sufficient statistic for $\mathscr{P}$. Then a necessary and sufficient condition for all similar tests to have Neyman structure with respect to T is that the family $\mathscr{P}^T$ of distributions of T is boundedly complete.*

Proof. Suppose first that $\mathscr{P}^T$ is boundedly complete, and let $\phi(X)$ be similar with respect to $\mathscr{P}$. Then

$$E[\phi(X) - \alpha] = 0 \quad \text{for all} \quad P \in \mathscr{P}$$

and hence, if $\psi(t)$ denotes the conditional expectation of $\phi(X) - \alpha$ given t,

$$E\psi(T) = 0 \quad \text{for all} \quad P^T \in \mathscr{P}^T.$$

Since $\psi(t)$ can be taken to be bounded by Lemma 3 of Chapter 2, it follows from the bounded completeness of $\mathscr{P}^T$ that $\psi(t) = 0$ and hence $E[\phi(X)|t] = \alpha$ a.e. $\mathscr{P}^T$, as was to be proved.

Conversely suppose that $\mathscr{P}^T$ is not boundedly complete. Then there exists a function f such that $|f(t)| \leq M$ for some M, that $Ef(T) = 0$ for all $P^T \in \mathscr{P}^T$, and $f(T) \neq 0$ with positive probability for some $P^T \in \mathscr{P}^T$. Let $\phi(t) = cf(t) + \alpha$ where $c = \min(\alpha, 1 - \alpha)/M$. Then ϕ is a critical function since $0 \leq \phi(t) \leq 1$, and it is a similar test since $E\phi(T) = \alpha$ for all $P^T \in \mathscr{P}^T$. But ϕ does not have Neyman structure since $\phi(T) \neq \alpha$ with positive probability for at least some distribution in $\mathscr{P}^T$.

4. UMP UNBIASED TESTS FOR MULTIPARAMETER EXPONENTIAL FAMILIES

An important class of hypotheses concerns a real-valued parameter in an exponential family, with the remaining parameters occurring as unspecified nuisance parameters. In many of these cases, UMP unbiased tests exist and can be constructed by means of the theory of the preceding section.

Let X be distributed according to

$$(10) \quad dP^X_{\theta,\vartheta}(x) = C(\theta, \vartheta) \exp\left[\theta U(x) + \sum_{i=1}^{k} \vartheta_i T_i(x)\right] d\mu(x), \qquad (\theta, \vartheta) \in \Omega$$

and let $\vartheta = (\vartheta_1, \cdots, \vartheta_k)$ and $T = (T_1, \cdots, T_k)$. We shall consider the problems of testing the following hypotheses H_j against the alternatives $K_j, j = 1, \cdots 4$:

$$
\begin{array}{ll}
H_1: \theta \leqq \theta_0 & K_1: \theta > \theta_0 \\
H_2: \theta \leqq \theta_1 \text{ or } \theta \geqq \theta_2 & K_2: \theta_1 < \theta < \theta_2 \\
H_3: \theta_1 \leqq \theta \leqq \theta_2 & K_3: \theta < \theta_1 \text{ or } \theta > \theta_2 \\
H_4: \theta = \theta_0 & K_4: \theta \neq \theta_0.
\end{array}
$$

We shall assume that the parameter space Ω is convex, and that it has dimension $k + 1$, that is, that it is not contained in a linear space of dimension $<k + 1$. This is the case in particular when Ω is the natural parameter space of the exponential family. We shall also assume that there are points in Ω with θ both $<$ and $>\theta_0$, θ_1, and θ_2 respectively.

Attention can be restricted to the sufficient statistics (U, T) which have the joint distribution

$$
(11) \quad dP_{\theta,\vartheta}^{U,T}(u, t) = C(\theta, \vartheta) \exp\left(\theta u + \sum_{i=1}^{k} \vartheta_i t_i\right) d\nu(u, t), \qquad (\theta, \vartheta) \in \Omega.
$$

When $T = t$ is given, U is the only remaining variable and by Lemma 8 of Chapter 2 the conditional distribution of U given t constitutes an exponential family

$$
dP_\theta^{U|t}(u) = C_t(\theta) \exp(\theta u)\, d\nu_t(u).
$$

In this conditional situation there exists by Corollary 2 of Chapter 3 a UMP test for testing H_1 with critical function ϕ_1 satisfying

$$
(12) \qquad \phi(u, t) = \begin{cases} 1 & \text{when} \quad u > C_0(t) \\ \gamma_0(t) & \text{when} \quad u = C_0(t) \\ 0 & \text{when} \quad u < C_0(t) \end{cases}
$$

where the functions C_0 and γ_0 are determined by

$$
(13) \qquad E_{\theta_0}[\phi_1(U, T)|t] = \alpha \quad \text{for all} \quad t.
$$

For testing H_2 in the conditional family there exists by Theorem 6 of Chapter 3 a UMP test with critical function

$$
(14) \qquad \phi(u, t) = \begin{cases} 1 & \text{when} \quad C_1(t) < u < C_2(t) \\ \gamma_i(t) & \text{when} \quad u = C_i(t), \qquad i = 1, 2 \\ 0 & \text{when} \quad u < C_1(t) \text{ or } > C_2(t) \end{cases}
$$

where the C's and γ's are determined by

$$
(15) \qquad E_{\theta_1}[\phi_2(U, T)|t] = E_{\theta_2}[\phi_2(U, T)|t] = \alpha.
$$

Consider next the test ϕ_3 satisfying

$$\phi(u, t) = \begin{cases} 1 & \text{when} \quad u < C_1(t) \text{ or } > C_2(t) \\ \gamma_i(t) & \text{when} \quad u = C_i(t), \qquad i = 1, 2 \\ 0 & \text{when} \quad C_1(t) < u < C_2(t) \end{cases} \tag{16}$$

with the C's and γ's determined by

$$E_{\theta_1}[\phi_3(U, T)|t] = E_{\theta_2}[\phi_3(U, T)|t] = \alpha. \tag{17}$$

When $T = t$ is given, this is by Section 2 of the present chapter UMP unbiased for testing H_3 and UMP among all tests satisfying (17).

Finally, let ϕ_4 be a critical function satisfying (16) with the C's and γ's determined by

$$E_{\theta_0}[\phi_4(U, T)|t] = \alpha \tag{18}$$

and

$$E_{\theta_0}[U\phi_4(U, T)|t] = \alpha E_{\theta_0}[U|t]. \tag{19}$$

Then given $T = t$, it follows again from the results of Section 2 that ϕ_4 is UMP unbiased for testing H_4 and UMP among all tests satisfying (18) and (19).

So far, the critical functions ϕ_j have been considered as conditional tests given $T = t$. Reinterpreting them now as tests depending on U and T for the hypotheses concerning the distribution of X (or the joint distribution of U and T) as originally stated, we have the following main theorem.

Theorem 3. *Define the critical functions:* ϕ_1 *by* (12) *and* (13); ϕ_2 *by* (14) *and* (15); ϕ_3 *by* (16) *and* (17); ϕ_4 *by* (16), (18), *and* (19). *These constitute UMP unbiased level* α *tests for testing the hypotheses* $H_1, \cdots, H_4$ *respectively when the joint distribution of* U *and* T *is given by* (11).

Proof. The statistic T is sufficient for ϑ if θ has any fixed value, and hence T is sufficient for each

$$\omega_j = \{(\theta, \vartheta):\ (\theta, \vartheta) \in \Omega,\ \theta = \theta_j\}, \qquad j = 0, 1, 2.$$

By Lemma 8 of Chapter 2, the associated family of distributions of T is given by

$$dP^T_{\theta_j,\vartheta}(t) = C(\theta_j, \vartheta) \exp\left(\sum_{i=1}^{k} \vartheta_i t_i\right) d\nu_\theta(t), \qquad (\theta_j, \vartheta) \in \omega_j; \quad j = 0, 1, 2.$$

Since by assumption Ω is convex and of dimension $k + 1$ and contains points on both sides of $\theta = \theta_j$, it follows that ω_j is convex and of dimension

k. Thus ω_j contains a k-dimensional rectangle; by Theorem 1 the family

$$\mathscr{P}_j^T = \{P_{\theta_j,\vartheta}^T: (\theta, \vartheta) \in \omega_j\}$$

is complete; and similarity of a test ϕ on ω_j implies

$$E_{\theta_j}[\phi(U, T)|t] = \alpha.$$

(1) Consider first H_1. By Theorem 6 of Chapter 2 the power function of all tests is continuous for an exponential family. It is therefore enough to prove ϕ_1 to be UMP among all tests that are similar on ω_0 (Lemma 1), and hence among those satisfying (13). On the other hand, the over-all power of a test ϕ against an alternative (θ, ϑ) is

$$E_{\theta,\vartheta}[\phi(U, T)] = \int \left[\int \varphi(u, t)\, dP_\theta^{U|t}(u) \right] dP_{\theta,\vartheta}^T(t). \tag{20}$$

One therefore maximizes the over-all power by maximizing the power of the conditional test, given by the expression in brackets, separately for each t. Since ϕ_1 has the property of maximizing the conditional power against any $\theta > \theta_0$ subject to (13), this establishes the desired result.

(2) The proof for H_2 and H_3 is completely analogous. By Lemma 1, it is enough to prove ϕ_2 and ϕ_3 to be UMP among all tests that are similar on both ω_1 and ω_2, and hence among all tests satisfying (15). For each t, ϕ_2 and ϕ_3 maximize the conditional power for their respective problems subject to this condition and therefore also the unconditional power.

(3) Unbiasedness of a test of H_4 implies similarity on ω_0 and

$$\frac{\partial}{\partial\theta}[E_{\theta,\vartheta}\phi(U, T)] = 0 \quad \text{on} \quad \omega_0.$$

The differentiation on the left-hand side of this equation can be carried out under the expectation sign, and by the computation which earlier led to (6), the equation is seen to be equivalent to

$$E_{\theta,\vartheta}[U\phi(U, T) - \alpha U] = 0 \quad \text{on} \quad \omega_0.$$

Therefore, since $\mathscr{P}_0^T$ is complete, unbiasedness implies (18) and (19). As in the preceding cases the test, which in addition satisfies (16), is UMP among all tests satisfying these two conditions. That it is UMP unbiased now follows, as in the proof of Lemma 1, by comparison with the test $\phi(u, t) \equiv \alpha$.

(4) The functions $\phi_1, \cdots, \phi_4$ were obtained above for each fixed t as a function of u. To complete the proof it is necessary to show that they

are jointly measurable in u and t, so that the expectation (20) exists. We shall prove this here for the case of ϕ_1; the proof for the other cases is sketched in Problems 14 and 15. To establish the measurability of ϕ_1, one needs to show that the functions $C_0(t)$ and $\gamma_0(t)$ defined by (12) and (13) are t-measurable. Omitting the subscript 0, and denoting the conditional distribution function of U given $T = t$ and for $\theta = \theta_0$ by

$$F_t(u) = P_{\theta_0}\{U \leqq u | t\}$$

one can rewrite (13) as

$$F_t(C) - \gamma[F_t(C) - F_t(C - 0)] = 1 - \alpha.$$

Here $C = C(t)$ is such that $F_t(C - 0) \leqq 1 - \alpha \leqq F_t(C)$, and hence

$$C(t) = F_t^{-1}(1 - \alpha)$$

where $F_t^{-1}(y) = \inf\{u: F_t(u) \geqq y\}$. It follows that $C(t)$ and $\gamma(t)$ will both be measurable provided $F_t(u)$ and $F_t(u - 0)$ are jointly measurable in u and t and $F_t^{-1}(1 - \alpha)$ is measurable in t.

For each fixed u the function $F_t(u)$ is a measurable function of t, and for each fixed t it is a cumulative distribution function and therefore in particular nondecreasing and continuous on the right. From the second property it follows that $F_t(u) \geqq c$ if and only if for each n there exists a rational number r such that $u \leqq r < u + 1/n$ and $F_t(r) \geqq c$. Therefore, if the rationals are denoted by $r_1, r_2, \cdots$,

$$\{(u, t): F_t(u) \geqq c\} = \bigcap_n \bigcup_i \{(u, t): 0 \leqq r_i - u < 1/n, F_t(r_i) \geqq c\}.$$

This shows that $F_t(u)$ is jointly measurable in u and t. The proof for $F_t(u - 0)$ is completely analogous. Since $F_t^{-1}(y) \leqq u$ if and only if $F_t(u) \geqq y$, $F_t^{-1}(y)$ is t-measurable for any fixed y and this completes the proof.

The test ϕ_1 of the above theorem is also UMP unbiased if Ω is replaced by the set $\Omega' = \Omega \cap \{(\theta, \vartheta): \theta \geqq \theta_0\}$ and hence for testing $H': \theta = \theta_0$ against $\theta > \theta_0$. The assumption that Ω should contain points with $\theta < \theta_0$ was in fact used only to prove that the boundary set ω_0 contains a k-dimensional rectangle, and this remains valid if Ω is replaced by Ω'.

The remainder of this chapter as well as the next chapter will be concerned mainly with applications of the preceding theorem to various statistical problems. While this provides the most expeditious proof that the tests in all these cases are UMP unbiased, there is available also a variation of the approach, which is more elementary. The proof of Theorem 3 is quite elementary except for the following points: (i) the fact that the conditional distributions of U given $T = t$ constitute an exponential family, (ii) that the family of distributions of T is complete, (iii) that the derivative of $E_{\theta,\vartheta}\,\phi(U, T)$ exists and can be computed by differentiating under the expectation sign, (iv) that the functions $\phi_1, \cdots, \phi_4$ are measurable. Instead of verifying (i) through (iv) in general, as was done in the above proof, it is possible in applications of the theorem to check these conditions directly for each specific problem, which in some cases is quite easy.

Through a transformation of parameters, Theorem 3 can be extended to cover hypotheses concerning parameters of the form

$$\theta^* = a_0\theta + \sum_{i=1}^{k} a_i\vartheta_i, \qquad a_0 \neq 0.$$

This transformation is formally given by the following lemma, the proof of which is immediate.

Lemma 2. *The exponential family of distributions* (10) *can also be written as*

$$dP^X_{\theta,\vartheta}(x) = K(\theta^*, \vartheta) \exp\left[\theta^* U^*(x) + \Sigma\vartheta_i T_i^*(x)\right] d\mu(x)$$

where

$$U^* = \frac{U}{a_0}, \qquad T_i^* = T_i - \frac{a_i}{a_0}\, U.$$

Application of Theorem 3 to the form of the distributions given in the lemma leads to UMP unbiased tests of the hypothesis $H_1^*: \theta^* \leqq \theta_0$ and the analogously defined hypotheses H_2^*, H_3^*, H_4^*.

When testing one of the hypotheses H_j one is frequently interested in the power $\beta(\theta', \vartheta)$ of ϕ_j against some alternative θ'. As is indicated by the notation and is seen from (20), this power will usually depend on the unknown nuisance parameters ϑ. On the other hand, the power of the conditional test given $T = t$,

$$\beta(\theta' \mid t) = E_{\theta'}[\phi(U, T) \mid t],$$

is independent of ϑ and therefore has a known value.

The quantity $\beta(\theta' \mid t)$ can be interpreted in two ways. (i) It is the probability of rejecting H when $T = t$. Once T has been observed to have the value t it may be felt, at least in certain problems, that this is a more appropriate expression of the power in the given situation than $\beta(\theta', \vartheta)$, which is obtained by averaging $\beta(\theta' \mid t)$ with respect to other values of t not relevant to the situation at hand. This argument leads to difficulties since in many cases the conditioning could be carried even further and it is not clear where the process should stop. (ii) A more clear-cut interpretation is obtained by considering $\beta(\theta'|t)$ as an estimate of $\beta(\theta', \vartheta)$. Since

$$E_{\theta',\vartheta}[\beta(\theta'|T)] = \beta(\theta', \vartheta),$$

this estimate is unbiased in the sense of Chapter 1, equation (11). It follows further from the theory of unbiased estimation and the completeness of the exponential family that among all unbiased estimates of $\beta(\theta', \vartheta)$ the present one has the smallest variance.*

Regardless of the interpretation, $\beta(\theta'|t)$ has the disadvantage compared with an unconditional power that it becomes available only after the observations have been taken. It therefore cannot be used to plan the experiment and in particular to determine the sample size, if this must be done prior to the experiment. On the other hand, a simple sequential procedure guaranteeing a specified power β against the alternatives $\theta = \theta'$ is obtained by continuing taking observations until the conditional power $\beta(\theta'|t)$ is $\geqq \beta$.

5. COMPARING TWO POISSON OR BINOMIAL POPULATIONS

A problem arising in many different contexts is the comparison of two treatments or of one treatment with a control situation in which no treatment is applied. If the observations consist of the number of successes in a sequence of trials for each treatment, for example the number of cures of a certain disease, the problem becomes that of testing the equality of two binomial probabilities. If the basic distributions are Poisson, for example in a comparison of the radioactivity of two substances, one will be testing the equality of two Poisson distributions.

When testing whether a treatment has a beneficial effect by comparing it with the control situation of no treatment, the problem is of the one-sided type. If ξ_2 and ξ_1 denote the parameter values when the treatment is or is not applied, the class of alternatives is K: $\xi_2 > \xi_1$. The hypothesis is $\xi_2 = \xi_1$ if it is known a priori that there is either no effect or a beneficial one; it is $\xi_2 \leqq \xi_1$ if the possibility is admitted that the treatment may actually be harmful. Since the test is the same for the two hypotheses, the second somewhat safer hypothesis would seem preferable in most cases.

A one-sided formulation is sometimes appropriate also when a new treatment or process is being compared with a standard one, where the new treatment is of interest only if it presents an improvement. On the

* See Theorem 5.1 of Lehmann and Scheffé, "Completeness, similar regions and unbiased estimates," *Sankhyā*, Vol. 10 (1950), pp. 305–340.

other hand, if the two treatments are on an equal footing, the hypothesis $\xi_2 = \xi_1$ of equality of two treatments is tested against the two-sided alternatives $\xi_2 \neq \xi_1$. The formulation of this problem as one of hypothesis testing is usually quite artificial since in case of rejection of the hypothesis one will obviously wish to know which of the treatments is better.* Such two-sided tests do, however, have important applications to the problem of obtaining confidence limits for the extent by which one treatment is better than the other.

To apply Theorem 3 to this comparison problem it is necessary to express the distributions in an exponential form with $\theta = f(\xi_1, \xi_2)$, for example $\theta = \xi_2 - \xi_1$ or ξ_2/ξ_1 such that the hypotheses of interest become equivalent to those of Theorem 3. In the present section the problem will be considered for Poisson and binomial distributions; the case of normal distributions will be taken up in Chapter 5.

We consider first the Poisson problem in which X and Y are independently distributed according to $P(\lambda)$ and $P(\mu)$ so that their joint distribution can be written as

$$P\{X = x, Y = y\} = \frac{e^{-(\lambda+\mu)}}{x!y!} \exp\left[y \log \frac{\mu}{\lambda} + (x + y) \log \lambda\right].$$

By Theorem 3 there exist UMP unbiased tests of the four hypotheses $H_1, \cdots, H_4$ concerning the parameter $\theta = \log(\mu/\lambda)$ or equivalently concerning the ratio $\rho = \mu/\lambda$. This includes in particular the hypotheses $\mu \leq \lambda$ (or $\mu = \lambda$) against the alternatives $\mu > \lambda$, and $\mu = \lambda$ against $\mu \neq \lambda$. Comparing the distribution of (X, Y) with (10), one has $U = Y$ and $T = X + Y$, and by Theorem 3 the tests are performed conditionally on the integer points of the line segment $X + Y = t$ in the positive quadrant of the x, y-plane. The conditional distribution of Y given $X + Y = t$ is (Problem 12 of Chapter 2)

$$P\{Y = y \mid X + Y = t\} = \binom{t}{y}\left(\frac{\mu}{\lambda + \mu}\right)^y \left(\frac{\lambda}{\lambda + \mu}\right)^{t-y} \qquad y = 0, 1, \cdots, t,$$

the binomial distribution corresponding to t trials and probability $p = \mu/(\lambda + \mu)$ of success. The original hypotheses therefore reduce to the corresponding ones about the parameter p of a binomial distribution. The hypothesis $H: \mu \leq a\lambda$, for example, becomes $H: p \leq a/(a + 1)$,

* For a discussion of the comparison of two treatments as a three-decision problem, see Bahadur, "A property of the t-statistic," *Sankhyā*, Vol. 12 (1952), pp. 79–88, and Lehmann, "A theory of some multiple decision procedures," *Ann. Math. Stat.*, Vol. 28 (1957), pp. 1–25, 547–572.

which is rejected when Y is too large. The cutoff point depends of course, in addition to a, also on t. It can be determined from tables of the binomial, and for large t approximately from tables of the normal distribution.

In many applications the ratio $\rho = \mu/\lambda$ is a reasonable measure of the extent to which the two Poisson populations differ, since the parameters λ and μ measure the rates (in time or space) at which two Poisson processes produce the events in question. One might therefore hope that the power of the above tests depends only on this ratio, but this is not the case. On the contrary, for each fixed value of ρ corresponding to an alternative to the hypothesis being tested, the power $\beta(\lambda, \mu) = \beta(\lambda, \rho\lambda)$ is an increasing function of λ, which tends to 1 as $\lambda \to \infty$ and to α as $\lambda \to 0$. To see this consider the power $\beta(\rho|t)$ of the conditional test given t. This is an increasing function of t since it is the power of the optimum test based on t binomial trials. The conditioning variable T has a Poisson distribution with parameter $\lambda(1 + \rho)$, and its distribution for varying λ forms an exponential family. It follows (Lemma 2 of Chapter 3) that the over-all power $E[\beta(\rho|T)]$ is an increasing function of λ. As $\lambda \to 0$ or ∞, T tends in probability to 0 or ∞, and the power against a fixed alternative ρ tends to α or 1.

The above test is also applicable to samples $X_1, \cdots, X_m$ and $Y_1, \cdots, Y_n$ from two Poisson distributions. The statistics $X = \sum_{i=1}^m X_i$ and $Y = \sum_{j=1}^n Y_j$ are then sufficient for λ and μ, and have Poisson distributions with parameters $m\lambda$ and $n\mu$ respectively. In planning an experiment one might wish to determine $m = n$ so large that the test of, say, $H: \rho \leq \rho_0$ has power against a specified alternative ρ_1 greater than or equal to some preassigned β. However, it follows from the discussion of the power function for $n = 1$ which applies equally to any other n, that this cannot be achieved for any fixed n no matter how large. This is seen more directly by noting that as $\lambda \to 0$, for both $\rho = \rho_0$ and $\rho = \rho_1$ the probability of the event $X = Y = 0$ tends to 1. Therefore, the power of any level α test against $\rho = \rho_1$ and for varying λ cannot be bounded away from α. This difficulty can be overcome only by permitting observations to be taken sequentially. One can for example determine t_0 so large that the test of the hypothesis $p \leq \rho_0/(1 + \rho_0)$ on the basis of t_0 binomial trials has power $\geq \beta$ against the alternative $p_1 = \rho_1/(1 + \rho_1)$. By observing $(X_1, Y_1), (X_2, Y_2), \cdots$ and continuing until $\Sigma(X_i + Y_i) \geq t_0$, one obtains a test with power $\geq \beta$ against all alternatives with $\rho \geq \rho_1$.*

The corresponding comparison of two binomial probabilities is quite

* A discussion of this and alternative procedures for achieving the same aim is given by Birnbaum, "Statistical methods for Poisson processes and exponential populations," *J. Am. Stat. Assoc.*, Vol. 49, pp. 254–266.

similar. Let X and Y be independent binomial variables with joint distribution

$$P\{X = x, Y = y\} = \binom{m}{x} p_1^x q_1^{m-x} \binom{n}{y} p_2^y q_2^{n-y}$$
$$= \binom{m}{x}\binom{n}{y} q_1^m q_2^n \exp\left[y\left(\log\frac{p_2}{q_2} - \log\frac{p_1}{q_1}\right) + (x+y)\log\frac{p_1}{q_1}\right].$$

The four hypotheses $H_1, \cdots, H_4$ can then be tested concerning the parameter $\theta = \log\left(\frac{p_2}{q_2}\Big/\frac{p_1}{q_1}\right)$ or equivalently concerning the ratio $\rho = \frac{p_2}{q_2}\Big/\frac{p_1}{q_1}$. This includes in particular the problems of testing $H_1': p_2 \leq p_1$ against $p_2 > p_1$ and $H_4': p_2 = p_1$ against $p_2 \neq p_1$. As in the Poisson case, $U = Y$ and $T = X + Y$, and the test is carried out in terms of the conditional distribution of Y on the line segment $X + Y = t$. This distribution is given by

$$(21)\quad P\{Y = y \mid X + Y = t\} = C_t(\rho)\binom{m}{t-y}\binom{n}{y}\rho^y, \qquad y = 0, 1, \cdots, t,$$

where $C_t(\rho) = 1/\sum_{y'=0}^{t}\binom{m}{t-y'}\binom{n}{y'}\rho^{y'}$. In the particular case of the hypotheses H_1' and H_4', the boundary value θ_0 of (13), (18), and (19) is 0, and the corresponding value of ρ is $\rho_0 = 1$. The conditional distribution then reduces to

$$P\{Y = y \mid X + Y = t\} = \frac{\binom{m}{t-y}\binom{n}{y}}{\binom{m+n}{t}}, \qquad y = 0, \cdots, t,$$

which is the hypergeometric distribution.*

6. TESTING FOR INDEPENDENCE IN A 2 × 2 TABLE

The problem of deciding whether two characteristics A and B are independent in a population was discussed in Section 4 of Chapter 3

* Tables facilitating the tests in this case are given, among others, by Mainland, Herrera, and Sutcliffe, *Tables for Use with Binomial Samples*, New York, Department of Medical Statistics, N. Y. Univ. College of Medicine, 1956, and by Armsen, "Tables for significance tests of 2 × 2 contingency tables," *Biometrika*, Vol. 42 (1955), pp. 494–511.

(Example 4), under the assumption that the marginal probabilities $p(A)$ and $p(B)$ are known. The most informative sample of size s was found to be one selected entirely from that one of the four categories A, $\tilde{A}$, B, or $\tilde{B}$, say A, which is rarest in the population. The problem then reduces to testing the hypothesis $H: p = p(B)$ in a binomial distribution $b(p, s)$.

In the more usual situation that $p(A)$ and $p(B)$ are not known, a sample from one of the categories such as A does not provide a basis for distinguishing between the hypothesis and the alternatives. This follows from the fact that the number in the sample possessing characteristic B then constitutes a binomial variable with probability $p(B|A)$, which is completely unknown both when the hypothesis is true and when it is false. The hypothesis can, however, be tested if samples are taken both from categories A and $\tilde{A}$ or both from B and $\tilde{B}$. In the latter case, for example, if the sample sizes are m and n, the numbers of cases possessing characteristic A in the two samples constitute independent variables with binomial distributions $b(p_1, m)$ and $b(p_2, n)$ respectively where $p_1 = P(A|B)$ and $p_2 = P(A|\tilde{B})$. The hypothesis of independence of the two characteristics: $p(A|B) = p(A)$, is then equivalent to the hypothesis $p_1 = p_2$, and the problem reduces to that treated in the preceding section.

Instead of selecting samples from two of the categories, it is frequently more convenient to take the sample at random from the population as a whole. The results of such a sample can be summarized in the following 2×2 contingency table, the entries of which give the numbers in the various categories.

	A	$\tilde{A}$	
B	X	X'	M
$\tilde{B}$	Y	Y'	N
	T	T'	s

The joint distribution of the variables X, X', Y, and Y' is multinomial, and is given by

$$P\{X = x, X' = x', Y = y, Y' = y'\} = \frac{s!}{x!x'!y!y'!} p_{AB}^{x} p_{\tilde{A}B}^{x'} p_{A\tilde{B}}^{y} p_{\tilde{A}\tilde{B}}^{y'}$$

$$= \frac{s!}{x!x'!y!y'!} p_{\tilde{A}\tilde{B}}^{s} \exp\left(x \log \frac{p_{AB}}{p_{\tilde{A}\tilde{B}}} + x' \log \frac{p_{\tilde{A}B}}{p_{\tilde{A}\tilde{B}}} + y \log \frac{p_{A\tilde{B}}}{p_{\tilde{A}\tilde{B}}}\right).$$

Lemma 2 and Theorem 3 are therefore applicable to any parameter of the form

$$\theta^* = a_0 \log \frac{p_{AB}}{p_{\tilde{A}\tilde{B}}} + a_1 \log \frac{p_{\tilde{A}B}}{p_{\tilde{A}\tilde{B}}} + a_2 \log \frac{p_{A\tilde{B}}}{p_{\tilde{A}\tilde{B}}}.$$

Putting $a_1 = a_2 = 1$, $a_0 = -1$, $\Delta = e^{\theta^*} = (p_{\tilde{A}B}p_{A\tilde{B}})/(p_{AB}p_{\tilde{A}\tilde{B}})$, and denoting the probabilities of A and B in the population by $p_A = p_{AB} + p_{A\tilde{B}}$, $p_B = p_{AB} + p_{\tilde{A}B}$, one finds

$$p_{AB} = p_A p_B + \frac{1-\Delta}{\Delta} p_{\tilde{A}B}p_{A\tilde{B}}$$

$$p_{\tilde{A}B} = p_{\tilde{A}} p_B - \frac{1-\Delta}{\Delta} p_{\tilde{A}B}p_{A\tilde{B}}$$

$$p_{A\tilde{B}} = p_A p_{\tilde{B}} - \frac{1-\Delta}{\Delta} p_{\tilde{A}B}p_{A\tilde{B}}$$

$$p_{\tilde{A}\tilde{B}} = p_{\tilde{A}} p_{\tilde{B}} + \frac{1-\Delta}{\Delta} p_{\tilde{A}B}p_{A\tilde{B}}.$$

Independence of A and B is therefore equivalent to $\Delta = 1$, and $\Delta < 1$ and $\Delta > 1$ correspond to positive and negative dependence respectively.†

The test of the hypothesis of independence, or any of the four hypotheses concerning Δ, is carried out in terms of the conditional distribution of X given $X + X' = m$, $X + Y = t$. Instead of computing this distribution directly, consider first the conditional distribution subject only to the condition $X + X' = m$, and hence $Y + Y' = s - m = n$. This is seen to be

$$P\{X = x, Y = y \mid X + X' = m\}$$
$$= \binom{m}{x}\binom{n}{y}\left(\frac{p_{AB}}{p_B}\right)^x\left(\frac{p_{\tilde{A}B}}{p_B}\right)^{m-x}\left(\frac{p_{A\tilde{B}}}{p_{\tilde{B}}}\right)^y\left(\frac{p_{\tilde{A}\tilde{B}}}{p_{\tilde{B}}}\right)^{n-y},$$

which is the distribution of two independent binomial variables, the number of successes in m and n trials with probability $p_1 = p_{AB}/p_B$ and $p_2 = p_{A\tilde{B}}/p_{\tilde{B}}$. Actually, this is clear without computation since we are now dealing with samples of fixed size m and n from the subpopulations B and $\tilde{B}$, and the probability of A in these subpopulations is p_1 and p_2. If now the additional restriction $X + Y = t$ is imposed, the conditional distribution of X subject to the two conditions $X + X' = m$ and $X + Y = t$ is the same as that of X given $X + Y = t$ in the case of two independent binomials considered in the previous section. It is therefore given by

$$P\{X = x \mid X + X' = m, X + Y = t\} = C_t(\rho)\binom{m}{x}\binom{n}{t-x}\rho^{t-x},$$
$$x = 0, \cdots, t,$$

† Δ is equivalent to Yule's measure of association, which is $Q = (1 - \Delta)/(1 + \Delta)$. For a discussion of this and related measures see Goodman and Kruskal, "Measures of association for error classifications," *J. Am. Stat. Assoc.*, Vol. 49 (1954), pp. 732–764.

that is, by (21) expressed in terms of x instead of y. (Here the choice of X as testing variable is quite arbitrary; we could equally well again have chosen Y.) For the parameter ρ one finds

$$\rho = \frac{p_2}{q_2}\bigg/\frac{p_1}{q_1} = \frac{p_{\tilde{A}B}p_{A\tilde{B}}}{p_{AB}p_{\tilde{A}\tilde{B}}} = \Delta.$$

From these considerations it follows that the conditional test given $X + X' = m$, $X + Y = t$, for testing any of the hypotheses concerning Δ is identical with the conditional test given $X + Y = t$ of the same hypothesis concerning $\rho = \Delta$ in the preceding section, in which $X + X' = m$ was given a priori. In particular, the conditional test for testing the hypothesis of independence $\Delta = 1$, the *Fisher–Irwin test*, is the same as that of testing the equality of two binomial p's and is therefore given in terms of the hypergeometric distribution.

At the beginning of the section it was pointed out that the hypothesis of independence can be tested on the basis of samples obtained in a number of different ways. Either samples of fixed size can be taken from A and $\tilde{A}$ or from B and $\tilde{B}$, or the sample can be selected at random from the population at large. Which of these designs is most efficient depends on the cost of sampling from the various categories and from the population at large, and also on the cost of performing the necessary classification of a selected individual with respect to the characteristics in question. Suppose, however, for a moment that these considerations are neglected and that the designs are compared solely in terms of the power that the resulting tests achieve against a common alternative. Then the following results* can be shown to hold asymptotically as the total sample size s tends to infinity.

(i) If samples of size m and n ($m + n = s$) are taken from B and $\tilde{B}$ or from A and $\tilde{A}$, the best choice of m and n is $m = n = s/2$.

(ii) It is better to select samples of equal size $s/2$ from B and $\tilde{B}$ than from A and $\tilde{A}$ provided $|p_B - 1/2| > |p_A - 1/2|$.

(iii) Selecting the sample at random from the population at large is worse than taking equal samples from either A and $\tilde{A}$ or from B and $\tilde{B}$.

These statements, which we shall not prove here, can be established by using the normal approximation for the distribution of the binomial variables X and Y when m and n are fixed, and by noting that under random sampling from the population at large, M/s and N/s tend in probability to p_B and $p_{\tilde{B}}$ respectively.

* These results were conjectured by Berkson and proved by Neyman in a course on χ^2.

7. THE SIGN TEST

To test consumer preferences between two products, a sample of n subjects is asked to state their preferences. Each subject is recorded as plus or minus as it favors product A or B. The total number Y of plus signs is then a binomial variable with distribution $b(p, n)$. Consider the problem of testing the hypothesis $p = 1/2$ of no difference against the alternatives $p \neq 1/2$. (As in previous such problems we disregard here that in case of rejection it will be necessary to decide which of the two products is preferred.) The appropriate test is the two-sided *sign test*, which rejects when $|Y - \frac{1}{2}n|$ is too large. This is UMP unbiased (Section 2).

Sometimes the subjects are also given the possibility of declaring themselves as undecided. If p_+, p_-, and p_0 denote the probabilities of preference for product A, product B, and of no preference respectively, the numbers X, Y, and Z of decisions in favor of these three possibilities are distributed according to the multinomial distribution

$$\frac{n!}{x!y!z!} p_-^x p_+^y p_0^z \qquad (x + y + z = n), \tag{22}$$

and the hypothesis to be tested is $H: p_+ = p_-$. The distribution (22) can also be written as

$$\frac{n!}{x!y!z!} \left(\frac{p_+}{1 - p_0 - p_+}\right)^y \left(\frac{p_0}{1 - p_0 - p_+}\right)^z (1 - p_0 - p_+)^n,$$

and is then seen to constitute an exponential family with $U = Y$, $T = Z$, $\theta = \log [p_+/(1 - p_0 - p_+)]$, $\vartheta = \log [p_0/(1 - p_0 - p_+)]$. Rewriting the hypothesis H as $p_+ = 1 - p_0 - p_+$, it is seen to be equivalent to $\theta = 0$. There exists therefore a UMP unbiased test of H, which is obtained by considering z as fixed and determining the best unbiased conditional test of H given $Z = z$. Since the conditional distribution of Y given z is a binomial distribution $b(p, n - z)$ with $p = p_+/(p_+ + p_-)$, the problem reduces to that of testing the hypothesis $p = 1/2$ in a binomial distribution with $n - z$ trials, for which the rejection region is $|Y - \frac{1}{2}(n - z)| > C(z)$. The UMP unbiased test is therefore obtained by disregarding the number of cases in which no preference is expressed (the number of *ties*), and applying the sign test to the remaining data.

The power of the test depends strongly on p_0, which governs the distribution of Z. For large p_0, the number $n - z$ of trials in the conditional binomial distribution can be expected to be small, and the test will thus have little power. This may be an advantage in the present case, since a sufficiently high value of p_0, regardless of the value of p_+/p_-, implies that the population as a whole is largely indifferent with respect to the products.

The above conditional sign test applies to any situation in which the observations are the result of n independent trials, each of which is either a success (+), a failure (−), or a tie. As an alternative treatment of ties, it is sometimes proposed to assign each tie at random (with probability 1/2 each) to either plus or minus. The total number Y' of plus signs after the ties have been broken is then a binomial variable with distribution $b(\pi, n)$ where $\pi = p_+ + \frac{1}{2}p_0$. The hypothesis H becomes $\pi = 1/2$, and is rejected when $|Y' - \frac{1}{2}n| > C$, where the probability of rejection is α when $\pi = 1/2$. This test can be viewed also as a randomized test based on X, Y, and Z, and it is unbiased for testing H in its original form since p_+ is $=$ or $\neq p_-$ as π is $=$ or $\neq 1/2$. Since the test involves randomization other than on the boundaries of the rejection region, it is less powerful than the UMP unbiased test for this situation, so that the random breaking of ties results in a loss of power.

This remark might be thought to throw some light on the question of whether in the determination of consumer preferences it is better to permit the subject to remain undecided or to force an expression of preference. However, here the assumption of a completely random assignment in case of a tie does not apply. Even when the subject is not conscious of a definite preference, there will usually be a slight inclination toward one of the two possibilities, which in a majority of the cases will be brought out by a forced decision. This will be balanced in part by the fact that such forced decisions are more variable than those reached voluntarily. Which of these two factors dominates depends on the strength of the preference.

Frequently, the question of preference arises between a standard product and a possible modification or a new product. If each subject is required to express a definite preference, the hypothesis of interest is usually the one-sided hypothesis $p_+ \leqq p_-$, where + denotes a preference for the modification. However, if an expression of indifference is permitted, the hypothesis to be tested is not $p_+ \leqq p_-$ but rather $p_+ \leqq p_0 + p_-$ since typically the modification is of interest only if it is actually preferred. As was shown in Chapter 3, Example 8, the one-sided sign test which rejects when the number of plus signs is too large is UMP for this problem.

In some investigations, the subject is asked not only to express a preference but to give a more detailed evaluation, such as a score on some numerical scale. Depending on the situation, the hypothesis can then take on one of two forms. One may be interested in the hypothesis that there is no difference in the consumer's reaction to the two products. Formally, this states that the distribution of the scores $X_1, \cdots, X_n$ expressing the degree of preference of the n subjects for the modified product is symmetric about the origin. This problem, for which a UMP unbiased test does not exist without further assumptions, will be considered in Chapter 6, Section 9.

Alternatively, the hypothesis of interest may continue to be $H: p_+ = p_-$. Since $p_- = P\{X < 0\}$ and $p_+ = P\{X > 0\}$, this now becomes

$$H: P\{X > 0\} = P\{X < 0\}.$$

Here symmetry of X is no longer assumed even when $P\{X < 0\} = P\{X > 0\}$. If no assumptions are made concerning the distribution of X beyond the fact that the set of its possible values is given, the sign test based on the number of X's that are positive and negative continues to be UMP unbiased.

To see this, note that any distribution of X can be specified by the probabilities

$$p_- = P\{X < 0\}, \qquad p_+ = P\{X > 0\}, \qquad p_0 = P\{X = 0\},$$

and the conditional distributions F_- and F_+ of X given $X < 0$ and $X > 0$ respectively. Consider any fixed distributions F'_-, F'_+ and denote by $\mathscr{F}_0$ the family of all distributions with $F_- = F'_-$, $F_+ = F'_+$ and arbitrary p_-, p_+, p_0. Any test that is unbiased for testing H in the original family of distributions $\mathscr{F}$ in which F_- and F_+ are unknown is also unbiased for testing H in the smaller family $\mathscr{F}_0$. We shall show below that there exists a UMP unbiased test ϕ_0 of H in $\mathscr{F}_0$. It turns out that ϕ_0 is also unbiased for testing H in $\mathscr{F}$ and is independent of F'_-, F'_+. Let ϕ be any other unbiased test of H in $\mathscr{F}$, and consider any fixed alternative which without loss of generality can be assumed to be in $\mathscr{F}_0$. Since ϕ is unbiased for $\mathscr{F}$, it is unbiased for testing $p_+ = p_-$ in $\mathscr{F}_0$; the power of ϕ_0 against the particular alternative is therefore at least as good as that of ϕ. Hence ϕ_0 is UMP unbiased.

To determine the UMP unbiased test of H in $\mathscr{F}_0$, let the densities of F'_- and F'_+ with respect to some measure μ be f'_- and f'_+. The joint density of the X's at a point $(x_1, \cdots, x_n)$ with

$$x_{i_1}, \cdots, x_{i_r} < 0 = x_{j_1} = \cdots = x_{j_s} < x_{k_1}, \cdots, x_{k_m}$$

is

$$p_-^r p_0^s p_+^m f'_-(x_{i_1}) \cdots f'_-(x_{i_r}) f'_+(x_{k_1}) \cdots f'_+(x_{k_m}).$$

The set of statistics (r, s, m) is sufficient for (p_-, p_0, p_+) and its distribution is given by (22) with $x = r$, $y = m$, $z = s$. The sign test is therefore seen to be UMP unbiased as before.

8. PROBLEMS

Section 1

1. *Admissibility.* Any UMP unbiased test ϕ_0 is admissible in the sense that there cannot exist another test ϕ_1, which is at least as powerful as ϕ_0 against all alternatives and more powerful against some.

[If ϕ is unbiased and ϕ' is uniformly at least as powerful as ϕ, then ϕ' is also unbiased.]

2. *Critical levels.* Consider a family of tests of $H: \theta = \theta_0$ (or $\theta \leqq \theta_0$), with level α rejection regions S_α such that (a) $P_{\theta_0}\{X \in S_\alpha\} = \alpha$ for all $0 < \alpha < 1$, and (b) $S_{\alpha_0} = \bigcap_{\alpha > \alpha_0} S_\alpha$ for all $0 < \alpha_0 < 1$, which in particular implies $S_\alpha \subset S_{\alpha'}$, for $\alpha < \alpha'$.

(i) Then the critical level $\hat{\alpha}$ is given by $\hat{\alpha} = \hat{\alpha}(x) = \inf\{\alpha\colon\, x \in S_\alpha\}$.

(ii) When $\theta = \theta_0$, the distribution of $\hat{\alpha}$ is the uniform distribution over (0, 1).

(iii) If the tests S_α are unbiased, the distribution of $\hat{\alpha}$ under any alternative θ satisfies

$$P_\theta\{\hat{\alpha} \leqq \alpha\} \geqq P_{\theta_0}\{\hat{\alpha} \leqq \alpha\} = \alpha,$$

so that it is shifted toward the origin.

If the critical values are available from a number of independent experiments, they can be combined by (ii) and (iii) to provide an over-all test* of the hypothesis.

[$\hat{\alpha} \leqq \alpha$ if and only if $x \in S_\alpha$, and hence $P_\theta\{\hat{\alpha} \leqq \alpha\} = P_\theta\{X \in S_\alpha\} = \beta_\alpha(\theta)$, which is α for $\theta = \theta_0$ and $\geqq \alpha$ if θ is an alternative to H.]

Section 2

3. Let X have the binomial distribution $b(p, n)$ and consider the hypothesis $H: p = p_0$ at level of significance α. Determine the boundary values of the UMP unbiased test for $n = 10$, $\alpha = .1$, $p_0 = .2$ and $\alpha = .05$, $p_0 = .4$, and in each case graph the power functions of both the unbiased and the equal tails test.

4. Let X have the Poisson distribution $P(\tau)$, and consider the hypothesis $H: \tau = \tau_0$. Then condition (6) reduces to

$$\sum_{x=C_1+1}^{C_2-1} \frac{\tau_0^{x-1}}{(x-1)!} e^{-\tau_0} + \sum_{i=1}^{2} (1 - \gamma_i) \frac{\tau_0^{C_i - 1}}{(C_i - 1)!} e^{-\tau_0} = 1 - \alpha.$$

5. Let T_n/θ have a χ^2-distribution with n degrees of freedom. For testing $H: \theta = 1$ at level of significance $\alpha = .05$, find n so large that the power of the UMP unbiased test is $\geqq .9$ against both $\theta \geqq 2$ and $\theta \leqq \frac{1}{2}$. How large does n have to be if the test is not required to be unbiased?

* For a discussion of a number of such tests see Wallis "Compounding probabilities from independent significance tests," *Econometrica*, Vol. 10 (1942), pp. 229–248, and Birnbaum, "Combining independent tests of significance," *J. Am. Stat. Assoc.*, Vol. 49 (1954), pp. 559–574.

6. Let X and Y be independently distributed according to one-parameter exponential families, so that their joint distribution is given by

$$dP_{\theta_1,\theta_2}(x, y) = C(\theta_1)e^{\theta_1 T(x)}\,d\mu(x)K(\theta_2)e^{\theta_2 U(y)}\,d\nu(y).$$

Then a UMP unbiased test does not exist for testing H: $\theta_1 = a$, $\theta_2 = b$ against the alternatives $\theta_1 \neq a$ or $\theta_2 \neq b$.

[The most powerful unbiased test against the alternatives $\theta_1 \neq a$, $\theta_2 = b$ and $\theta_1 = a$, $\theta_2 \neq b$ have acceptance regions $C_1 < T(x) < C_2$ and $K_1 < U(y) < K_2$ respectively. These tests are also unbiased against the wider class of alternatives K: $\theta_1 \neq a$ or $\theta_2 \neq b$ or both.]

7. Let (X, Y) be distributed according to the exponential family

$$dP_{\theta_1,\theta_2}(x, y) = C(\theta_1, \theta_2)e^{\theta_1 x + \theta_2 y}\,d\mu(x, y).$$

The only unbiased test for testing H: $\theta_1 \leq a$, $\theta_2 \leq b$ against K: $\theta_1 > a$ or $\theta_2 > b$ or both is $\phi(x, y) \equiv \alpha$.

[Take $a = b = 0$, and let $\beta(\theta_1, \theta_2)$ be the power function of any level α test. Unbiasedness implies $\beta(0, \theta_2) = \alpha$ for $\theta_2 < 0$ and hence for all θ_2 since $\beta(0, \theta_2)$ is an analytic function of θ_2. For fixed $\theta_2 > 0$, $\beta(\theta_1, \theta_2)$ considered as a function of θ_1 therefore has a minimum at $\theta_1 = 0$, so that $\partial\beta(\theta_1, \theta_2)/\partial\theta_1$ vanishes at $\theta_1 = 0$ for all positive θ_2, and hence for all θ_2. By considering alternatively positive and negative values of θ_2 and using the fact that the partial derivatives of all orders of $\beta(\theta_1, \theta_2)$ with respect to θ_1 are analytic, one finds that for each fixed θ_2 these derivatives all vanish at $\theta_1 = 0$ and hence that the function β must be a constant. Because of the completeness of (X, Y), $\beta(\theta_1, \theta_2) \equiv \alpha$ implies $\phi(x, y) \equiv \alpha$.]

8. For testing the hypothesis H: $\theta = \theta_0$ in the one-parameter exponential family of Section 2, let $\mathscr{C}$ be the totality of tests satisfying (3) and (5) for some $-\infty \leq C_1 \leq C_2 \leq \infty$ and $0 \leq \gamma_1, \gamma_2 \leq 1$.

(i) $\mathscr{C}$ is complete in the sense that given any level α test ϕ_0 of H there exists $\phi \in \mathscr{C}$ such that ϕ is uniformly at least as powerful as ϕ_0.

(ii) If $\phi_1, \phi_2 \in \mathscr{C}$, then neither of the two tests is uniformly more powerful than the other.

(iii) Let the problem be considered as a two-decision problem, with decisions d_0 and d_1 corresponding to acceptance and rejection of H, and with loss function $L(\theta, d_i) = L_i(\theta)$, $i = 0, 1$. Then $\mathscr{C}$ is minimal essentially complete provided $L_1(\theta) < L_0(\theta)$ for all $\theta \neq \theta_0$.

(iv) Extend the result of part (iii) to the hypothesis H': $\theta_1 \leq \theta \leq \theta_2$.

[(i) Let the derivative of the power function of ϕ_0 at θ_0 be $\beta'_{\phi_0}(\theta_0) = \rho$. Then there exists $\phi \in \mathscr{C}$ such that $\beta'_{\phi}(\theta_0) = \rho$ and ϕ is UMP among all tests satisfying this condition.

(ii) See Chapter 3, end of Section 7.

(iii) See Chapter 3, proof of Theorem 3.]

Section 3

9. Let $X_1, \cdots, X_n$ be a sample from (i) the normal distribution $N(a\sigma, \sigma^2)$, with a fixed and $0 < \sigma < \infty$; (ii) the uniform distribution $R(\theta - \frac{1}{2}, \theta + \frac{1}{2})$, $-\infty < \theta < \infty$; (iii) the uniform distribution $R(\theta_1, \theta_2)$, $-\infty < \theta_1 < \theta_2 < \infty$. For these three families of distributions the following statistics are sufficient: (i) $T = (\Sigma X_i, \Sigma X_i^2)$, (ii) and (iii) $T = (\min(X_1, \cdots, X_n), \max(X_1, \cdots, X_n))$.

The family of distributions of T is complete for case (iii), but for (i) and (ii) it is not complete or even boundedly complete.

[(i) The distribution of $\Sigma X_i / \sqrt{\Sigma X_i^2}$ does not depend on σ.]

10. Let $X_1, \cdots, X_m$ and $Y_1, \cdots, Y_n$ be samples from $N(\xi, \sigma^2)$ and $N(\xi, \tau^2)$. Then $T = (\Sigma X_i, \Sigma Y_j, \Sigma X_i^2, \Sigma Y_j^2)$, which in Example 5 was seen not to be complete, is also not boundedly complete.

[Let $f(t)$ be 1 or -1 as $\bar{y} - \bar{x}$ is positive or not.]

11. *Counterexample.* Let X be a random variable taking on the values $-1, 0, 1, 2, \cdots$ with probabilities

$$P_\theta\{X = -1\} = \theta; \qquad P_\theta\{X = x\} = (1 - \theta)^2\theta^x, \qquad x = 0, 1, \cdots.$$

Then $\mathscr{P} = \{P_\theta, 0 < \theta < 1\}$ is boundedly complete but not complete.

12. Let $\mathscr{P} = \{P\}$ be a family of distributions with the property that for any $P, Q \in \mathscr{P}$, there exists $0 < p < 1$ such that $pP + (1 - p)Q \in \mathscr{P}$. Suppose that $h(x_1, \cdots, x_n)$ is a symmetric function satisfying

$$\int h(x_1, \cdots, x_n)\, dP(x_1) \cdots dP(x_n) = 0 \quad \text{for all} \quad P \in \mathscr{P}. \tag{23}$$

Then

$$\int h(x_1, \cdots, x_n)\, dP_1(x_1) \cdots dP_n(x_n) = 0 \quad \text{for all} \quad P_1, \cdots, P_n \in \mathscr{P}. \tag{24}$$

[(1) If $P_1, \cdots, P_k \in \mathscr{P}$ there exist probabilities $p_1, \cdots, p_k$, positive and adding up to 1, such that $(p_1P_1 + \cdots + p_iP_i)/(p_1 + \cdots + p_i) \in \mathscr{P}$ for all $i = 1, \cdots, k$.

(2) For any integers $1 \leq i_1 < i_2 < \cdots < i_k \leq n$, let $\alpha(i_1, \cdots, i_k)$ be the set of all n-tuples $(j_1, \cdots, j_n)$ such that (a) every component is one of the integers $i_1, \cdots, i_k$, and (b) each of these integers occurs at least once among $(j_1, \cdots, j_n)$. If

$$I(j_1, \cdots, j_n) = p_{j_1} \cdots p_{j_n} \int h(x_1, \cdots, x_n)\, dP_{j_1}(x_1) \cdots dP_{j_n}(x_n),$$

then (23) implies

$$\sum_{\alpha(i_1, \cdots, i_k)} I(j_1, \cdots, j_n) = 0$$

for all $(i_1, \cdots, i_k)$ with $k \leq n$.

This is proved by induction over k. For $k = 1$ it is a direct consequence of (23). To prove for example that $\Sigma_{\alpha(1,2)} I(j_1, \cdots, j_n) = 0$, let $P = (p_1P_1 + p_2P_2)/(p_1 + p_2)$ be the element of $\mathscr{P}$ guaranteed by (1). Then

$$0 = \int h(x_1, \cdots, x_n)\, dP(x_1) \cdots dP(x_n)(p_1 + p_2)^n$$

$$= \sum_{\alpha(1,2)} I(j_1, \cdots, j_n) + \sum_{\alpha(1) \cup \alpha(2)} I(j_1, \cdots, j_n),$$

and the result follows since the second term on the right-hand side has already been shown to be zero.

(3) It follows from (2) with $k = n$ that $\Sigma I(j_1, \cdots, j_n) = 0$ when the summation extends over all permutations $(j_1, \cdots, j_n)$ of $(1, \cdots, n)$. Since I is symmetric in its n arguments, this shows that $I(1, \cdots, n) = 0$, as was to be proved.]

13. *Continuation.* Let $\mathscr{C}$ be the class of uniform distributions over finite intervals, and let $\mathscr{P}$ be the class of convex combinations of a finite number of distributions from $\mathscr{C}$. If $X_1, \cdots, X_n$ are identically and independently distributed according to $P \in \mathscr{P}$, the set of order statistics $T = (X^{(1)}, \cdots, X^{(n)})$ is sufficient for $\mathscr{P}$, and the family $\mathscr{P}^T$ of distributions of T is complete.

[That T is sufficient follows from Example 7 of Chapter 2. Completeness of $\mathscr{P}^T$ is seen by applying the preceding problem to the equation $E_P h(T) = 0$ for all $P \in \mathscr{P}$.]

Section 4

14. *Measurability of tests of Theorem 3.* The function ϕ_3 defined by (16) and (17) is jointly measurable in u and t.

[With $C_1 = v$ and $C_2 = w$, the determining equations for v, w, γ_1, γ_2 are

$$(25)\quad F_t(v-) + [1 - F_t(w)] + \gamma_1[F_t(v) - F_t(v-)] + \gamma_2[F_t(w) - F_t(w-)] = \alpha$$

and

$$(26)\quad G_t(v-) + [1 - G_t(w)] + \gamma_1[G_t(v) - G_t(v-)] + \gamma_2[G_t(w) - G_t(w-)] = \alpha$$

where

$$(27)\qquad F_t(u) = \int_{-\infty}^{u} C_t(\theta_1)e^{\theta_1 y}\,d\nu_t(y), \qquad G_t(u) = \int_{-\infty}^{u} C_t(\theta_2)e^{\theta_2 y}\,d\nu_t(y)$$

denote the conditional cumulative distribution function of U given t when $\theta = \theta_1$ and $\theta = \theta_2$ respectively.

(1) For each $0 \leqq y \leqq \alpha$ let $v(y, t) = F_t^{-1}(y)$ and $w(y, t) = F_t^{-1}(1 - \alpha + y)$, where the inverse function is defined as in the proof of Theorem 3. Define $\gamma_1(y, t)$ and $\gamma_2(y, t)$ so that for $v = v(y, t)$ and $w = w(y, t)$,

$$F_t(v-) + \gamma_1[F_t(v) - F_t(v-)] = y$$

$$1 - F_t(w) + \gamma_2[F_t(w) - F_t(w-)] = \alpha - y.$$

(2) Let $H(y, t)$ denote the left-hand side of (26), with $v = v(y, t)$, etc. Then $H(0, t) > \alpha$ and $H(\alpha, t) < \alpha$. This follows by Theorem 2 of Chapter 3 from the fact that $v(0, t) = -\infty$ and $w(\alpha, t) = \infty$ (which shows the conditional tests corresponding to $y = 0$ and $y = \alpha$ to be one-sided), and that the left-hand side of (26) for any y is the power of this conditional test.

(3) For fixed t, the functions

$$H_1(y, t) = G_t(v-) + \gamma_1[G_t(v) - G_t(v-)]$$

and

$$H_2(y, t) = 1 - G_t(w) + \gamma_2[G_t(w) - G_t(w-)]$$

are continuous functions of y. This is a consequence of the fact, which follows from (27), that a.e. $\mathscr{P}^T$ the discontinuities and flat stretches of F_t and G_t coincide.

(4) The function $H(y, t)$ is jointly measurable in y and t. This follows from the continuity of H by an argument similar to the proof of measurability of $F_t(u)$ in the text. Define

$$y(t) = \inf \{y\colon\ H(y, t) < \alpha\},$$

and let $v(t) = v[y(t), t]$, etc. Then (25) and (26) are satisfied for all t. The measurability of $v(t)$, $w(t)$, $\gamma_1(t)$, and $\gamma_2(t)$ defined in this manner will follow from measurability in t of $y(t)$ and $F_t^{-1}[y(t)]$. This is a consequence of the relations, which hold for all real c,

$$\{t\colon\ y(t) < c\} = \bigcup_{r<c} \{t\colon\ H(r, t) < \alpha\}$$

where r indicates a rational, and

$$\{t\colon\ F_t^{-1}[y(t)] \leqq c\} = \{t\colon\ y(t) - F_t(c) \leqq 0\}.]$$

15. *Continuation.* The function ϕ_4 defined by (16), (18), and (19) is jointly measurable in u and t.

[The proof, which otherwise is essentially like that outlined in the preceding problem, requires the measurability in z and t of the integral

$$g(z, t) = \int_{-\infty}^{z-} u\, dF_t(u).$$

This integral is absolutely convergent for all t since F_t is a distribution belonging to an exponential family. For any $z < \infty$, $g(z, t) = \lim g_n(z, t)$, where

$$g_n(z, t) = \sum_{j=1}^{\infty} \left(z - \frac{j}{2^n}\right)\left[F_t\left(z - \frac{j-1}{2^n} - 0\right) - F_t\left(z - \frac{j}{2^n} - 0\right)\right],$$

and the measurability of g follows from that of the functions g_n. The inequalities corresponding to those obtained in step (2) of the preceding problem result from the property of the conditional one-sided tests established in Problem 18 of Chapter 3.]

16. The UMP unbiased tests of the hypotheses $H_1, \cdots, H_4$ of Theorem 3 are unique if attention is restricted to tests depending on U and the T's.

Section 5

17. Let X and Y be independently distributed with Poisson distributions $P(\lambda)$ and $P(\mu)$. Find the power of the UMP unbiased test of $H\colon\ \mu \leqq \lambda$, against the alternatives $\lambda = .1,\ \mu = .2;\ \ \lambda = 1,\ \mu = 2;\ \ \lambda = 10,\ \mu = 20;\ \ \lambda = .1,\ \mu = .4$; at level of significance $\alpha = .1$.

[Since $T = X + Y$ has the Poisson distribution $P(\lambda + \mu)$, the power is

$$\beta = \sum_{t=0}^{\infty} \beta(t) \frac{(\lambda + \mu)^t}{t!} e^{-(\lambda+\mu)}$$

where $\beta(t)$ is the power of the conditional test given t against the alternative in question.]

18. *Sequential comparison of two binomials.* Consider two sequences of binomial trials with probabilities of success p_1 and p_2 respectively, and let $\rho = (p_2/q_2) \div (p_1/q_1)$.

(i) If $\alpha < \beta$, no test with fixed numbers of trials m and n for testing H: $\rho = \rho_0$ can have power $\geqq \beta$ against all alternatives with $\rho = \rho_1$.

(ii) The following is a simple sequential sampling scheme leading to the desired result. Let the trials be performed in pairs of one of each kind, and restrict attention to those pairs in which one of the trials is a success and the other a failure. If experimentation is continued until N such pairs have been observed, the number of pairs in which the successful trial belonged to the first series has the binomial distribution $b(\pi, N)$ with $\pi = p_1q_2/(p_1q_2 + p_2q_1) = 1/(1 + \rho)$. A test of arbitrarily high power against ρ_1 is therefore obtained by taking N large enough.

(iii) The pairs of trials to which attention is restricted in (ii) constitute independent binomial trials with probability π of success. An alternative procedure for testing H: $\pi = \pi_0$ (or $\pi \leqq \pi_0$) to that given in (ii) is the sequential probability ratio test, based on a sequence of such pairs, for testing $\pi = \pi_0$ against $\pi = \pi_1$.

Section 6

19. *Runs.* Consider a sequence of N dependent trials, and let X_i be 1 or 0 as the ith trial is a success or failure. Suppose that the sequence has the *Markov* property*

$$P\{X_i = 1|x_1, \cdots, x_{i-1}\} = P\{X_i = 1|x_{i-1}\}$$

and the property of *stationarity* according to which $P\{X_i = 1\}$ and $P\{X_i = 1|x_{i-1}\}$ are independent of i. The distribution of the X's is then specified by the probabilities

$$p_1 = P\{X_i = 1|X_{i-1} = 1\} \quad \text{and} \quad p_0 = P\{X_i = 1|X_{i-1} = 0\}$$

and by the initial probabilities

$$\pi_1 = P\{X_1 = 1\} \quad \text{and} \quad \pi_0 = 1 - \pi_1 = P\{X_1 = 0\}.$$

(i) Stationarity implies that

$$\pi_1 = p_0/(p_0 + q_1), \qquad \pi_0 = q_1/(p_0 + q_1).$$

(ii) A set of successive outcomes $x_i, x_{i+1}, \cdots, x_{i+j}$ is said to form a *run* of zeros if $x_i = x_{i+1} = \cdots = x_{i+j} = 0$, and $x_{i-1} = 1$ or $i = 1$, and $x_{i+j+1} = 1$ or $i + j = N$. A run of ones is defined analogously. The probability of any particular sequence of outcomes $(x_1, \cdots, x_N)$ is

$$\frac{1}{p_0 + q_1} p_0^v p_1^{n-v} q_1^u q_0^{m-u},$$

where m and n denote the number of zeros and ones, and u and v the number of runs of zeros and ones in the sequence.

20. *Continuation.* For testing the hypothesis of independence of the X's, H: $p_0 = p_1$, against the alternatives K: $p_0 < p_1$, consider the *run test*, which rejects H when the total number of runs $R = U + V$ is less than a constant

* For a recent discussion of statistical problems in more complex Markov chains, see Anderson and Goodman, "Statistical inference about Markov chains," *Ann. Math. Stat.*, Vol. 28 (1957), pp. 89–110, and Goodman, "Simplified runs tests and likelihood ratio tests for Markoff chains," *Biometrika*, Vol. 45 (1958), pp. 181–197.

$C(m)$ depending on the number m of zeros in the sequence. When $R = C(m)$, the hypothesis is rejected with probability $\gamma(m)$, where C and γ are determined by

$$P_H\{R < C(m)|m\} + \gamma(m)P_H\{R = C(m)|m\} = \alpha.$$

(i) Against any alternative of K the most powerful similar test (which is at least as powerful as the most powerful unbiased test) coincides with the run test in that it rejects H when $R < C(m)$. Only the supplementary rule for bringing the conditional probability of rejection (given m) up to α depends on the specific alternative under consideration.

(ii) The run test is unbiased against the alternatives K.

(iii) The conditional distribution of R given m, when H is true, is*

$$P\{R = 2r\} = \frac{2\binom{m-1}{r-1}\binom{n-1}{r-1}}{\binom{m+n}{m}};$$

$$P\{R = 2r+1\} = \frac{\binom{m-1}{r-1}\binom{n-1}{r} + \binom{m-1}{r}\binom{n-1}{r-1}}{\binom{m+n}{m}}.$$

[(i) Unbiasedness implies that the conditional probability of rejection given m is α for all m. The most powerful conditional level α test rejects H for those sample sequences for which $\Delta(u, v) = (p_0/p_1)^v(q_1/q_0)^u$ is too large. Since $p_0 < p_1$ and $q_1 < q_0$ and since $|v - u|$ can only take on the values 0 and 1, it follows that

$$\Delta(1, 1) > \Delta(1, 2), \Delta(2, 1) > \Delta(2, 2) > \Delta(2, 3), \Delta(3, 2) > \cdots.$$

Thus only the relation between $\Delta(i, i+1)$ and $\Delta(i+1, i)$ depends on the specific alternative, and this establishes the desired result.

(ii) That the above conditional test is unbiased for each m is seen by writing its power as

$$\beta(p_0, p_1|m) = (1 - \gamma)P\{R < C(m)|m\} + \gamma P\{R \leqq C(m)|m\},$$

since by (i) the rejection regions $R < C(m)$ and $R < C(m) + 1$ are both UMP at their respective conditional levels.

(iii) When H is true, the conditional probability given m of any set of m zeros and n ones is $1\Big/\binom{m+n}{m}$. The number of ways of dividing n ones into r groups is $\binom{n-1}{r-1}$ and that of dividing m zeros into $r+1$ groups is $\binom{m-1}{r}$. The conditional probability of getting $r+1$ runs of zeros and r runs of ones is therefore $\binom{m-1}{r}\binom{n-1}{r-1}\Big/\binom{m+n}{m}$. To complete the proof, note that the total number of runs is $2r+1$ if and only if there are either $r+1$ runs of zeros and r runs of ones or r runs of zeros and $r+1$ runs of ones.]

* This distribution is tabled by Swed and Eisenhart, "Tables for testing randomness of grouping in a sequence of alternatives," *Ann. Math. Stat.*, Vol. 14 (1943), pp. 66–87. For further discussion of the run test see Wolfowitz, "On the theory of runs with some applications to quality control," *Ann. Math. Stat.*, Vol. 14 (1943), pp. 280–288.

21. *Rank-sum test.* Let $Y_1, \cdots, Y_N$ be independently distributed according to the binomial distributions $b(p_i, n_i)$, $i = 1, \cdots, N$, where

$$p_i = 1/[1 + e^{-(\alpha+\beta x_i)}].$$

This is the model frequently assumed in bio-assay, where x_i denotes the dose, or some function of the dose such as its logarithm, of a drug given to n_i experimental subjects, and where Y_i is the number among these subjects which respond to the drug at level x_i. Here the x_i are known, and α and β are unknown parameters.

(i) The joint distribution of the Y's constitutes an exponential family, and UMP unbiased tests exist for the four hypotheses of Theorem 3, concerning both α and β.

(ii) Suppose in particular that $x_i = \Delta i$, where Δ is known, and that $n_i = 1$ for all i. Let n be the number of successes in the N trials, and let these successes occur in the s_1st, s_2nd, $\cdots$, s_nth trial where $s_1 < s_2 < \cdots < s_n$. Then the UMP unbiased test for testing H: $\beta = 0$ against the alternatives $\beta > 0$ is carried out conditionally, given n, and rejects when the *rank-sum* $\sum_{i=1}^n s_i$ is too large.*

(iii) Let $Y_1, \cdots, Y_M$ and $Z_1, \cdots, Z_N$ be two independent sets of experiments of the type described at the beginning of the problem, corresponding, say, to two different drugs. If Y_i is distributed as $b(p_i, m_i)$ and Z_j as $b(\pi_j, n_j)$, with

$$p_i = 1/[1 + e^{-(\alpha+\beta u_i)}], \qquad \pi_j = 1/[1 + e^{-(\gamma+\delta v_j)}],$$

UMP unbiased tests exist for the four hypotheses concerning $\gamma - \alpha$ and $\delta - \beta$.

9. REFERENCES

Bartlett, M. S.

(1937) "Properties of sufficiency and statistical tests," *Proc. Roy. Soc. London, Ser. A*, Vol. 160, pp. 268–282.

[Points out that *exact* (that is, similar) tests can be obtained by combining the conditional tests given the different values of a sufficient statistic. Applications.]

David, F. N.

(1947) "A power function for tests of randomness in a sequence," *Biometrika*, Vol. 34, pp. 335–339.

[Discusses the run test in connection with the model of Problem 19.]

Feller, W.

(1936) "Note on regions similar to the sample space," *Stat. Res. Mem.*, Vol. II, pp. 117–125.

[Obtains a result which implies the completeness of order statistics.]

Fisher, R. A.

(1934) *Statistical Methods for Research Workers*, Edinburgh, Oliver and Boyd, 5th and subsequent editions, Section 21.02.

[Proposes the conditional test for the hypothesis of independence in a 2 × 2 table.]

* For tables of this test, which is formally equivalent to the two-sample Wilcoxon test discussed in Chapter 6, Section 8, see Fix and Hodges, "Significance probabilities of the Wilcoxon test," *Ann. Math. Stat.*, Vol. 26 (1955), pp. 301–312.

Fraser, D. A. S.

(1953) "Completeness of order statistics," *Canad. J. Math.*, Vol. 6, pp. 42–45. [Problems 12 and 13.]

Ghosh, M. N.

(1948) "On the problem of similar regions," *Sankhyā*, Vol. 8, pp. 329–338. [Theorem 1.]

Girshick, M. A., Frederick Mosteller, and L. J. Savage

(1946) "Unbiased estimates for certain binomial sampling problems with applications," *Ann. Math. Stat.*, Vol. 17, pp. 13–23. [Problem 11.]

Haldane, J. B. S., and C. A. B. Smith

(1948) "A simple exact test for birth-order effect," *Ann. Eugenics*, Vol. 14, pp. 117–124.

[Proposes the rank-sum test in a setting similar to that of Problem 21.]

Hoel, Paul G.

(1945) "Testing the homogeneity of Poisson frequencies," *Ann. Math. Stat.*, Vol. 16, pp. 362–368.

[First example of Section 5.]

(1948) "On the uniqueness of similar regions," *Ann. Math. Stat.*, Vol. 19, pp. 66–71. [Theorem 1 under regularity assumptions.]

Irwin, J. O.

(1935) "Tests of significance for differences between percentages based on small numbers," *Metron*, Vol. 12, pp. 83–94.

[Proposes the conditional test for the hypothesis of independence in a 2 × 2 table, which was also proposed by Fisher (cf. Yates, "Contingency tables involving small numbers and the χ^2 test," *J. Roy. Stat. Soc., Suppl.*, Vol. 1 (1934), pp. 217–235).]

Kruskal, William H.

(1957) "Historical notes on the Wilcoxon unpaired two-sample test," *J. Am. Stat. Assoc.*, Vol. 52, pp. 356–360.

[Gives the early history of the rank-sum test of Problem 21.]

Lehmann, E. L.

(1947) "On families of admissible tests," *Ann. Math. Stat.*, Vol. 18, pp. 97–104. [Problem 8.]

(1950) "Some principles of the theory of testing hypotheses," *Ann. Math. Stat.*, Vol. 21, pp. 1–26.

[Lemma 1.]

(1952) "Testing multiparameter hypotheses," *Ann. Math. Stat.*, Vol. 23, pp. 541–552. [Problem 7.]

Lehmann, E. L., and Henry Scheffé

(1950, 1955) "Completeness, similar regions, and unbiased estimation," *Sankhyā*, Vol. 10, pp. 305–340; Vol. 15, pp. 219–236.

[Introduces the concept of completeness. Theorem 3 and applications.]

Nandi, H. K.

(1951) "On type B_1 and type B regions," *Sankhyā*, Vol. 11, pp. 13–22.

[One of the cases of Theorem 3, under regularity assumptions.]

Neyman, J.

(1935) "Sur la vérification des hypothèses statistiques composées," *Bull. soc. math. France*, Vol. 63, pp. 246–266.

[Theory of tests of composite hypotheses that are locally unbiased and locally most powerful.]

(1941) "On a statistical problem arising in routine analyses and in sampling inspection of mass distributions," *Ann. Math. Stat.*, Vol. 12, pp. 46–76.

Neyman, J., and E. S. Pearson

(1933) "On the problem of the most efficient tests of statistical hypotheses," *Phil. Trans. Roy. Soc., Ser. A*, Vol. 231, pp. 289–337.

[Introduces the concept of similarity and develops a method for determining the totality of similar regions.]

(1936, 1938) "Contributions to the theory of testing statistical hypotheses," *Stat. Res. Mem.*, Vol. I, pp. 1–37; Vol. II, pp. 25–57.

[Defines unbiasedness and determines both locally and UMP unbiased tests of certain classes of simple hypotheses.]

Przyborowski, J., and H. Wilenski

(1939) "Homogeneity of results in testing samples from Poisson series," *Biometrika*, Vol. 31, pp. 313–323.

[Derives the UMP similar test for the equality of two Poisson parameters.]

Putter, Joseph

(1955) "The treatment of ties in some nonparametric tests," *Ann. Math. Stat.*, Vol. 26, pp. 368–386.

[Discusses the treatment of ties in the sign test.]

Scheffé, Henry

(1943) "On a measure problem arising in the theory of non-parametric tests," *Ann. Math. Stat.*, Vol. 14, pp. 227–233.

[Proves the completeness of order statistics.]

Sverdrup, Erling

(1953) "Similarity, unbiasedness, minimaxibility and admissibility of statistical test procedures," *Skand. Aktuar. Tidskrift*, Vol. 36, pp. 64–86.

[Theorem 1 and results of the type of Theorem 3. Applications including the 2×2 table.]

Tocher, K. D.

(1950) "Extension of Neyman-Pearson theory of tests to discontinuous variates," *Biometrika*, Vol. 37, pp. 130–144.

[Proves the optimum property of the test of Fisher and Irwin given in Section 6.]

Wald, Abraham

(1947) *Sequential Analysis*, New York, John Wiley & Sons, Section 6.3.

[Problem 18(iii).]

Walsh, John E.

(1949) "Some significance tests for the median which are valid under very general conditions," *Ann. Math. Stat.*, Vol. 20, pp. 64–81.

[Contains a result related to Problem 12.]

CHAPTER 5

Unbiasedness: Applications to Normal Distributions; Confidence Intervals

1. STATISTICS INDEPENDENT OF A SUFFICIENT STATISTIC

A general expression for the UMP unbiased tests of the hypotheses $H_1: \theta \leqq \theta_0$ and $H_4: \theta = \theta_0$ in the exponential family

$$dP_{\theta,\vartheta}(x) = C(\theta, \vartheta) \exp [\theta U(x) + \Sigma\vartheta_i T_i(x)]\, d\mu(x) \tag{1}$$

was given in Theorem 3 of the preceding chapter. However, this turns out to be inconvenient in the applications to normal and certain other families of continuous distributions, with which we shall be concerned in the present chapter. In these applications, the tests can be given a more convenient form, in which they no longer appear as conditional tests in terms of U given t but are expressed in terms of a single test statistic.

This reduction depends on the existence of a statistic $V = h(U, T)$ which is independent of T when $\theta = \theta_0$, and which for each fixed t is monotone in U for H_1 and linear in U for H_4. The critical function ϕ_1 for testing H_1 then satisfies

$$\phi(v) = \begin{cases} 1 & \text{when } v > C_0 \\ \gamma_0 & \text{when } v = C_0 \\ 0 & \text{when } v < C_0 \end{cases} \tag{2}$$

where C_0 and γ_0 are no longer dependent on t, and are determined by

$$E_{\theta_0} \phi_1(V) = \alpha. \tag{3}$$

Similarly the test ϕ_4 of H_4 reduces to

$$\phi(v) = \begin{cases} 1 & \text{when } v < C_1 \text{ or } v > C_2 \\ \gamma_i & \text{when } v = C_i, \qquad i = 1, 2 \\ 0 & \text{when } C_1 < v < C_2 \end{cases} \tag{4}$$

where the C's and γ's are determined by

$$E_{\theta_0}[\phi_4(V)] = \alpha \tag{5}$$

and

$$E_{\theta_0}[V\phi_4(V)] = \alpha E_{\theta_0}(V). \tag{6}$$

The corresponding reduction for the hypotheses $H_2: \theta \leqq \theta_1$ or $\theta \geqq \theta_2$ and $H_3: \theta_1 \leqq \theta \leqq \theta_2$ requires that V be monotone in U for each fixed t, and be independent of T when $\theta = \theta_1$ and $\theta = \theta_2$. The test ϕ_3 is then given by (4) with the C's and γ's determined by

$$E_{\theta_1} \phi_3(V) = E_{\theta_2} \phi_3(V) = \alpha. \tag{7}$$

The test for H_2 as before has the critical function

$$\phi_2(v; \alpha) = 1 - \phi_3(v; 1 - \alpha).$$

This is summarized in the following theorem.

Theorem 1. *Suppose that the distribution of X is given by* (1), *and that $V = h(U, T)$ is independent of T when $\theta = \theta_0$. Then ϕ_1 is UMP unbiased for testing H_1 provided the function h is increasing in u for each t, and ϕ_4 is UMP unbiased for H_4 provided*

$$h(u, t) = a(t)u + b(t) \quad \text{with} \quad a(t) > 0.$$

The tests ϕ_2 and ϕ_3 are UMP unbiased for H_2 and H_3 if V is independent of T when $\theta = \theta_1$ and θ_2, and if h is increasing in u for each t.

Proof. The test of H_1 defined by (12) and (13) of Chapter 4 is equivalent to that given by (2), with the constants determined by

$$P_{\theta_0}\{V > C_0(t) | t\} + \gamma_0(t) P_{\theta_0}\{V = C_0(t) | t\} = \alpha.$$

By assumption, V is independent of T when $\theta = \theta_0$, and C_0 and γ_0 therefore do not depend on t. This completes the proof for H_1, and that for H_2 and H_3 is quite analogous.

The test of H_4 given in Section 4 of Chapter 4 is equivalent to that defined by (4) with the constants C_i and γ_i determined by $E_{\theta_0}[\phi_4(V, t) | t] = \alpha$ and

$$E_{\theta_0}\left[\phi_4(V, t)\, \frac{V - b(t)}{a(t)} \,\Big|\, t\right] = \alpha E_{\theta_0}\left[\frac{V - b(t)}{a(t)} \,\Big|\, t\right],$$

which reduces to

$$E_{\theta_0}[V\phi_4(V, t) | t] = \alpha E_{\theta_0}[V | t].$$

Since V is independent of T for $\theta = \theta_0$, so are the C's and γ's, as was to be proved.

To prove the required independence of V and T in applications of Theorem 1 to special cases, the standard methods of distribution theory are available: transformation of variables, characteristic functions, and the geometric method. Frequently, an alternative approach, which is particularly useful also in determining a suitable statistic V, is provided by the following theorem.

Theorem 2. *Let the family of possible distributions of X be $\mathscr{P} = \{P_\vartheta, \vartheta \in \omega\}$, let T be sufficient for $\mathscr{P}$, and suppose that the family $\mathscr{P}^T$ of distributions of T is boundedly complete. If V is any statistic whose distribution does not depend on ϑ, then V is independent of T.*

Proof. For any critical function ϕ, the expectation $E_\vartheta\phi(V)$ is by assumption independent of ϑ. It therefore follows from Theorem 2 of Chapter 4 that $E[\phi(V)|t]$ is constant (a.e. $\mathscr{P}^T$) for every critical function ϕ and hence that V is independent of T.

Corollary 1. *Let $\mathscr{P}$ be the exponential family obtained from* (1) *by letting θ have some fixed value. Then a statistic V is independent of T for all ϑ provided the distribution of V does not depend on ϑ.*

Proof. It follows from Theorem 1 of Chapter 4 that $\mathscr{P}^T$ is complete and hence boundedly complete, and the preceding theorem is therefore applicable.

Example 1. Let $X_1, \cdots, X_n$ be independently, normally distributed with mean ξ and variance σ^2. Suppose first that σ^2 is fixed at σ_0^2. Then the assumptions of Corollary 1 hold with $T = \bar{X} = \Sigma X_i/n$ and ϑ proportional to ξ. Let f be any function satisfying

$$f(x_1 + c, \cdots, x_n + c) = f(x_1, \cdots, x_n) \quad \text{for all real} \quad c.$$

If

$$V = f(X_1, \cdots, X_n),$$

then also $V = f(X_1 - \xi, \cdots, X_n - \xi)$. Since the variables $X_i - \xi$ are distributed as $N(0, \sigma_0^2)$ which does not involve ξ, the distribution of V does not depend on ξ. It follows from Corollary 1 that any such statistic V, and therefore in particular $V = \Sigma(X_i - \bar{X})^2$, is independent of $\bar{X}$. This is true for all σ.

Suppose, on the other hand, that ξ is fixed at ξ_0. Then Corollary 1 applies with $T = \Sigma(X_i - \xi_0)^2$ and $\vartheta = -1/2\sigma^2$. Let f be any function such that

$$f(cx_1, \cdots, cx_n) = f(x_1, \cdots, x_n) \quad \text{for all} \quad c > 0,$$

and let

$$V = f(X_1 - \xi_0, \cdots, X_n - \xi_0).$$

Then V is unchanged if each $X_i - \xi_0$ is replaced by $(X_i - \xi_0)/\sigma$ and since these variables are normally distributed with zero mean and unit variance, the distribution of V does not depend on σ. It follows that all such statistics V, and hence for example

$$(\bar{X} - \xi_0)/\sqrt{\Sigma(X_i - \bar{X})^2} \quad \text{and} \quad (\bar{X} - \xi_0)/\sqrt{\Sigma(X_i - \xi_0)^2},$$

are independent of $\Sigma(X_i - \xi_0)^2$. This, however, does not hold for all ξ, but only when $\xi = \xi_0$.

Example 2. Let U_1/σ_1^2 and U_2/σ_2^2 be independently distributed according to χ^2-distributions with f_1 and f_2 degrees of freedom respectively, and suppose that $\sigma_2^2/\sigma_1^2 = a$. The joint density of the U's is then

$$Cu_1^{\frac{1}{2}f_1-1}u_2^{\frac{1}{2}f_2-1} \exp\left[-\frac{1}{2\sigma_2^2}(au_1 + u_2)\right]$$

so that Corollary 1 is applicable with $T = aU_1 + U_2$ and $\vartheta = -1/2\sigma_2^2$. Since the distribution of

$$V = \frac{U_2}{U_1} = a\frac{U_2/\sigma_2^2}{U_1/\sigma_1^2}$$

does not depend on σ_2, V is independent of $aU_1 + U_2$. For the particular case that $\sigma_2 = \sigma_1$, this proves the independence of U_2/U_1 and $U_1 + U_2$.

Example 3. Let $(X_1, \cdots, X_n)$ and $(Y_1, \cdots, Y_n)$ be samples from normal distributions $N(\xi, \sigma^2)$ and $N(\eta, \tau^2)$ respectively. Then $T = (\bar{X}, \Sigma X_i^2, \bar{Y}, \Sigma Y_i^2)$ is sufficient for $(\xi, \sigma^2, \eta, \tau^2)$ and the family of distributions of T is complete. Since

$$V = \frac{\Sigma(X_i - \bar{X})(Y_i - \bar{Y})}{\sqrt{\Sigma(X_i - \bar{X})^2\Sigma(Y_i - \bar{Y})^2}}$$

is unchanged when X_i and Y_i are replaced by $(X_i - \xi)/\sigma$ and $(Y_i - \eta)/\tau$, the distribution of V does not depend on any of the parameters, and Theorem 2 shows V to be independent of T.

2. TESTING THE PARAMETERS OF A NORMAL DISTRIBUTION

The four hypotheses $\sigma \leq \sigma_0$, $\sigma \geq \sigma_0$, $\xi \leq \xi_0$, $\xi \geq \xi_0$ concerning the variance σ^2 and mean ξ of a normal distribution were discussed in Chapter 3, Section 9, and it was pointed out there that at the usual significance levels there exists a UMP test only for the first one. We shall now show that the standard (likelihood ratio) tests are UMP unbiased for the above four hypotheses as well as for some of the corresponding two-sided problems.

For varying ξ and σ, the densities

$$(8)\qquad (2\pi\sigma^2)^{-n/2} \exp\left(-\frac{n\xi^2}{2\sigma^2}\right) \exp\left(-\frac{1}{2\sigma^2}\Sigma x_i^2 + \frac{\xi}{\sigma^2}\Sigma x_i\right)$$

of a sample $X_1, \cdots, X_n$ from $N(\xi, \sigma^2)$ constitute a two-parameter exponential family, which coincides with (1) for

$$\theta = -1/2\sigma^2, \qquad \vartheta = n\xi/\sigma^2, \qquad U(x) = \Sigma x_i^2, \qquad T(x) = \bar{x} = \Sigma x_i/n.$$

By Theorem 3 of Chapter 4 there exists therefore a UMP unbiased test of the hypothesis $\theta \geqq \theta_0$, which for $\theta_0 = -1/2\sigma_0^2$ is equivalent to $H: \sigma \geqq \sigma_0$. The rejection region of this test can be obtained from (12) of Chapter 4, with the inequalities reversed since the hypothesis is now $\theta \geqq \theta_0$. In the present case this becomes

$$\Sigma x_i^2 \leqq C_0(\bar{x})$$

where

$$P_{\sigma_0}\{\Sigma X_i^2 \leqq C_0(\bar{x}) | \bar{x}\} = \alpha.$$

If this is written as

$$\Sigma x_i^2 - n\bar{x}^2 \leqq C_0'(\bar{x}),$$

it follows from the independence of $\Sigma X_i^2 - n\bar{X}^2 = \Sigma(X_i - \bar{X})^2$ and $\bar{X}$ (Example 1) that $C_0'(\bar{x})$ does not depend on $\bar{x}$. The test therefore rejects when $\Sigma(x_i - \bar{x})^2 \leqq C_0'$, or equivalently when

$$\Sigma(x_i - \bar{x})^2/\sigma_0^2 \leqq C_0, \tag{9}$$

with C_0 determined by $P_{\sigma_0}\{\Sigma(X_i - \bar{X})^2/\sigma_0^2 \leqq C_0\} = \alpha$. Since $\Sigma(X_i - \bar{X})^2/\sigma_0^2$ has a χ^2-distribution with $n - 1$ degrees of freedom, the determining condition for C_0 is

$$\int_0^{C_0} \chi_{n-1}^2(y)\, dy = \alpha \tag{10}$$

where χ_{n-1}^2 denotes the density of a χ^2 variable with $n - 1$ degrees of freedom.

The same result can be obtained through Theorem 1. A statistic $V = h(U, T)$ of the kind required by the theorem—that is, independent of $\bar{X}$ for $\sigma = \sigma_0$ and all ξ—is

$$V = \Sigma(X_i - \bar{X})^2 = U - nT^2.$$

This is in fact independent of $\bar{X}$ for all ξ and σ^2. Since $h(u, t)$ is an increasing function of u for each t, it follows that the UMP unbiased test has a rejection region of the form $V \leqq C_0'$.

This derivation also shows that the UMP unbiased rejection region for $H: \sigma \leqq \sigma_1$ or $\sigma \geqq \sigma_2$ is

$$C_1 < \Sigma(x_i - \bar{x})^2 < C_2 \tag{11}$$

where the C's are given by

$$\int_{C_1/\sigma_1^2}^{C_2/\sigma_1^2} \chi_{n-1}^2(y)\, dy = \int_{C_1/\sigma_2^2}^{C_2/\sigma_2^2} \chi_{n-1}^2(y)\, dy = \alpha. \tag{12}$$

Since $h(u, t)$ is linear in u, it is further seen that the UMP unbiased test of $H: \sigma = \sigma_0$ has the acceptance region

$$C_1' < \Sigma(x_i - \bar{x})^2/\sigma_0^2 < C_2' \tag{13}$$

with the constants determined by

$$\int_{C_1'}^{C_2'} \chi_{n-1}^2(y)\, dy = \frac{1}{n-1} \int_{C_1'}^{C_2'} y\chi_{n-1}^2(y)\, dy = 1 - \alpha. \tag{14}$$

This is just the test obtained in Example 2 of Chapter 4 with $\Sigma(x_i - \bar{x})^2$ in place of Σx_i^2 and $n - 1$ degrees of freedom instead of n, as could have been foreseen. Theorem 1 shows for this and the other hypotheses considered that the UMP unbiased test depends only on V. Since the distributions of V do not depend on ξ, and constitute an exponential family in σ, the problems are thereby reduced to the corresponding ones for a one-parameter exponential family which were solved previously.

The power of the above tests can be obtained explicitly in terms of the χ^2-distribution. In the case of the one-sided test (9) for example, it is given by

$$\beta(\sigma) = P_\sigma\left\{\frac{\Sigma(X_i - \bar{X})^2}{\sigma^2} \leqq \frac{C_0\sigma_0^2}{\sigma^2}\right\} = \int_0^{C_0\sigma_0^2/\sigma^2} \chi_{n-1}^2(y)\, dy.$$

The same method can be applied to the problems of testing the hypotheses $\xi \leqq \xi_0$ against $\xi > \xi_0$ and $\xi = \xi_0$ against $\xi \neq \xi_0$. As is seen by transforming to the variables $X_i - \xi_0$, there is no loss of generality in assuming that $\xi_0 = 0$. It is convenient here to make the identification of (8) with (1) through the correspondence

$$\theta = n\xi/\sigma^2, \qquad \vartheta = -1/2\sigma^2, \qquad U(x) = \bar{x}, \qquad T(x) = \Sigma x_i^2.$$

Theorem 3 of Chapter 4 then shows that UMP unbiased tests exist for the hypotheses $\theta \leqq 0$ and $\theta = 0$, which are equivalent to $\xi \leqq 0$ and $\xi = 0$. Since

$$V = \bar{X}/\sqrt{\Sigma(X_i - \bar{X})^2} = U/\sqrt{T - nU^2}$$

is independent of $T = \Sigma X_i^2$ when $\xi = 0$ (Example 1), it follows from Theorem 1 that the UMP unbiased rejection region for $H: \xi \leqq 0$ is $V \geqq C_0'$ or equivalently

$$t(x) \geqq C_0 \tag{15}$$

where

$$t(x) = \frac{\sqrt{n}\bar{x}}{\sqrt{\dfrac{1}{n-1}\Sigma(x_i - \bar{x})^2}}. \tag{16}$$

In order to apply the theorem to H': $\xi = 0$, let $W = \bar{X}/\sqrt{\Sigma X_i^2}$. This is also independent of ΣX_i^2 when $\xi = 0$, and in addition is linear in $U = \bar{X}$. The distribution of W is symmetric about 0 when $\xi = 0$, and conditions (4), (5), (6) with W in place of V are therefore satisfied for the rejection region $|w| \geqq C'$ with $P_{\xi=0}\{|W| \geqq C'\} = \alpha$. Since

$$t(x) = \frac{\sqrt{(n-1)n}\,W(x)}{\sqrt{1 - nW^2(x)}},$$

the absolute value of $t(x)$ is an increasing function of $|W(x)|$, and the rejection region is equivalent to

$$|t(x)| \geqq C. \tag{17}$$

From (16) it is seen that $t(X)$ is the ratio of the two independent random variables $\sqrt{n}\bar{X}/\sigma$ and $\sqrt{\Sigma(X_i - \bar{X})^2/(n-1)\sigma^2}$. The denominator is distributed as the square root of a χ^2 variable with $n - 1$ degrees of freedom, divided by $n - 1$; the distribution of the numerator, when $\xi = 0$ is the normal distribution $N(0, 1)$. The distribution of such a ratio is *Student's* t-distribution with $n - 1$ degrees of freedom, which has probability density

$$t_{n-1}(y) = \frac{1}{\sqrt{\pi(n-1)}} \frac{\Gamma(\frac{1}{2}n)}{\Gamma[\frac{1}{2}(n-1)]} \frac{1}{\left(1 + \frac{y^2}{n-1}\right)^{\frac{1}{2}n}}. \tag{18}$$

The distribution is symmetric about 0, and the constants C_0 and C of the one- and two-sided tests are determined by

$$\int_{C_0}^{\infty} t_{n-1}(y)\,dy = \alpha \quad \text{and} \quad \int_{C}^{\infty} t_{n-1}(y)\,dy = \frac{\alpha}{2}. \tag{19}$$

For $\xi \neq 0$, the distribution of $t(X)$ is the so-called *noncentral* t-distribution, which is derived in Problem 3. Some properties of the power function of the one- and two-sided t-test are given in Problems 1, 2, and 4. We note here that the distribution of $t(X)$, and therefore the power of the above tests, depends only on the noncentrality parameter $\delta = \sqrt{n}\xi/\sigma$. This is seen from the expression of the probability density given in Problem 3, but can also be shown by the following direct argument. Suppose that $\xi'/\sigma' = \xi/\sigma \neq 0$, and denote the common value of ξ'/ξ and σ'/σ by c, which is then also different from zero. If $X_i' = cX_i$ and the X_i are distributed as $N(\xi, \sigma^2)$, the variables X_i' have distribution $N(\xi', \sigma'^2)$. Also $t(X) = t(X')$, and hence $t(X')$ has the same distribution as $t(X)$, as was to be proved.

If ξ_1 denotes any alternative value to $\xi = 0$, the power $\beta(\xi, \sigma) = f(\delta)$ depends on σ. As $\sigma \to \infty$, $\delta \to 0$, and

$$\beta(\xi_1, \sigma) \to f(0) = \beta(0, \sigma) = \alpha,$$

since f is continuous by Theorem 6 of Chapter 2. Therefore, regardless of the sample size the probability of detecting the hypothesis to be false when $\xi \geqq \xi_1 > 0$ cannot be made $\geqq \beta > \alpha$ for all σ. This is not surprising since the distributions $N(0, \sigma^2)$ and $N(\xi_1, \sigma^2)$ become practically indistinguishable when σ is sufficiently large. To obtain a procedure with guaranteed power for $\xi \geqq \xi_1$, the sample size must be made to depend on σ. This can be achieved by a sequential procedure, with the stopping rule depending on an estimate of σ, but not with a procedure of fixed sample size. (See Problems 15 and 17.)

The tests of the more general hypotheses $\xi \leqq \xi_0$ and $\xi = \xi_0$ are reduced to those above by transforming to the variables $X_i - \xi_0$. The rejection regions for these hypotheses are given as before by (15), (17), and (19), but now with

$$t(x) = \frac{\sqrt{n}(\bar{x} - \xi_0)}{\sqrt{\dfrac{1}{n-1}\Sigma(x_i - \bar{x})^2}}.$$

It is seen from the representation of (8) as an exponential family with $\theta = n\xi/\sigma^2$ that there exists a UMP unbiased test of the hypothesis $a \leqq \xi/\sigma^2 \leqq b$, but the method does not apply to the more interesting hypothesis $a \leqq \xi \leqq b$;* nor is it applicable to the corresponding hypothesis for the mean expressed in σ-units: $a \leqq \xi/\sigma \leqq b$, which will be discussed in Chapter 6.

The tests for mean and variance, which above were proved to be UMP unbiased, in one important respect behave very differently. If the variables $X_1, \cdots, X_n$ constitute a sample from any distribution with finite variance and zero mean and if the sample size n is sufficiently large, the distribution of the statistic (16) will be approximately the normal distribution $N(0, 1)$. This follows from the central limit theorem, according to which $\sqrt{n}\bar{X}/\sigma$ has the limiting distribution $N(0, 1)$, and the fact that $\Sigma(X_i - \bar{X})^2/(n-1)\sigma^2$ tends to one in probability, by a convergence theorem of Cramér.† As a consequence, at least for large samples,

* This problem is discussed in Section 3 of Hodges and Lehmann, "Testing the approximate validity of statistical hypotheses," *J. Roy. Stat. Soc., Ser. B.*, Vol. 16 (1954), pp. 261–268.

† For a statement and proof of this theorem see Cramér, *Mathematical Methods of Statistics*, Princeton Univ. Press, 1946, p. 254.

the size of the t-test will be approximately equal to the stated significance level even when the underlying distribution is not normal.*

On the other hand, the limiting distribution of $\Sigma(X_i - \bar{X})^2/\sqrt{n}\sigma^2$ is not independent of the underlying distribution of the X_i but depends on the fourth moment $E(X_i^4)$. To see this, suppose without loss of generality that $E(X_i) = 0$ since $\Sigma(X_i - \bar{X})^2$ does not depend on the mean of the X_i. Then $\sqrt{n}\bar{X}$ has the limiting distribution $N(0, \sigma^2)$ and $n\bar{X}^2/\sqrt{n}$ tends to zero in probability. It follows that $[\Sigma(X_i - \bar{X})^2 - n\sigma^2]/\sqrt{n}$ has the same limiting distribution as $[\Sigma X_i^2 - n\sigma^2]/\sqrt{n}$; namely by the central limit theorem the normal distribution $N(0, \tau^2)$ where τ^2 is the variance of the variables X_i^2. As a consequence, the size of the variance tests (9) and (11) may be far from the stated significance levels even for large samples when the underlying distribution is not normal.

3. COMPARING THE MEANS AND VARIANCES OF TWO NORMAL DISTRIBUTIONS

The problem of comparing the parameters of two normal distributions arises in the comparison of two treatments, products, etc., under conditions similar to those discussed in Chapter 4 at the beginning of Section 5. We consider first the comparison of two variances σ^2 and τ^2, which occurs for example when one is concerned with the variability of analyses made by two different laboratories or by two different methods, and specifically the hypotheses $H: \tau^2/\sigma^2 \leqq \Delta_0$ and $H': \tau^2/\sigma^2 = \Delta_0$.

Let $X = (X_1, \cdots, X_m)$ and $Y = (Y_1, \cdots, Y_n)$ be samples from the normal distributions $N(\xi, \sigma^2)$ and $N(\eta, \tau^2)$ with joint density

$$C(\xi, \eta, \sigma, \tau) \exp\left(-\frac{1}{2\sigma^2}\Sigma x_i^2 - \frac{1}{2\tau^2}\Sigma y_j^2 + \frac{m\xi}{\sigma^2}\bar{x} + \frac{n\eta}{\tau^2}\bar{y}\right).$$

This is an exponential family with the four parameters

$$\theta = -1/2\tau^2, \qquad \vartheta_1 = -1/2\sigma^2, \qquad \vartheta_2 = n\eta/\tau^2, \qquad \vartheta_3 = m\xi/\sigma^2$$

* More detailed investigations of the behavior of the t-test for non-normal distributions were carried out by Gayen, "The distribution of Student's t in random samples of any size from non-normal universes," *Biometrika*, Vol. 36 (1949), pp. 353–369, and by Geary, "The distribution of Student's ratio for non-normal samples, *Suppl. J. Roy. Stat. Soc.*, Vol. 3 (1936), pp. 178–184. In particular, it is shown there that the limiting behavior takes over much sooner for the two-sided than for the one-sided test, and that in fact the one-sided test for small samples is quite sensitive to departures from normality. See also Tukey, "Some elementary problems of importance to small sample practice," *Human Biology*, Vol. 20 (1948), pp. 205–214, and the survey paper by Wallace, "Asymptotic approximations to distributions," *Ann. Math. Stat.*, Vol. 29 (1958), pp. 635–654.

and the sufficient statistics

$$U = \Sigma Y_j^2, \qquad T_1 = \Sigma X_i^2, \qquad T_2 = \bar{Y}, \qquad T_3 = \bar{X}.$$

It can be expressed equivalently (see Lemma 2 of Chapter 4), in terms of the parameters

$$\theta^* = (-1/2\tau^2) + (1/2\Delta_0\sigma^2), \qquad \vartheta_i^* = \vartheta_i \qquad (i = 1, 2, 3)$$

and the statistics

$$U^* = \Sigma Y_j^2, \qquad T_1^* = \Sigma X_i^2 + (1/\Delta_0)\Sigma Y_j^2, \qquad T_2^* = \bar{Y}, \qquad T_3^* = \bar{X}.$$

The hypotheses $\theta^* \leqq 0$ and $\theta^* = 0$, which are equivalent to H and H' respectively, therefore possess UMP unbiased tests by Theorem 3 of Chapter 4.

When $\tau^2 = \Delta_0\sigma^2$, the distribution of the statistic

$$V = \frac{\Sigma(Y_j - \bar{Y})^2/\Delta_0}{\Sigma(X_i - \bar{X})^2} = \frac{\Sigma(Y_j - \bar{Y})^2/\tau^2}{\Sigma(X_i - \bar{X})^2/\sigma^2}$$

does not depend on σ, ξ, or η, and it follows from Corollary 1 that V is independent of (T_1^*, T_2^*, T_3^*). The UMP unbiased test of H is therefore given by (2) and (3), so that the rejection region can be written as

$$\frac{\Sigma(Y_j - \bar{Y})^2/\Delta_0(n-1)}{\Sigma(X_i - \bar{X})^2/(m-1)} \geqq C_0. \tag{20}$$

When $\tau^2 = \Delta_0\sigma^2$, the statistic on the left-hand side of (20) is the ratio of the two independent χ^2 variables $\Sigma(Y_j - \bar{Y})^2/\tau^2$ and $\Sigma(X_i - \bar{X})^2/\sigma^2$, each divided by the number of its degrees of freedom. The distribution of such a ratio is the *F-distribution* with $n - 1$ and $m - 1$ degrees of freedom, which has the density

$$(21) \quad F_{n-1,m-1}(y) = \frac{\Gamma[\frac{1}{2}(m+n-2)]}{\Gamma[\frac{1}{2}(m-1)]\Gamma[\frac{1}{2}(n-1)]}\left(\frac{n-1}{m-1}\right)^{\frac{1}{2}(n-1)} \frac{y^{\frac{1}{2}(n-1)-1}}{\left(1 + \dfrac{n-1}{m-1}y\right)^{\frac{1}{2}(m+n-2)}}.$$

The constant C_0 of (20) is then determined by

$$\int_{C_0}^{\infty} F_{n-1,m-1}(y)\,dy = \alpha. \tag{22}$$

In order to apply Theorem 1 to H' let

$$W = \frac{\Sigma(Y_j - \bar{Y})^2/\Delta_0}{\Sigma(X_i - \bar{X})^2 + (1/\Delta_0)\Sigma(Y_j - \bar{Y})^2}.$$

This is also independent of $T^* = (T_1^*, T_2^*, T_3^*)$ when $\tau^2 = \Delta_0\sigma^2$, and is linear in U^*. The UMP unbiased acceptance region of H' is therefore

$$C_1 \leqq W \leqq C_2 \tag{23}$$

with the constants determined by (5) and (6) where V is replaced by W. On dividing numerator and denominator of W by σ^2 it is seen that for $\tau^2 = \Delta_0\sigma^2$, the statistic W is a ratio of the form $W_1/(W_1 + W_2)$ where W_1 and W_2 are independent χ^2 variables with $n - 1$ and $m - 1$ degrees of freedom respectively. Equivalently, $W = Y/(1 + Y)$ where $Y = W_1/W_2$ and where $(m - 1)Y/(n - 1)$ has the distribution $F_{n-1,m-1}$. The distribution of W is the *beta-distribution*† with density

$$\begin{aligned}(24)\quad & B_{\frac{1}{2}(n-1),\frac{1}{2}(m-1)}(w) \\ &= \frac{\Gamma[\frac{1}{2}(m+n-2)]}{\Gamma[\frac{1}{2}(m-1)]\Gamma[\frac{1}{2}(n-1)]} w^{\frac{1}{2}(n-3)}(1-w)^{\frac{1}{2}(m-3)}, \qquad 0 < w < 1.\end{aligned}$$

Conditions (5) and (6), by means of the relations

$$E(W) = \frac{n-1}{m+n-2}$$

and

$$wB_{\frac{1}{2}(n-1),\frac{1}{2}(m-1)}(w) = \frac{n-1}{m+n-2} B_{\frac{1}{2}(n+1),\frac{1}{2}(m-1)}(w)$$

become

$$\int_{C_1}^{C_2} B_{\frac{1}{2}(n-1),\frac{1}{2}(m-1)}(w)\,dw = \int_{C_1}^{C_2} B_{\frac{1}{2}(n+1),\frac{1}{2}(m-1)}(w)\,dw = 1 - \alpha. \tag{25}$$

The definition of V shows that its distribution depends only on the ratio τ^2/σ^2, and so does the distribution of W. The power of the tests (20) and (23) is therefore also a function only of the variable $\Delta = \tau^2/\sigma^2$; it can be expressed explicitly in terms of the F-distribution, for example in the first case by

$$\beta(\Delta) = P\left\{\frac{\Sigma(Y_j - \bar{Y})^2/\tau^2(n-1)}{\Sigma(X_i - \bar{X})^2/\sigma^2(m-1)} \geqq \frac{C_0\Delta_0}{\Delta}\right\} = \int_{C_0\Delta_0/\Delta}^{\infty} F_{n-1,m-1}(y)\,dy.$$

The hypothesis of equality of the means ξ, η of two normal distributions with unknown variances σ^2 and τ^2, the so-called *Behrens-Fisher problem* is not accessible by the present method. (See Example 5 of Chapter 4;

† The relationship $W = Y/(1 + Y)$ shows the F- and beta-distributions to be equivalent. An advantage of the latter are the extensive tables of its cumulative distribution function, *Tables of the Incomplete Beta Function*, Cambridge Univ. Press, 1932, edited by Karl Pearson.

for a possible approach to this problem see Chapter 6, Section 6.) We shall therefore consider only the simpler case in which the two variances are assumed to be equal. The joint density of the X's and Y's is then

$$(26) \qquad C(\xi, \eta, \sigma) \exp\left[-\frac{1}{2\sigma^2}(\Sigma x_i^2 + \Sigma y_j^2) + \frac{\xi}{\sigma^2}\Sigma x_i + \frac{\eta}{\sigma^2}\Sigma y_j\right],$$

which is an exponential family with parameters

$$\theta = \eta/\sigma^2, \qquad \vartheta_1 = \xi/\sigma^2, \qquad \vartheta_2 = -1/2\sigma^2$$

and the sufficient statistics

$$U = \Sigma Y_j, \qquad T_1 = \Sigma X_i, \qquad T_2 = \Sigma X_i^2 + \Sigma Y_j^2.$$

For testing the hypotheses

$$H: \eta - \xi \leqq 0 \qquad \text{and} \qquad H': \eta - \xi = 0$$

it is more convenient to represent the densities as an exponential family with the parameters

$$\theta^* = \frac{\eta - \xi}{\left(\frac{1}{m} + \frac{1}{n}\right)\sigma^2}, \qquad \vartheta_1^* = \frac{m\xi + n\eta}{(m+n)\,\sigma^2}, \qquad \vartheta_2^* = \vartheta_2$$

and the sufficient statistics

$$U^* = \bar{Y} - \bar{X}, \qquad T_1^* = m\bar{X} + n\bar{Y}, \qquad T_2^* = \Sigma X_i^2 + \Sigma Y_j^2.$$

That this is possible is seen from the identity

$$m\xi\bar{x} + n\eta\bar{y} = \frac{(\bar{y} - \bar{x})(\eta - \xi)}{\frac{1}{m} + \frac{1}{n}} + \frac{(m\bar{x} + n\bar{y})(m\xi + n\eta)}{m + n}.$$

It follows from Theorem 3 of Chapter 4 that UMP unbiased tests exist for the hypotheses $\theta^* \leqq 0$ and $\theta^* = 0$, and hence for H and H'.

When $\eta = \xi$, the distribution of

$$V = \frac{\bar{Y} - \bar{X}}{\sqrt{\Sigma(X_i - \bar{X})^2 + \Sigma(Y_j - \bar{Y})^2}} = \frac{U^*}{\sqrt{T_2^* - \frac{1}{m+n}T_1^{*2} - \frac{mn}{m+n}U^{*2}}}$$

does not depend on the common mean ξ or on σ, as is seen by replacing X_i by $(X_i - \xi)/\sigma$ and Y_j by $(Y_j - \xi)/\sigma$ in the expression for V, and V

is independent of (T_1^*, T_2^*). The rejection region of the UMP unbiased test of H can therefore be written as $V \geqq C_0'$ or

$$t(X, Y) \geqq C_0, \tag{27}$$

where

$$t(X, Y) = \frac{(\bar{Y} - \bar{X}) \Big/ \sqrt{\frac{1}{m} + \frac{1}{n}}}{\sqrt{[\Sigma(X_i - \bar{X})^2 + \Sigma(Y_j - \bar{Y})^2]/(m + n - 2)}}. \tag{28}$$

The statistic $t(X, Y)$ is the ratio of the two independent variables

$$\frac{\bar{Y} - \bar{X}}{\sqrt{\left(\frac{1}{m} + \frac{1}{n}\right)\sigma^2}} \quad \text{and} \quad \sqrt{\frac{\Sigma(X_i - \bar{X})^2 + \Sigma(Y_j - \bar{Y})^2}{(m + n - 2)\sigma^2}}.$$

The numerator is normally distributed with mean $(\eta - \xi)/\sqrt{m^{-1} + n^{-1}}\sigma$ and unit variance; the denominator as the square root of a χ^2 variable with $(m + n - 2)$ degrees of freedom, divided by $(m + n - 2)$. Hence $t(X, Y)$ has a noncentral t-distribution with $(m + n - 2)$ degrees of freedom and noncentrality parameter

$$\delta = \frac{\eta - \xi}{\sqrt{\frac{1}{m} + \frac{1}{n}}\sigma}.$$

When in particular $\eta - \xi = 0$, the distribution of $t(X, Y)$ is Student's t-distribution, and the constant C_0 is determined by

$$\int_{C_0}^{\infty} t_{m+n-2}(y)\, dy = \alpha. \tag{29}$$

As before, the assumptions required by Theorem 1 for H' are not satisfied by V itself but by a function of V,

$$W = \frac{\bar{Y} - \bar{X}}{\sqrt{\Sigma X_i^2 + \Sigma Y_j^2 - \frac{(\Sigma X_i + \Sigma Y_j)^2}{m + n}}}$$

which is related to V through

$$V = \frac{W}{\sqrt{1 - \frac{mn}{m + n} W^2}}.$$

Since W is a function of V it is also independent of (T_1^*, T_2^*) when $\eta = \xi$;

in addition it is a linear function of U^* with coefficients dependent only on T^*. The distribution of W being symmetric about 0 when $\eta = \xi$, it follows, as in the derivation of the corresponding rejection region (17) for the one-sample problem, that the UMP unbiased test of H' rejects when $|W|$ is too large or equivalently when

$$|t(X, Y)| > C. \tag{30}$$

The constant C is determined by

$$\int_C^\infty t_{m+n-2}(y)\, dy = \frac{\alpha}{2}. \tag{31}$$

The power of the tests (27) and (30) depends only on $(\eta - \xi)/\sigma$ and is given in terms of the noncentral t-distribution. Its properties are analogous to those of the one-sample t-test (Problems 1, 2, and 4).

As in the corresponding one-sample problem, the tests based on the t-statistic (28) are insensitive to departures from normality while this is not the case for tests based on the F-ratio (20).† The result follows in both cases by applying the argument given in the one-sample problem. The robustness of the t-test will be seen from a different point of view in Section 8, where a modified test is discussed, the size of which is exactly independent of the underlying distribution.

4. CONFIDENCE INTERVALS AND FAMILIES OF TESTS

Confidence bounds for a parameter θ corresponding to a confidence level $1 - \alpha$ were defined in Chapter 3, Section 5, for the case that the distribution of the random variable X depends only on θ. When nuisance parameters ϑ are present the defining condition for a lower confidence bound $\underline{\theta}$ becomes

$$P_{\theta,\vartheta}\{\underline{\theta}(X) \leqq \theta\} \geqq 1 - \alpha \quad \text{for all} \quad \theta, \vartheta. \tag{32}$$

Similarly, confidence intervals for θ at confidence level $1 - \alpha$ are defined as a set of random intervals with end points $\underline{\theta}(X)$, $\bar{\theta}(X)$ such that

$$P_{\theta,\vartheta}\{\underline{\theta}(X) \leqq \theta \leqq \bar{\theta}(X)\} \geqq 1 - \alpha \quad \text{for all} \quad \theta, \vartheta. \tag{33}$$

† Tests for two or more variances which do not suffer from this disadvantage are discussed by Box, "Non-normality and tests of variances," *Biometrika*, Vol. 40 (1953), pp. 318–335, and by Box and Andersen, "Permutation theory in the derivation of robust criteria and the study of departures from assumptions," *J. Roy. Stat. Soc., Ser. B*, Vol. 17 (1955), pp. 1–34.

The infimum over (θ, ϑ) of the left-hand side of (32) and (33) is the *confidence coefficient* associated with these statements.

As was already indicated in Chapter 3, confidence statements permit a dual interpretation. Directly, they provide bounds for the unknown parameter θ and thereby a solution to the problem of estimating θ. The statement $\underline{\theta} \leqq \theta \leqq \bar{\theta}$ is not as precise as a point estimate, but it has the advantage that the probability of it being correct can be guaranteed to be at least $1 - \alpha$. Similarly, a lower confidence bound can be thought of as an estimate $\underline{\theta}$, which overestimates the true parameter value with probability $\leqq \alpha$. In particular for $\alpha = \frac{1}{2}$, if $\underline{\theta}$ satisfies

$$P_{\theta,\vartheta}\{\underline{\theta} \leqq \theta\} = P_{\theta,\vartheta}\{\underline{\theta} \geqq \theta\} = \tfrac{1}{2},$$

the estimate is as likely to underestimate as to overestimate and is then said to be *median unbiased*. (See Chapter 1, Problem 3, for the relation of this property to a more general concept of unbiasedness.)

Alternatively, as was shown in Chapter 3, confidence statements can be viewed as equivalent to a family of tests. The following is essentially a review of the discussion of this relationship in Chapter 3, made slightly more specific by restricting attention to the two-sided case. For each θ_0 let $A(\theta_0)$ denote the acceptance region of a level α test (assumed for the moment to be nonrandomized) of the hypothesis $H(\theta_0)\colon \theta = \theta_0$. If

$$S(x) = \{\theta\colon x \in A(\theta)\}$$

then

$$\theta \in S(x) \quad \text{if and only if} \quad x \in A(\theta), \tag{34}$$

and hence

$$P_{\theta,\vartheta}\{\theta \in S(X)\} \geqq 1 - \alpha \quad \text{for all} \quad \theta, \vartheta. \tag{35}$$

Thus any family of level α acceptance regions, through the correspondence (34), leads to a family of confidence sets at confidence level $1 - \alpha$.

Conversely, given any class of confidence sets $S(x)$ satisfying (35), let

$$A(\theta) = \{x\colon \theta \in S(x)\}. \tag{36}$$

Then the sets $A(\theta_0)$ are level α acceptance regions for testing the hypotheses $H(\theta_0)\colon \theta = \theta_0$, and the confidence sets $S(x)$ show for each θ_0 whether for the particular x observed the hypothesis $\theta = \theta_0$ is accepted or rejected at level α.

Exactly the same arguments apply if the sets $A(\theta_0)$ are acceptance regions for the hypotheses $\theta \leqq \theta_0$. As will be seen below, one- and two-sided tests typically, although not always, lead to one-sided confidence bounds and to confidence intervals respectively.

Example 4. Confidence intervals for the mean ξ of a normal distribution with unknown variance can be obtained from the acceptance regions $A(\xi_0)$ of the hypotheses $H: \xi = \xi_0$. These are given by

$$\frac{|\sqrt{n}(\bar{x} - \xi_0)|}{\sqrt{\Sigma(x_i - \bar{x})^2/(n-1)}} \leq C,$$

where C is determined from the t-distribution so that the probability of this inequality is $1 - \alpha$ when $\xi = \xi_0$. [See (17) and (19) of Section 2.] The set $S(x)$ is then the set of ξ's satisfying this inequality with $\xi = \xi_0$, that is, the interval

$$(37) \quad \bar{x} - \frac{C}{\sqrt{n}}\sqrt{\frac{1}{n-1}\Sigma(x_i - \bar{x})^2} \leq \xi \leq \bar{x} + \frac{C}{\sqrt{n}}\sqrt{\frac{1}{n-1}\Sigma(x_i - \bar{x})^2}.$$

The class of these intervals therefore constitutes confidence intervals for ξ with confidence coefficient $1 - \alpha$.

The length of the intervals (37) is proportional to $\sqrt{\Sigma(x_i - \bar{x})^2}$ and their expected length to σ. For large σ, the intervals will therefore provide only little information concerning the unknown ξ. This is a consequence of the fact, which led to similar difficulties for the corresponding testing problem, that two normal distributions $N(\xi_0, \sigma^2)$ and $N(\xi_1, \sigma^2)$ with fixed difference of means become indistinguishable as σ tends to infinity. In order to obtain confidence intervals for ξ whose length does not tend to infinity with σ, it is necessary to determine the number of observations sequentially so that it can be adjusted to σ. A sequential procedure leading to confidence intervals of prescribed length is given in Problems 15 and 16.

However, even such a sequential procedure does not really dispose of the difficulty, but only shifts the lack of control from the length of the interval to the number of observations. As $\sigma \to \infty$, the number of observations required to obtain confidence intervals of bounded length also tends to infinity. Actually, in practice one will frequently have an idea of the order of magnitude of σ. With a sample either of fixed size or obtained sequentially, it is then necessary to establish a balance between the desired confidence $1 - \alpha$, the accuracy given by the length l of the interval, and the number of observations n one is willing to expend. In such an arrangement two of the three quantities $1 - \alpha$, l, and n will be fixed while the third is a random variable whose distribution depends on σ, so that it will be less well controlled than the others. If $1 - \alpha$ is taken as fixed, the choice between a sequential scheme and one of fixed sample size thus depends essentially on whether it is more important to control l or n.

To obtain lower confidence limits for ξ, consider the acceptance regions

$$\frac{\sqrt{n}(\bar{x} - \xi_0)}{\sqrt{\Sigma(x_i - \bar{x})^2/(n-1)}} \leq C_0$$

for testing $\xi \leq \xi_0$ against $\xi > \xi_0$. The sets $S(x)$ are then the one-sided intervals

$$\bar{x} - \frac{C_0}{\sqrt{n}}\sqrt{\frac{1}{n-1}\Sigma(x_i - \bar{x})^2} \leq \xi,$$

the left-hand sides of which therefore constitute the desired lower bounds $\underline{\xi}$. If $\alpha = \frac{1}{2}$, the constant C_0 is 0; the resulting confidence bound $\underline{\xi} = \bar{X}$ is a median unbiased estimate of ξ and among all such estimates it uniformly maximizes

$$P\{-\Delta_1 \leqq \underline{\xi} - \xi \leqq \Delta_2\} \quad \text{for all} \quad \Delta_1, \Delta_2 \geqq 0.$$

(For a proof see p. 83.)

5. UNBIASED CONFIDENCE SETS

Confidence sets can be viewed as a family of tests of the hypotheses $\theta \in H(\theta')$ against alternatives $\theta \in K(\theta')$ for varying θ'. A confidence level of $1 - \alpha$ then simply expresses the fact that all the tests are to be at level α, and the condition therefore becomes

$$(38) \qquad P_{\theta,\vartheta}\{\theta' \in S(X)\} \geqq 1 - \alpha \quad \text{for all} \quad \theta \in H(\theta') \quad \text{and all} \quad \vartheta.$$

In the case that $H(\theta')$ is the hypothesis $\theta = \theta'$ and $S(X)$ is the interval $[\underline{\theta}(X), \bar{\theta}(X)]$, this agrees with (33). In the one-sided case in which $H(\theta')$ is the hypothesis $\theta \leqq \theta'$ and $S(X) = \{\theta: \underline{\theta}(X) \leqq \theta\}$, the condition reduces to $P_{\theta,\vartheta}\{\underline{\theta}(X) \leqq \theta'\} \geqq 1 - \alpha$ for all $\theta' \geqq \theta$, and this is seen to be equivalent to (32). With this interpretation of confidence sets, the probabilities

$$(39) \qquad P_{\theta,\vartheta}\{\theta' \in S(X)\}, \qquad \theta \in K(\theta')$$

are the probabilities of false acceptance of $H(\theta')$ (error of the second kind). The smaller these probabilities are, the more desirable are the tests.

From the point of view of estimation, on the other hand, (39) is the probability of covering the wrong value θ'. With a controlled probability of covering the true value, the confidence sets will be more informative the less likely they are to cover false values of the parameter. In this sense the probabilities (39) provide a measure of the accuracy of the confidence sets. A justification of (39) in terms of loss functions was given for the one-sided case in Chapter 3, Section 5.

In the presence of nuisance parameters, UMP tests usually do not exist and this implies the nonexistence of confidence sets that are uniformly most accurate in the sense of minimizing (39) for all θ' such that $\theta \in K(\theta')$ and for all ϑ. This suggests restricting attention to confidence sets which in a suitable sense are unbiased. In analogy with the corresponding definition for tests, a family of confidence sets at confidence level $1 - \alpha$ is said to be *unbiased* if

$$(40) \qquad P_{\theta,\vartheta}\{\theta' \in S(X)\} \leqq 1 - \alpha \quad \text{for all} \quad \theta' \quad \text{such that} \quad \theta \in K(\theta') \quad \text{and for all } \vartheta \text{ and } \theta,$$

so that the probability of covering these false values does not exceed the confidence level.

In the two- and one-sided cases mentioned above, condition (40) reduces to

$$P_{\theta,\vartheta}\{\underline{\theta} \leqq \theta' \leqq \bar{\theta}\} \leqq 1 - \alpha \quad \text{for all} \quad \theta' \neq \theta \quad \text{and all} \quad \vartheta$$

and

$$P_{\theta,\vartheta}\{\underline{\theta} \leqq \theta'\} \leqq 1 - \alpha \quad \text{for all} \quad \theta' < \theta \quad \text{and all} \quad \vartheta.$$

With this definition of unbiasedness, unbiased families of tests lead to unbiased confidence sets and conversely. A family of confidence sets is uniformly most accurate unbiased at confidence level $1 - \alpha$ if it minimizes the probabilities

$$P_{\theta,\vartheta}\{\theta' \in S(X)\} \quad \text{for all} \quad \theta' \quad \text{such that} \quad \theta \in K(\theta') \quad \text{and for all } \vartheta \text{ and } \theta,$$

subject to (38) and (40). The confidence sets obtained on the basis of the UMP unbiased tests of the present and preceding chapter are therefore uniformly most accurate unbiased. This applies in particular to the confidence intervals obtained in the preceding section. Some further examples are the following.

Example 5. If $X_1, \cdots, X_n$ is a sample from $N(\xi, \sigma^2)$, the UMP unbiased test of the hypothesis $\sigma = \sigma_0$ is given by the acceptance region (13)

$$C_1' \leqq \Sigma(x_i - \bar{x})^2/\sigma_0^2 \leqq C_2'$$

where C_1' and C_2' are determined by (14). The most accurate unbiased confidence intervals for σ^2 are therefore

$$\frac{1}{C_2'}\Sigma(x_i - \bar{x})^2 \leqq \sigma^2 \leqq \frac{1}{C_1'}\Sigma(x_i - \bar{x})^2.$$

Similarly, from (9) and (10) the most accurate unbiased upper confidence limits for σ^2 are

$$\sigma^2 \leqq \frac{1}{C_0}\Sigma(x_i - \bar{x})^2$$

where

$$\int_{C_0}^{\infty} \chi_{n-1}^2(y)\,dy = 1 - \alpha.$$

The corresponding lower confidence limits are uniformly most accurate (without the restriction of unbiasedness) by Chapter 3, Section 9.

Example 6. Confidence intervals for the difference $\Delta = \eta - \xi$ of the means of two normal distributions with common variance are obtained from tests of the hypothesis $\eta - \xi = \Delta_0$. If $X_1, \cdots, X_m$ and $Y_1, \cdots, Y_n$ are distributed as $N(\xi, \sigma^2)$ and $N(\eta, \sigma^2)$ respectively, and if $Y_j' = Y_j - \Delta_0$, $\eta' = \eta - \Delta_0$, the hypothesis can be expressed in terms of the variables X_i and Y_j' as $\eta' - \xi = 0$. From (28) and (30) the UMP unbiased acceptance region is then seen to be

$$\frac{|(\bar{y} - \bar{x} - \Delta_0)| \Big/ \sqrt{\dfrac{1}{m} + \dfrac{1}{n}}}{\sqrt{[\Sigma(x_i - \bar{x})^2 + \Sigma(y_j - \bar{y})^2]/(m + n - 2)}} \leqq C,$$

where C is determined by (31). The most accurate unbiased confidence intervals for $\eta - \xi$ are therefore

$$(\bar{y} - \bar{x}) - CS \leqq \eta - \xi \leqq (\bar{y} - \bar{x}) + CS \tag{41}$$

where

$$S^2 = \left(\frac{1}{m} + \frac{1}{n}\right) \frac{\Sigma(x_i - \bar{x})^2 + \Sigma(y_j - \bar{y})^2}{m + n - 2}.$$

The one-sided intervals are obtained analogously.

Example 7. If $X_1, \cdots, X_m$ and $Y_1, \cdots, Y_n$ are samples from $N(\xi, \sigma^2)$ and $N(\eta, \tau^2)$, most accurate unbiased confidence intervals for $\Delta = \tau^2/\sigma^2$ are derived from the acceptance region (23) as

$$\frac{1 - C_2}{C_2} \frac{\Sigma(y_j - \bar{y})^2}{\Sigma(x_i - \bar{x})^2} \leqq \frac{\tau^2}{\sigma^2} \leqq \frac{1 - C_1}{C_1} \frac{\Sigma(y_j - \bar{y})^2}{\Sigma(x_i - \bar{x})^2} \tag{42}$$

where C_1 and C_2 are determined from (25).* In the particular case that $m = n$, the intervals take on the simpler form

$$\frac{1}{k} \frac{\Sigma(y_j - \bar{y})^2}{\Sigma(x_i - \bar{x})^2} \leqq \frac{\tau^2}{\sigma^2} \leqq k \frac{\Sigma(y_j - \bar{y})^2}{\Sigma(x_i - \bar{x})^2} \tag{43}$$

where k is determined from the F-distribution. Most accurate unbiased lower confidence limits for the variance ratio are

$$\underline{\Delta} = \frac{1}{C_0} \frac{\Sigma(y_j - \bar{y})^2/(n - 1)}{\Sigma(x_i - \bar{x})^2/(m - 1)} \leqq \frac{\tau^2}{\sigma^2} \tag{44}$$

with C_0 given by (22). If in (22) α is taken to be $\frac{1}{2}$, this lower confidence limit $\underline{\Delta}$ becomes a median unbiased estimate of τ^2/σ^2. Among all such estimates it uniformly minimizes

$$P\left\{-\Delta_1 \leqq \frac{\tau^2}{\sigma^2} - \underline{\Delta} \leqq \Delta_2\right\} \quad \text{for all} \quad \Delta_1, \Delta_2 \geqq 0.$$

(For a proof see p. 83.)

So far it has been assumed that the tests from which the confidence sets are obtained are nonrandomized. The modifications that are necessary when this assumption is not satisfied were discussed in Chapter 3. The randomized tests can then be interpreted as being nonrandomized in the space of X and an auxiliary variable V which is uniformly distributed on the unit interval. If in particular X is integer-valued as in the binomial or Poisson case, the tests can be represented in terms of the continuous variable $X + V$. In this way, most accurate unbiased confidence intervals can be obtained, for example, for a binomial probability p from the UMP unbiased tests of $H: p = p_0$ (Example 1 of Chapter 4). It

* A comparison of these limits with those obtained from the equal tails test is given by Scheffé, "On the ratio of the variances of two normal populations," *Ann. Math. Stat.*, Vol. 13 (1942), pp. 371–388.

is not clear a priori that the resulting confidence sets for p will necessarily be intervals. This is, however, a consequence of the following lemma.

Lemma 1. *Let X be a real-valued random variable with probability density $p_\theta(x)$ which has monotone likelihood ratio in x. Suppose that UMP unbiased tests of the hypotheses $H(\theta_0)$: $\theta = \theta_0$ exist and are given by the acceptance regions*

$$C_1(\theta_0) \leqq x \leqq C_2(\theta_0),$$

and that they are strictly unbiased. Then the functions $C_i(\theta)$ are strictly increasing in θ, and the most accurate unbiased confidence intervals for θ are

$$C_2^{-1}(x) \leqq \theta \leqq C_1^{-1}(x).$$

Proof. Let $\theta_0 < \theta_1$ and let $\beta_0(\theta)$ and $\beta_1(\theta)$ denote the power functions

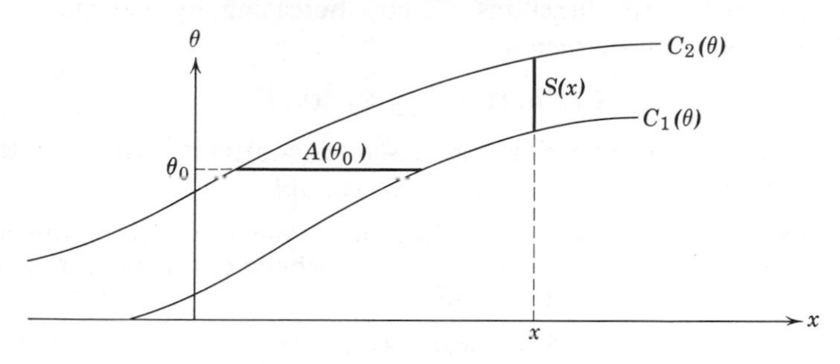

Figure 1

of the above tests ϕ_0 and ϕ_1 for testing $\theta = \theta_0$ and $\theta = \theta_1$. It follows from the strict unbiasedness of the tests that

$$E_{\theta_0}[\phi_1(X) - \phi_0(X)] = \beta_1(\theta_0) - \alpha > 0 > \alpha - \beta_0(\theta_1)$$
$$= E_{\theta_1}[\phi_1(X) - \phi_0(X)].$$

Thus neither of the two intervals $[C_1(\theta_i), C_2(\theta_i)]$ $(i = 0, 1)$ contains the other, and it is seen from Lemma 2(ii) of Chapter 3 that $C_i(\theta_0) < C_i(\theta_1)$ for $i = 1, 2$. The functions C_i therefore have inverses, and the inequalities defining the acceptance region for $H(\theta)$ are equivalent to $C_2^{-1}(x) \leqq \theta \leqq C_1^{-1}(x)$, as was to be proved.

The situation is indicated in Figure 1. From the boundaries $x = C_1(\theta)$ and $x = C_2(\theta)$ of the acceptance regions $A(\theta)$ one obtains for each fixed value of x the confidence set $S(x)$ as the interval of θ's for which $C_1(\theta) \leqq x \leqq C_2(\theta)$.

By Section 2 of Chapter 4, the conditions of the lemma are satisfied in particular for a one-parameter exponential family, provided the tests are nonrandomized. In cases such as that of binomial or Poisson distributions, where the family is exponential but X is integer-valued so that randomization is required, the intervals can be obtained by applying the lemma to the variable $X + V$ instead of X, where V is independent of X and uniformly distributed over (0, 1).

In Lemma 1, the distribution of X was assumed to depend only on θ. Consider now the exponential family (1) in which nuisance parameters are present in addition to θ. The UMP unbiased tests of $\theta = \theta_0$ are then performed as conditional tests given $T = t$, and the confidence intervals for θ will as a consequence also be obtained conditionally. If the conditional distributions are continuous, the acceptance regions will be of the form

$$C_1(\theta; t) \leqq u \leqq C_2(\theta; t)$$

where for each t the functions C_i are increasing by Lemma 1. The confidence intervals are then

$$C_2^{-1}(u; t) \leqq \theta \leqq C_1^{-1}(u; t).$$

If the conditional distributions are discrete, continuity can be obtained as before through addition of a uniform variable.

Example 8. Let X and Y be independent Poisson variables with means λ and μ, and let $\rho = \mu/\lambda$. The conditional distribution of Y given $X + Y = t$ is the binomial distribution $b(p, t)$ with

$$p = \rho/(1 + \rho).$$

The UMP unbiased test $\phi(y, t)$ of the hypothesis $\rho = \rho_0$ is defined for each t as the UMP unbiased conditional test of the hypothesis $p = \rho_0/(1 + \rho_0)$. If

$$\underline{p}(t) \leqq p \leqq \bar{p}(t)$$

are the associated most accurate unbiased confidence intervals for p given t, it follows that the most accurate unbiased confidence intervals for μ/λ are

$$\underline{p}(t)/[1 - \underline{p}(t)] \leqq \mu/\lambda \leqq \bar{p}(t)/[1 - \bar{p}(t)].$$

The binomial tests which determine the functions $\underline{p}(t)$ and $\bar{p}(t)$ are discussed in Example 1 of Chapter 4.

6. REGRESSION

The relation between two variables X and Y can be studied by drawing an unrestricted sample and observing the two variables for each subject, obtaining n pairs of measurements $(X_1, Y_1), \cdots, (X_n, Y_n)$ (see Section 11 and Chapter 6, Problem 11). Alternatively, it is frequently possible to control one of the variables such as the age of a subject, the temperature at which an experiment is performed, or the strength of the treatment that is being applied. Observations $Y_1, \cdots, Y_n$ of Y can then be obtained at a number of predetermined levels $x_1, \cdots, x_n$ of x. Suppose that for fixed x the distribution of Y is normal with constant variance σ^2 and a mean which is a function of x, *the regression of Y on x*, and which is assumed to be linear,

$$E[Y|x] = \alpha + \beta x.$$

Putting $v_i = (x_i - \bar{x})/\sqrt{\Sigma(x_j - \bar{x})^2}$ and $\gamma + \delta v_i = \alpha + \beta x_i$, so that $\Sigma v_i = 0$, $\Sigma v_i^2 = 1$, and

$$\alpha = \gamma - \delta \frac{\bar{x}}{\sqrt{\Sigma(x_j - \bar{x})^2}}, \qquad \beta = \frac{\delta}{\sqrt{\Sigma(x_j - \bar{x})^2}},$$

the joint density of $Y_1, \cdots, Y_n$ is

$$\frac{1}{(\sqrt{2\pi}\,\sigma)^n} \exp\left[-\frac{1}{2\sigma^2}\Sigma(y_i - \gamma - \delta v_i)^2\right].$$

These densities constitute an exponential family (1) with

$$\begin{aligned} &U = \Sigma v_i Y_i, \qquad && T_1 = \Sigma Y_i^2, \qquad && T_2 = \Sigma Y_i \\ &\theta = \delta/\sigma^2, && \vartheta_1 = -1/2\sigma^2, && \vartheta_2 = \gamma/\sigma^2. \end{aligned}$$

This representation implies the existence of UMP unbiased tests of the hypotheses $a\gamma + b\delta = c$ where a, b, and c are given constants, and therefore of most accurate unbiased confidence intervals for the parameter

$$\rho = a\gamma + b\delta.$$

To obtain these confidence intervals explicitly, one requires the UMP unbiased test of $H: \rho = \rho_0$, which is given by the acceptance region

$$\frac{|b\Sigma v_i Y_i + a\bar{Y} - \rho_0|/\sqrt{(a^2/n) + b^2}}{\sqrt{[\Sigma(Y_i - \bar{Y})^2 - (\Sigma v_i Y_i)^2]/(n-2)}} \leq C \tag{45}$$

where

$$\int_{-C}^{C} t_{n-2}(y)\,dy = 1 - \alpha.$$

(See Problem 20 and Chapter 7, Section 6.) The resulting confidence intervals for ρ are centered at $b\Sigma v_i Y_i + a\bar{Y}$ and their length is

$$L = 2C\sqrt{[(a^2/n) + b^2][\Sigma(Y_i - \bar{Y})^2 - (\Sigma v_i Y_i)^2]/(n-2)}.$$

It follows from the transformations given in Problem 20 that $[\Sigma(Y_i - \bar{Y})^2 - (\Sigma v_i Y_i)^2]/\sigma^2$ has a χ^2-distribution with $n - 2$ degrees of freedom and hence that the expected length of the intervals is

$$E(L) = 2C_n\sigma\sqrt{(a^2/n) + b^2}.$$

In particular applications, a and b typically are functions of the x's. If these are at the disposal of the experimenter and there is therefore some choice with respect to a and b, the expected length of L is minimized by minimizing $(a^2/n) + b^2$. Actually, it is not clear that the expected length is a good criterion for the accuracy of confidence intervals, since short intervals are desirable when they cover the true parameter value but not necessarily otherwise. However, the same result holds for other criteria such as the expected value of $(\bar{\rho} - \rho)^2 + (\rho - \underline{\rho})^2$ or more generally of $f_1(|\bar{\rho} - \rho|) + f_2(|\rho - \underline{\rho}|)$, where f_1 and f_2 are increasing functions of their arguments. (See Problem 20.) Furthermore, the same choice of a and b also minimizes the probability of the intervals covering any false value of the parameter. We shall therefore consider $(a^2/n) + b^2$ as an inverse measure of the accuracy of the intervals.

Example 9. Confidence intervals for the slope $\beta = \delta/\sqrt{\Sigma(x_j - \bar{x})^2}$ are obtained from the above intervals by letting $a = 0$ and $b = 1/\sqrt{\Sigma(x_j - \bar{x})^2}$. Here the accuracy increases with $\Sigma(x_j - \bar{x})^2$ and, if the x_j must be chosen from an interval $[C_0, C_1]$, it is maximized by putting half of the values at each end point. However, from a practical point of view, this is frequently not a good design since it permits no check of the linearity of the regression.

Example 10. Another parameter of interest is the value $\alpha + \beta x_0$ to be expected from an observation Y at $x = x_0$. Since

$$\alpha + \beta x_0 = \gamma + \delta(x_0 - \bar{x})/\sqrt{\Sigma(x_j - \bar{x})^2},$$

the constants a and b are $a = 1$, $b = (x_0 - \bar{x})/\sqrt{\Sigma(x_j - \bar{x})^2}$. The maximum accuracy is obtained by minimizing $|\bar{x} - x_0|$ and if $\bar{x} = x_0$ cannot be achieved exactly also maximizing $\Sigma(x_j - \bar{x})^2$.

Example 11. Frequently it is of interest to estimate the point x at which $\alpha + \beta x$ has a preassigned value. One may for example wish to find the dosage $x = -\alpha/\beta$ at which $E(Y|x) = 0$, or equivalently the value $v = (x - \bar{x})/\sqrt{\Sigma(x_j - \bar{x})^2}$ at which $\gamma + \delta v = 0$. Most accurate unbiased confidence sets for the solution $-\gamma/\delta$ of this equation can be obtained from the UMP unbiased tests of the hypotheses $-\gamma/\delta = v_0$. The acceptance regions of these tests are given by (45) with $a = 1$, $b = v_0$, and $\rho_0 = 0$, and the resulting confidence sets for v are the sets of values v satisfying

$$v^2[C^2S^2 - (\Sigma v_i Y_i)^2] - 2v\bar{Y}(\Sigma v_i Y_i) + \frac{1}{n}(C^2S^2 - n\bar{Y}^2) \geqq 0$$

where $S^2 = [\Sigma(Y_i - \bar{Y})^2 - (\Sigma v_i Y_i)^2]/(n - 2)$. If the associated quadratic equation in v has roots $\underline{v}, \bar{v}$ the confidence statement becomes

$$\underline{v} \leqq v \leqq \bar{v} \quad \text{when} \quad |\Sigma v_i Y_i|/S > C$$

and

$$v \leqq \underline{v} \quad \text{or} \quad v \geqq \bar{v} \quad \text{when} \quad |\Sigma v_i Y_i|/S < C.$$

The somewhat surprising possibility that the confidence sets may be the outside of an interval actually is quite appropriate here. When the line

$y = \gamma + \delta v$ is nearly parallel to the v-axis, the intercept with the v-axis will be large in absolute value, but its sign can be changed by a very small change in angle. There is the further possibility that the discriminant of the quadratic polynomial is negative,

$$n\bar{Y}^2 + (\Sigma v_i Y_i)^2 < C^2S^2,$$

in which case the associated quadratic equation has no solutions. This condition implies that the leading coefficient of the quadratic polynomial is positive, so that the confidence set in this case becomes the whole real axis. The fact that the confidence sets are not necessarily finite intervals has led to the suggestion that their use be restricted to the cases in which they do have this form. Such usage will however affect the probability with which the sets cover the true value and hence the validity of the reported confidence coefficient.*

7. PERMUTATION TESTS

For the comparison of a treatment with a control situation in which no treatment is given, it was shown in Section 3 that the one-sided t-test is UMP unbiased for testing $H: \eta = \xi$ against $\eta - \xi = \Delta > 0$ when the measurements $X_1, \cdots, X_m$ and $Y_1, \cdots, Y_n$ are samples from normal populations $N(\xi, \sigma^2)$ and $N(\eta, \sigma^2)$. We shall now consider this problem without the assumption of normality, supposing instead that the X's and Y's are samples from distributions with densities $f(x)$ and $f(y - \Delta)$ where f is assumed to be continuous a.e. but otherwise unknown. In this nonparametric formulation, the joint density of the variables is

$$f(x_1) \cdots f(x_m) f(y_1 - \Delta) \cdots f(y_n - \Delta), \qquad f \in \mathscr{F}, \tag{46}$$

where $\mathscr{F}$ is the family of all probability densities that are continuous a.e.

If there is much variation in the population being sampled, the sensitivity of the experiment can frequently be increased by dividing the population into more homogeneous subgroups, defined for example by some characteristic such as age or sex. A sample of size N_i $(i = 1, \cdots, c)$ is then taken from the ith subpopulation, m_i to serve as controls and the other $n_i = N_i - m_i$ to receive the treatment. If the observations in the ith subgroup of such a *stratified sample* are denoted by

$$(X_{i1}, \cdots, X_{im_i}; Y_{i1}, \cdots, Y_{in_i}) = (Z_{i1}, \cdots, Z_{iN_i}),$$

the density of $Z = (Z_{11}, \cdots, Z_{cN_c})$ is

$$p_\Delta(z) = \prod_{i=1}^{c} [f_i(x_{i1}) \cdots f_i(x_{im_i}) f_i(y_{i1} - \Delta) \cdots f_i(y_{in_i} - \Delta)]. \tag{47}$$

* A method for obtaining the size of this effect has been developed by Neyman and tables have been computed on its basis by Fix. This work is reported by Bennett, "On the performance characteristic of certain methods of determining confidence limits," *Sankhyā*, Vol. 18 (1957), pp. 1–12.

Unbiasedness of a test ϕ for testing $\Delta = 0$ against $\Delta > 0$ implies that for all $f_1, \cdots, f_c$

$$\int \phi(z) p_0(z)\, dz = \alpha \qquad (dz = dz_{11} \cdots dz_{cN_c}). \tag{48}$$

Theorem 3. *If $\mathscr{F}$ is the family of all probability densities f that are continuous a.e., then* (48) *holds for all $f_1, \cdots, f_c \in \mathscr{F}$ if and only if*

$$\frac{1}{N_1! \cdots N_c!} \sum_{z' \in S(z)} \phi(z') = \alpha \qquad \text{a.e.} \tag{49}$$

where $S(z)$ is the set of points obtained from z by permuting for each $i = 1, \cdots, c$ the coordinates z_{ij} $(j = 1, \cdots, N_i)$ within the ith subgroup, in all $N_1! \cdots N_c!$ possible ways.

Proof. To prove the result for the case $c = 1$, note that the set of order statistics $T(Z) = (Z^{(1)}, \cdots, Z^{(N)})$ is a complete sufficient statistic for $\mathscr{F}$ (Chapter 4, Example 6). A necessary and sufficient condition for (48) is therefore

$$E[\phi(Z) \mid T(z)] = \alpha \qquad \text{a.e.} \tag{50}$$

The set $S(z)$ in the present case $(c = 1)$ consists of the $N!$ points obtained from z through permutation of coordinates, so that $S(z) = \{z' : T(z') = T(z)\}$. It follows from Section 4 of Chapter 2 that the conditional distribution of Z given $T(z)$ assigns probability $1/N!$ to each of the $N!$ points of $S(z)$. Thus (50) is equivalent to

$$\frac{1}{N!} \sum_{z' \in S(z)} \phi(z') = \alpha \qquad \text{a.e.,} \tag{51}$$

as was to be proved. The proof for general c is completely analogous and is left as an exercise (Problem 21).

The tests satisfying (49) are called *permutation tests.* An extension of this definition is given in Problem 31.

8. MOST POWERFUL PERMUTATION TESTS

For the problem of testing the hypothesis $H: \Delta = 0$ of no treatment effect on the basis of a stratified sample with density (47) it was shown in the preceding section that unbiasedness implies (49). We shall now determine the test which, subject to (49), maximizes the power against a fixed alternative (47) or more generally against an alternative with arbitrary fixed density $h(z)$.

The power of a test ϕ against an alternative h is

$$\int \phi(z)h(z)\,dz = \int E[\phi(Z)|t]\,dP^T(t).$$

Let $t = T(z) = (z^{(1)}, \cdots, z^{(N)})$ so that $S(z) = S(t)$. As was seen in Example 7 and Problem 4 of Chapter 2, the conditional expectation of $\phi(Z)$ given $T(Z) = t$ is

$$\psi(t) = \frac{\sum_{z \in S(t)} \phi(z)h(z)}{\sum_{z \in S(t)} h(z)}.$$

To maximize the power of ϕ subject to (49) it is therefore necessary to maximize $\psi(t)$ for each t subject to this condition. The problem thus reduces to the determination of a function ϕ which, subject to

$$\sum_{z \in S(t)} \phi(z) \frac{1}{N_1! \cdots N_c!} = \alpha,$$

maximizes

$$\sum_{z \in S(t)} \phi(z) \frac{h(z)}{\sum_{z' \in S(t)} h(z')}.$$

By the Neyman-Pearson fundamental lemma, this is achieved by rejecting H for those points z of $S(t)$ for which the ratio

$$\frac{h(z) N_1! \cdots N_c!}{\sum_{z' \in S(t)} h(z')}$$

is too large. Thus the most powerful test is given by the critical function

$$\phi(z) = \begin{cases} 1 & \text{when} \quad h(z) > C[T(z)] \\ \gamma & \text{when} \quad h(z) = C[T(z)] \\ 0 & \text{when} \quad h(z) < C[T(z)]. \end{cases} \tag{52}$$

To carry out the test, the $N_1! \cdots N_c!$ points of each set $S(z)$ are ordered according to the values of the density h. The hypothesis is rejected for the k largest values and with probability γ for the $(k + 1)$st value, where k and γ are defined by

$$k + \gamma = \alpha N_1! \cdots N_c!.$$

Consider now in particular the alternatives (47). The most powerful permutation test is seen to depend on Δ and the f_i, and is therefore not UMP.

Of special interest is the class of normal alternatives with common variance:

$$f_i = N(\xi_i, \sigma^2).$$

The most powerful test against these alternatives, which turns out to be independent of the ξ_i, σ^2, and Δ, is appropriate when approximate normality is suspected but the assumption is not felt to be reliable. It may then be desirable to control the size of the test at level α regardless of the form of the densities f_i and to have the test unbiased against all alternatives (47). However, among the class of tests satisfying these broad restrictions it is natural to make the selection so as to maximize the power against the type of alternative one expects to encounter, that is, against the normal alternatives.

With the above choice of f_i, (47) becomes

$$(53) \qquad h(z) = (\sqrt{2\pi}\,\sigma)^{-N} \exp\left[-\frac{1}{2\sigma^2}\sum_{i=1}^{c}\left(\sum_{j=1}^{m_i}(z_{ij}-\xi_i)^2 + \sum_{j=m_i+1}^{N_i}(z_{ij}-\xi_i-\Delta)^2\right)\right].$$

Since the factor $\exp[-\sum_i\sum_{j=1}^{N_i}(z_{ij}-\xi_i)^2/2\sigma^2]$ is constant over $S(t)$, the test (52) therefore rejects H when $\exp(\Delta\sum_i\sum_{j=m_i+1}^{N_i}z_{ij}) > C[T(z)]$ and hence when

$$(54) \qquad \sum_{i=1}^{c}\sum_{j=1}^{n_i} y_{ij} = \sum_{i=1}^{c}\sum_{j=m_i+1}^{N_i} z_{ij} > C[T(z)].$$

Of the $N_1! \cdots N_c!$ values that the test statistic takes on over $S(t)$, only $\binom{N_1}{n_1}\cdots\binom{N_c}{n_c}$ are distinct since the value of the statistic is the same for any two points z' and z'' for which $(z'_{i1}, \cdots, z'_{im_i})$ and $(z''_{i1}, \cdots, z''_{im_i})$ are permutations of each other for each i. It is therefore enough to compare these distinct values, and to reject H for the k' largest ones and with probability γ' for the $(k'+1)$st, where

$$k' + \gamma' = \alpha\binom{N_1}{n_1}\cdots\binom{N_c}{n_c}.$$

The test (54) is most powerful against the normal alternatives under consideration among all tests which are unbiased and of level α for testing $H: \Delta = 0$ in the original family (47) with $f_1, \cdots, f_c \in \mathscr{F}$. To complete the proof of this statement it is still necessary to prove the test unbiased against the alternatives (47). We shall show more generally that it is unbiased against all alternatives for which X_{ij} $(j = 1, \cdots, m_i)$, Y_{ik} $(k = 1, \cdots, n_i)$ are independently distributed with cumulative distribution functions F_i, G_i respectively such that Y_{ik} is stochastically larger than X_{ij}, that is, such that $G_i(z) \leqq F_i(z)$ for all z. This is a consequence of the following lemma.

Lemma 2. *Let $X_1, \cdots, X_m$; $Y_1, \cdots, Y_n$ be samples from continuous distributions F, G and let $\phi(x_1, \cdots, x_m;\ y_1, \cdots, y_n)$ be a critical function such that* (a) *its expectation is α whenever $G = F$, and* (b) *$y_i \leqq y_i'$ for $i = 1, \cdots, n$ implies*

$$\phi(x_1, \cdots, x_m; y_1, \cdots, y_n) \leqq \phi(x_1, \cdots, x_m; y_1', \cdots, y_n').$$

Then the expectation $\beta = \beta(F, G)$ of ϕ is $\geqq \alpha$ for all pairs of distributions for which Y is stochastically larger than X; it is $\leqq \alpha$ if X is stochastically larger than Y.

Proof. By Lemma 1 of Chapter 3 there exist functions f, g and independent random variables $V_1, \cdots, V_{m+n}$ such that the distributions of $f(V_i)$ and $g(V_i)$ are F and G respectively and that $f(z) \leqq g(z)$ for all z. Then

$$E\phi[f(V_1), \cdots, f(V_m); f(V_{m+1}), \cdots, f(V_{m+n})] = \alpha$$

and

$$E\phi[f(V_1), \cdots, f(V_m); g(V_{m+1}), \cdots, g(V_{m+n})] = \beta.$$

Since for all $(v_1, \cdots, v_{m+n})$,

$$\phi[f(v_1), \cdots, f(v_m); f(v_{m+1}), \cdots, f(v_{m+n})]$$
$$\leqq \phi[f(v_1), \cdots, f(v_m); g(v_{m+1}), \cdots, g(v_{m+n})]$$

the same inequality holds for the expectations of both sides and hence $\alpha \leqq \beta$.

The proof for the case that X is stochastically larger than Y is completely analogous. The lemma also generalizes to the case of c vectors $(X_{i1}, \cdots, X_{im_i}; Y_{i1}, \cdots, Y_{in_i})$ with distributions (F_i, G_i). If the expectation of a function ϕ is then α when $F_i = G_i$ and ϕ is nondecreasing in each y_{ij} when all other variables are held fixed, it follows as before that the expectation of ϕ is $\geqq \alpha$ when the random variables with distribution G_i are stochastically larger than those with distribution F_i.

In applying the lemma to the permutation test (54) it is enough to consider the case $c = 1$, the argument in the more general case being completely analogous. Since the rejection probability of the test (54) is α whenever $F = G$, it is only necessary to show that the critical function ϕ of the test satisfies (b). Now $\phi = 1$ if $\sum_{i=m+1}^{m+n} z_i$ exceeds sufficiently many of the sums $\sum_{i=m+1}^{m+n} z_{j_i}$, and hence if sufficiently many of the differences

$$\sum_{i=m+1}^{m+n} z_i - \sum_{i=m+1}^{m+n} z_{j_i}$$

are positive. For a particular permutation $(j_1, \cdots, j_{m+n})$

$$\sum_{i=m+1}^{m+n} z_i - \sum_{i=m+1}^{m+n} z_{j_i} = \sum_{i=1}^{p} z_{s_i} - \sum_{i=1}^{p} z_{r_i}$$

where $r_1 < \cdots < r_p$ denote those of the integers $j_{m+1}, \cdots, j_{m+n}$ that are $\leqq m$, and $s_1 < \cdots < s_p$ those of the integers $m+1, \cdots, m+n$ not included in the set $(j_{m+1}, \cdots, j_{m+n})$. If $\Sigma z_{s_i} - \Sigma z_{r_i}$ is positive and $y_i \leqq y_i'$, that is, $z_i \leqq z_i'$ for $i = m+1, \cdots, m+n$, then the difference $\Sigma z_{s_i}' - \Sigma z_{r_i}$ is also positive and hence ϕ satisfies (b).

The same argument also shows that the rejection probability of the test is $\leqq \alpha$ when the density of the variables is given by (47) with $\Delta \leqq 0$. The test is therefore equally appropriate if the hypothesis $\Delta = 0$ is replaced by $\Delta \leqq 0$.

Except for small values of the sample sizes N_i, an exact application of the permutation test (54) is impracticable since the amount of computation very quickly becomes prohibitive. In the case $c = 1$, for example, determination of the cutoff point $C[T(z)]$ requires finding the sets of subscripts $(j_1, \cdots, j_n)$ giving rise to the k largest values of $\sum_{i=1}^{n} z_{j_i}$ where k is the largest integer not exceeding $\alpha\binom{N}{n}$. If $\alpha = .05$, $k = 12$ for $m = n = 5$ but already exceeds 9000 for $m = n = 10$. There is however available a very convenient large sample approximation. On multiplying both sides of the inequality

$$\Sigma y_j > C[T(z)]$$

by $[(1/m) + (1/n)]$ and subtracting $(\Sigma x_i + \Sigma y_j)/m$, the rejection region for $c = 1$ becomes $\bar{y} - \bar{x} > C[T(z)]$ or $W = (\bar{y} - \bar{x})/\sqrt{\sum_{i=1}^{N} (z_i - \bar{z})^2} > C[T(z)]$ since the denominator of W is constant over $S(z)$ and hence depends only on $T(z)$. As was seen at the end of Section 3, this is equivalent to

$$\frac{(\bar{y} - \bar{x})\Big/\sqrt{\dfrac{1}{m} + \dfrac{1}{n}}}{\sqrt{[\Sigma(x_i - \bar{x})^2 + \Sigma(y_j - \bar{y})^2]/(m + n - 2)}} > C[T(z)]. \tag{55}$$

The rejection region therefore has the form of a t-test in which the constant cutoff point C_0 of (27) has been replaced by a random one. It turns out that when the hypothesis is true, so that the Z's are identically and independently distributed, and if $E|Z|^3 < \infty$ and m/n is bounded away from zero and infinity as m and n tend to infinity, the difference between the random cutoff point $C[T(Z)]$ and C_0 tends to zero in probability.* In the limit, the permutation test therefore becomes equivalent to the t-test given by (27)–(29). It follows that the *permutation test can be approximated for large samples by the standard t-test.* An exactly analogous result holds for $c > 1$; the appropriate t-test is given in Chapter 7, Problem 7.

9. RANDOMIZATION AS A BASIS FOR INFERENCE

The problem of testing for the effect of a treatment was considered in Section 3 under the assumption that the treatment and control measurements $X_1, \cdots, X_m$ and $Y_1, \cdots, Y_n$ constitute samples from normal distributions and in Sections 7 and 8 without relying on the assumption of normality. We shall now consider in somewhat more detail the structure of the experiment from which the data are obtained, resuming for the moment the assumption that the distributions involved are normal.

Suppose that the experimental material consists of $m + n$ patients, plants, pieces of material, etc., drawn at random from the population to which the treatment could be applied. The treatment is given to n of these while the other m serve as controls. The characteristic that is to be influenced by the treatment is then measured in each case, leading to observations $X_1, \cdots, X_m$; $Y_1, \cdots, Y_n$.

To be specific, suppose that the treatment is carried out by injecting a drug and that $m + n$ ampules are assigned to the $m + n$ patients. The ith measurement can be considered as the sum of two components. One, say U_i, is associated with the ith patient; the other, V_i, with the ith ampule and the circumstances under which it is administered and under which the measurements are taken. The variables U_i and V_i are assumed to be independently distributed, the V's with normal distribution $N(\eta, \sigma^2)$ or $N(\xi, \sigma^2)$ as the ampule contains the drug or is one of those used for control. If in addition the U's are assumed to constitute a random sample from $N(\mu, \sigma_1^2)$ it follows that the X's and Y's are independently normally distributed with common variance $\sigma^2 + \sigma_1^2$ and means

$$E(X) = \mu + \xi, \qquad E(Y) = \mu + \eta.$$

* An account of the required limit theorems and references to the original work of Dwass, Hoeffding, Noether, and Wald and Wolfowitz is given in Chapter 6, Section 6, of Fraser, *Nonparametric Methods in Statistics*, New York, John Wiley & Sons, 1957. For a discussion of more precise approximations to permutation tests see Box and Andersen, "Permutation theory in the derivation of robust criteria and the study of departures from assumption," *J. Roy. Stat. Soc., Ser. B*, Vol. XVII (1955), pp. 1–34.

Except for a change of notation their joint distribution is then given by (26), and the hypothesis $\eta = \xi$ can be tested by the standard t-test.

Unfortunately, under actual experimental conditions, it is frequently not possible to ensure that the patients or other experimental units constitute a random sample from the population of such units. They may be patients in a certain hospital at a given time, or prisoners volunteering for the experiment, and may constitute a haphazard rather than a random sample. In this case the U's would have to be considered as unknown constants since they are not obtained by any definite sampling procedure. This assumption is appropriate also in a different context. Suppose that the experimental units are all the machines in a shop or fields on a farm. If the experiment is performed only to determine the best method for this particular shop or farm, these experimental units are the only relevant ones; that is, a replication of the experiment would consist in comparing the two treatments again for the same machines or fields rather than for a new batch drawn at random from a large population. In this case the units themselves, and therefore the u's, are constant.

Under the above assumptions the joint density of the $m + n$ measurements is

$$\frac{1}{(\sqrt{2\pi}\sigma)^{m+n}} \exp\left[-\frac{1}{2\sigma^2}\left(\sum_{i=1}^{m}(x_i - u_i - \xi)^2 + \sum_{j=1}^{n}(y_j - u_{m+j} - \eta)^2\right)\right].$$

Since the u's are completely arbitrary, it is clearly impossible to distinguish between $H: \eta = \xi$ and the alternatives $K: \eta > \xi$. In fact, every distribution of K also belongs to H and vice versa, and the most powerful level α test for testing H against any simple alternative specifying ξ, η, σ, and the u's rejects H with probability α regardless of the observations.

Data which could serve as a basis for testing whether or not the treatment has an effect can be obtained through the fundamental device of *randomization*. Suppose that the $N = m + n$ patients are assigned to the N ampules at random, that is, in such a way that each of the $N!$ possible assignments has probability $1/N!$ of being chosen. Then for a given assignment the N measurements are independently normally distributed with variance σ^2 and means $\xi + u_{j_i}$ $(i = 1, \cdots, m)$ and $\eta + u_{j_i}$ $(i = m + 1, \cdots, m + n)$. The over-all joint density of the variables

$$(Z_1, \cdots, Z_N) = (X_1, \cdots, X_m;\ Y_1, \cdots, Y_n)$$

is therefore

$$(56)\quad \frac{1}{N!}\sum_{(j_1,\cdots,j_N)} \frac{1}{(\sqrt{2\pi}\sigma)^N} \times \exp\left[\frac{-1}{2\sigma^2}\left(\sum_{i=1}^{m}(x_i - u_{j_i} - \xi)^2 + \sum_{i=1}^{n}(y_i - u_{j_{m+i}} - \eta)^2\right)\right]$$

where the outer summation extends over all $N!$ permutations $(j_1, \cdots, j_N)$ of $(1, \cdots, N)$. Under the hypothesis $\eta = \xi$ this density can be written as

$$\frac{1}{N!} \sum_{(j_1,\cdots,j_N)} \frac{1}{(\sqrt{2\pi}\sigma)^N} \exp\left[-\frac{1}{2\sigma^2} \sum_{i=1}^{N} (z_i - \zeta_{j_i})^2\right], \tag{57}$$

where $\zeta_{j_i} = u_{j_i} + \xi = u_{j_i} + \eta$.

Without randomization, a set of y's which is large relative to the x-values could be explained entirely in terms of the unit effects u_i. However, if these are assigned to the y's at random, they will on the average balance those assigned to the x's. As a consequence, a marked superiority of the second sample becomes very unlikely under the hypothesis, and must therefore be put down to the effectiveness of the treatment.

The method of assigning the treatments to the experimental units completely at random permits the construction of a level α test of the hypothesis $\eta = \xi$, whose power exceeds α against all alternatives $\eta - \xi > 0$. The actual power of such a test will however depend not only on the alternative value of $\eta - \xi$, which measures the effect of the treatment, but also on the unit effects u_i. In particular, if there is excessive variation among the u's, this will swamp the treatment effect (much in the same way as an increase in the variance σ^2 would), and the test will accordingly have little power to detect any given alternative $\eta - \xi$.

In such cases the sensitivity of the experiment can be increased by an approach exactly analogous to the method of stratified sampling discussed in Section 7. In the present case this means replacing the process of complete randomization described above by a more restricted randomization procedure. The experimental material is divided into subgroups, which are more homogeneous than the material as a whole, so that within each group the differences among the u's are small. In animal experiments, for example, this can frequently be achieved by a division into litters. Randomization is then applied only within each group. If the ith group contains N_i units, n_i of these are selected at random to receive the treatment, and the remaining $m_i = N_i - n_i$ serve as controls ($\Sigma N_i = N$, $\Sigma m_i = m$, $\Sigma n_i = n$).

An example of this approach is the method of *matched pairs*. Here the experimental units are divided into pairs, which are as like each other as possible with respect to all relevant properties, so that within each pair the difference of the u's will be as small as possible. Suppose that the material consists of n such pairs, and denote the associated unit effects (the U's of the previous discussion) by $U_1, U_1'; \cdots; U_n, U_n'$. Let the first and second member of each pair receive the treatment or serve as control respectively, and let the observations for the ith pair be X_i and Y_i.

If the matching is completely successful, as may be the case, for example, when the same patient is used twice in the investigation of a sleeping drug, or when identical twins are used, then $U_i' = U_i$ for all i, and the density of the X's and Y's is

$$\text{(58)} \qquad \frac{1}{(\sqrt{2\pi}\sigma)^{2n}} \exp\left[-\frac{1}{2\sigma^2}[\Sigma(x_i - \xi - u_i)^2 + \Sigma(y_i - \eta - u_i)^2]\right].$$

The UMP unbiased test for testing $H: \eta = \xi$ against $\eta > \xi$ is then given in terms of the differences $W_i = Y_i - X_i$ by the rejection region

$$\text{(59)} \qquad \sqrt{n}\bar{w} \bigg/ \sqrt{\frac{1}{n-1}\Sigma(w_i - \bar{w})^2} > C.$$

(See Problem 25.)

However, usually one is not willing to trust the assumption $u_i' = u_i$ even after matching, and it again becomes necessary to randomize. Since as a result of the matching the variability of the u's within each pair is presumably considerably smaller than the over-all variation, randomization is carried out only within each pair. For each pair, one of the units is selected with probability 1/2 to receive the treatment, while the other serves as control. The density of the X's and Y's is then

$$\text{(60)} \qquad \frac{1}{2^n}\frac{1}{(\sqrt{2\pi}\sigma)^{2n}} \prod_{i=1}^{n} \left\{\exp\left[-\frac{1}{2\sigma^2}[(x_i - \xi - u_i)^2 + (y_i - \eta - u_i')^2]\right] + \exp\left[-\frac{1}{2\sigma^2}[(x_i - \xi - u_i')^2 + (y_i - \eta - u_i)^2]\right]\right\}.$$

Under the hypothesis $\eta = \xi$, and writing

$$z_{i1} = x_i, \qquad z_{i2} = y_i, \qquad \zeta_{i1} = \xi + u_i, \qquad \zeta_{i2} = \eta + u_i' \qquad (i = 1, \cdots, n),$$

this becomes

$$\text{(61)} \qquad \frac{1}{2^n} \Sigma \frac{1}{(\sqrt{2\pi}\sigma)^{2n}} \exp\left[-\frac{1}{2\sigma^2}\sum_{i=1}^{n}\sum_{j=1}^{2}(z_{ij} - \zeta_{ij}')^2\right].$$

Here the outer summation extends over the 2^n points $\zeta' = (\zeta_{11}', \cdots, \zeta_{n2}')$ for which $(\zeta_{i1}', \zeta_{i2}')$ is either (ζ_{i1}, ζ_{i2}) or (ζ_{i2}, ζ_{i1}).

10. PERMUTATION TESTS AND RANDOMIZATION

It was shown in the preceding section that randomization provides a basis for testing the hypothesis $\eta = \xi$ of no treatment effect, without any assumptions concerning the experimental units. In the present section, a specific test will be derived for this problem. When the experimental units are treated as constants, the probability density of the observations is given by (56) in the case of complete randomization and by (60) in the case of matched pairs. More generally, let the experimental material be divided into c subgroups, let the randomization be applied within each subgroup, and let the observations in the ith subgroup be

$$(Z_{i1}, \cdots, Z_{iN_i}) = (X_{i1}, \cdots, X_{im_i};\ Y_{i1}, \cdots, Y_{in_i}).$$

For any point $u = (u_{11}, \cdots, u_{cN_c})$, let $S(u)$ denote as before the set of $N_1! \cdots N_c!$ points obtained from u by permuting the coordinates within each subgroup in all $N_1! \cdots N_c!$ possible ways. Then the joint density of the Z's is

$$(62) \quad \frac{1}{N_1! \cdots N_c!} \sum_{u' \in S(u)} \frac{1}{(\sqrt{2\pi}\sigma)^N} \times \exp\left[-\frac{1}{2\sigma^2} \sum_{i=1}^{c} \left(\sum_{j=1}^{m_i} (z_{ij} - \xi - u'_{ij})^2 + \sum_{j=m_i+1}^{N_i} (z_{ij} - \eta - u'_{ij})^2 \right)\right]$$

and under the hypothesis of no treatment effect

$$(63) \quad p_{\sigma,\zeta}(z) = \frac{1}{N_1! \cdots N_c!} \sum_{\zeta' \in S(\zeta)} \frac{1}{(\sqrt{2\pi}\sigma)^N} \exp\left[-\frac{1}{2\sigma^2} \sum_{i=1}^{c} \sum_{j=1}^{N_i} (z_{ij} - \zeta'_{ij})^2\right].$$

It may happen that the coordinates of u or ζ are not distinct. If then some of the points of $S(u)$ or $S(\zeta)$ also coincide, each should be counted with its proper multiplicity. More precisely, if the $N_1! \cdots N_c!$ relevant permutations of $N_1 + \cdots + N_c$ coordinates are denoted by g_k, $k = 1, \cdots, N_1! \cdots N_c!$, $S(\zeta)$ can be taken to be the ordered set of points $g_k\zeta$, $k = 1, \cdots, N_1! \cdots N_c!$, and (63), for example, becomes

$$p_{\sigma,\zeta}(z) = \frac{1}{N_1! \cdots N_c!} \sum_{k=1}^{N_1! \cdots N_c!} \frac{1}{(\sqrt{2\pi}\sigma)^N} \exp\left(-\frac{1}{2\sigma^2} |z - g_k\zeta|^2\right)$$

where $|u^2|$ stands for $\sum_{i=1}^{c} \sum_{j=1}^{N_i} u_{ij}^2$.

Theorem 4. *A necessary and sufficient condition for a critical function ϕ to satisfy*

$$(64) \quad \int \phi(z) p_{\sigma,\zeta}(z)\, dz \leqq \alpha \qquad (dz = dz_{11} \cdots dz_{cN_c})$$

for all $\sigma > 0$ and all vectors ζ is that

$$(65) \quad \frac{1}{N_1! \cdots N_c!} \sum_{z' \in S(z)} \phi(z') \leqq \alpha \qquad \text{a.e.}$$

The proof will be based on the following lemma.

Lemma 3. *Let A be a set in N-space with positive Lebesgue measure $\mu(A)$. Then for any $\epsilon > 0$ there exist real numbers $\sigma > 0$* and $\xi_1, \cdots, \xi_N$ such that

$$P\{(X_1, \cdots, X_N) \in A\} \geqq 1 - \epsilon$$

where the X's are independently normally distributed with means $E(X_i) = \xi_i$ and variance $\sigma^2_{X_i} = \sigma^2$.

Proof. Suppose without loss of generality that $\mu(A) < \infty$. Given any $\eta > 0$, there exists a square Q such that

$$\mu(Q \cap \tilde{A}) \leqq \eta\mu(Q).$$

This follows from the fact that almost every point of A has metric density 1,* or from the more elementary fact that a measurable set can be approximated in measure by unions of disjoint squares. Let a be such that

$$\frac{1}{\sqrt{2\pi}} \int_{-a}^{a} \exp(-t^2/2)\, dt = \left(1 - \frac{\epsilon}{2}\right)^{1/N}$$

and let

$$\eta = \frac{\epsilon}{2}\left(\frac{\sqrt{2\pi}}{2a}\right)^N.$$

If $(\xi_1, \cdots, \xi_N)$ is the center of Q, and if $\sigma = b/a = (1/2a)[\mu(Q)]^{1/N}$ where $2b$ is the length of the side of Q, then

$$\frac{1}{(\sqrt{2\pi}\sigma)^N} \int_{\tilde{A} \cap \tilde{Q}} \exp\left[-\frac{1}{2\sigma^2} \Sigma(x_i - \xi_i)^2\right] dx_1 \cdots dx_N$$
$$\leqq \frac{1}{(\sqrt{2\pi}\sigma)^N} \int_{\tilde{Q}} \exp\left[-\frac{1}{2\sigma^2} \Sigma(x_i - \xi_i)^2\right] dx_1 \cdots dx_N$$
$$= 1 - \left[\frac{1}{\sqrt{2\pi}} \int_{-a}^{a} \exp(-t^2/2)\, dt\right]^N = \frac{\epsilon}{2}.$$

On the other hand,

$$\frac{1}{(\sqrt{2\pi}\sigma)^N} \int_{\tilde{A} \cap Q} \exp\left[-\frac{1}{2\sigma^2} \Sigma(x_i - \xi_i)^2\right] dx_1 \cdots dx_N$$
$$\leqq \frac{1}{(\sqrt{2\pi}\sigma)^N} \mu(\tilde{A} \cap Q) < \frac{\epsilon}{2},$$

and by adding the two inequalities one obtains the desired result.

* See for example Hobson, *Theory of Functions of a Real Variable*, Vol. 1, Cambridge Univ. Press, 3rd ed., 1927, p. 194.

Proof of the theorem. Let ϕ be any critical function, and let

$$\psi(z) = \frac{1}{N_1! \cdots N_c!} \sum_{z' \in S(z)} \phi(z').$$

If (65) does not hold, there exists $\eta > 0$ such that $\psi(z) > \alpha + \eta$ on a set A of positive measure. By the lemma there exists $\sigma > 0$ and $\zeta = (\zeta_{11}, \cdots, \zeta_{cN_c})$ such that $P\{Z \in A\} > 1 - \eta$ when $Z_{11}, \cdots, Z_{cN_c}$ are independently normally distributed with common variance σ^2 and means $E(Z_{ij}) = \zeta_{ij}$. It follows that

$$(66) \quad \int \phi(z) p_{\sigma,\zeta}(z)\, dz = \int \psi(z) p_{\sigma,\zeta}(z)\, dz \geqq \int_A \psi(z) \frac{1}{(\sqrt{2\pi}\sigma)^N}$$

$$\times \exp\left[-\frac{1}{2\sigma^2} \Sigma\Sigma(z_{ij} - \zeta_{ij})^2\right] dz > (\alpha + \eta)(1 - \eta),$$

which is $> \alpha$ since $\alpha + \eta < 1$. This proves that (64) implies (65). The converse follows from the first equality in (66).

Corollary 2. *Let H be the class of densities*

$$\{p_{\sigma,\zeta}(z):\ \sigma > 0,\ -\infty < \zeta_{ij} < \infty\}.$$

A complete family of tests for H at level of significance α is the class of tests $\mathscr{C}$ satisfying

$$(67) \qquad \frac{1}{N_1! \cdots N_c!} \sum_{z' \in S(z)} \phi(z') = \alpha \qquad \text{a.e.}$$

Proof. The corollary states that for any given level α test ϕ_0 there exists an element ϕ of $\mathscr{C}$ which is uniformly at least as powerful as ϕ_0. By the preceding theorem the average value of ϕ_0 over each set $S(z)$ is $\leqq \alpha$. On the sets for which this inequality is strict, one can increase ϕ_0 to obtain a critical function ϕ satisfying (67), and such that $\phi_0(z) \leqq \phi(z)$ for all z. Since against all alternatives the power of ϕ is at least that of ϕ_0, this establishes the result. An explicit construction of ϕ, which shows that it can be chosen to be measurable, is given in Problem 28.

This corollary shows that the normal randomization model (62) leads exactly to the class of tests that was previously found to be relevant when the U's constitute a sample but the assumption of normality was not imposed. It therefore follows from Section 8 that the most powerful level α test for testing (63) against a simple alternative (62) is given by (52) with $h(z)$ equal to the probability density (62). If $\eta - \xi = \Delta$, the rejection region of this test reduces to

$$(68) \quad \sum_{u \in S(u)} \exp\left[\frac{1}{\sigma^2} \sum_{i=1}^{c} \left(\sum_{j=1}^{N_i} z_{ij} u'_{ij} + \Delta \sum_{j=m_i+1}^{N_i} (z_{ij} - u'_{ij})\right)\right] > C[T(z)],$$

since both $\Sigma\Sigma z_{ij}$ and $\Sigma\Sigma z_{ij}^2$ are constant on $S(z)$ and therefore functions only of $T(z)$. It is seen that this test depends on Δ and the unit effects u_{ij}, so that a UMP test does not exist.

Among the alternatives (62) a subclass occupies a central position and is of particular interest. This is the class of alternatives specified by the assumption that the unit effects u_i constitute a sample from a normal distribution. Although this assumption cannot be expected to hold exactly—in fact, it was just as a safeguard against the possibility of its breakdown that randomization was introduced—it is in many cases reasonable to suppose that it holds at least approximately. The resulting subclass of alternatives is given by the probability densities

$$(69)\quad \frac{1}{(\sqrt{2\pi}\sigma)^N} \times \exp\left[-\frac{1}{2\sigma^2}\sum_{i=1}^{c}\left(\sum_{j=1}^{m_i}(z_{ij}-u_i-\xi)^2+\sum_{j=m_i+1}^{N_i}(z_{ij}-u_i-\eta)^2\right)\right].$$

These alternatives are suggestive also from a slightly different point of view. The procedure of assigning the experimental units to the treatments at random within each subgroup was seen to be appropriate when the variation of the u's is small within these groups and is employed when this is believed to be the case. This suggests, at least as an approximation, the assumption of constant $u_{ij} = u_i$, which is the limiting case of a normal distribution as the variance tends to zero, and for which the density is also given by (69).

Since the alternatives (69) are the same as the alternatives (53) of Section 8 with $u_i - \xi = \xi_i$, $u_i - \eta = \xi_i - \Delta$, *the permutation test* (54) *is seen to be most powerful for testing the hypothesis* $\eta = \xi$ *in the normal randomization model* (62) *against the alternatives* (69) *with* $\eta - \xi > 0$. The test retains this property in the still more general setting in which neither normality nor the sample property of the U's is assumed to hold. Let the joint density of the variables be

$$\sum_{u'\in S(u)}\prod_{i=1}^{c}\left[\prod_{j=1}^{m_i} f_i(z_{ij}-u'_{ij}-\xi)\prod_{j=m_i+1}^{N_i} f_i(z_{ij}-u'_{ij}-\eta)\right],$$

with f_i continuous a.e. but otherwise unspecified.* Under the hypothesis H: $\eta = \xi$, this density is symmetric in the variables $(z_{i1}, \cdots, z_{iN_i})$ of the ith subgroup for each i, so that any permutation test (49) has rejection

* Actually, all that is needed is that $f_1, \cdots, f_c \in \mathscr{F}$ where $\mathscr{F}$ is any family containing all normal distributions.

probability α for all distributions of H. By Corollary 2, these permutation tests therefore constitute a complete class, and the result follows.

11. TESTING FOR INDEPENDENCE IN A BIVARIATE NORMAL DISTRIBUTION

So far, the methods of the present chapter have been illustrated mainly by the two-sample problem. As a further example, we shall now apply two of the formulations that have been discussed, the normal model of Section 3 and the nonparametric one of Section 7, to the hypothesis of independence in a bivariate distribution.

The probability density of a sample $(X_1, Y_1), \cdots, (X_n, Y_n)$ from a bivariate normal distribution is

$$\frac{1}{(2\pi\sigma\tau\sqrt{1-\rho^2})^n} \exp\left[-\frac{1}{2(1-\rho^2)}\left(\frac{1}{\sigma^2}\Sigma(x_i-\xi)^2 - \frac{2\rho}{\sigma\tau}\Sigma(x_i-\xi)(y_i-\eta) + \frac{1}{\tau^2}\Sigma(y_i-\eta)^2\right)\right]. \tag{70}$$

Here (ξ, σ^2) and (η, τ^2) are the mean and variance of X and Y respectively, and ρ is the correlation coefficient between X and Y. The hypotheses $\rho \leqq \rho_0$ and $\rho = \rho_0$ for arbitrary ρ_0 cannot be treated by the methods of the present chapter, and will be taken up in Chapter 6. For the present, we shall consider only the hypothesis $\rho = 0$ that X and Y are independent, and the corresponding one-sided hypothesis $\rho \leqq 0$.

The family of densities (70) is of the exponential form (1) with

$$U = \Sigma X_i Y_i, \quad T_1 = \Sigma X_i^2, \quad T_2 = \Sigma Y_i^2, \quad T_3 = \Sigma X_i, \quad T_4 = \Sigma Y_i$$

and

$$\theta = \frac{\rho}{\sigma\tau(1-\rho^2)}, \quad \vartheta_1 = \frac{-1}{2\sigma^2(1-\rho^2)}, \quad \vartheta_2 = \frac{-1}{2\tau^2(1-\rho^2)},$$

$$\vartheta_3 = \frac{1}{1-\rho^2}\left(\frac{\xi}{\sigma^2} - \frac{\eta\rho}{\sigma\tau}\right), \quad \vartheta_4 = \frac{1}{1-\rho^2}\left(\frac{\eta}{\tau^2} - \frac{\xi\rho}{\sigma\tau}\right).$$

The hypothesis $H: \rho \leqq 0$ is equivalent to $\theta \leqq 0$. Since the sample correlation coefficient

$$R = \frac{\Sigma(X_i - \bar{X})(Y_i - \bar{Y})}{\sqrt{\Sigma(X_i - \bar{X})^2 \Sigma(Y_i - \bar{Y})^2}}$$

is unchanged when the X_i and Y_i are replaced by $(X_i - \xi)/\sigma$ and $(Y_i - \eta)/\tau$, the distribution of R does not depend on ξ, η, σ, or τ, but only on ρ. For $\theta = 0$ it therefore does not depend on $\vartheta_1, \cdots, \vartheta_4$ and hence by Theorem 2, R is independent of $(T_1, \cdots, T_4)$ when $\theta = 0$. It follows from Theorem 1 that the UMP unbiased test of H rejects when

$$R \geqq C_0, \tag{71}$$

or equivalently when

$$\frac{R}{\sqrt{(1 - R^2)/(n - 2)}} > K_0. \tag{72}$$

The statistic R is linear in U, and its distribution for $\rho = 0$ is symmetric about 0. The UMP unbiased test of the hypothesis $\rho = 0$ against the alternatives $\rho \neq 0$ therefore rejects when

$$\frac{|R|}{\sqrt{(1 - R^2)/(n - 2)}} > K_1. \tag{73}$$

Since $\sqrt{n - 2}\ R/\sqrt{1 - R^2}$ has the t-distribution with $n - 2$ degrees of freedom when $\rho = 0$ (Problem 32), the constants K_0 and K_1 in the above tests are given by

$$\int_{K_0}^{\infty} t_{n-2}(y)\, dy = \alpha \qquad \text{and} \qquad \int_{K_1}^{\infty} t_{n-2}(y)\, dy = \frac{\alpha}{2}. \tag{74}$$

Since the distribution of R depends only on the correlation coefficient ρ, the same is true of the power of these tests.

We next consider the problem without the assumption of normality, in a nonparametric formulation. For any bivariate distribution of (X, Y), let Y_x denote a random variable whose distribution is the conditional distribution of Y given x. We shall say that there is *positive dependence* between X and Y if for any $x < x'$ the variable $Y_{x'}$ is stochastically larger than Y_x. Generally speaking, larger values of Y will then correspond to larger values of X; this is the intuitive meaning of positive dependence. An example is furnished by any normal bivariate distribution with $\rho > 0$. (See Problem 36.)

Consider now the hypothesis of independence against the alternatives of positive dependence in a general bivariate distribution possessing a probability density with respect to Lebesgue measure. Unbiasedness of a test ϕ implies that the rejection probability is α when X and Y are independent, and hence that

$$\int \phi(x_1, \cdots, x_n; y_1, \cdots, y_n) f_1(x_1) \cdots f_1(x_n) f_2(y_1) \cdots f_2(y_n)\, dx\, dy = \alpha$$

for all probability densities f_1 and f_2. By Theorem 3 this in turn implies

$$\frac{1}{(n!)^2} \Sigma \phi(x_{i_1}, \cdots, x_{i_n}; y_{j_1}, \cdots, y_{j_n}) = \alpha.$$

Here the summation extends over the $(n!)^2$ points of the set $S(x, y)$, which is obtained from a fixed point (x, y) with $x = (x_1, \cdots, x_n)$, $y = (y_1, \cdots, y_n)$ by permuting the x-coordinates and the y-coordinates, each among themselves in all possible ways.

Among all tests satisfying this condition, the most powerful one against the normal alternatives (70) with $\rho > 0$ rejects for the k' largest values of (70) in each set $S(x, y)$, where $k'/(n!)^2 = \alpha$. Since Σx_i^2, Σy_i^2, Σx_i, Σy_i are all constant on $S(x, y)$, the test equivalently rejects for the k' largest values of $\Sigma x_i y_i$ in each $S(x, y)$.

Of the $(n!)^2$ values that the statistic $\Sigma X_i Y_i$ takes on over $S(x, y)$, only $n!$ are distinct since the statistic remains unchanged if the X's and Y's are subjected to the same permutation. A simpler form of the test is therefore obtained, for example by rejecting H for the k largest values of $\Sigma x^{(i)} y_{j_i}$ of each set $S(x, y)$, where $x^{(1)} < \cdots < x^{(n)}$ and $k/n! = \alpha$. The test can be shown to be unbiased against all alternatives with positive dependence. (See Problem 41 of Chapter 6.)

In order to obtain a comparison of the permutation test with the standard normal test based on the sample correlation coefficient R, let $T(X, Y)$ denote the set of ordered X's and Y's,

$$T(X, Y) = (X^{(1)}, \cdots, X^{(n)};\ Y^{(1)}, \cdots, Y^{(n)}).$$

The rejection region of the permutation test can then be written as

$$\Sigma X_i Y_i > C[T(X, Y)],$$

or equivalently as

$$R > K[T(X, Y)].$$

It again turns* out that the difference between $K[T(X, Y)]$ and the cutoff point C_0 of the corresponding normal test (71) tends to zero, and that the two tests become equivalent in the limit as n tends to infinity. Sufficient conditions for this are that $\sigma_X^2, \sigma_Y^2 > 0$ and $E(|X|^3), E(|Y|^3) < \infty$. For large n, the standard normal test (71) therefore serves as an approximation for the permutation test which is impractical except for small sample sizes.

12. PROBLEMS

Section 2

1. Let $X_1, \cdots, X_n$ be a sample from $N(\xi, \sigma^2)$. The power of Student's t-test is an increasing function of ξ/σ in the one-sided case H: $\xi \leq 0$, K: $\xi > 0$, and of $|\xi|/\sigma$ in the two-sided case H: $\xi = 0$, K: $\xi \neq 0$.

* For a proof see the book by Fraser, to which reference is made in Section 8.

[If

$$S = \sqrt{\frac{1}{n-1}\Sigma(X_i - \bar{X})^2},$$

the power in the two-sided case is given by

$$1 - P\left\{-\frac{CS}{\sigma} - \frac{\sqrt{n}\xi}{\sigma} \leq \frac{\sqrt{n}(\bar{X} - \xi)}{\sigma} \leq \frac{CS}{\sigma} - \frac{\sqrt{n}\xi}{\sigma}\right\}$$

and the result follows from the fact that it holds conditionally for each fixed value of S/σ.]

2. In the situation of the previous problem there exists no test for testing $H: \xi = 0$ at level α, which for all σ has power $\geq \beta > \alpha$ against the alternatives (ξ, σ) with $\xi = \xi_1 > 0$.

[Let $\beta(\xi_1, \sigma)$ be the power of any level α test of H, and let $\beta(\sigma)$ denote the power of the most powerful test for testing $\xi = 0$ against $\xi = \xi_1$ when σ is known. Then $\inf \beta_\sigma(\xi_1, \sigma) \leq \inf_\sigma \beta(\sigma) = \alpha$.]

3. (i) Let Z and V be independently distributed as $N(\delta, 1)$ and χ^2 with f degrees of freedom respectively. Then the ratio $Z \div \sqrt{V/f}$ has the noncentral t-distribution with f degrees of freedom and noncentrality parameter δ, the probability density of which is†

$$(75)\quad p_\delta(t) = \frac{1}{2^{\frac{1}{2}(f+1)}\Gamma(\frac{1}{2}f)\sqrt{\pi f}}\int_0^\infty y^{\frac{1}{2}(f-1)} \exp\left(-\tfrac{1}{2}y\right) \exp\left[-\frac{1}{2}\left(t\sqrt{\frac{y}{f}} - \delta\right)^2\right] dy$$

or equivalently

$$(76)\quad p_\delta(t) = \frac{1}{2^{\frac{1}{2}(f-1)}\Gamma(\frac{1}{2}f)\sqrt{\pi f}} \exp\left(-\frac{1}{2}\frac{f\delta^2}{f+t^2}\right) \left(\frac{f}{f+t^2}\right)^{\frac{1}{2}(f+1)} \int_0^\infty v^f \exp\left[-\frac{1}{2}\left(v - \frac{\delta t}{\sqrt{f+t^2}}\right)^2\right] dv.$$

Another form is obtained by making the substitution $w = t\sqrt{y}/\sqrt{f}$ in (75).

(ii) If $X_1, \cdots, X_n$ are independently distributed as $N(\xi, \sigma^2)$, then $\sqrt{n}\bar{X} \div \sqrt{\Sigma(X_i - \bar{X})^2/(n-1)}$ has the noncentral t-distribution with $n - 1$ degrees of freedom and noncentrality parameter $\delta = \sqrt{n}\,\xi/\sigma$.

[(i) The first expression is obtained from the joint density of Z and V by transforming to $t = z \div \sqrt{v/f}$ and v.]

4. Let $X_1, \cdots, X_n$ be a sample from $N(\xi, \sigma^2)$. Denote the power of the one-sided t-test of H: $\xi \leq 0$ against the alternative ξ/σ by $\beta(\xi/\sigma)$, and by $\beta^*(\xi/\sigma)$ the power of the test appropriate when σ is known. Determine $\beta(\xi/\sigma)$ for

† The cumulative distribution function as well as the probability density of this distribution has been tabled by Resnikoff and Lieberman, *Tables of the Non-central t-distribution*, Stanford Univ. Press, 1957. See also Merrington and Pearson, "An approximation to the distribution of non-central t," *Biometrika*, Vol. 45 (1958), pp. 484–491.

$n = 5, 10, 15$, $\alpha = .05$, $\xi/\sigma = .7, .8, .9, 1.0, 1.1, 1.2$ and in each case compare it with $\beta^*(\xi/\sigma)$. Do the same for the two-sided case.

5. Let $Z_1, \cdots, Z_n$ be independently normally distributed with common variance σ^2 and means $E(Z_i) = \zeta_i (i = 1, \cdots, s)$, $E(Z_i) = 0$ $(i = s + 1, \cdots, n)$. There exist UMP unbiased tests for testing $\zeta_1 \leqq \zeta_1^0$ and $\zeta_1 = \zeta_1^0$ given by the rejection regions

$$\frac{Z_1 - \zeta_1^0}{\sqrt{\sum_{i=s+1}^{n} Z_i^2/(n-s)}} > C_0 \quad \text{and} \quad \frac{|Z_1 - \zeta_1^0|}{\sqrt{\sum_{i=s+1}^{n} Z_i^2/(n-s)}} > C.$$

When $\zeta_1 = \zeta_1^0$, the test statistic has the t-distribution with $n - s$ degrees of freedom.

6. Let $X_1, \cdots, X_n$ be independently normally distributed with common variance σ^2 and means $\xi_1, \cdots, \xi_n$, and let $Z_i = \Sigma_{j=1}^n a_{ij} X_j$ be an orthogonal transformation (that is, $\Sigma_{i=1}^n a_{ij} a_{ik} = 1$ or 0 as $j = k$ or $j \neq k$). The Z's are normally distributed with common variance σ^2 and means $\zeta_i = \Sigma a_{ij} \xi_j$.

[The density of the Z's is obtained from that of the X's by substituting $x_i = \Sigma b_{ij} z_j$ where (b_{ij}) is the inverse of the matrix (a_{ij}), and multiplying by the Jacobian which is 1.]

7. If $X_1, \cdots, X_n$ is a sample from $N(\xi, \sigma^2)$, the UMP unbiased test of $\xi \leqq 0$ and $\xi = 0$ can be obtained from Problems 5 and 6 by making an orthogonal transformation to variables $Z_1, \cdots, Z_n$ such that $Z_1 = \sqrt{n}\,\bar{X}$.

[Then

$$\sum_{i=2}^{n} Z_i^2 = \sum_{i=1}^{n} Z_i^2 - Z_1^2 = \sum_{i=1}^{n} X_i^2 - n\bar{X}^2 = \sum_{i=1}^{n} (X_i - \bar{X})^2.]$$

8. Let $X_1, X_2, \cdots$ be a sequence of independent variables distributed as $N(\xi, \sigma^2)$ and let $Y_n = [nX_{n+1} - (X_1 + \cdots + X_n)]/\sqrt{n(n+1)}$.

(i) The variables $Y_1, Y_2, \cdots$ are independently distributed as $N(0, \sigma^2)$.

(ii) On the basis of the Y's, the hypothesis $\sigma = \sigma_0$ can be tested against $\sigma = \sigma_1$ by means of a sequential probability ratio test.

Section 3

9. Let $X_1, \cdots, X_n$ and $Y_1, \cdots, Y_n$ be independent samples from $N(\xi, \sigma^2)$ and $N(\eta, \tau^2)$ respectively. Determine the sample size necessary to obtain power $\geqq \beta$ against the alternatives $\tau/\sigma > \Delta$ when $\alpha = .05$, $\beta = .9$, $\Delta = 1.5, 2, 3$, and the hypothesis being tested is H: $\tau/\sigma \leqq 1$.

10. If $m = n$, the acceptance region (23) can be written as

$$\max (S_Y^2/\Delta_0 S_X^2, \Delta_0 S_X^2/S_Y^2) \leqq (1 - C)/C,$$

where $S_X^2 = \Sigma(X_i - \bar{X})^2$, $S_Y^2 = \Sigma(Y_i - \bar{Y})^2$ and where C is determined by

$$\int_0^C B_{n-1,\, n-1}(w)\, dw = \frac{\alpha}{2}.$$

11. Let $X_1, \cdots, X_m$ and $Y_1, \cdots, Y_n$ be samples from $N(\xi, \sigma^2)$ and $N(\eta, \sigma^2)$. The UMP unbiased test for testing $\eta - \xi = 0$ can be obtained through Problems

5 and 6 by making an orthogonal transformation from $(X_1, \cdots, X_m, Y_1, \cdots, Y_n)$ to $(Z_1, \cdots, Z_{m+n})$ such that $Z_1 = (\bar{Y} - \bar{X})/\sqrt{(1/m) + (1/n)}$, $Z_2 = (\Sigma X_i + \Sigma Y_j)/\sqrt{m+n}$.

12. *Exponential densities.* Let $X_1, \cdots, X_n$ be a sample from a distribution with exponential density $a^{-1}e^{-(x-b)/a}$ for $x \geqq b$.

(i) For testing $a = 1$ there exists a UMP unbiased test given by the acceptance region

$$C_1 \leqq 2\Sigma[x_i - \min(x_1, \cdots, x_n)] \leqq C_2$$

where the test statistic has a χ^2-distribution with $2n - 2$ degrees of freedom when $a = 1$, and C_1, C_2 are determined by

$$\int_{C_1}^{C_2} \chi^2_{2n-2}(y)\, dy = \int_{C_1}^{C_2} \chi^2_{2n}(y)\, dy = 1 - \alpha.$$

(ii) For testing $b = 0$ there exists a UMP unbiased test given by the acceptance region

$$0 \leqq \frac{n \min(x_1, \cdots, x_n)}{\Sigma[x_i - \min(x_1, \cdots, x_n)]} \leqq C.$$

When $b = 0$, the test statistic has probability density

$$p(u) = (n-1)/(1+u)^n, \qquad u \geqq 0.$$

[These distributions for varying b do not constitute an exponential family and Theorem 3 is therefore not directly applicable.

(i) One can restrict attention to the ordered variables $X^{(1)} < \cdots < X^{(n)}$ since these are sufficient for a and b, and transform to new variables $Z_1 = nX^{(1)}$, $Z_i = (n - i + 1)[X^{(i)} - X^{(i-1)}]$ for $i = 2, \cdots, n$ as in Problem 13 of Chapter 2. When $a = 1$, Z_1 is a complete sufficient statistic for b, and the test is therefore obtained by considering the conditional problem given z_1. Since $\Sigma_{i=2}^n Z_i$ is independent of Z_1, the conditional UMP unbiased test has the acceptance region $C_1 \leqq \Sigma_{i=2}^n Z_i \leqq C_2$ for each z_1, and the result follows.

(ii) When $b = 0$, $\Sigma_{i=1}^n Z_i$ is a complete sufficient statistic for a, and the test is therefore obtained by considering the conditional problem given $\Sigma_{i=1}^n z_i$. The remainder of the argument uses the fact that $Z_1/\Sigma_{i=1}^n Z_i$ is independent of $\Sigma_{i=1}^n Z_i$ when $b = 0$, and otherwise is similar to that used to prove Theorem 1.]

13. Extend the results of the preceding problem to the case, considered in Problem 8, Chapter 3, that observation is continued only until $X^{(1)}, \cdots, X^{(r)}$ have been observed.

Section 4

14. On the basis of a sample $X = (X_1, \cdots, X_n)$ of fixed size from $N(\xi, \sigma^2)$ there do not exist confidence intervals for ξ with positive confidence coefficient and of bounded length.

[Consider any family of confidence intervals $\delta(X) \pm L/2$ of constant length L. Let $\xi_1, \cdots, \xi_{2N}$ be such that $|\xi_i - \xi_j| > L$ whenever $i \neq j$. Then the sets $S_i = \{x\colon |\delta(x) - \xi_i| \leqq L/2\}$ $(i = 1, \cdots, 2N)$ are mutually exclusive. Also, there exists $\sigma_0 > 0$ such that

$$|P_{\xi_i,\sigma}\{X \in S_i\} - P_{\xi_1,\sigma}\{X \in S_i\}| \leqq 1/2N \quad \text{for} \quad \sigma > \sigma_0$$

as is seen by transforming to new variables $Y_j = (X_j - \xi_1)/\sigma$ and applying Lemmas 2 and 4 of the Appendix. Since $\min_i P_{\xi_1,\sigma}\{X \in S_i\} \leq 1/2N$, it follows for $\sigma > \sigma_0$ that $\min_i P_{\xi_i,\sigma}\{X \in S_i\} \leq 1/N$, and hence that

$$\inf_{\xi,\sigma} P_{\xi,\sigma}\{|\delta(X) - \xi| \leq L/2\} \leq 1/N.$$

The confidence coefficient associated with the intervals $\delta(X) \pm L/2$ is therefore zero, and the same must be true a fortiori of any set of confidence intervals of length $\leq L$.]

15. *Stein's two-stage procedure.* (i) If mS^2/σ^2 has a χ^2-distribution with m degrees of freedom, and if the conditional distribution of Y given $S = s$ is $N(0, \sigma^2/S^2)$, then Y has Student's t-distribution with m degrees of freedom.

(ii) Let $X_1, X_2, \cdots$ be independently distributed as $N(\xi, \sigma^2)$. Let $\bar{X}_0 = \Sigma_{i=1}^{n_0} X_i/n_0$, $S^2 = \Sigma_{i=1}^{n_0}(X_i - \bar{X}_0)^2/(n_0 - 1)$, and let $a_1 = \cdots = a_{n_0} = a$, $a_{n_0+1} = \cdots = a_n = b$ and $n \geq n_0$ be measurable functions of S. Then

$$Y = \frac{\sum_{i=1}^{n} a_i(X_i - \xi)}{\sqrt{S^2 \sum_{i=1}^{n} a_i^2}}$$

has Student's distribution with $n_0 - 1$ degrees of freedom.

(iii) Consider a two-stage sampling scheme Π_1, in which S^2 is computed from an initial sample of size n_0, and then $n - n_0$ additional observations are taken. The size of the second sample is such that

$$n = \max\left\{n_0 + 1, \left[\frac{S^2}{c}\right] + 1\right\}$$

where c is any given constant and where $[y]$ denotes the largest integer $\leq y$. There then exist numbers $a_1, \cdots, a_n$ such that $a_1 = \cdots = a_{n_0}, a_{n_0+1} = \cdots = a_n$, $\Sigma_{i=1}^n a_i = 1$, $\Sigma_{i=1}^n a_i^2 = c/S^2$. It follows from (ii) that $\Sigma_{i=1}^n a_i(X_i - \xi)/\sqrt{c}$ has Student's t-distribution with $n_0 - 1$ degrees of freedom.

(iv) The following sampling scheme Π_2, which does not require that the second sample contain at least one observation, is slightly more efficient than Π_1 for the applications to be made in Problems 16 and 17. Let n_0, S^2, and c be defined as before, let

$$n = \max\left\{n_0, \left[\frac{S^2}{c}\right] + 1\right\},$$

$a_i = 1/n$ $(i = 1, \cdots, n)$ and $\bar{X} = \Sigma_{i=1}^n a_i X_i$. Then $\sqrt{n}(\bar{X} - \xi)/S$ has again the t-distribution with $n_0 - 1$ degrees of freedom.

[(ii) Given $S = s$, the quantities a, b, and n are constants, $\Sigma_{i=1}^{n_0} a_i(X_i - \xi) = n_0 a(\bar{X}_0 - \xi)$ is distributed as $N(0, n_0 a^2\sigma^2)$, and the numerator of Y is therefore normally distributed with zero mean and variance $\sigma^2 \Sigma_{i=1}^n a_i^2$. The result now follows from (i).]

16. *Confidence intervals of fixed length for a normal mean.* (i) In the two-stage procedure Π_1 defined in part (iii) of the preceding problem, let the number c be determined for any given $L > 0$ and $0 < \gamma < 1$ by

$$\int_{-L/2\sqrt{c}}^{L/2\sqrt{c}} t_{n_0-1}(y)\, dy = \gamma$$

where t_{n_0-1} denotes the density of the t-distribution with $n_0 - 1$ degrees of freedom. Then the intervals $\Sigma_{i=1}^{n} a_i X_i \pm L/2$ are confidence intervals for ξ of length L and with confidence coefficient γ.

(ii) Let c be defined as in (i), and let the sampling procedure be Π_2 as defined in part (iv) of Problem 15. The intervals $\bar{X} \pm L/2$ are then confidence intervals of length L for ξ with confidence coefficient $\geqq \gamma$, while the expected number of observations required is slightly lower than under Π_1.

[(i) The probability that the intervals cover ξ equals

$$P_{\xi,\sigma}\left\{-\frac{L}{2\sqrt{c}} \leqq \frac{\sum_{i=1}^{n} a_i(X_i - \xi)}{\sqrt{c}} \leqq \frac{L}{2\sqrt{c}}\right\} = \gamma.$$

(ii) The probability that the intervals cover ξ equals

$$P_{\xi,\sigma}\left\{\frac{\sqrt{n}|\bar{X} - \xi|}{S} \leqq \frac{\sqrt{n}L}{2S}\right\} \geqq P_{\xi,\sigma}\left\{\frac{\sqrt{n}|\bar{X} - \xi|}{S} \leqq \frac{L}{2\sqrt{c}}\right\} = \gamma.$$

17. *Two-stage t-tests with power independent of σ.* (i) For the procedure Π_1 with any given c, let C be defined by

$$\int_{C}^{\infty} t_{n_0-1}(y)\, dy = \alpha.$$

Then the rejection region $(\Sigma_{i=1}^{n} a_i X_i - \xi_0)/\sqrt{c} > C$ defines a level α test of $H: \xi \leqq \xi_0$ with strictly increasing power function $\beta_c(\xi)$ depending only on ξ.

(ii) Given any alternative ξ_1 and any $\alpha < \beta < 1$, the number c can be chosen so that $\beta_c(\xi_1) = \beta$.

(iii) The test with rejection region $\sqrt{n}(\bar{X} - \xi_0)/S > C$ based on Π_2 and the same c as in (i) is a level α test of H which is uniformly more powerful than the test given in (i).

(iv) Extend parts (i)–(iii) to the problem of testing $\xi = \xi_0$ against $\xi \neq \xi_0$.

[(i) and (ii) The power of the test is

$$\beta_c(\xi) = \int_{C - \frac{\xi - \xi_0}{\sqrt{c}}}^{\infty} t_{n_0-1}(y)\, dy.$$

(iii) This follows from the inequality $\sqrt{n}\,|\xi - \xi_0|/S \geqq |\xi - \xi_0|/\sqrt{c}$.]

Section 5

18. Let $X_1, \cdots, X_n$ be distributed as in Problem 12. Then the most accurate unbiased confidence intervals for the scale parameter a are

$$\frac{2}{C_2}\Sigma[x_i - \min(x_1, \cdots, x_n)] \leqq a \leqq \frac{2}{C_1}\Sigma[x_i - \min(x_1, \cdots, x_n)].$$

19. Most accurate unbiased confidence intervals exist in the following situations:

(i) If X, Y are independent with binomial distributions $b(p_1, m)$ and $b(p_2, n)$, for the parameter p_1q_2/p_2q_1.

(ii) In a 2×2 table, for the parameter Δ of Chapter 4, Section 6.

Section 6

20. (i) Under the assumptions made at the beginning of Section 6, the UMP unbiased test of $H: \rho = \rho_0$ is given by (45).

(ii) Let $(\underline{\rho}, \bar{\rho})$ be the associated most accurate unbiased confidence intervals for $\rho = a\gamma + b\delta$ where $\underline{\rho} = \underline{\rho}(a, b)$, $\bar{\rho} = \bar{\rho}(a, b)$. Then if f_1 and f_2 are increasing functions, the expected value of $f_1(|\bar{\rho} - \rho|) + f_2(|\rho - \underline{\rho}|)$ is an increasing function of $a^2/n + b^2$.

[(i) Make any orthogonal transformation from $y_1, \cdots, y_n$ to new variables $z_1, \cdots, z_n$ such that $z_1 = \Sigma_i[bv_i + (a/n)]y_i/\sqrt{(a^2/n) + b^2}$, $z_2 = \Sigma_i(av_i - b)y_i/\sqrt{a^2 + nb^2}$, and apply Problems 5 and 6.

(ii) If $a_1^2/n + b_1^2 < a_2^2/n + b_2^2$, the random variable $|\bar{\rho}(a_2, b_2) - \rho|$ is stochastically larger than $|\bar{\rho}(a_1, b_1) - \rho|$, and analogously for $\underline{\rho}$.]

Section 7

21. Prove Theorem 3 for arbitrary values of c.

Section 8

22. If $c = 1$, $m = n = 4$, $\alpha = .1$, and the ordered coordinates $z^{(1)}, \cdots, z^{(N)}$ of a point z are 1.97, 2.19, 2.61, 2.79, 2.88, 3.02, 3.28, 3.41, determine the points of $S(z)$ belonging to the rejection region (54).

23. *Confidence intervals for a shift.* Let $X_1, \cdots, X_m$; $Y_1, \cdots, Y_n$ be independently distributed according to continuous distributions $F(x)$ and $G(y) = F(y - \Delta)$ respectively. Without any further assumptions concerning F, confidence intervals for Δ can be obtained from permutation tests of the hypotheses $H(\Delta_0)$: $\Delta = \Delta_0$. Specifically, consider the point $(z_1, \cdots, z_{m+n}) = (x_1, \cdots, x_m, y_1 - \Delta, \cdots, y_n - \Delta)$ and the $\binom{m+n}{m}$ permutations $i_1 < \cdots < i_m$; $i_{m+1} < \cdots < i_{m+n}$ of the integers $1, \cdots, m + n$. Suppose that the hypothesis $H(\Delta)$ is accepted for the k of these permutations which lead to the smallest values of

$$\left| \sum_{j=m+1}^{m+n} z_{i_j}/n - \sum_{j=1}^{m} z_{i_j}/m \right|$$

where $k = (1 - \alpha)\binom{m+n}{m}$. Then the totality of values Δ for which $H(\Delta)$ is accepted constitute an interval, and these intervals are confidence intervals for Δ at confidence level $1 - \alpha$.

[A point is in the acceptance region for $H(\Delta)$ if

$$|\Sigma(y_j - \Delta)/n - \Sigma x_i/m| = |\bar{y} - \bar{x} - \Delta|$$

is exceeded by at least $\binom{m+n}{m} - k$ of the quantities $|\bar{y}' - \bar{x}' - \gamma\Delta|$ where

$(x_1', \cdots, x_m', y_1', \cdots, y_n')$ is a permutation of $(x_1, \cdots, x_m, y_1, \cdots, y_n)$, the quantity γ is determined by this permutation, and $|\gamma| \leqq 1$. The desired result now follows from the fact that if

$$(\bar{y} - \bar{x} - \Delta)^2 \leqq (\bar{y}' - \bar{x}' - \gamma\Delta)^2$$

or more generally if $(a - \Delta)^2 \leqq (b - \gamma\Delta)^2$ for some a and b both when $\Delta = \Delta_0$ and when $\Delta = \Delta_1$, then the same inequality holds for any Δ between Δ_0 and Δ_1.]

Section 9

24. In the matched pairs experiment for testing the effect of a treatment, suppose that only the differences $Z_i = Y_i - X_i$ are observable. The Z's are assumed to be a sample from an unknown continuous distribution, which under the hypothesis of no treatment effect is symmetric with respect to the origin. Under the alternatives it is symmetric with respect to a point $\zeta > 0$. Determine the test which among all unbiased tests maximizes the power against the alternatives that the Z's are a sample from $N(\zeta, \sigma^2)$ with $\zeta > 0$.

[Under the hypothesis, the set of statistics $(\Sigma_{i=1}^n Z_i^2, \cdots, \Sigma_{i=1}^n Z_i^{2n})$ is sufficient; that it is complete is shown as the corresponding result in Theorem 3. The remainder of the argument follows the lines of Section 8.]

25. (i) If $X_1, \cdots, X_n$; $Y_1, \cdots, Y_n$ are independent normal variables with common variance σ^2 and means $E(X_i) = \xi_i$, $E(Y_i) = \xi_i + \Delta$ the UMP unbiased test of $\Delta = 0$ against $\Delta > 0$ is given by (59).

(ii) Determine the most accurate unbiased confidence intervals for Δ.

[(i) The structure of the problem becomes clear if one makes the orthogonal transformation $X_i' = (Y_i - X_i)/\sqrt{2}$, $Y_i' = (X_i + Y_i)/\sqrt{2}$.]

26. *Comparison of two designs.* Under the assumptions made at the beginning of Section 9, one has the following comparison of the methods of complete randomization and matched pairs. The unit effects and experimental effects U_i and V_i are independently normally distributed with variances σ_1^2, σ^2 and means $E(U_i) = \mu$ and $E(V_i) = \xi$ or η as V_i corresponds to a control or treatment. With complete randomization, the observations are $X_i = U_i + V_i$ $(i = 1, \cdots, n)$ for the controls and $Y_i = U_{n+i} + V_{n+i}$ $(i = 1, \cdots, n)$ for the treated cases, with $E(X_i) = \mu + \xi$, $E(Y_i) = \mu + \eta$. For the matched pairs, if the matching is assumed to be perfect, the X's are as before but $Y_i = U_i + V_{n+i}$. UMP unbiased tests are given by (27) for complete randomization and by (59) for matched pairs. The distribution of the test statistic under an alternative $\Delta = \eta - \xi$ is the noncentral t-distribution with noncentrality parameter $\sqrt{n}\,\Delta/\sqrt{2(\sigma^2 + \sigma_1^2)}$ and $(2n - 2)$ degrees of freedom in the first case, and with noncentrality parameter $\sqrt{n}\,\Delta/\sqrt{2}\sigma$ and $n - 1$ degrees of freedom in the second one. Thus the method of matched pairs has the disadvantage of a smaller number of degrees of freedom and the advantage of a larger noncentrality parameter. For $\alpha = .05$ and $\Delta = 4$, compare the power of the two methods as a function of n when $\sigma_1 = 1$, $\sigma = 2$ and when $\sigma_1 = 2$, $\sigma = 1$.

27. *Continuation.* An alternative comparison of the two designs is obtained by considering the expected length of the most accurate unbiased confidence intervals for $\Delta = \eta - \xi$ in each case. Carry this out for varying n and confidence coefficient $1 - \alpha = .95$ when $\sigma_1 = 1$, $\sigma = 2$ and when $\sigma_1 = 2$, $\sigma = 1$.

Section 10

28. Suppose that a critical function ϕ_0 satisfies (65) but not (67) and let $\alpha < \frac{1}{2}$. Then the following construction provides a measurable critical function ϕ satisfying (67) and such that $\phi_0(z) \leqq \phi(z)$ for all z. Inductively, sequences of functions $\phi_1, \phi_2, \cdots$ and $\psi_0, \psi_1, \cdots$ are defined through the relations

$$\psi_m(z) = \sum_{z' \in S(z)} \phi_m(z')/N_1! \cdots N_c!, \qquad m = 0, 1, \cdots$$

and

$$\phi_m(z) = \begin{cases} \phi_{m-1}(z) + [\alpha - \psi_{m-1}(z)] & \text{if both } \phi_{m-1}(z) \text{ and } \psi_{m-1}(z) \text{ are } < \alpha \\ \phi_{m-1}(z) & \text{otherwise.} \end{cases}$$

The function $\phi(z) = \lim \phi_m(z)$ then satisfies the required conditions.

[The functions ϕ_m are nondecreasing and between 0 and 1. It is further seen by induction that $0 \leqq \alpha - \psi_m(z) \leqq (1 - \gamma)^m[\alpha - \psi_0(z)]$ where $\gamma = 1/N_1! \cdots N_c!$.]

29. Consider the problem of testing H: $\eta = \xi$ in the family of densities (62) when it is given that $\sigma > c > 0$ and that the point $(\zeta_{11}, \cdots, \zeta_{cN_c})$ of (63) lies in a bounded region R containing a rectangle, where c and R are known. Then Theorem 4 is no longer applicable. However, unbiasedness of a test ϕ of H implies (67), and therefore reduces the problem to the class of permutation tests.

[Unbiasedness implies $\int\phi(z)p_{\sigma,\zeta}(z)\,dz = \alpha$ and hence

$$\alpha = \int \psi(z) p_{\sigma,\zeta}(z)\,dz = \int \psi(z) \frac{1}{(\sqrt{2\pi}\sigma)^N} \exp\left[-\frac{1}{2\sigma^2}\Sigma\Sigma(z_{ij} - \zeta_{ij})^2\right]$$

for all $\sigma > c$ and ζ in R. The result follows from completeness of this last family.]

30. To generalize Theorem 4 to other designs, let $Z = (Z_1, \cdots, Z_N)$ and let $G = \{g_1, \cdots, g_r\}$ be a group of permutations of N coordinates or more generally a group of orthogonal transformations of N-space. If

$$p_{\sigma,\zeta}(z) = \frac{1}{r}\sum_{k=1}^{r} \frac{1}{(\sqrt{2\pi}\sigma)^N} \exp\left(-\frac{1}{2\sigma^2}|z - g_k\zeta|^2\right) \tag{77}$$

where $|z|^2 = \Sigma z_i^2$, then $\int\phi(z)p_{\sigma,\zeta}(z)\,dz \leqq \alpha$ for all $\sigma > 0$ and all ζ implies

$$\frac{1}{r}\sum_{z' \in S(z)} \phi(z') \leqq \alpha \qquad \text{a.e.} \tag{78}$$

where $S(z)$ is the set of points in N-space obtained from z by applying to it all the transformations g_k, $k = 1, \cdots, r$.

31. *Generalization of Corollary* 2. Let H be the class of densities (77) with $\sigma > 0$ and $-\infty < \zeta_i < \infty$ $(i = 1, \cdots, N)$. A complete family of tests of H at level of significance α is the class of *permutation tests* satisfying

$$\frac{1}{r}\sum_{z' \in S(z)} \phi(z') = \alpha \qquad \text{a.e.} \tag{79}$$

Section 11

32. (i) If the joint distribution of X and Y is the bivariate normal distribution (70), then the conditional distribution of Y given x is the normal distribution with variance $\tau^2(1 - \rho^2)$ and mean $\eta + (\rho\tau/\sigma)(x - \xi)$.

(ii) Let $(X_1, Y_1), \cdots, (X_n, Y_n)$ be a sample from a bivariate normal distribution, let R be the sample correlation coefficient, and suppose that $\rho = 0$. Then the conditional distribution of $\sqrt{n-2}\, R/\sqrt{1 - R^2}$ given $x_1, \cdots, x_n$ is Student's t-distribution with $n - 2$ degrees of freedom provided $\Sigma(x_i - \bar{x})^2 > 0$. This is therefore also the unconditional distribution of this statistic.

(iii) The probability density of R itself is then

$$p(r) = \frac{1}{\sqrt{\pi}} \frac{\Gamma[\frac{1}{2}(n-1)]}{\Gamma[\frac{1}{2}(n-2)]} (1 - r^2)^{\frac{1}{2}n-2}. \tag{80}$$

[(ii) If $v_i = (x_i - \bar{x})/\sqrt{\Sigma(x_j - \bar{x})^2}$ so that $\Sigma v_i = 0$, $\Sigma v_i^2 = 1$, the statistic can be written as

$$\frac{\Sigma v_i Y_i}{\sqrt{[\Sigma Y_i^2 - n\bar{Y}^2 - (\Sigma v_i Y_i)^2]/(n-2)}}.$$

Since its distribution depends only on ρ one can assume $\eta = 0$, $\tau = 1$. The desired result follows from Problem 6 by making an orthogonal transformation from $(Y_1, \cdots, Y_n)$ to $(Z_1 \cdots, Z_n)$ such that $Z_1 = \sqrt{n}\,\bar{Y}$, $Z_2 = \Sigma v_i Y_i$.]

33. (i) Let $(X_1, Y_1), \cdots, (X_n, Y_n)$ be a sample from the bivariate normal distribution (70), and let $S_1^2 = \Sigma(X_i - \bar{X})^2$, $S_2^2 = \Sigma(Y_i - \bar{Y})^2$, $S_{12} = \Sigma(X_i - \bar{X})(Y_i - \bar{Y})$. There exists a UMP unbiased test for testing the hypothesis $\tau/\sigma = \Delta$. Its acceptance region is

$$\frac{|\Delta^2 S_1^2 - S_2^2|}{\sqrt{(\Delta^2 S_1^2 + S_2^2)^2 - 4\Delta^2 S_{12}^2}} \leq C,$$

and the probability density of the test statistic is given by (80) when the hypothesis is true.

(ii) Under the assumption $\tau = \sigma$, there exists a UMP unbiased test for testing $\eta = \xi$, with acceptance region $|\bar{Y} - \bar{X}|/\sqrt{S_1^2 + S_2^2 - S_{12}} \leq C$. On multiplication by a suitable constant the test statistic has Student's t-distribution with $n - 1$ degrees of freedom when $\eta = \xi$. (Without the assumption $\tau = \sigma$, this hypothesis is a special case of the one considered in Chapter 7, Example 11.)

[(i) The transformation $U = \Delta X + Y$, $V = X - (1/\Delta)Y$ reduces the problem to that of testing that the correlation coefficient in a bivariate normal distribution is zero.

(ii) Transform to new variables $V_i = Y_i - X_i$, $U_i = Y_i + X_i$.]

34. Let $(X_1, Y_1), \cdots, (X_n, Y_n)$ be a sample from the bivariate normal distribution (70), and let $S_1^2 = \Sigma(X_i - \bar{X})^2$, $S_{12} = \Sigma(X_i - \bar{X})(Y_i - \bar{Y})$, $S_2^2 = \Sigma(Y_i - \bar{Y})^2$.

(i) Then (S_1^2, S_{12}, S_2^2) are independently distributed of $(\bar{X}, \bar{Y})$, and their joint distribution is the same as that of $(\Sigma_{i=1}^{n-1} X_i'^2, \Sigma_{i=1}^{n-1} X_i' Y_i', \Sigma_{i=1}^{n-1} Y_i'^2)$ where (X_i', Y_i'), $i = 1, \cdots, n - 1$, are a sample from the distribution (70) with $\xi = \eta = 0$.

(ii) Let $X_1, \cdots, X_m$ and $Y_1, \cdots, Y_m$ be two samples from $N(0, 1)$. Then the joint density of $S_1^2 = \Sigma X_i^2$, $S_{12} = \Sigma X_i Y_i$, $S_2^2 = \Sigma Y_i^2$ is

$$\frac{1}{4\pi\Gamma(m-1)}(s_1^2 s_2^2 - s_{12}^2)^{\frac{1}{2}(m-3)} \exp\left[-\tfrac{1}{2}(s_1^2 + s_2^2)\right]$$

for $s_{12}^2 \leq s_1^2 s_2^2$, and zero elsewhere.

(iii) The joint density of the statistics (S_1^2, S_{12}, S_2^2) of part (i) is

$$\frac{(s_1^2 s_2^2 - s_{12}^2)^{\frac{1}{2}(n-4)}}{4\pi\Gamma(n-2)(\sigma\tau\sqrt{1-\rho^2})^{n-1}} \exp\left[-\frac{1}{2(1-\rho^2)}\left(\frac{s_1^2}{\sigma^2} - \frac{2\rho s_{12}}{\sigma\tau} + \frac{s_2^2}{\tau^2}\right)\right] \tag{81}$$

for $s_{12}^2 \leq s_1^2 s_2^2$, and zero elsewhere.

[(i) Make an orthogonal transformation from $X_1, \cdots, X_n$ to $X_1', \cdots, X_n'$ such that $X_n' = \sqrt{n}\bar{X}$, and apply the same orthogonal transformation also to $Y_1, \cdots, Y_n$. Then $Y_n' = \sqrt{n}\bar{Y}$,

$$\sum_{i=1}^{n-1} X_i'^2 = \sum_{i=1}^{n}(X_i - \bar{X})^2, \quad \sum_{i=1}^{n-1} X_i' Y_i' = \sum_{i=1}^{n}(X_i - \bar{X})(Y_i - \bar{Y}),$$

$$\sum_{i=1}^{n-1} Y_i'^2 = \sum_{i=1}^{n}(Y_i - \bar{Y})^2.$$

The pairs of variables $(X_1', Y_1'), \cdots, (X_n', Y_n')$ are independent, each with a bivariate normal distribution with the same variances and correlation as those of (X, Y) and with means $E(X_i') = E(Y_i') = 0$ for $i = 1, \cdots, n-1$.

(ii) Consider first the joint distribution of $S_{12} = \Sigma x_i Y_i$ and $S_2^2 = \Sigma Y_i^2$ given $x_1, \cdots, x_m$. Letting $Z_1 = S_{12}/\sqrt{\Sigma x_i^2}$ and making an orthogonal transformation from $Y_1, \cdots, Y_m$ to $Z_1, \cdots, Z_m$ so that $S_2^2 = \Sigma_{i=1}^m Z_i^2$, the variables Z_1 and $\Sigma_{i=2}^m Z_i^2 = S_2^2 - Z_1^2$ are independently distributed as $N(0, 1)$ and χ_{m-1}^2 respectively. From this the joint conditional density of $S_{12} = s_1 Z_1$ and S_2^2 is obtained by a simple transformation of variables. Since the conditional distribution depends on the x's only through s_1^2, the joint density of S_1^2, S_{12}, S_2^2 is found by multiplying the above conditional density by the marginal one of S_1^2, which is χ_m^2. The proof is completed through use of the identity

$$\Gamma[\tfrac{1}{2}(m-1)]\Gamma(\tfrac{1}{2}m) = \frac{\sqrt{\pi}\,\Gamma(m-1)}{2^{m-2}}.$$

(iii) If $(X', Y') = (X_1', Y_1'; \cdots; X_m', Y_m')$ is a sample from a bivariate normal distribution with $\xi = \eta = 0$, then $T = (\Sigma X_i'^2, \Sigma X_i' Y_i', \Sigma Y_i'^2)$ is sufficient for $\theta = (\sigma, \rho, \tau)$, and the density of T is obtained from that given in part (ii) for $\theta_0 = (1, 0, 1)$ through the identity [Chapter 3, Problem 10(i)]

$$p_\theta^T(t) = p_{\theta_0}^T(t)[p_\theta^{X',Y'}(x', y')/p_{\theta_0}^{X',Y'}(x', y')].$$

The result now follows from part (i) with $m = n - 1$.]

35. If $(X_1, Y_1), \cdots, (X_n, Y_n)$ is a sample from a bivariate normal distribution, the probability density of the sample correlation coefficient R is*

* This density and the associated cumulative distribution function are tabled by David, *Tables of the Correlation Coefficient*, Cambridge Univ. Press, 1938.

$$(82)\quad p_\rho(r) = \frac{2^{n-3}}{\pi(n-3)!}(1-\rho^2)^{\frac{1}{2}(n-1)}(1-r^2)^{\frac{1}{2}(n-4)} \sum_{k=0}^{\infty} \Gamma^2[\tfrac{1}{2}(n+k-1)]\frac{(2\rho r)^k}{k!}$$

or alternatively

$$(83)\quad p_\rho(r) = \frac{n-2}{\pi}(1-\rho^2)^{\frac{1}{2}(n-1)}(1-r^2)^{\frac{1}{2}(n-4)} \int_0^1 \frac{t^{n-2}}{(1-\rho r t)^{n-1}} \frac{1}{\sqrt{1-t^2}}\,dt.$$

Another form is obtained by making the transformation $t = (1-v)/(1-\rho r v)$ in the integral on the right-hand side of (83). The integral then becomes

$$(84)\quad \frac{1}{(1-\rho r)^{\frac{1}{2}(2n-3)}} \int_0^1 \frac{(1-v)^{n-2}}{\sqrt{2v}}[1-\tfrac{1}{2}v(1+\rho r)]^{-\frac{1}{2}}\,dv.$$

Expanding the last factor in powers of v, the density becomes

$$(85)\quad \frac{n-2}{\sqrt{2\pi}}\frac{\Gamma(n-1)}{\Gamma(n-\frac{1}{2})}(1-\rho^2)^{\frac{1}{2}(n-1)}(1-r^2)^{\frac{1}{2}(n-4)}(1-\rho r)^{-n+\frac{3}{2}} F\left(\tfrac{1}{2};\tfrac{1}{2};n-\tfrac{1}{2};\frac{1+\rho r}{2}\right)$$

where

$$(86)\quad F(a,b,c,x) = \sum_{j=0}^{\infty} \frac{\Gamma(a+j)}{\Gamma(a)}\frac{\Gamma(b+j)}{\Gamma(b)}\frac{\Gamma(c)}{\Gamma(c+j)}\frac{x^j}{j!}$$

is a hypergeometric function.

[To obtain the first expression make a transformation from (S_1^2, S_2^2, S_{12}) with density (81) to (S_1^2, S_2^2, R) and expand the factor $\exp\{\rho s_{12}/(1-\rho^2)\sigma\tau\} = \exp\{\rho r s_1 s_2/(1-\rho^2)\sigma\tau\}$ into a power series. The resulting series can be integrated term by term with respect to s_1^2 and s_2^2. The equivalence with the second expression is seen by expanding the factor $(1-\rho r t)^{-(n-1)}$ under the integral in (83) and integrating term by term.]

36. If X and Y have a bivariate normal distribution with correlation coefficient $\rho > 0$, they are positively dependent in the sense of Section 11.

[The conditional distribution of Y given x is normal with mean $\eta + \rho\tau\sigma^{-1}(x-\xi)$ and variance $\tau^2(1-\rho^2)$. Through addition to such a variable of the positive quantity $\rho\tau\sigma^{-1}(x'-x)$ it is transformed into one with the conditional distribution of Y given $x' > x$.]

13. REFERENCES

Basu, D.
(1955) "On statistics independent of a complete sufficient statistic," *Sankhyā*, Vol. 15, pp. 377–380 and Vol. 20 (1958), pp. 223–226.
[Theorem 2.]

Chapman, Douglas G.
(1950) "Some two-sample tests," *Ann. Math. Stat.*, Vol. 21, pp. 601–606.
[Extends Problems 15–17 to the comparison of two means.]

Fisher, R. A.
(1915) "Frequency distribution of the values of the correlation coefficient in samples from an indefinitely large population," *Biometrika*, Vol. 10, pp. 507–521.
[Derives the distribution of the sample correlation coefficient from a bivariate normal distribution.]
(1931) "Properties of the [*Hh*] functions," *Brit. Assoc. Math. Tables*, Vol. 1 (3rd ed., 1951, pp. xxviii–xxxvii).
[Derivation of noncentral t-distribution.]
(1935) *The Design of Experiments*, Edinburgh, Oliver and Boyd.
[Contains the basic ideas concerning permutation tests. In particular, points out how randomization provides a basis for inference and proposes the permutation version of the t-test as not requiring the assumption of normality.]

Girshick, M. A.
(1946) "Contributions to the theory of sequential analysis. I." *Ann. Math. Stat.*, Vol. 17, pp. 123–143.
[Problem 8.]

Helmert, F. R.
(1876) "Die Genauigkeit der Formel von Peters zur Berechnung des wahrscheinlichen Beobachtungsfehlers direkter Beobachtungen gleicher Genauigkeit," *Astron. Nachrichten*, Vol. 88, No. 2096–97, pp. 113–132.
[Obtains the distribution of $\Sigma(X_i - \bar{X})^2$ when the X's are independently, normally distributed.]

Hsu, C. T.
(1940), "On samples from a normal bivariate population," *Ann. Math. Stat.*, Vol. 11, pp. 410–426.
[Problem 33(ii).]

Lehmann, E. L.
(1947) "On optimum tests of composite hypotheses with one constraint," *Ann. Math. Stat.*, Vol. 18, pp. 473–494.
[Determines best similar regions for a number of problems including Problem 12.]

Lehmann, E. L., and Stein, C.
(1949) "On the theory of some non-parametric hypotheses," *Ann. Math. Stat.*, Vol. 20, pp. 28–45.
[Develops the theory of optimum permutation tests.]

Morgan, W. A.
(1939) "A test for the significance of the difference between the two variances in a sample from a normal bivariate population," *Biometrika*, Vol. 31, pp. 13–19.
[Problem 33(i).]

Neyman, J.
(1938) "On statistics the distribution of which is independent of the parameters involved in the original probability law of the observed variables," *Stat. Res. Mem.*, Vol. II, pp. 58–59.
[Essentially Theorem 2 under regularity assumptions.]

Paulson, Edward
(1941) "On certain likelihood ratio tests associated with the exponential distribution," *Ann. Math. Stat.*, Vol. 12, pp. 301–306.
[Discusses the power of the tests of Problem 12.]

Pitman, E. J. G.
(1937/38) "Significance tests which may be applied to samples from any population," *J. Roy. Stat. Soc. Suppl.*, Vol. 4, pp. 119–130, pp. 225–232, and *Biometrika*, Vol. 29, pp. 322–335.
[Develops the theory of randomization tests with many applications.]
(1939) "A note on normal correlation," *Biometrika*, Vol. 31, pp. 9–12.
[Problem 33(i).]

Stein, Charles
(1945) "A two-sample test for a linear hypothesis whose power is independent of the variance," *Ann. Math. Stat.*, Vol. 16, pp. 243–258.
[Problems 15–17.]

Student (W. S. Gosset)
(1908) "On the probable error of a mean," *Biometrika*, Vol. 6, pp. 1–25.
[Obtains the distribution of the t-statistic when the X's are a sample from $N(0, \sigma^2)$. A rigorous proof was given by R. A. Fisher, "Note on Dr. Burnside's recent paper on error of observation," *Proc. Camb. Phil. Soc.*, Vol. 21 (1923), pp. 655–658.]

CHAPTER 6

Invariance

1. SYMMETRY AND INVARIANCE

Many statistical problems exhibit symmetries, which provide natural restrictions to impose on the statistical procedures that are to be employed. Suppose, for example, that $X_1, \cdots, X_n$ are independently distributed with probability densities $p_{\theta_1}(x_1), \cdots, p_{\theta_n}(x_n)$. For testing the hypothesis $H: \theta_1 = \cdots = \theta_n$ against the alternative that the θ's are not all equal, the test should be symmetric in $x_1, \cdots, x_n$ since otherwise the acceptance or rejection of the hypothesis would depend on the (presumably quite irrelevant) numbering of these variables.

As another example consider a circular target with center O, on which are marked the impacts of a number of shots. Suppose that the points of impact are independent observations on a bivariate normal distribution centered on O. In testing this distribution for circular symmetry with respect to O, it seems reasonable to require that the test itself exhibit such symmetry. For if it lacks this feature, a two-dimensional (for example, Cartesian) coordinate system is required to describe the test, and acceptance or rejection will depend on the choice of this system, which under the assumptions made is quite arbitrary and has no bearing on the problem.

The mathematical expression of symmetry is invariance under a suitable group of transformations. In the first of the two examples above the group is that of all permutations of the variables $x_1, \cdots, x_n$ since a function of n variables is symmetric if and only if it remains invariant under all permutations of these variables. In the second example, circular symmetry with respect to the center O is equivalent to invariance under all rotations about O.

In general, let X be distributed according to a probability distribution P_θ, $\theta \in \Omega$, and let g be a transformation of the sample space $\mathscr{X}$. All such transformations considered in connection with invariance will be assumed to be 1 : 1 transformations of $\mathscr{X}$ onto itself. Denote by gX the random variable that takes on the value gx when $X = x$, and suppose that when the distribution of X is P_θ, $\theta \in \Omega$, the distribution of gX is $P_{\theta'}$ with θ'

also in Ω. The element θ' of Ω which is associated with θ in this manner will be denoted by $\bar{g}\theta$ so that

$$(1) \qquad P_\theta\{gX \in A\} = P_{\bar{g}\theta}\{X \in A\}.$$

Here the subscript θ on the left member of (1) indicates the distribution of X, not that of gX. Equation (1) can also be written as $P_\theta(g^{-1}A) = P_{\bar{g}\theta}(A)$ and hence as

$$(2) \qquad P_{\bar{g}\theta}(gA) = P_\theta(A).$$

The parameter set Ω remains invariant under g (or is preserved by g) if $\bar{g}\theta \in \Omega$ for all $\theta \in \Omega$, and if in addition for any $\theta' \in \Omega$ there exists $\theta \in \Omega$ such that $\bar{g}\theta = \theta'$. These two conditions can be expressed by the equation

$$(3) \qquad \bar{g}\Omega = \Omega.$$

The transformation $\bar{g}$ of Ω onto itself defined in this way is 1 : 1 provided the distributions P_θ corresponding to different values of θ are distinct. To see this let $\bar{g}\theta_1 = \bar{g}\theta_2$. Then $P_{\bar{g}\theta_1}(gA) = P_{\bar{g}\theta_2}(gA)$ and therefore $P_{\theta_1}(A) = P_{\theta_2}(A)$ for all A, so that $\theta_1 = \theta_2$.

Lemma 1. *Let g, g' be two transformations preserving Ω. Then the transformations $g'g$ and g^{-1} defined by*

$$(g'g)x = g'(gx) \qquad \text{and} \qquad g(g^{-1}x) = x \text{ for all } x \in \mathscr{X}$$

also preserve Ω and satisfy

$$(4) \qquad \overline{g'g} = \bar{g}' \cdot \bar{g} \qquad \text{and} \qquad \overline{(g^{-1})} = (\bar{g})^{-1}.$$

Proof. If the distribution of X is P_θ, that of gX is $P_{\bar{g}\theta}$ and that of $g'gX = g'(gX)$ is therefore $P_{\bar{g}'\cdot\bar{g}\theta}$. This establishes the first equation of (4); the proof of the second one is analogous.

We shall say that *the problem of testing $H: \theta \in \Omega_H$ against $K: \theta \in \Omega_K$ remains invariant* under a transformation g if $\bar{g}$ preserves both Ω_H and Ω_K, so that the equation

$$(5) \qquad \bar{g}\Omega_H = \Omega_H$$

holds in addition to (3). Let $\mathscr{C}$ be a class of transformations satisfying these two conditions, and let G be the smallest class of transformations containing $\mathscr{C}$ and such that $g, g' \in G$ implies that $g'g$ and g^{-1} belong to G. Then G is a group of transformations, all of which by Lemma 1 preserve both Ω and Ω_H. Any class $\mathscr{C}$ of transformations leaving the problem invariant can therefore be extended to a group G. It follows further from Lemma 1 that the class of induced transformations $\bar{g}$ form a group $\bar{G}$. The two equations (4) express the fact that $\bar{G}$ is a homomorphism of G.

In the presence of symmetries in both sample and parameter space represented by the groups G and $\bar{G}$, it is natural to restrict attention to tests ϕ which are also symmetric, that is, which satisfy

$$\phi(gx) = \phi(x) \quad \text{for all} \quad x \in X \quad \text{and} \quad g \in G. \tag{6}$$

A test ϕ satisfying (6) is said to be *invariant under G*. The restriction to invariant tests is a particular case of the principle of invariance formulated in Section 5 of Chapter 1. As was indicated there and in the examples above, a transformation g can be interpreted as a change of coordinates. From this point of view, a test is invariant if it is independent of the particular coordinate system in which the data are expressed.

A transformation g, in order to leave a problem invariant, must in particular preserve the class $\mathcal{A}$ of measurable sets over which the distributions P_θ are defined. This means that any set $A \in \mathcal{A}$ is transformed into a set of $\mathcal{A}$ and is the image of such a set, so that gA and $g^{-1}A$ both belong to $\mathcal{A}$. Any transformation satisfying this condition is said to be *bimeasurable*. Since a group with each element g also contains g^{-1}, its elements are automatically bimeasurable if all of them are measurable. If g' and g are bimeasurable, so are $g'g$ and g^{-1}. The transformations of the group G above generated by a class $\mathcal{C}$ are therefore all bimeasurable provided this is the case for the transformations of $\mathcal{C}$.

2. MAXIMAL INVARIANTS

If a problem is invariant under a group of transformations, the *principle of invariance* restricts attention to invariant tests. In order to obtain the best of these, it is convenient first to characterize the totality of invariant tests.

Let two points x_1, x_2 be considered equivalent under G

$$x_1 \sim x_2 \pmod{G}$$

if there exists a transformation $g \in G$ for which $x_2 = gx_1$. This is a true equivalence relation since G is a group and the sets of equivalent points, the *orbits* of G, therefore constitute a partition of the sample space. (Cf. Appendix, Section 1.) A point x traces out an orbit as all transformations g of G are applied to it; this means that the orbit containing x consists of the totality of points gx with $g \in G$. It follows from the definition of invariance that a function is invariant if and only if it is constant on each orbit.

A function T is said to be *maximal invariant* if it is invariant and if

$$T(x_1) = T(x_2) \quad \text{implies} \quad x_2 = gx_1 \quad \text{for some} \quad g \in G, \tag{7}$$

that is, if it is constant on the orbits but for each orbit takes on a different value. All maximal invariants are equivalent in the sense that their sets of constancy coincide.

Theorem 1. *Let $T(x)$ be a maximal invariant with respect to G. Then a necessary and sufficient condition for ϕ to be invariant is that it depends on x only through $T(x)$, that is, that there exists a function h for which $\phi(x) = h[T(x)]$ for all x.*

Proof. If $\phi(x) = h[T(x)]$ for all x, then $\phi(gx) = h[T(gx)] = h[T(x)] = \phi(x)$ so that ϕ is invariant. On the other hand, if ϕ is invariant and if $T(x_1) = T(x_2)$, then $x_2 = gx_1$ for some g and therefore $\phi(x_2) = \phi(x_1)$.

Example 1. (i) Let $x = (x_1, \cdots, x_n)$, and let G be the group of translations

$$gx = (x_1 + c, \cdots, x_n + c), \qquad -\infty < c < \infty.$$

Then the set of differences $y = (x_1 - x_n, \cdots, x_{n-1} - x_n)$ is invariant under G. To see that it is maximal invariant suppose that $x_i - x_n = x_i' - x_n'$ for $i = 1, \cdots, n - 1$. Putting $x_n' - x_n = c$, one has $x_i' = x_i + c$ for all i, as was to be shown. The function y is of course only one representation of the maximal invariant. Others are for example $(x_1 - x_2, x_2 - x_3, \cdots, x_{n-1} - x_n)$ or the redundant $(x_1 - \bar{x}, \cdots, x_n - \bar{x})$. In the particular case that $n = 1$, there are no invariants. The whole space is a single orbit so that for any two points there exists a transformation of G taking one into the other. In such a case the transformation group G is said to be *transitive*. The only invariant functions are then the constant functions $\phi(x) \equiv c$.

(ii) If G is the group of transformations

$$gx = (cx_1, \cdots, cx_n), \qquad c \neq 0,$$

a special role is played by any zero coordinates. However, in statistical applications the set of points for which none of the coordinates is zero typically has probability 1; attention can then be restricted to this part of the sample space and the set of ratios $x_1/x_n, \cdots, x_{n-1}/x_n$ is a maximal invariant. Without this restriction, two points x, x' are equivalent with respect to the maximal invariant partition if among their coordinates there is the same number of zeros (if any), if these occur at the same places and if for any two nonzero coordinates x_i, x_j the ratios x_j/x_i and x_j'/x_i' are equal.

(iii) Let $x = (x_1, \cdots, x_n)$ and let G be the group of all orthogonal transformations $x' = \Gamma x$ of n-space. Then Σx_i^2 is maximal invariant, that is, two points x and x^* can be transformed into each other by an orthogonal transformation if and only if they have the same distance from the origin. The proof of this is immediate if one restricts attention to the plane containing the points x, x^* and the origin.

Example 2. Let $x = (x_1, \cdots, x_n)$ and let G be the set of $n!$ permutations of the coordinates of x. Then the set of ordered coordinates (*order statistics*) $x^{(1)} \leq \cdots \leq x^{(n)}$ is maximal invariant. A permutation of the x_i obviously does not change the set of values of the coordinates and therefore not the $x^{(i)}$.

On the other hand, two points with the same set of ordered coordinates can be obtained from each other through a permutation of coordinates.

Example 3. Let G be the totality of transformations $x_i' = f(x_i)$, $i = 1, \cdots, n$, such that f is continuous and strictly increasing, and suppose that attention can be restricted to the points all of whose n coordinates are distinct. If the x_i are considered as n points on the real line, any such transformation preserves their order. Conversely, if $x_1, \cdots, x_n$ and $x_1', \cdots, x_n'$ are two sets of points in the same order, say $x_{i_1} < \cdots < x_{i_n}$ and $x'_{i_1} < \cdots < x'_{i_n}$, there exists a transformation f satisfying the required conditions and such that $x_i' = f(x_i)$ for all i. It can be defined for example as $f(x) = x + (x'_{i_1} - x_{i_1})$ for $x \leqq x_{i_1}$, $f(x) = x + (x'_{i_n} - x_{i_n})$ for $x \geqq x_{i_n}$, and to be linear between x_{i_k} and $x_{i_{k+1}}$ for $k = 1, \cdots, n-1$. A formal expression for the maximal invariant in this case is the set of *ranks* $(r_1, \cdots, r_n)$ of $(x_1, \cdots, x_n)$. Here the rank r_i of x_i is defined through

$$x_i = x^{(r_i)}$$

so that r_i is the number of x's $\leqq x_i$. In particular $r_i = 1$ if x_i is the smallest x, $r_i = 2$ if it is the second smallest, etc.

Frequently, it is convenient to obtain a maximal invariant in a number of steps, each corresponding to a subgroup of G. To illustrate the process and a difficulty that may arise in its application, let $x = (x_1, \cdots, x_n)$, suppose that the coordinates are distinct, and consider the group of transformations

$$gx = (ax_1 + b, \cdots, ax_n + b), \qquad a \neq 0, \qquad -\infty < b < \infty.$$

Applying first the subgroup of translations $x_i' = x_i + b$ a maximal invariant is $y = (y_1, \cdots, y_{n-1})$ with $y_i = x_i - x_n$. Another subgroup consists of the scale changes $x_i'' = ax_i$. This induces a corresponding change of scale in the y's: $y_i'' = ay_i$, and a maximal invariant with respect to this group acting on the y-space is $z = (z_1, \cdots, z_{n-2})$ with $z_i = y_i/y_{n-1}$. Expressing this in terms of the x's we get $z_i = (x_i - x_n)/(x_{n-1} - x_n)$, which is maximal invariant with respect to G.

Suppose now the process were carried out in the reverse order. Application first of the subgroup $x_i'' = ax_i$ yields as maximal invariant $u = (u_1, \cdots, u_{n-1})$ with $u_i = x_i/x_n$. However, the translations $x_i' = x_i + b$ do not induce transformations in u-space since $(x_i + b)/(x_n + b)$ is not a function of x_i/x_n.

Quite generally, let a transformation group G be *generated* by two subgroups D and E in the sense that it is the smallest group containing D and E. Then G consists of the totality of products $e_m d_m \cdots e_1 d_1$ for $m = 1, 2, \cdots$, with $d_i \in D$, $e_i \in E$ $(i = 1, \cdots, m)$.* The following theorem shows that whenever the process of determining a maximal invariant in steps can be carried out at all, it leads to a maximal invariant with respect to G.

* See Section 1 of the Appendix.

Theorem 2. *Let G be a group of transformations, and let D and E be two subgroups generating G. Suppose that $y = s(x)$ is maximal invariant with respect to D, and that for any $e \in E$*

$$s(x_1) = s(x_2) \quad \text{implies} \quad s(ex_1) = s(ex_2). \tag{8}$$

If $z = t(y)$ is maximal invariant under the group E^* of transformations e^* defined by

$$e^*y = s(ex) \quad \text{when} \quad y = s(x),$$

then $z = t[s(x)]$ is maximal invariant with respect to G.

Proof. To show that $t[s(x)]$ is invariant, let $x' = gx$, $g = e_m d_m \cdots e_1 d_1$. Then

$$t[s(x')] = t[s(e_m d_m \cdots e_1 d_1 x)] = t[e_m^* s(d_m \cdots e_1 d_1 x)]$$
$$= t[s(e_{m-1} d_{m-1} \cdots e_1 d_1 x)],$$

and the last expression can be reduced by induction to $t[s(x)]$. To see that $t[s(x)]$ is in fact maximal invariant, suppose that $t[s(x')] = t[s(x)]$. Setting $y' = s(x')$, $y = s(x)$ one has $t(y') = t(y)$, and since $t(y)$ is maximal invariant with respect to E^* there exists e^* such that $y' = e^*y$. Then $s(x') = e^*s(x) = s(ex)$, and by the maximal invariance of $s(x)$ with respect to D there exists $d \in D$ such that $x' = dex$. Since de is an element of G this completes the proof.

3. MOST POWERFUL INVARIANT TESTS

The class of all invariant functions can be obtained as the totality of functions of a maximal invariant $T(x)$. Therefore, in particular the class of all invariant tests is the totality of tests depending only on the maximal invariant statistic T. The latter statement, while correct for all the usual situations, actually requires certain qualifications regarding the class of measurable sets in T-space. These conditions will be discussed at the end of the section; they are satisfied in the examples below.

Example 4. Let $X = (X_1, \cdots, X_n)$, and suppose that the density of X is $f_i(x_1 - \theta, \cdots, x_n - \theta)$ under $H_i (i = 0, 1)$ where θ ranges from $-\infty$ to ∞. The problem of testing H_0 against H_1 is invariant under the group G of transformations

$$gx = (x_1 + c, \cdots, x_n + c), \qquad -\infty < c < \infty,$$

which in the parameter space induces the transformations

$$\bar{g}\theta = \theta + c.$$

By Example 1, a maximal invariant under G is $Y = (X_1 - X_n, \cdots, X_{n-1} - X_n)$. The distribution of Y is independent of θ and under H_i has the density

$$\int_{-\infty}^{\infty} f_i(y_1 + z, \cdots, y_{n-1} + z, z)\, dz.$$

When referred to Y, the problem of testing H_0 against H_1 therefore becomes one of testing a simple hypothesis against a simple alternative. The most powerful test is then independent of θ, and therefore UMP among all invariant tests. Its rejection region by the Neyman-Pearson lemma is

$$\frac{\int_{-\infty}^{\infty} f_1(y_1 + z, \cdots, y_{n-1} + z, z)\, dz}{\int_{-\infty}^{\infty} f_0(y_1 + z, \cdots, y_{n-1} + z, z)\, dz} = \frac{\int_{-\infty}^{\infty} f_1(x_1 + u, \cdots, x_n + u)\, du}{\int_{-\infty}^{\infty} f_0(x_1 + u, \cdots, x_n + u)\, du} > C.$$

Example 5. If $X_1, \cdots, X_n$ is a sample from $N(\xi, \sigma^2)$, the hypothesis H: $\sigma \geqq \sigma_0$ remains invariant under the transformations $X_i' = X_i + c$, $-\infty < c < \infty$. In terms of the sufficient statistics $Y = \bar{X}$, $S^2 = \Sigma(X_i - \bar{X})^2$ these transformations become $Y' = Y + c$, $(S^2)' = S^2$, and a maximal invariant is S^2. The class of invariant tests is therefore the class of tests depending on S^2. It follows from Theorem 2 of Chapter 3 that there exists a UMP invariant test, with rejection region $\Sigma(X_i - \bar{X})^2 \leqq C$. This coincides with the UMP unbiased test (9) of Chapter 5.

Example 6. If $X_1, \cdots, X_m$ and $Y_1, \cdots, Y_n$ are samples from $N(\xi, \sigma^2)$ and $N(\eta, \tau^2)$, a set of sufficient statistics is $T_1 = \bar{X}$, $T_2 = \bar{Y}$, $T_3 = \sqrt{\Sigma(X_i - \bar{X})^2}$, and $T_4 = \sqrt{\Sigma(Y_j - \bar{Y})^2}$. The problem of testing H: $\tau^2/\sigma^2 \leqq \Delta_0$ remains invariant under the transformations $T_1' = T_1 + c_1$, $T_2' = T_2 + c_2$, $T_3' = T_3$, $T_4' = T_4$, $-\infty < c_1, c_2 < \infty$, and also under a common change of scale of all four variables. A maximal invariant with respect to the first group is (T_3, T_4). In the space of this maximal invariant, the group of scale changes induces the transformations $T_3'' = cT_3$, $T_4'' = cT_4$, $0 < c$, which has as maximal invariant the ratio T_4/T_3. The statistic $Z = T_4^2/(n-1) \div T_3^2/(m-1)$ on division by $\Delta = \tau^2/\sigma^2$ has an F-distribution with density given by (21) of Chapter 5, so that the density of Z is

$$\frac{C(\Delta) z^{\frac{1}{2}(n-3)}}{\left(\Delta + \dfrac{n-1}{m-1} z\right)^{\frac{1}{2}(m+n-2)}}, \qquad z > 0.$$

For varying Δ, these densities constitute a family with monotone likelihood ratio, so that among all tests of H based on Z, and therefore among all invariant tests, there exists a UMP one given by the rejection region $Z > C$. This coincides with the UMP unbiased test (20) of Chapter 5.

Example 7. In the method of *paired comparisons* for testing whether a treatment has a beneficial effect, the experimental material consists of n pairs of subjects. From each pair, a subject is selected at random for treatment while the other serves as control. Let X_i be 1 or 0 as for the ith pair the experiment turns out in favor of the treated subject or the control, and let $p_i = P\{X_i = 1\}$. The hypothesis of no effect, H: $p_i = 1/2$ for $i = 1, \cdots, n$, is to be tested against the alternatives that $p_i > 1/2$ for all i.

The problem remains invariant under all permutations of the n variables $X_1, \cdots, X_n$, and a maximal invariant under this group is the total number of successes $X = X_1 + \cdots + X_n$. The distribution of X is

$$P\{X = k\} = q_1 \cdots q_n \Sigma \frac{p_{i_1}}{q_{i_1}} \cdots \frac{p_{i_k}}{q_{i_k}}$$

where $q_i = 1 - p_i$ and where the summation extends over all $\binom{n}{k}$ choices of subscripts $i_1 < \cdots < i_k$. The most powerful invariant test against an alternative $(p'_1, \cdots, p'_n)$ rejects H when

$$f(k) = \frac{1}{\binom{n}{k}} \Sigma \frac{p'_{i_1}}{q'_{i_1}} \cdots \frac{p'_{i_k}}{q'_{i_k}} > C.$$

To see that f is an increasing function of k, note that $a_i = p'_i/q'_i > 1$, and that

$$\sum_j \sum a_j a_{i_1} \cdots a_{i_k} = (k+1)\Sigma a_{i_1} \cdots a_{i_{k+1}}$$

and

$$\sum_j \sum a_{i_1} \cdots a_{i_k} = (n-k)\Sigma a_{i_1} \cdots a_{i_k}.$$

Here in both equations, the second summation on the left-hand side extends over all subscripts $i_1 < \cdots < i_k$ of which none is equal to j, and the summation on the right-hand side extends over all subscripts $i_1 < \cdots < i_k$ and $i_1 < \cdots < i_{k+1}$ respectively without restriction. Then

$$f(k+1) = \frac{1}{\binom{n}{k+1}} \Sigma a_{i_1} \cdots a_{i_{k+1}} = \frac{1}{(n-k)\binom{n}{k}} \sum_j \sum a_j a_{i_1} \cdots a_{i_k}$$

$$> \frac{1}{\binom{n}{k}} \Sigma a_{i_1} \cdots a_{i_k} = f(k),$$

as was to be shown. Regardless of the alternative chosen, the test therefore rejects when $k > C$, and hence is UMP invariant. If the ith comparison is considered as plus or minus as X_i is 1 or 0, this is seen to be another example of the sign test. (Cf. Chapter 3, Example 8, and Chapter 4, Section 7.)

Sufficient statistics provide a simplification of a problem by reducing the sample space; this process involves no change in the parameter space. Invariance, on the other hand, by reducing the data to a maximal invariant statistic T, whose distribution may depend only on a function of the parameter, typically also shrinks the parameter space. The details are given in the following theorem.

Theorem 3. *If $T(x)$ is invariant under G, and if $v(\theta)$ is maximal invariant under the induced group $\bar{G}$, then the distribution of $T(X)$ depends only on $v(\theta)$.*

Proof. Let $v(\theta_1) = v(\theta_2)$. Then $\theta_2 = \bar{g}\theta_1$, and hence

$$P_{\theta_1}\{T(X) \in B\} = P_{\theta_1}\{T(gX) \in B\} = P_{\bar{g}\theta_1}\{T(X) \in B\} = P_{\theta_2}\{T(X) \in B\}.$$

This result can be paraphrased by saying that the principle of invariance identifies all parameter points that are equivalent with respect to $\bar{G}$.

In applications, for instance in Examples 5 and 6, the maximal invariants

$T(x)$ and $\delta = v(\theta)$ under G and $\bar{G}$ are frequently real-valued, and the family of probability densities $p_\delta(t)$ of T has monotone likelihood ratio. For testing the hypothesis $H: \delta \leqq \delta_0$ there exists then a UMP test among those depending only on T, and hence a UMP invariant test. Its rejection region is $t \geqq C$ where

$$\int_C^\infty p_{\delta_0}(t)\, dt = \alpha. \tag{9}$$

Consider this problem now as a two-decision problem with decisions d_0 and d_1 of accepting or rejecting H, and a loss function $L(\theta, d_i) = L_i(\theta)$. Suppose that $L_i(\theta)$ depends only on the parameter δ, $L_i(\theta) = L_i'(\delta)$ say, and satisfies

$$L_1'(\delta) - L_0'(\delta) \gtrless 0 \quad \text{as} \quad \delta \lessgtr \delta_0. \tag{10}$$

It then follows from Theorem 3 of Chapter 3 that the family of rejection regions $t \geqq C(\alpha)$, as α varies from 0 to 1, forms a complete family of decision procedures among those depending only on t, and hence a complete family of invariant procedures. As before, the choice of a particular significance level α can be considered as a convenient way of specifying a test from this family.

At the beginning of the section it was stated that the class of invariant tests coincides with the class of tests based on a maximal invariant statistic $T = T(X)$. However, a statistic is not completely specified by a function but requires also specification of a class $\mathscr{B}$ of measurable sets. If in the present case $\mathscr{B}$ is the class of all sets B for which $T^{-1}(B) \in \mathscr{A}$, the desired statement is correct. For let $\phi(x) = \psi[T(x)]$ and ϕ be $\mathscr{A}$-measurable, and let C be a Borel set on the line. Then $\phi^{-1}(C) = T^{-1}[\psi^{-1}(C)] \in \mathscr{A}$ and hence $\psi^{-1}(C) \in \mathscr{B}$, so that ψ is $\mathscr{B}$-measurable and $\phi(x) = \psi[T(x)]$ is a test based on the statistic T.

In most applications, $T(x)$ is a measurable function taking on values in a Euclidean space and it is convenient to take $\mathscr{B}$ as the class of Borel sets. If $\phi(x) = \psi[T(x)]$ is then an arbitrary measurable function depending only on $T(x)$, it is not clear that $\psi(t)$ is necessarily $\mathscr{B}$-measurable. This measurability can be concluded if $\mathscr{X}$ is also Euclidean with $\mathscr{A}$ the class of Borel sets, and if the range of T is a Borel set. We shall prove it here only under the additional assumption (which in applications is usually obvious, and which will not be verified explicitly in each case) that there exists a vector-valued Borel-measurable function $Y(x)$ such that $[T(x), Y(x)]$ maps $\mathscr{X}$ onto a Borel subset of the product space $\mathscr{T} \times \mathscr{Y}$, that this mapping is 1 : 1, and that the inverse mapping is also Borel-measurable. Given any measurable function ϕ of x there exists then a measurable function ϕ' of (t, y) such that $\phi(x) \equiv \phi'[T(x), Y(x)]$. If ϕ depends only on $T(x)$, ϕ' depends only on t so that $\phi'(t, y) = \psi(t)$ say, and ψ is a measurable function of t.* In Example 1(i) for instance, where $x = (x_1, \cdots, x_n)$ and $T(x) = (x_1 - x_n, \cdots, x_{n-1} - x_n)$, the function $Y(x)$ can be taken as $Y(x) = x_n$.

4. SAMPLE INSPECTION BY VARIABLES

A sample is drawn from a lot of some manufactured product in order to decide whether the lot is of acceptable quality. In the simplest case, each sample item is classified directly as satisfactory or defective (*inspection by attributes*), and the decision is based on the total number of defectives. More generally, the quality of an item is characterized by a variable Y (*inspection by variables*), and an item is considered satisfactory if Y exceeds a given constant u. The probability of a defective is then

$$p = P\{Y \leqq u\}$$

and the problem becomes that of testing the hypothesis $H: p \geqq p_0$.

As was seen in Example 8 of Chapter 3, no use can be made of the actual value of Y unless something is known concerning the distribution of Y. In the absence of such information, the decision will be based as before simply on the number of defectives in the sample. We shall consider the problem now under the assumption that the measurements $Y_1, \cdots, Y_n$ constitute a sample from $N(\eta, \sigma^2)$. Then

$$p = \int_{-\infty}^{u} \frac{1}{\sqrt{2\pi}\sigma} \exp\left[-\frac{1}{2\sigma^2}(y - \eta)^2\right] dy = \Phi\left(\frac{u - \eta}{\sigma}\right)$$

where

$$\Phi(y) = \int_{-\infty}^{y} \frac{1}{\sqrt{2\pi}} \exp\left(-\tfrac{1}{2}t^2\right) dt$$

denotes the cumulative distribution function of a standard normal distribution, and the hypothesis H becomes $(u - \eta)/\sigma \geqq \Phi^{-1}(p_0)$. In terms of the variables $X_i = Y_i - u$, which have mean $\xi = \eta - u$ and variance σ^2, this reduces to

$$H: \xi/\sigma \leqq \theta_0$$

with $\theta_0 = -\Phi^{-1}(p_0)$. This hypothesis, which was considered in Chapter 5, Section 2, for $\theta_0 = 0$, occurs also in other contexts. It is appropriate when one is interested in the mean ξ of a normal distribution, expressed in σ-units rather than on a fixed scale.

For testing H, attention can be restricted to the pair of variables $\bar{X}$ and $S = \sqrt{\Sigma(X_i - \bar{X})^2}$ since they form a set of sufficient statistics for (ξ, σ).

* The last statement is an immediate consequence, for example, of Theorem B, Section 34, of Halmos' *Measure Theory*, New York, D. Van Nostrand Co., 1950.

These variables are independent, the distribution of $\bar{X}$ being $N(\xi, \sigma^2/n)$ and that of S/σ being χ_{n-1}. Multiplication of $\bar{X}$ and S by a common constant $c > 0$ transforms the parameters into $\xi' = c\xi$, $\sigma' = c\sigma$, so that ξ/σ and hence the problem of testing H remains invariant. A maximal invariant under these transformations is $\bar{x}/s$ or

$$t = \frac{\sqrt{n}\,\bar{x}}{s/\sqrt{n-1}},$$

the distribution of which depends only on the maximal invariant in the parameter space $\theta = \xi/\sigma$ (cf. Chapter 5, Section 2). Thus, the invariant tests are those depending only on t, and it remains to find the most powerful test of $H\colon \theta \leq \theta_0$ within this class.

The probability density of t is (Chapter 5, Problem 3)

$$p_\delta(t) = C\int_0^\infty \exp\left[-\frac{1}{2}\left(t\sqrt{\frac{w}{n-1}} - \delta\right)^2\right] w^{\frac{1}{2}(n-2)} \exp\left(-\tfrac{1}{2}w\right) dw,$$

where $\delta = \sqrt{n}\,\theta$ is the noncentrality parameter, and this will now be shown to constitute a family with monotone likelihood ratio. To see that the ratio

$$r(t) = \frac{\displaystyle\int_0^\infty \exp\left[-\frac{1}{2}\left(t\sqrt{\frac{w}{n-1}} - \delta_1\right)^2\right] w^{\frac{1}{2}(n-2)} \exp\left(-\tfrac{1}{2}w\right) dw}{\displaystyle\int_0^\infty \exp\left[-\frac{1}{2}\left(t\sqrt{\frac{w}{n-1}} - \delta_0\right)^2\right] w^{\frac{1}{2}(n-2)} \exp\left(-\tfrac{1}{2}w\right) dw}$$

is an increasing function of t for $\delta_0 < \delta_1$, suppose first that $t < 0$ and let $v = -t\sqrt{w/(n-1)}$. The ratio then becomes proportional to

$$\frac{\displaystyle\int_0^\infty f(v) \exp\left[-(\delta_1 - \delta_0)v - (n-1)v^2/2t^2\right] dv}{\displaystyle\int_0^\infty f(v) \exp\left[-(n-1)v^2/2t^2\right] dv} = \int \exp\left[-(\delta_1 - \delta_0)v\right] g_{t^2}(v)\, dv$$

where
$$f(v) = \exp(-\delta_0 v) v^{n-1} \exp(-v^2/2)$$

and
$$g_{t^2}(v) = \frac{f(v) \exp\left[-(n-1)v^2/2t^2\right]}{\displaystyle\int_0^\infty f(z) \exp\left[-(n-1)z^2/2t^2\right] dz}.$$

Since the family of probability densities $g_{t^2}(v)$ is a family with monotone likelihood ratio, the integral of $\exp[-(\delta_1 - \delta_0)v]$ with respect to this density is a decreasing function of t^2 (Problem 10 of Chapter 3), and hence an increasing function of t for $t < 0$. Similarly one finds that $r(t)$ is an increasing function of t for $t > 0$ by making the transformation $v = t\sqrt{w/(n-1)}$. By continuity it is then an increasing function of t for all t.

There exists therefore a UMP invariant test of $H: \xi/\sigma \leqq \theta_0$, which rejects when $t > C$, where C is determined by (9). In terms of the original variables Y_i the rejection region of the UMP invariant test of $H: p \geqq p_0$ becomes

$$\frac{\sqrt{n}(\bar{y} - u)}{\sqrt{\Sigma(y_i - \bar{y})^2/(n-1)}} > C. \tag{11}$$

If the problem is considered as a two-decision problem with losses $L_0(p)$ and $L_1(p)$ for accepting or rejecting $p \geqq p_0$, which depend only on p and satisfy the condition corresponding to (10), the class of tests (11) constitutes a complete family of invariant procedures as C varies from $-\infty$ to ∞.

Consider next the comparison of two products on the basis of samples $X_1, \cdots, X_m$; $Y_1, \cdots, Y_n$ from $N(\xi, \sigma^2)$ and $N(\eta, \sigma^2)$. If

$$p = \Phi\left(\frac{u - \xi}{\sigma}\right), \qquad \pi = \Phi\left(\frac{u - \eta}{\sigma}\right),$$

one wishes to test the hypothesis $p \leqq \pi$, which is equivalent to

$$H: \eta \leqq \xi.$$

The statistics $\bar{X}$, $\bar{Y}$, and $S = \sqrt{\Sigma(X_i - \bar{X})^2 + \Sigma(Y_j - \bar{Y})^2}$ are a set of sufficient statistics for ξ, η, σ. The problem remains invariant under the addition of an arbitrary common constant to $\bar{X}$ and $\bar{Y}$, which leaves $\bar{Y} - \bar{X}$ and S as maximal invariants. It is also invariant under multiplication of $\bar{X}$, $\bar{Y}$, and S, and hence of $\bar{Y} - \bar{X}$ and S, by a common positive constant, which reduces the data to the maximal invariant $(\bar{Y} - \bar{X})/S$. Since

$$t = \frac{(\bar{y} - \bar{x})\Big/\sqrt{\dfrac{1}{m} + \dfrac{1}{n}}}{s/\sqrt{m + n - 2}}$$

has a noncentral t-distribution with noncentrality parameter $\delta = \sqrt{mn}(\eta - \xi)/\sqrt{m+n}\,\sigma$, the UMP invariant test of $H: \eta - \xi \leqq 0$ rejects when $t > C$. This coincides with the UMP unbiased test (27) of Chapter 5, Section 3. Analogously the corresponding two-sided test (30) of Chapter 5, with rejection region $|t| \geqq C$, is UMP invariant for testing the hypothesis $p = \pi$ against the alternatives $p \neq \pi$ (Problem 10).

5. ALMOST INVARIANCE

Let G be a group of transformations leaving a family $\mathscr{P} = \{P_\theta, \theta \in \Omega\}$ of distributions of X invariant. A test ϕ is said to be *equivalent to an invariant test* if there exists an invariant test ψ such that $\phi(x) = \psi(x)$ for all x except possibly on a $\mathscr{P}$-null set N; ϕ is said to be *almost invariant with respect to G if*

$$\phi(gx) = \phi(x) \quad \text{for all} \quad x \in \mathscr{X} - N_g, \qquad g \in G \tag{12}$$

where the exceptional null set N_g is permitted to depend on g. This concept is required for investigating the relationship of invariance to unbiasedness and to certain other optimum properties. In this connection it is important to know whether a UMP invariant test is also UMP among almost invariant tests. This turns out to be the case under assumptions which are made precise in Theorem 4 below and which are satisfied in all the usual applications.

If ϕ is equivalent to an invariant test then $\phi(gx) = \phi(x)$ for all $x \notin N \cup g^{-1}N$. Since $P_\theta(g^{-1}N) = P_{\bar{g}\theta}(N) = 0$, it follows that ϕ is then almost invariant. The following theorem gives conditions under which conversely any almost invariant test is equivalent to an invariant one.

Theorem 4. *Let G be a group of transformations of $\mathscr{X}$, and let $\mathscr{A}$ and $\mathscr{B}$ be σ-fields of subsets of $\mathscr{X}$ and G such that for any set $A \in \mathscr{A}$ the set of pairs (x, g) for which $gx \in A$ is measurable $\mathscr{A} \times \mathscr{B}$. Suppose further that there exists a σ-finite measure ν over G such that $\nu(B) = 0$ implies $\nu(Bg) = 0$ for all $g \in G$. Then any measurable function that is almost invariant under G (where "almost" refers to some σ-finite measure μ) is equivalent to an invariant function.*

Proof. Because of the measurability assumptions, the function $\phi(gx)$ considered as a function of the two variables x and g is measurable $\mathscr{A} \times \mathscr{B}$. It follows that $\phi(gx) - \phi(x)$ is measurable $\mathscr{A} \times \mathscr{B}$, and so therefore is the set S of points (x, g) with $\phi(gx) \neq \phi(x)$. If ϕ is almost invariant, any section of S with fixed g is a μ-null set. By Fubini's theorem (Theorem 3 of Chapter 2) there exists therefore a μ-null set N such that for all $x \in \mathscr{X} - N$

$$\phi(gx) = \phi(x) \qquad \text{a.e. } \nu.$$

Without loss of generality suppose that $\nu(G) = 1$, and let A be the set of points x for which

$$\int \phi(g'x)\, d\nu(g') = \phi(gx) \qquad \text{a.e. } \nu.$$

If

$$f(x, g) = \left| \int \phi(g'x)\, d\nu(g') - \phi(gx) \right|,$$

then A is the set of points x for which

$$\int f(x, g)\, d\nu(g) = 0.$$

Since this integral is a measurable function of x, it follows that A is measurable. Let

$$\psi(x) = \begin{cases} \int \phi(gx)\, d\nu(g) & \text{if} \quad x \in A \\ 0 \quad \text{if} \quad x \notin A. \end{cases}$$

Then ψ is measurable and $\psi(x) = \phi(x)$ for $x \notin N$ since $\phi(gx) = \phi(x)$ a.e. ν implies that $\int\phi(g'x)\, d\nu(g') = \phi(x)$ and that $x \in A$. To show that ψ is invariant it is enough to prove that the set A is invariant. For any point $x \in A$, the function $\phi(gx)$ is constant except on a null-subset N_x of G. Then $\phi(ghx)$ has the same constant value for all $g \notin N_x h^{-1}$ which by assumption is again a ν-null set; and hence $hx \in A$, which completes the proof.

Corollary 1. *Suppose that the problem of testing $H: \theta \in \omega$ against $K: \theta \in \Omega - \omega$ remains invariant under G and that the assumptions of Theorem 4 hold. Then if ϕ_0 is UMP invariant, it is also UMP within the class of almost invariant tests.*

Proof. If ϕ is almost invariant, it is equivalent to an invariant test ψ by Theorem 4. The tests ϕ and ψ have the same power function, and hence ϕ_0 is uniformly at least as powerful as ϕ.

In applications, $\mathscr{P}$ is usually a dominated family, and μ any σ-finite measure equivalent to $\mathscr{P}$ (which exists by Theorem 2 of the Appendix). If ϕ is almost invariant with respect to $\mathscr{P}$ it is then almost invariant with respect to μ and hence equivalent to an invariant test. Typically, the sample space $\mathscr{X}$ is an n-dimensional Euclidean space, $\mathscr{A}$ is the class of Borel sets, and the elements of G are transformations of the form $y = f(x, \tau)$ where τ ranges over a set of positive measure in an m-dimensional space and f is a Borel measurable vector-valued function of $m + n$ variables. If $\mathscr{B}$ is taken as the class of Borel sets in m-space, the measurability conditions of the theorem are satisfied.

The requirement that for all $g \in G$ and $B \in \mathscr{B}$

$$\nu(B) = 0 \quad \text{implies} \quad \nu(Bg) = 0 \tag{13}$$

is satisfied in particular when

$$\nu(Bg) = \nu(B) \quad \text{for all} \quad g \in G,\ B \in \mathscr{B}. \tag{14}$$

The existence of such a *right invariant measure* is guaranteed for a large class of groups by the theory of Haar measure. Alternatively, it is usually not difficult to check condition (13) directly.

Example 8. Let G be the group of all nonsingular linear transformations of n-space. Relative to a fixed coordinate system the elements of G can be represented by nonsingular $n \times n$ matrices $A = (a_{ij})$, $A' = (a'_{ij})$, $\cdots$ with the matrix product serving as the group product of two such elements. The σ-field $\mathscr{B}$ can be taken to be the class of Borel sets in the space of the n^2 elements of the matrices, and the measure ν can be taken as Lebesgue measure over $\mathscr{B}$. Consider now a set S of matrices with $\nu(S) = 0$, and the set S^* of matrices $A'A$ with $A' \in S$ and A fixed. If $a = \max |a_{ij}|$, $C' = A'A$, and $C'' = A''A$, the inequalities $|a''_{ij} - a'_{ij}| \leqq \varepsilon$ for all i, j imply $|c''_{ij} - c'_{ij}| \leqq na\varepsilon$. Since a set has ν-measure zero if and only if it can be covered by a union of rectangles whose total measure does not exceed any given $\varepsilon > 0$, it follows that $\nu(S^*) = 0$, as was to be proved.

In the preceding chapters, tests were compared purely in terms of their power functions (possibly weighted according to the seriousness of the losses involved). Since the restriction to invariant tests is a departure from this point of view, it is of interest to consider the implications of applying invariance to the power functions rather than to the tests themselves. Any test that is invariant or almost invariant under a group G has a power function, which is invariant under the group $\bar{G}$ induced by G in the parameter space.

To see that the converse is in general not true, let X_1, X_2, X_3 be independently, normally distributed with mean ξ and variance σ^2, and consider the hypothesis $\sigma \geqq \sigma_0$. The test with rejection region

$$|X_2 - X_1| > k \quad \text{when} \quad \bar{X} < 0$$
$$|X_3 - X_2| > k \quad \text{when} \quad \bar{X} \geqq 0$$

is not invariant under the group G of transformations $X'_i = X_i + c$ but its power function is invariant under the associated group $\bar{G}$.

The two properties, almost invariance of a test ϕ and invariance of its power function, become equivalent if before the application of invariance considerations the problem is reduced to a sufficient statistic whose distributions constitute a boundedly complete family.

Lemma 2. *Let the family $\mathscr{P}^T = \{P_\theta^T, \theta \in \Omega\}$ of distributions of T be boundedly complete, and let the problem of testing $H: \theta \in \Omega_H$ remain invariant under a group G of transformations of T for all θ. Then a necessary and sufficient condition for the power function of a test $\psi(t)$ to be invariant under the induced group $\bar{G}$ over Ω is that $\psi(t)$ be almost invariant under G.*

Proof. For all $\theta \in \Omega$ we have $E_{\bar{g}\theta}\psi(T) = E_\theta\psi(gT)$. If ψ is almost invariant, $E_\theta\psi(T) = E_\theta\psi(gT)$ and hence $E_{\bar{g}\theta}\psi(T) = E_\theta\psi(T)$ so that the power function of ψ is invariant. Conversely, if $E_\theta\psi(T) = E_{\bar{g}\theta}\psi(T)$, then $E_\theta\psi(T) = E_\theta\psi(gT)$, and it follows from the bounded completeness of $\mathscr{P}^T$ that $\psi(gt) = \psi(t)$ a.e. $\mathscr{P}^T$.

As a consequence, it is seen that UMP almost invariant tests also possess the following optimum property.

Theorem 5. *Under the assumptions of Lemma 2, let $v(\theta)$ be maximal invariant with respect to $\bar{G}$ and suppose that among the tests of H based on the sufficient statistic T there exists a UMP almost invariant one, say $\psi_0(t)$. Then $\psi_0(t)$ is UMP in the class of all tests based on the original observations X, whose power function depends only on $v(\theta)$.*

Proof. Let $\phi(x)$ be any such test, and let $\psi(t) = E[\phi(X)|t]$. The power function of $\psi(t)$, being identical with that of $\phi(x)$, depends then only on $v(\theta)$, and hence is invariant under $\bar{G}$. It follows from Lemma 2 that $\psi(t)$ is almost invariant under G, and $\psi_0(t)$ is uniformly at least as powerful as $\psi(t)$ and therefore as $\phi(x)$.

Example 9. For the hypothesis $\tau^2 \leqq \sigma^2$ concerning the variances of two normal distributions, the statistics $(\bar{X}, \bar{Y}, S_X^2, S_Y^2)$ constitute a complete set of sufficient statistics. It was shown in Example 6 that there exists a UMP invariant test with respect to a suitable group G, which has rejection region $S_Y^2/S_X^2 > C_0$. Since in the present case almost invariance of a test with respect to G implies that it is equivalent to an invariant one (Problem 13), Theorem 5 is applicable with $v(\theta) = \Delta = \tau^2/\sigma^2$, and the test is therefore UMP among all tests whose power function depends only on Δ.

6. UNBIASEDNESS AND INVARIANCE

The principles of unbiasedness and invariance complement each other in that each is successful in cases where the other is not. For example, there exist UMP unbiased tests for the comparison of two binomial or Poisson distributions, problems to which invariance considerations are not applicable. UMP unbiased tests also exist for testing the hypothesis $\sigma = \sigma_0$ against $\sigma \neq \sigma_0$ in a normal distribution, while invariance does not reduce this problem sufficiently far. Conversely, there exist UMP invariant tests of hypotheses specifying the values of more than one parameter (to be considered in Chapter 7) but for which the class of unbiased tests has no UMP member. There are also hypotheses, for example the one-sided hypothesis $\xi/\sigma \leqq \theta_0$ in a univariate normal distribution or $\rho \leqq \rho_0$ in a bivariate one (Problem 11) with $\theta_0, \rho_0 \neq 0$, where a UMP invariant test exists but the existence of a UMP unbiased test does not follow by the methods of Chapter 5 and is still an open question.

On the other hand, to some problems both principles have been applied successfully. These include Student's hypotheses $\xi \leqq \xi_0$ and $\xi = \xi_0$ concerning the mean of a normal distribution, and the corresponding two-sample problems $\eta - \xi \leqq \Delta_0$ and $\eta - \xi = \Delta_0$ when the variances of the two samples are assumed equal. Other examples are the one-sided hypotheses $\sigma^2 \geqq \sigma_0^2$ and $\tau^2/\sigma^2 \geqq \Delta_0$ concerning the variances of one or two normal distributions. The hypothesis of independence $\rho = 0$ in a bivariate normal distribution is still another case in point (Problem 11). In all these examples the two optimum procedures coincide. We shall now show that this is not accidental but is the case whenever the UMP invariant test is UMP also among all almost invariant tests and the UMP unbiased test is unique. In this sense, the principles of unbiasedness and of almost invariance are consistent.

Theorem 6. *Suppose that for a given testing problem there exists a UMP unbiased test ϕ^* which is unique (up to sets of measure zero), and that there also exists a UMP almost invariant test with respect to some group G. Then the latter is also unique (up to sets of measure zero), and the two tests coincide a.e.*

Proof. If $U(\alpha)$ is the class of unbiased level α tests, and if $g \in G$, then $\phi \in U(\alpha)$ if and only if $\phi g \in U(\alpha)$.† Denoting the power function of the test ϕ by $\beta_\phi(\theta)$, we thus have

$$\beta_{\phi^* g}(\theta) = \beta_{\phi^*}(\bar{g}\theta) = \sup_{\phi \in U(\alpha)} \beta_\phi(\bar{g}\theta) = \sup_{\phi \in U(\alpha)} \beta_{\phi g}(\theta) = \sup_{\phi g \in U(\alpha)} \beta_{\phi g}(\theta) = \beta_{\phi^*}(\theta).$$

It follows that ϕ^* and $\phi^* g$ have the same power function, and, because of the uniqueness assumption, that ϕ^* is almost invariant. Therefore, if ϕ' is UMP almost invariant, we have $\beta_{\phi'}(\theta) \geqq \beta_{\phi^*}(\theta)$ for all θ. On the other hand, ϕ' is unbiased as is seen by comparing it with the invariant test $\phi(x) \equiv \alpha$, and hence $\beta_{\phi'}(\theta) \leqq \beta_{\phi^*}(\theta)$ for all θ. Since ϕ' and ϕ^* therefore have the same power function, they are equal a.e. because of the uniqueness of ϕ^*, as was to be proved.

This theorem provides an alternative derivation for some of the tests of Chapter 5. In Theorem 3 of Chapter 4, the existence of UMP unbiased tests was established for one- and two-sided hypotheses concerning the parameter θ of the exponential family (10) of Chapter 4. For this family, the statistics (U, T) are sufficient and complete, and in terms of these statistics the UMP unbiased test is therefore unique. Convenient explicit expressions for some of these tests, which were derived in Chapter 5, can instead be obtained by noting that when a UMP almost invariant test exists, the same test by Theorem 6 must also be UMP unbiased. This proves for example that the tests of Examples 5 and 6 of the present chapter are UMP unbiased.

The principles of unbiasedness and invariance can be used to supplement each other in cases where neither principle alone leads to a solution but where they do so when applied in conjunction. As an example consider a sample $X_1, \cdots, X_n$ from $N(\xi, \sigma^2)$ and the problem of testing

† ϕg denotes the critical function which assigns to x the value $\phi(gx)$.

$H: \xi/\sigma = \theta_0 \neq 0$ against the two-sided alternatives that $\xi/\sigma \neq \theta_0$. Here sufficiency and invariance reduce the problem to the consideration of $t = \sqrt{n}\,\bar{x}/\sqrt{\Sigma(x_i - \bar{x})^2/(n-1)}$. The distribution of this statistic is the noncentral t-distribution with noncentrality parameter $\delta = \sqrt{n}\xi/\sigma$, and $n - 1$ degrees of freedom. For varying δ, the family of these distributions can be shown to be strictly of Polya type and hence in particular of type 3.* It follows as in Chapter 3, Problem 25, that among all tests of H based on t, there exists a UMP unbiased one with acceptance region $C_1 \leqq t \leqq C_2$ where C_1, C_2 are determined by the conditions

$$P_{\delta_0}\{C_1 \leqq t \leqq C_2\} = 1 - \alpha \quad \text{and} \quad \partial P_\delta\{C_1 \leqq t \leqq C_2\}/\partial\delta|_{\delta=\delta_0} = 0.$$

In terms of the original observations, this test then has the property of being UMP among all tests that are unbiased and invariant. Whether it is also UMP unbiased without the restriction to invariant tests is an open problem.

Another case in which the combination of invariance and unbiasedness appears to offer a promising approach is the so-called *Behrens-Fisher problem.* Let $X_1, \cdots, X_m$ and $Y_1, \cdots, Y_n$ be samples from normal distributions $N(\xi, \sigma^2)$ and $N(\eta, \tau^2)$ respectively. The problem is that of testing $H: \eta \leqq \xi$ (or $\eta = \xi$) without assuming equality of the variances σ^2 and τ^2. A set of sufficient statistics for $(\xi, \eta, \sigma, \tau)$ is then $(\bar{X}, \bar{Y}, S_X^2, S_Y^2)$ where $S_X^2 = \Sigma(X_i - \bar{X})^2$ and $S_Y^2 = \Sigma(Y_j - \bar{Y})^2$. Adding the same constant to $\bar{X}$ and $\bar{Y}$ reduces the problem to $\bar{Y} - \bar{X}, S_X^2, S_Y^2$, and multiplication of all variables by a common positive constant to $(\bar{Y} - \bar{X})/\sqrt{S_X^2 + S_Y^2}$ and S_Y^2/S_X^2. One would expect any reasonable invariant rejection region to be of the form

$$\frac{\bar{Y} - \bar{X}}{\sqrt{S_X^2 + S_Y^2}} \geqq g\left(\frac{S_Y^2}{S_X^2}\right) \tag{15}$$

for some suitable function g. If this test is also to be unbiased, the probability of (15) must equal α when $\eta = \xi$ for all values of τ/σ. Whether there exists a function g with this property is an open question. However, an approximate solution is available, which has been tabled† and which for practical purposes provides a satisfactory test.

* Karlin, "Decision theory for Pólya type distributions. Case of two actions, I.," *Proc. Third Berkeley Symposium on Mathematical Statistics and Probability*, Vol. 1, Berkeley, Univ. Calif. Press, pp. 115–129.

† Welch, "The generalization of Student's problem when several different population variances are involved," *Biometrika*, Vol. 34 (1947), pp. 28–35; Aspin, "Tables for use in comparisons whose accuracy involves two variances," *Biometrika*, Vol. 36 (1949), pp. 290–296. See also Chernoff, "Asymptotic studentization in testing of hypotheses," *Ann. Math. Stat.*, Vol. 20 (1949), pp. 268–278, and Wallace, "Asymptotic approximations to distributions," *Ann. Math. Stat.*, Vol. 29 (1958), pp. 635–654, Section 8.

Any UMP unbiased test has the important property of admissibility (Problem 1 of Chapter 4), so that there cannot exist another test which is uniformly at least as powerful and against some alternatives actually more powerful than the given one. The corresponding property does not necessarily hold for UMP invariant tests as is shown by the following example.

Example 10.* Let (X_{11}, X_{12}) and (X_{21}, X_{22}) have bivariate normal distributions with zero means and covariance matrices

$$\begin{pmatrix} \sigma_1^2 & \rho\sigma_1\sigma_2 \\ \rho\sigma_1\sigma_2 & \sigma_2^2 \end{pmatrix} \quad \text{and} \quad \begin{pmatrix} \Delta\sigma_1^2 & \Delta\rho\sigma_1\sigma_2 \\ \Delta\rho\sigma_1\sigma_2 & \Delta\sigma_2^2 \end{pmatrix}.$$

Suppose that these matrices are nonsingular, or equivalently that $|\rho| \neq 1$, but that σ_1, σ_2, ρ, and Δ are otherwise unknown. The problem of testing $\Delta = 1$ against $\Delta > 1$ remains invariant under the group G of all common nonsingular transformations

$$\begin{aligned} X'_{i1} &= a_{11}X_{i1} + a_{12}X_{i2} \\ X'_{i2} &= a_{21}X_{i1} + a_{22}X_{i2} \end{aligned} \qquad (i = 1, 2).$$

Since the probability is 0 that $X_{11}X_{22} = X_{12}X_{21}$, the 2×2 matrix (X_{ij}) is nonsingular with probability 1, and the sample space can therefore be restricted to be the set of all nonsingular such matrices. Given any two sample points $Z = (X_{ij})$ and $Z' = (X'_{ij})$ there exists a nonsingular linear transformation A such that $Z' = AZ$. There are therefore no invariants under G, and the only invariant size α test is $\phi = \alpha$. It follows vacuously that this is UMP invariant although its power is $\beta(\Delta) \equiv \alpha$. On the other hand, X_{11} and X_{21} are independently distributed as $N(0, \sigma_1^2)$ and $N(0, \Delta\sigma_1^2)$. On the basis of these observations there exists a UMP test for testing $\Delta = 1$ against $\Delta > 1$ with rejection region $X_{21}^2/X_{11}^2 > C$ (Problem 33, Chapter 3). The power function of this test is strictly increasing in Δ and hence $> \alpha$ for all $\Delta > 1$.

Admissibility of optimum invariant tests therefore cannot be taken for granted but must be established separately for each case. Let $\delta = v(\theta)$ be maximal invariant under $\bar{G}$ and suppose in order to be specific that the hypothesis to be tested is $\delta \leqq \delta_0$. To prove admissibility of a level α test ϕ_0, it is sufficient to show for some subset Ω' of alternatives that if ϕ is any level α test, then $E_\theta\,\phi(X) \geqq E_\theta\,\phi_0(X)$ for all $\theta \in \Omega'$ implies $E_\theta\,\phi(X) = E_\theta\,\phi_0(X)$ for all θ. Admissibility proofs typically fall into one of three categories as they establish this (a) locally, that is, for all θ satisfying $\delta_0 < v(\theta) < \delta_1$ for some $\delta_1 > \delta_0$; (b) for all sufficiently distant alternatives, that is, all alternatives satisfying $v(\theta) > \delta_2$ for some $\delta_2 > \delta_0$; (c) for all alternatives at any given distance δ, that is, satisfying $v(\theta) = \delta$. Proofs of type (a) or (b) are not entirely satisfactory since they do not

* This example was communicated to me by Professor C. M. Stein

rule out the existence of a test with better power for all alternatives of practical importance and worse only when both tests have power very close to 1 or at alternatives so close to the hypothesis that the value of the power there is immaterial.

As an example consider the UMP unbiased test ϕ_1 of Theorem 3, Chapter 4, for testing $H: \theta \leqq \theta_0$ against $\theta > \theta_0$ in the presence of nuisance parameters ϑ. To show that this is locally admissible, let ϕ be any other level α test of H. If $E_{\theta_0,\vartheta}\,\phi(X) < \alpha$ for some ϑ then by continuity there exists $\theta_1 > \theta_0$ such that for $\theta_0 < \theta < \theta_1$, $E_{\theta,\vartheta}\,\phi(X) < \alpha < E_{\theta,\vartheta}\,\phi_1(X)$, and it follows that locally ϕ is not uniformly as powerful as ϕ_1. If on the other hand $E_{\theta_0,\vartheta}\,\phi(X) = \alpha$ for all ϑ then $E_{\theta,\vartheta}\,\phi(X) \leqq E_{\theta,\vartheta}\,\phi_1(X)$ for all $\theta > \theta_0$ and all ϑ since in the proof of Theorem 3, ϕ_1 was shown to be UMP among all tests that are similar on the boundary. This argument does not however eliminate the possibility of a test which is biased near H but uniformly more powerful than ϕ_0 against all alternatives being at least a certain distance from H. Admissibility against distant alternatives has been proved for certain hypotheses concerning exponential families,* and against alternatives at any given distance for some location parameter problems† including that of testing $\xi/\sigma \leqq \theta_0$ against $\xi/\sigma = \theta$, in a normal distribution.

7. RANK TESTS

One of the basic problems of statistics is the two-sample problem of testing the equality of two distributions. A typical example is the comparison of a treatment with a control, where the hypothesis of no treatment effect is tested against the alternatives of a beneficial effect. This was considered in Chapters 4 and 5 under the assumption of normality, and the appropriate test was seen to be based on Student's t. It was also shown that when approximate normality is suspected but the assumption cannot be trusted, one is led to replacing the t-test by its permutation analogue, which in turn can be approximated by the original t-test.

We shall consider the same problem below without, at least for the moment, making any assumptions concerning even the approximate form of the underlying distributions, assuming only that they are continuous. The observations then consist of samples $X_1, \cdots, X_m$ and $Y_1, \cdots, Y_n$

* Birnbaum, "Characterizations of complete classes of tests of some multiparameter hypotheses with applications to likelihood ratio tests," *Ann. Math. Stat.*, Vol. 26 (1955), pp. 21–36, and Stein, "The admissibility of Hotelling's T^2-test," *Ann. Math. Stat.*, Vol. 27 (1956), pp. 616–623.

† Lehmann and Stein, "The admissibility of certain invariant statistical tests involving a translation parameter," *Ann. Math. Stat.*, Vol. 24 (1953), pp. 473–479.

from two distributions with continuous cumulative distribution functions F and G, and the problem becomes that of testing the hypothesis

$$H_1: G = F.$$

If the treatment effect is assumed to be additive, the alternatives are $G(y) = F(y - \Delta)$. We shall here consider the more general possibility that the size of the effect may depend on the value of y (so that Δ becomes a nonnegative function of y) and therefore test H_1 against the one-sided alternatives that the Y's are stochastically larger than the X's,

$$K_1: G(z) \leqq F(z) \quad \text{for all} \quad z, \qquad \text{and} \qquad G \neq F.$$

An alternative experiment that can be performed to test the effect of a treatment consists of the comparison of N pairs of subjects, which have been matched so as to eliminate as far as possible any differences not due to the treatment. One member of each pair is chosen at random to receive the treatment while the other serves as control. If the normality assumption of Chapter 5, Section 4, is dropped and the pairs of subjects can be considered to constitute a sample, the observations $(X_1, Y_1), \cdots, (X_N, Y_N)$ are a sample from a continuous bivariate distribution F. The hypothesis of no effect is then equivalent to the assumption that F is symmetric with respect to the line $y = x$

$$H_2: F(x, y) = F(y, x).$$

Another basic problem, which occurs in many different contexts, concerns the dependence or independence of two variables. In particular, if $(X_1, Y_1), \cdots, (X_N, Y_N)$ is a sample from a bivariate distribution F, one will be interested in the hypothesis

$$H_3: F(x, y) = G_1(x)H_2(y)$$

that X and Y are independent, which was considered for normal distributions in Section 9 of Chapter 5. The alternatives of interest may, for example, be that X and Y are positively dependent (cf. Chapter 5, Section 11). An alternative formulation results when x, instead of being random, can be selected for the experiment. If the chosen values are $x_1 < \cdots < x_N$ and F_i denotes the distribution of Y given x_i, the Y's are independently distributed with continuous cumulative distribution functions $F_1, \cdots, F_N$. The hypothesis of independence of Y from x becomes

$$H_4: F_1 = \cdots = F_N$$

while under the alternatives of positive dependence the variables Y_i are stochastically increasing with i.

In these and other similar problems, invariance reduces the data so completely that the actual values of the observations are discarded and only certain order relations between different groups of variables are retained. It is nevertheless possible on this basis to test the various hypotheses in question, and the resulting tests frequently are nearly as powerful as the standard normal tests. We shall now carry out this reduction for the four problems above.

The two-sample problem of testing H_1 against K_1 remains invariant under the group G of all transformations

$$x_i' = f(x_i), \qquad y_j' = f(y_j) \qquad (i = 1, \cdots, m; j = 1, \cdots, n)$$

such that f is continuous and strictly increasing. This follows from the fact that these transformations preserve both the continuity of a distribution and the property of two variables being either identically distributed or one being stochastically larger than the other. As was seen (with a different notation) in Example 3, a maximal invariant under G is the set of ranks

$$(R'; S') = (R_1', \cdots, R_m'; S_1', \cdots, S_n')$$

of $X_1, \cdots, X_m; Y_1, \cdots, Y_n$ in the combined sample. Since the distribution of $(R_1', \cdots, R_m'; S_1', \cdots, S_n')$ is symmetric in the first m and in the last n variables for all distributions F and G, a set of sufficient statistics for (R', S') is the set of the X-ranks and that of the Y-ranks without regard to the subscripts of the X's and Y's. This can be represented by the ordered X-ranks and Y-ranks

$$R_1 < \cdots < R_m \quad \text{and} \quad S_1 < \cdots < S_n,$$

and therefore by one of these sets alone since each of them determines the other. Any invariant test is thus a *rank test*, that is, it depends only on the ranks of the observations, for example on $(S_1, \cdots, S_n)$.

To obtain a similar reduction for H_2, it is convenient first to make the transformation $Z_i = Y_i - X_i$, $W_i = X_i + Y_i$. The pairs of variables (Z_i, W_i) are then again a sample from a continuous bivariate distribution. Under the hypothesis this distribution is symmetric with respect to the w-axis, while under the alternatives the distribution is shifted in the direction of the positive z-axis. The problem is unchanged if all the w's are subjected to the same transformation $w_i' = g(w_i)$ where g is $1:1$ and has at most a finite number of discontinuities, and $(Z_1, \cdots, Z_N)$ constitutes a maximal invariant under this group. [Cf. Problem 2(ii).]

The Z's are a sample from a continuous univariate distribution D, for which the hypothesis of symmetry with respect to the origin

$$H_2': D(z) + D(-z) = 1 \quad \text{for all} \quad z$$

is to be tested against the alternatives that the distribution is shifted toward positive z-values. This problem is invariant under the group G of all transformations

$$z_i' = f(z_i) \qquad (i = 1, \cdots, N)$$

such that f is continuous, odd, and strictly increasing. If $z_{i_1}, \cdots, z_{i_m} < 0 < z_{j_1}, \cdots, z_{j_n}$ where $i_1 < \cdots < i_m$ and $j_1 < \cdots < j_n$, let $s_1', \cdots, s_n'$ denote the ranks of $z_{j_1}, \cdots, z_{j_n}$ among the absolute values $|z_1|, \cdots, |z_N|$ and $r_1', \cdots, r_m'$ the ranks of $|z_{i_1}|, \cdots, |z_{i_m}|$ among $|z_1|, \cdots, |z_N|$. The transformations f preserve the sign of each observation, and hence in particular also the numbers m and n. Since f is a continuous, strictly increasing function of $|z|$, it leaves the order of the absolute values invariant and therefore the ranks r_i' and s_j'. To see that the latter are maximal invariant, let $(z_1, \cdots, z_N)$ and $(z_1', \cdots, z_N')$ be two sets of points with $m' = m$, $n' = n$ and the same r_i' and s_j'. There exists a continuous, strictly increasing function on the positive real axis such that $|z_i'| = f(|z_i|)$ and $f(0) = 0$. If f is defined for negative z by $f(-z) = -f(z)$, it belongs to G and $z_i' = f(z_i)$ for all i, as was to be proved. As in the preceding problem, sufficiency permits the further reduction to the ordered ranks $r_1 < \cdots < r_m$ and $s_1 < \cdots < s_n$. This retains the information for the rank of each absolute value whether it belongs to a positive or negative observation, but not with which positive or negative observation it is associated.

The situation is very similar for the hypotheses H_3 and H_4. The problem of testing for independence in a bivariate distribution against the alternatives of positive dependence is unchanged if the X_i and Y_i are subjected to transformations $X_i' = f(X_i)$, $Y_i' = g(Y_i)$ such that f and g are continuous and strictly increasing. This leaves as maximal invariant the ranks $(R_1', \cdots, R_N')$ of $(X_1, \cdots, X_N)$ among the X's and the ranks $(S_1', \cdots, S_N')$ of $(Y_1, \cdots, Y_N)$ among the Y's. The distribution of $(R_1', S_1'), \cdots, (R_N', S_N')$ is symmetric in these N pairs for all distributions of (X, Y). It follows that a sufficient statistic is $(S_1, \cdots, S_N)$ where $(1, S_1), \cdots, (N, S_N)$ is a permutation of $(R_1', S_1'), \cdots, (R_N', S_N')$ and where therefore S_i is the rank of the variable Y associated with the ith smallest X.

The hypothesis H_4 that $Y_1, \cdots, Y_n$ constitutes a sample is to be tested against the alternatives K_4 that the Y_i are stochastically increasing with i. This problem is invariant under the group of transformations $y_i' = f(y_i)$ where f is continuous and strictly increasing. A maximal invariant under this group is the set of ranks $S_1, \cdots, S_N$ of $Y_1, \cdots, Y_N$.

Some invariant tests of the hypotheses H_1 and H_2 will be considered in the next two sections. Corresponding results concerning H_3 and H_4 are given in Problems 39–41.

8. THE TWO-SAMPLE PROBLEM

The principle of invariance reduces the problem of testing the two-sample hypothesis $H: G = F$ against the one-sided alternatives K that the Y's are stochastically larger than the X's, to the ranks $S_1 < \cdots < S_n$ of the Y's. The specification of the S_i is equivalent to specifying for each of the $N = m + n$ positions within the combined sample, the smallest, the next smallest, etc., whether it is occupied by an x or a y. Since for any set of observations n of the N positions are occupied by y's and since the $\binom{N}{n}$ possible assignments of n positions to the y's are all equally likely when $G = F$, the joint distribution of the S_i under H is

$$P\{S_1 = s_1, \cdots, S_n = s_n\} = 1 \Big/ \binom{N}{n} \tag{16}$$

for each set $1 \leq s_1 < s_2 < \cdots < s_n \leq N$. Any rank test of H of size $\alpha = k \Big/ \binom{N}{n}$ therefore has a rejection region consisting of exactly k points $(s_1, \cdots, s_n)$.

For testing H against K there does not exist a UMP rank test, and hence no UMP invariant test. This follows for example from a consideration of two of the standard tests for this problem, since each is most powerful among all rank tests against some alternative. The two tests in question have rejection regions of the form

$$h(s_1) + \cdots + h(s_n) > C. \tag{17}$$

One, the *Wilcoxon two-sample test*,* is obtained from (17) by letting $h(s) = s$, so that it rejects H when the sum of the y-ranks is too large. We shall show below that for sufficiently small Δ, this is most powerful against the alternatives that F is the logistic distribution $F(x) = 1/(1 + e^{-x})$ and that $G(y) = F(y - \Delta)$. The other test, the *Fisher-Yates test*, has the rejection region (17) with $h(s) = E(V^{(s)})$ where $V^{(1)} < \cdots < V^{(N)}$ is an ordered sample of size N from a standard normal distribution.† This is most powerful against the alternatives that F and G are normal distributions with common variance and means ξ and $\eta = \xi + \Delta$, when Δ is sufficiently small.

* For tables of this test cf. p. 157.

† Tables of the expected order statistics from a normal distribution are given in *Biometrika Tables for Statisticians*, Vol. 1, Cambridge Univ. Press, 1954, Table 28 (to 3 decimals for $N \leq 20$ and to 2 decimals for $N \leq 50$), and by Teichroew, "Tables of expected values of order statistics and products of order statistics · · ·," *Ann. Math. Stat.*, Vol. 27 (1956), pp. 410–426 (to 10 decimals for $N \leq 20$).

To prove that these tests have the stated properties it is necessary to know the distribution of $(S_1, \cdots, S_n)$ under the alternatives. If F and G have densities f and g such that f is positive whenever g is, the joint distribution of the S_i is given by

$$P\{S_1 = s_1, \cdots, S_n = s_n\} = E\left[\frac{g(V^{(s_1)})}{f(V^{(s_1)})} \cdots \frac{g(V^{(s_n)})}{f(V^{(s_n)})}\right] \Big/ \binom{N}{n} \tag{18}$$

where $V^{(1)} < \cdots < V^{(N)}$ is an ordered sample of size N from the distribution F. (See Problem 22.) Consider in particular the translation alternatives

$$g(y) = f(y - \Delta),$$

and the problem of maximizing the power for small values of Δ. Suppose that f is differentiable and that the probability (18), which is now a function of Δ, can be differentiated with respect to Δ under the expectation sign. The derivative of (18) at $\Delta = 0$ is then

$$\frac{\partial}{\partial \Delta} P_\Delta \{S_1 = s_1, \cdots, S_n = s_n\} \big|_{\Delta=0}$$
$$= -E\left[\frac{f'(V^{(s_1)})}{f(V^{(s_1)})} + \cdots + \frac{f'(V^{(s_n)})}{f(V^{(s_n)})}\right] \Big/ \binom{N}{n}.$$

Since under the hypothesis the probability of any ranking is given by (16), it follows from the Neyman-Pearson lemma in the extended form of Theorem 5, Chapter 3, that the derivative of the power function at $\Delta = 0$ is maximized by the rejection region

$$-\sum_{i=1}^{n} E\left[\frac{f'(V^{(s_i)})}{f(V^{(s_i)})}\right] > C. \tag{19}$$

The same test maximizes the power itself for sufficiently small Δ. To see this let s denote a general rank point $(s_1, \cdots, s_n)$, and denote by $s^{(j)}$ the rank point giving the jth largest value to the left-hand side of (19). If $\alpha = k \Big/ \binom{N}{n}$, the power of the test is then

$$\beta(\Delta) = \sum_{j=1}^{k} P_\Delta(s^{(j)}) = \sum_{j=1}^{k} \left[\frac{1}{\binom{N}{n}} + \Delta \frac{\partial}{\partial \Delta} P_\Delta(s^{(j)})\big|_{\Delta=0} + \cdots\right].$$

Since there is only a finite number of points s, there exists for each j a number $\Delta_j > 0$ such that the point $s^{(j)}$ also gives the jth largest value to $P_\Delta(s)$ for all $\Delta < \Delta_j$. If Δ is less than the smallest of the numbers Δ_j, $j = 1, \cdots, \binom{N}{n}$, the test also maximizes $\beta(\Delta)$.

If $f(x)$ is the normal density $N(\xi, \sigma^2)$,

$$-\frac{f'(x)}{f(x)} = -\frac{d}{dx}\log f(x) = \frac{x-\xi}{\sigma^2},$$

and the left-hand side of (19) becomes

$$\Sigma E\frac{V^{(s_i)}-\xi}{\sigma^2} = \frac{1}{\sigma}\Sigma E(W^{(s_i)})$$

where $W^{(1)} < \cdots < W^{(N)}$ is an ordered sample from $N(0, 1)$. The test that maximizes the power against these alternatives (for sufficiently small Δ) is therefore the Fisher-Yates test.

In the case of the logistic distribution,

$$F(x) = 1/(1+e^{-x}), \qquad f(x) = e^{-x}/(1+e^{-x})^2,$$

and hence

$$-f'(x)/f(x) = 2F(x) - 1.$$

The locally most powerful rank test therefore rejects when $\Sigma E[F(V^{(s_i)})] > C$. If V has the distribution F and $0 \leqq y \leqq 1$,

$$P\{F(V) \leqq y\} = P\{V \leqq F^{-1}(y)\} = F[F^{-1}(y)] = y,$$

so that $U = F(V)$ is uniformly distributed over $(0, 1)$.* The rejection region can therefore be written as $\Sigma E(U^{(s_i)}) > C$ where $U^{(1)} < \cdots < U^{(N)}$ is an ordered sample of size N from the uniform distribution $R(0, 1)$. Since $E(U^{(s_i)}) = s_i/(N+1)$, the test is seen to be the Wilcoxon test.

Both the Fisher-Yates test and the Wilcoxon test are unbiased against the one-sided alternatives K. In fact, let ϕ be the critical function of any test determined by (17) with h nondecreasing. Then ϕ is nondecreasing in the y's and the probability of rejection is α for all $F = G$. It follows from Lemma 2 of Chapter 5 that the test is unbiased against all alternatives of K.

It follows from the unbiasedness properties of these tests that the most powerful invariant tests in the two cases considered are also most powerful against their respective alternatives among all tests that are invariant and unbiased. The nonexistence of a UMP test is therefore not relieved by restricting the tests to be unbiased as well as invariant. Nor does the application of the unbiasedness principle alone lead to a solution, as was seen in the discussion of permutation tests in Chapter 5, Section 8. With the failure of these two principles, both singly and in conjunction, the problem is left not only without a solution but even

* This transformation, which takes a random variable with continuous distribution F into a uniformly distributed variable, is known as the *probability integral transformation*.

without a formulation. A possible formulation (stringency) will be discussed in Chapter 8. However, the determination of a most stringent test for the two-sample hypothesis is an open problem.

Although optimum properties have not yet been established for any two-sample test, both tests mentioned above appear to be very satisfactory in practice, as are others such as *van der Waerden's test*† which has the rejection region (17) with $h(s) = \Phi^{-1}(s/N + 1)$ where Φ is the cumulative distribution function of a standard normal distribution. Even when F and G are normal with common variance, these tests are nearly as powerful as the t-test.

To obtain a numerical comparison, suppose that the two samples are of equal size and consider the ratio n^*/n of the number of observations required by two tests to obtain the same power β against the same alternative. Let $m = n$ and $m^* = n^* = g(n)$ be the sample sizes required by one of the rank tests and the t-test respectively, and suppose (as is the case for the tests under consideration) that the ratio n^*/n tends to a limit e independent of α and β as $n \to \infty$. Then e is called the *asymptotic efficiency* of the rank test relative to the t-test. Thus, if in a particular case $e = \frac{1}{2}$, the rank test requires approximately twice as many observations as the t-test to achieve the same power.

In the particular case of the Wilcoxon test,‡ e turns out to be equal to $3/\pi \sim .95$ when F and G are normal distributions with equal variance. When F and G are not necessarily normal but differ only in location, e depends on the form of the distribution. It is always $\geq .864$, but may exceed 1 and can in fact be infinite. The situation is even more favorable for the Fisher-Yates test. Its asymptotic efficiency relative to the t-test is always ≥ 1 when F and G differ only in location; it is 1 in the particular case that F is normal. The same results hold for van der Waerden's test, which appears to be asymptotically equivalent to that of Fisher and Yates.

The above results do not depend on the assumption of equal sample sizes; they are also valid if m/n and m^*/n^* tend to a common limit ρ as $n \to \infty$ where $0 < \rho < \infty$. At least in the case that F is normal, the asymptotic results agree well with those found for very small samples.

† Tables facilitating this test are given by van der Waerden and Nievergelt, *Tables for Comparing Two Samples by X-Test and Sign Test*, Berlin, Springer Verlag, 1956.

‡ For a discussion of these and related efficiency results, see for example Hodges and Lehmann, "The efficiency of some nonparametric competitors of the t-test," *Ann. Math. Stat.*, Vol. 27 (1956), pp. 324–335; Chernoff and Savage, "Asymptotic normality and efficiency of certain nonparametric test statistics," *Ann. Math. Stat.*, Vol. 29 (1958), pp. 972–994; van der Waerden, "Order tests for the two-sample problem and their powers," *Koninkl. Ned. Akad. Wetenschap., Proc., Ser. A*, Vol. 55 (1952), pp. 435–458 and Vol. 56 (1953), pp. 303–316.

For testing $G = F$ against the two-sided alternatives that the Y's are either stochastically smaller or larger than the X's, two-sided versions of the above tests can be used. In particular, if $m = n$, (17) suggests the rejection region

$$|\Sigma h(s_j) - \Sigma h(r_i)| > C.$$

The theory here is in still less satisfactory state than in the one-sided case. Thus, for the two-sided Wilcoxon test obtained by putting $h(k) = k$, and other similar tests, it is not even known whether they are unbiased against the two-sided alternatives in question, or whether they are admissible within the class of all rank tests. On the other hand, the relative asymptotic efficiencies are the same as in the one-sided case.

The two-sample hypothesis $G = F$ can also be tested against the general alternatives $G \neq F$. This problem arises in deciding whether two products, two sets of data, etc., can be pooled when nothing is known about the underlying distributions. Since the alternatives are now unrestricted, the problem remains invariant under all transformations $x_i' = f(x_i)$, $y_j' = f(y_j)$, $i = 1, \cdots, m$; $j = 1, \cdots, n$ such that f has only a finite number of discontinuities. There are no invariants under this group, so that the only invariant test is $\phi(x, y) \equiv \alpha$. This is however not admissible since there do exist tests of H that are strictly unbiased against all alternatives $G \neq F$ (Problem 34). The test most commonly employed for this problem is the *Smirnov test*. Let the *sample cumulative distribution functions* of the two samples be defined by

$$S_{x_1, \cdots, x_m}(z) = a/m, \qquad S_{y_1, \cdots, y_n}(z) = b/n,$$

where a and b are the number of x's and y's less or equal to z respectively. Then H is rejected according to this test* when

$$\sup_z |S_{x_1, \cdots, x_m}(z) - S_{y_1, \cdots, y_n}(z)| > C.$$

9. THE HYPOTHESIS OF SYMMETRY

When the method of paired comparisons is used to test the hypothesis of no treatment effect, the problem was seen in Section 7 to reduce through invariance to that of testing the hypothesis

$$H_2': D(z) + D(-z) = 1 \quad \text{for all} \quad z,$$

* A survey dealing with the theory of this and related tests and containing references to the relevant tables is given by Darling, "The Kolmogorov-Smirnov, Cramer-von Mises tests," *Ann. Math. Stat.*, Vol. 28 (1957), pp. 823–838. A detailed study of the distribution of the test statistic under the hypothesis is presented by Hodges, "The significance probability of the Smirnov two-sample test," *Arkiv Mat.*, Vol. 3 (1957), pp. 469–486.

which states that the distribution D of the differences $Z_i = Y_i - X_i$ ($i = 1, \cdots, N$) is symmetric with respect to the origin. The distribution D can be specified by the triple (ρ, F, G) where

$$\rho = P\{Z \leqq 0\}, \qquad F(z) = P\{|Z| \leqq z | Z < 0\}, \qquad G(z) = P\{Z \leqq z | Z > 0\},$$

and the hypothesis of symmetry with respect to the origin then becomes

$$H: \rho = \tfrac{1}{2}, \qquad G = F.$$

Invariance and sufficiency were shown to reduce the data to the ranks $S_1 < \cdots < S_n$ of the positive Z's among the absolute values $|Z_1|, \cdots, |Z_N|$. The probability of $S_1 = s_1, \cdots, S_n = s_n$ is the probability of this event given that there are n positive observations multiplied by the probability that the number of positive observations is n. Hence

$$P\{S_1 = s_1, \cdots, S_n = s_n\} = \binom{N}{n}(1 - \rho)^n \rho^{N-n} P_{F,G}\{S_1 = s_1, \cdots, S_n = s_n | n\}$$

where the second factor is given by (18). Under H, this becomes

$$P\{S_1 = s_1, \cdots, S_n = s_n\} = 1/2^N$$

for each of the $\sum_{n=0}^{N}\binom{N}{n} = 2^N$ n-tuples $(s_1, \cdots, s_n)$ satisfying $1 \leqq s_1 < \cdots < s_n \leqq N$. Any rank test of size $\alpha = k/2^N$ therefore has a rejection region containing exactly k such points $(s_1, \cdots, s_n)$.

The alternatives K of a beneficial treatment effect are characterized by the fact that the variable Z being sampled is stochastically larger than some random variable which is symmetrically distributed about 0. It is again suggestive to use rejection regions of the form $h(s_1) + \cdots + h(s_n) > C$, where however n is no longer a constant as it was in the two-sample problem but depends on the observations. Two particular cases are the *Wilcoxon one-sample test*, which is obtained by putting $h(s) = s$, and the analogue of the Fisher-Yates test with $h(s) = E(W^{(s)})$ where $W^{(1)} < \cdots < W^{(N)}$ are the ordered values of $|V_1|, \cdots, |V_N|$, the V's being a sample from $N(0, 1)$. The W's are therefore an ordered sample of size N from a distribution with density $\sqrt{2/\pi}\, e^{-w^2/2}$ for $w \geqq 0$.

As in the two-sample problem, it can be shown that each of these tests is most powerful (among all invariant tests) against certain alternatives, and that they are both unbiased against the class K. Their asymptotic efficiencies relative to the t-test for testing that the mean of Z is zero have the same values $3/\pi$ and 1 as the corresponding two-sample tests, when the distribution of Z is normal.

In certain applications, for example where the various comparisons are made under different experimental conditions, or by different methods, it may be unrealistic to assume that the variables $Z_1, \cdots, Z_N$ have a common distribution. Suppose instead that the Z_i are still independently distributed but with arbitrary continuous distributions D_i. The hypothesis to be tested is that each of these distributions is symmetric with respect to the origin.

This problem remains invariant under all transformations $z_i' = f_i(z_i)$, $i = 1, \cdots, N$, such that each f_i is continuous, odd, and strictly increasing. A maximal invariant is then the number n of positive observations, and it follows from Example 7 that there exists a UMP invariant test, the *sign test*, which rejects when n is too large. This test reflects the fact that the magnitude of the observations or of their absolute values can be explained entirely in terms of the spread of the distributions D_i, so that only the signs of the Z's are relevant.

Frequently, it seems reasonable to assume that the Z's are identically distributed but the assumption cannot be trusted. One would then prefer to use the information provided by the ranks s_i but requires a test which controls the probability of false rejection even when the assumption fails. As is shown by the following lemma, this requirement is in fact satisfied for every (symmetric) rank test. Actually, the lemma will not require even the independence of the Z's; it will show that any symmetric rank test continues to correspond to the stated level of significance provided only the treatment is assigned at random within each pair.

Lemma 3. *Let $\phi(z_1, \cdots, z_N)$ be symmetric in its N variables and such that*

$$E_D\,\phi(Z_1, \cdots, Z_N) = \alpha \tag{20}$$

when the Z's are a sample from any continuous distribution D which is symmetric with respect to the origin. Then

$$E\phi(Z_1, \cdots, Z_N) = \alpha \tag{21}$$

if the joint distribution of the Z's is unchanged under the 2^N transformations $Z_1' = \pm Z_1, \cdots, Z_N' = \pm Z_N$.

Proof. Condition (20) implies

$$\sum_{(j_1,\cdots,j_N)} \sum \phi(\pm z_{j_1}, \cdots, \pm z_{j_N})/(2^N \cdot N!) = \alpha \qquad \text{a.e.} \tag{22}$$

where the outer summation extends over all $N!$ permutations $(j_1, \cdots, j_N)$ and the inner one over all 2^N possible choices of the signs $+$ and $-$. This is proved exactly as was Theorem 3 of Chapter 5. If in addition ϕ is symmetric, (22) implies

$$\Sigma\phi(\pm z_1, \cdots, \pm z_N)/2^N = \alpha. \tag{23}$$

Suppose that the distribution of the Z's is invariant under the 2^N transformations in question. Then the conditional probability of any sign combination of $Z_1, \cdots, Z_N$ given $|Z_1|, \cdots, |Z_N|$ is $1/2^N$. Hence (23) is equivalent to

(24) $$E[\phi(Z_1, \cdots, Z_N) \,|\, |Z_1|, \cdots, |Z_N|\,] = \alpha \quad \text{a.e.}$$

and this implies (21), as was to be proved.

10. INVARIANT CONFIDENCE SETS

Confidence sets for a parameter θ in the presence of nuisance parameters ϑ were discussed in Chapter 5 (Sections 4 and 5) under the assumption that θ is real-valued. The correspondence between acceptance regions $A(\theta_0)$ of the hypotheses $H(\theta_0)\colon \theta = \theta_0$ and confidence sets $S(x)$ for θ given by (34) and (35) of Chapter 5 is, however, independent of this assumption; it is valid regardless of whether θ is real-valued, vector-valued, or possibly a label for a completely unknown distribution function (in the latter case, confidence intervals become confidence bands for the distribution function). This correspondence, which can be summarized by the relationship

(25) $$\theta \in S(x) \quad \text{if and only if} \quad x \in A(\theta),$$

was the basis for deriving uniformly most accurate and uniformly most accurate unbiased confidence sets. In the present section, it will be used to obtain uniformly most accurate invariant confidence sets.

We begin by defining invariance for confidence sets. Let G be a group of transformations of the variable X preserving the family of distributions $\{P_{\theta,\vartheta}, (\theta, \vartheta) \in \Omega\}$ and let $\bar{G}$ be the induced group of transformations of Ω. If $\bar{g}(\theta, \vartheta) = (\theta', \vartheta')$ we shall suppose that θ' depends only on $\bar{g}$ and θ and not on ϑ, so that $\bar{g}$ induces a transformation in the space of θ. In order to keep the notation from becoming unnecessarily complex, it will then be convenient to write also $\theta' = \bar{g}\theta$. For each transformation $g \in G$, denote by g^* the transformation acting on sets S in θ-space and defined by

(26) $$g^*S = \{\bar{g}\theta\colon \theta \in S\},$$

so that g^*S is the set obtained by applying the transformation $\bar{g}$ to each point θ of S. A confidence procedure, given by a class of confidence sets $S(x)$ is then said to be *invariant under* G if

(27) $$g^*S(x) = S(gx) \quad \text{for all} \quad x \in \mathscr{X}, g \in G.$$

This definition is a particular case of the invariance concept discussed in Chapter 1. If the transformation g is interpreted as a change of coordinates, (27) means that the confidence statement does not depend on the coordinate system used to express the data. The statement that the transformed parameter $\bar{g}\theta$ lies in $S(gx)$ is equivalent to stating that $\theta \in g^{*-1}S(gx)$, which is equivalent to the original statement $\theta \in S(x)$ provided (27) holds.

Example 11. Let X, Y be independently normally distributed with means ξ, η and unit variance, and let G be the group of all rigid motions of the plane, which is generated by all translations and orthogonal transformations. Here $\bar{g} = g$ for all $g \in G$. An example of an invariant class of confidence sets is given by

$$S(x, y) = \{(\xi, \eta)\colon (x - \xi)^2 + (y - \eta)^2 \leqq C\},$$

the class of circles with radius $\sqrt{C}$ and center (x, y). The set $g^*S(x, y)$ is the set of all points $g(\xi, \eta)$ with $(\xi, \eta) \in S(x, y)$, and hence is obtained by subjecting $S(x, y)$ to the rigid motion g. The result is the circle with radius $\sqrt{C}$ and center $g(x, y)$, and (27) is therefore satisfied.

In accordance with the definitions given in Chapters 3 and 5, a class of confidence sets for θ will be said to be *uniformly most accurate invariant* at confidence level $1 - \alpha$ if among all invariant classes of sets at that level it minimizes the probability

$$P_{\theta,\vartheta}\{\theta' \in S(X)\} \quad \text{for all} \quad \theta' \neq \theta.$$

In order to derive confidence sets with this property from families of UMP invariant tests, we shall now investigate the relationship between invariance of confidence sets and of the associated tests.

Suppose that for each θ_0 there exists a group of transformations G_{θ_0}, which leaves invariant the problem of testing $H(\theta_0)\colon \theta = \theta_0$, and denote by G the group of transformations generated by the totality of groups G_θ.

Lemma 4. (i) *Let $S(x)$ be any class of confidence sets that is invariant under G and let $A(\theta) = \{x\colon \theta \in S(x)\}$; then the acceptance region $A(\theta)$ is invariant under G_θ for each θ.*

(ii) *If in addition, for each θ_0 the acceptance region $A(\theta_0)$ is UMP invariant for testing $H(\theta_0)$ at level α, the class of confidence sets $S(x)$ is uniformly most accurate among all invariant confidence sets at confidence level $1 - \alpha$.*

Proof. (i) Consider any fixed θ and let $g \in G_\theta$. Then

$$\begin{aligned} gA(\theta) &= \{gx\colon \theta \in S(x)\} = \{x\colon \theta \in S(g^{-1}x)\} = \{x\colon \theta \in g^{*-1}S(x)\} \\ &= \{x\colon \bar{g}\theta \in S(x)\} = \{x\colon \theta \in S(x)\} = A(\theta). \end{aligned}$$

Here the third equality holds since $S(x)$ is invariant, and the fifth one since $g \in G_\theta$ and therefore $\bar{g}\theta = \theta$.

(ii) If $S'(x)$ is any other invariant class of confidence sets at the prescribed level, the associated acceptance regions $A'(\theta)$ by (i) define invariant tests of the hypotheses $H(\theta)$. It follows that these tests are uniformly at most as powerful as those with acceptance regions $A(\theta)$ and hence that

$$P_{\theta,\vartheta}\{\theta' \in S(X)\} \leqq P_{\theta,\vartheta}\{\theta' \in S'(X)\} \quad \text{for all} \quad \theta' \neq \theta,$$

as was to be proved.

It is an immediate consequence of the lemma that if UMP invariant acceptance regions $A(\theta)$ have been found for each hypothesis $H(\theta)$ (invariant with respect to G_θ), and if the confidence sets $S(x) = \{\theta : x \in A(\theta)\}$ are invariant under G, then they are uniformly most accurate invariant.

Example 12. Under the assumptions of Example 11, the problem of testing $\xi = \xi_0, \eta = \eta_0$ is invariant under the group G_{ξ_0, η_0} of orthogonal transformations about the point (ξ_0, η_0):

$$X' - \xi_0 = a_{11}(X - \xi_0) + a_{12}(Y - \eta_0), \quad Y' - \eta_0 = a_{21}(X - \xi_0) + a_{22}(Y - \eta_0)$$

where the matrix (a_{ij}) is orthogonal. There exists under this group a UMP invariant test, which has the acceptance region (Problem 8 of Chapter 7)

$$(X - \xi_0)^2 + (Y - \eta_0)^2 \leqq C.$$

Let G_0 be the smallest group containing the groups $G_{\xi,\eta}$ for all ξ, η. Since this is a subgroup of the group G of Example 11 (the two groups actually coincide but this is immaterial for the argument), the confidence sets $(X - \xi)^2 + (Y - \eta)^2 \leqq C$ are invariant under G_0 and hence uniformly most accurate invariant.

Example 13. Let $X_1, \cdots, X_n$ be independently normally distributed with mean ξ and variance σ^2. Confidence intervals for ξ are based on the hypotheses $H(\xi_0)$: $\xi = \xi_0$, which are invariant under the groups G_{ξ_0} of transformations $X_i' = a(X_i - \xi_0) + \xi_0$ $(a \neq 0)$. The UMP invariant test of $H(\xi_0)$ has acceptance region

$$\sqrt{(n-1)n}\,|\bar{X} - \xi_0| / \sqrt{\Sigma(X_i - \bar{X})^2} \leqq C,$$

and the associated confidence intervals are

$$(28) \quad \bar{X} - \frac{C}{\sqrt{n(n-1)}}\sqrt{\Sigma(X_i - \bar{X})^2} \leqq \xi \leqq \bar{X} + \frac{C}{\sqrt{n(n-1)}}\sqrt{\Sigma(X_i - \bar{X})^2}.$$

The group G in the present case consists of all transformations g: $X_i' = aX_i + b$ $(a \neq 0)$, which on ξ induces the transformation $\bar{g}$: $\xi' = a\xi + b$. Application of the associated transformation g^* to the interval (28) takes it into the set of points $a\xi + b$ for which ξ satisfies (28), that is, into the interval with end points

$$a\bar{X} + b - |a|\, C\sqrt{\Sigma(X_i - \bar{X})^2/n(n-1)}$$

and

$$a\bar{X} + b + |a|\, C\sqrt{\Sigma(X_i - \bar{X})^2/n(n-1)}.$$

Since this coincides with the interval obtained by replacing X_i in (28) with $aX_i + b$, the confidence intervals (28) are invariant under G_0 and hence uniformly most accurate invariant.

11. CONFIDENCE BANDS FOR A DISTRIBUTION FUNCTION

Suppose that $X = (X_1, \cdots, X_n)$ is a sample from an unknown continuous cumulative distribution function F, and that lower and upper bounds L_X and M_X are to be determined such that with preassigned probability $1 - \alpha$ the inequalities

$$L_X(u) \leqq F(u) \leqq M_X(u) \quad \text{for all} \quad u$$

hold for all continuous cumulative distribution functions F. This problem is invariant under the group G of transformations

$$X_i' = g(X_i), \qquad i = 1, \cdots, n,$$

where g is any continuous strictly increasing function. The induced transformation in the parameter space is $\bar{g}F = F(g^{-1})$.

If $S(x)$ is the set of continuous cumulative distribution functions

$$S(x) = \{F: L_x(u) \leqq F(u) \leqq M_x(u) \quad \text{for all} \quad u\},$$

then

$$\begin{aligned} g^*S(x) &= \{\bar{g}F: L_x(u) \leqq F(u) \leqq M_x(u) \quad \text{for all} \quad u\} \\ &= \{F: L_x[g^{-1}(u)] \leqq F(u) \leqq M_x[g^{-1}(u)] \quad \text{for all} \quad u\}. \end{aligned}$$

For an invariant procedure, this must coincide with the set

$$S(gx) = \{F: L_{g(x_1),\cdots,g(x_n)}(u) \leqq F(u) \leqq M_{g(x_1),\cdots,g(x_n)}(u) \quad \text{for all} \quad u\}.$$

The condition of invariance is therefore

$$L_{g(x_1),\cdots,g(x_n)}[g(u)] = L_x(u); \qquad M_{g(x_1),\cdots,g(x_n)}[g(u)] = M_x(u) \quad \text{for all } x \text{ and } u.$$

To characterize the totality of invariant procedures, consider the sample cumulative distribution function T_x given by

$$T_x(u) = i/n \quad \text{for} \quad x^{(i)} \leqq u < x^{(i+1)}, \qquad i = 0, \cdots, n,$$

where $x^{(1)} < \cdots < x^{(n)}$ is the ordered sample and where $x^{(0)} = -\infty$, $x^{(n+1)} = \infty$. Then a necessary and sufficient condition for L and M to satisfy the above invariance condition is the existence of numbers $a_0, \cdots, a_n; a_0', \cdots, a_n'$ such that

$$L_x(u) = a_i, \qquad M_x(u) = a_i' \quad \text{for} \quad x^{(i)} < u < x^{(i+1)}.$$

That this condition is sufficient is immediate. To see that it is also necessary, let u, u' be any two points satisfying $x^{(i)} < u < u' < x^{(i+1)}$.

Given any $y_1, \cdots, y_n$ and v with $y^{(i)} < v < y^{(i+1)}$ there exist $g, g' \in G$ such that

$$g(y^{(i)}) = g'(y^{(i)}) = x^{(i)}, \qquad g(v) = u, \qquad g'(v) = u'.$$

If L_x, M_x are invariant, it then follows that $L_x(u') = L_y(v)$ and $L_x(u) = L_y(v)$, and hence that $L_x(u') = L_x(u)$ and similarly $M_x(u') = M_x(u)$, as was to be proved. This characterization shows L_x and M_x to be step functions whose discontinuity points are restricted to those of T_x.

Since any two continuous strictly increasing cumulative distribution functions can be transformed into one another through a transformation $\bar{g}$, it follows that all these distributions have the same probability of being covered by an invariant confidence band. (See Problem 48.) Suppose now that F is continuous but no longer strictly increasing. If I is any interval of constancy of F, there are no observations in I so that I is also an interval of constancy of the sample cumulative distribution function. It follows that the probability of the confidence band covering F is not affected by the presence of I and hence is the same for all continuous cumulative distribution functions F.

For any numbers a_i, a'_i let Δ_i, Δ'_i be determined by

$$a_i = (i/n) - \Delta_i, \qquad a'_i = (i/n) + \Delta'_i.$$

Then it was seen above that any numbers $\Delta_0, \cdots, \Delta_n; \Delta'_0, \cdots, \Delta'_n$ define a confidence band for F, which is invariant and hence has constant probability of covering the true F. From these confidence bands a test can be obtained of the hypothesis of *goodness of fit* $F = F_0$ that the unknown F equals a hypothetical distribution F_0. The hypothesis is accepted if F_0 lies entirely within the band, that is, if

$$-\Delta_i < F_0(u) - T_x(u) < \Delta'_i \quad \text{for all} \quad x^{(i)} < u < x^{(i+1)}$$
$$\text{and all} \quad i = 1, \cdots, n.$$

Within this class of tests there exists no UMP member and the most common choice of the Δ's is $\Delta_i = \Delta'_i = \Delta$ for all i. The acceptance region of the resulting *Kolmogorov test** can be written as

$$\sup_{-\infty < u < \infty} |F_0(u) - T_x(u)| < \Delta.$$

* A survey dealing with the theory of this and related tests (including tests for goodness of fit when the hypothesis specifies a parametric family rather than a single distribution) is given by Darling, "The Kolmogorov-Smirnov, Cramér-von Mises tests," *Ann. Math. Stat.*, Vol. 28 (1957), pp. 823–838. This paper contains in particular also references to the tables which are required to carry out the test. A discussion of some of the associated one-sided tests is given by Chapman, "A comparative study of several one-sided goodness-of-fit tests," *Ann. Math. Stat.*, Vol. 29 (1958), pp. 655–674.

This is the limiting case of the Smirnov two-sample test as the size of the second sample tends to infinity.

12. PROBLEMS

Section 1

1. Let G be a group of measurable transformations of $(\mathscr{X}, \mathscr{A})$ leaving $\mathscr{P} = \{P_\theta, \theta \in \Omega\}$ invariant, and let $T(x)$ be a measurable transformation to $(\mathscr{T}, \mathscr{B})$. Suppose that $T(x_1) = T(x_2)$ implies $T(gx_1) = T(gx_2)$ for all $g \in G$, so that G induces a group G^* on $\mathscr{T}$ through $g^*T(x) = T(gx)$, and suppose further that the induced transformations g^* are measurable $\mathscr{B}$. Then G^* leaves the family $\mathscr{P}^T = \{P_\theta^T, \theta \in \Omega\}$ of distributions of T invariant.

Section 2

2. (i) Let $\mathscr{X}$ be the totality of points $x = (x_1, \cdots, x_n)$ for which all coordinates are different from zero, and let G be the group of transformations $x_i' = cx_i$, $c > 0$. Then a maximal invariant under G is $(\text{sgn } x_n, x_1/x_n, \cdots, x_{n-1}/x_n)$ where $\text{sgn } x$ is 1 or -1 as x is positive or negative.

(ii) Let $\mathscr{X}$ be the space of points $x = (x_1, \cdots, x_n)$ for which all coordinates are distinct and let G be the group of all transformations $x_i' = f(x_i)$, $i = 1, \cdots, n$, such that f is a 1 : 1 transformation of the real line onto itself with at most a finite number of discontinuities. Then G is transitive over $\mathscr{X}$.

[(ii) Let $x = (x_1, \cdots, x_n)$ and $x' = (x_1', \cdots, x_n')$ be any two points of $\mathscr{X}$. Let $I_1, \cdots, I_n$ be a set of mutually exclusive open intervals which (together with their end points) cover the real line and such that $x_j \in I_j$. Let $I_1', \cdots, I_n'$ be a corresponding set of intervals for $x_1', \cdots, x_n'$. Then there exists a transformation f which maps each I_j continuously onto I_j', maps x_j onto x_j', and the set of $n - 1$ end points of $I_1, \cdots, I_n$ onto the set of end points of $I_1', \cdots, I_n'$.]

3. (i) A sufficient condition for (8) to hold is that D is a normal subgroup of G.

(ii) If G is the group of transformations $x' = ax + b$, $a \neq 0$, $-\infty < b < \infty$, then the subgroup of translations $x' = x + b$ is normal but the subgroup $x' = ax$ is not.

[The defining property of a normal subgroup is that given $d \in D$, $g \in G$, there exists $d' \in D$ such that $gd = d'g$. The equality $s(x_1) = s(x_2)$ implies $x_2 = dx_1$ for some $d \in D$, and hence $ex_2 = edx_1 = d'ex_1$. The result (i) now follows since s is invariant under D.]

Section 3

4. Let X, Y have the joint probability density $f(x, y)$. Then the integral $h(z) = \int_{-\infty}^{\infty} f(y - z, y)\, dy$ is finite for almost all z, and is the probability density of $Z = Y - X$.

[Since $P\{Z \leq b\} = \int_{-\infty}^{b} h(z)\, dz$, it is finite and hence h is finite almost everywhere.]

5. (i) Let $X = (X_1, \cdots, X_n)$ have probability density $(1/\theta^n)f[(x_1 - \xi)/\theta, \cdots, (x_n - \xi)/\theta]$ where $-\infty < \xi < \infty$, $0 < \theta$ are unknown, and where f is even.

The problem of testing $f = f_0$ against $f = f_1$ remains invariant under the transformations $x_i' = ax_i + b$ $(i = 1, \cdots, n)$, $a \neq 0$, $-\infty < b < \infty$, and the most powerful invariant test is given by the rejection region

$$\int_{-\infty}^{\infty} \int_0^{\infty} v^{n-2} f_1(vx_1 + u, \cdots, vx_n + u)\, dv\, du$$
$$> C \int_{-\infty}^{\infty} \int_0^{\infty} v^{n-2} f_0(vx_1 + u, \cdots, vx_n + u)\, dv\, du.$$

(ii) Let $X = (X_1, \cdots, X_n)$ have probability density $f(x_1 - \Sigma_{j=1}^k w_{1j}\beta_j, \cdots, x_n - \Sigma_{j=1}^k w_{nj}\beta_j)$ where $k < n$, the w's are given constants, the β's are unknown, and where we wish to test $f = f_0$ against $f = f_1$. The problem remains invariant under the transformations $x_i' = x_i + \Sigma_{j=1}^k w_{ij}\gamma_j$, $-\infty < \gamma_1, \cdots, \gamma_k < \infty$, and the most powerful invariant test is given by the rejection region

$$\frac{\int \cdots \int f_1(x_1 - \Sigma w_{1j}\beta_j, \cdots, x_n - \Sigma w_{nj}\beta_j)\, d\beta_1, \cdots, d\beta_k}{\int \cdots \int f_0(x_1 - \Sigma w_{1j}\beta_j, \cdots, x_n - \Sigma w_{nj}\beta_j)\, d\beta_1, \cdots, d\beta_k} > C.$$

[A maximal invariant is given by

$$y = \left(x_1 - \sum_{r=n-k+1}^{n} a_{1r}x_r,\ x_2 - \sum_{r=n-k+1}^{n} a_{2r}x_r,\ \cdots,\ x_{n-k} - \sum_{r=n-k+1}^{n} a_{n-k,r}x_r\right)$$

for suitably chosen constants a_{ir}.]

6. Let $X_1, \cdots, X_m$; $Y_1, \cdots, Y_n$ be samples from exponential distributions with densities $\sigma^{-1}e^{-(x-\xi)/\sigma}$ for $x \geqq \xi$, and $\tau^{-1}e^{-(y-\eta)/\tau}$ for $y \geqq \eta$.

(i) For testing $\tau/\sigma \leqq \Delta$ against $\tau/\sigma > \Delta$, there exists a UMP invariant test with respect to the group G: $X_i' = aX_i + b$, $Y_j' = aY_j + c$, $a > 0$, $-\infty < b$, $c < \infty$, and its rejection region is

$$\Sigma[y_j - \min(y_1, \cdots, y_n)]/\Sigma[x_i - \min(x_1, \cdots, x_m)] > C.$$

(ii) This test is also UMP unbiased.

(iii) Extend these results to the case that only the r smallest X's and the s smallest Y's are observed.

[(ii) See Problem 12 of Chapter 5.]

7. If $X_1, \cdots, X_n$ and $Y_1, \cdots, Y_n$ are samples from $N(\xi, \sigma^2)$ and $N(\eta, \tau^2)$ respectively, the problem of testing $\tau^2 = \sigma^2$ against the two-sided alternatives $\tau^2 \neq \sigma^2$ remains invariant under the group G generated by the transformations $X_i' = aX_i + b$, $Y_i' = aY_i + c$, $a \neq 0$, and $X_i' = Y_i$, $Y_i' = X_i$. There exists a UMP invariant test under G with rejection region

$$W = \max\{\Sigma(Y_i - \bar{Y})^2/\Sigma(X_i - \bar{X})^2, \Sigma(X_i - \bar{X})^2/\Sigma(Y_i - \bar{Y})^2\} \geqq k.$$

[The ratio of the probability densities of W for $\tau^2/\sigma^2 = \Delta$ and $\tau^2/\sigma^2 = 1$ is proportional to $[(1 + w)/(\Delta + w)]^{n-1} + [(1 + w)/(1 + \Delta w)]^{n-1}$ for $w \geqq 1$. The derivative of this expression is $\geqq 0$ for all Δ.]

Section 4

8. (i) When testing H: $p \leqq p_0$ against K: $p > p_0$ by means of the test corresponding to (11), determine the sample size required to obtain power

β against $p = p_1$, $\alpha = .05$, $\beta = .9$ for the cases $p_0 = .1$, $p_1 = .15$, .20, .25; $p_0 = .05$, $p_1 = .10$, .15, .20, .25; $p_0 = .01$, $p_1 = .02$, .05, .10, .15, .20.

(ii) Compare this with the sample size required if the inspection is by attributes, and the test is based on the total number of defectives, and with the expected sample size if the binomial sequential probability ratio test is used for testing p_0 against p_1.

9. *Sequential t-test.* The hypothesis $p \geqq p_0$ or equivalently $\xi/\sigma \leqq \delta_0$ of Section 4 can be tested by means of the following sequential t-test. Let $\delta_0 < \delta_1$ and $t_1 = \operatorname{sgn} x_1$, and for $n > 1$ let

$$s_n^2 = \sum_{i=1}^{n} \left(x_i - \frac{x_1 + \cdots + x_n}{n} \right)^2 \quad \text{and} \quad t_n = \frac{(x_1 + \cdots + x_n)/\sqrt{n}}{s_n/\sqrt{n-1}}.$$

If $p_\delta(t_1, \cdots, t_n)$ denotes the joint (generalized) density of $t_1, \cdots, t_n$, observation is continued as long as

$$A_0 < p_{\delta_1}(t_1, \cdots, t_n)/p_{\delta_0}(t_1, \cdots, t_n) < A_1,$$

and at the first violation of these inequalities the hypothesis is accepted or rejected as the probability ratio is $<A_0$ or $>A_1$.

(i) It can be shown* that this procedure terminates with probability 1. Use this to show that the inequalities (34) of Chapter 3 hold in the present case.

(ii) The procedure is greatly simplified and can be based on tables of the noncentral t density by noting that

$$\frac{p_{\delta_1}(t_1, \cdots, t_n)}{p_{\delta_0}(t_1, \cdots, t_n)} = \frac{p_{\delta_1}(t_n)}{p_{\delta_0}(t_n)}$$

where for $n > 1$, $p_\delta(t_n)$ is the density of the noncentral t-distribution given by (75) of Chapter 5 with $f = n$.

(iii) It is interesting to note that the probability ratio can be expressed as the ratio of the average densities of the original variables $X_1, \cdots, X_n$, averaged with respect to the scale-invariant measure $d\sigma/\sigma$; that is, it equals

$$\frac{\displaystyle\int_0^\infty \frac{1}{(\sqrt{2\pi}\sigma)^n} \exp\left[-\frac{1}{2\sigma^2} \Sigma(x_i - \delta_1\sigma)^2 \right] \frac{1}{\sigma}\, d\sigma}{\displaystyle\int_0^\infty \frac{1}{(\sqrt{2\pi}\sigma)^n} \exp\left[-\frac{1}{2\sigma^2} \Sigma(x_i - \delta_0\sigma)^2 \right] \frac{1}{\sigma}\, d\sigma}.$$

(i) The argument is the same as that used to prove (34) of Chapter 3.

(ii) To prove this result, which is equivalent to the statement that t_n is sufficient for δ on the basis of $t_1, \cdots, t_n$, it is enough to show that for fixed δ_0, the ratio $p_\delta(t_1, \cdots, t_n)/p_{\delta_0}(t_1, \cdots, t_n)$ is a function only of δ and t_n. If $y_i = x_i/|x_1|$ $(i = 1, \cdots, n)$, the density $p_\delta(t_1, \cdots, t_n)$ differs from the joint density $h_\delta(y_1, \cdots, y_n)$ of the y's by a factor independent of δ so that the ratio of the p_δ's equals the corresponding ratio of the h_δ's. The joint density of the y's is

$$h_\delta(y_1, \cdots, y_n) = \frac{1}{(\sqrt{2\pi}\sigma)^n} \int_0^\infty v^{n-1} \exp\left[-\frac{1}{2\sigma^2} \sum_{i=1}^{n} (vy_i - \delta\sigma)^2 \right] dv$$

* David and Kruskal, "The Wagr sequential t-test reaches a decision with probability one," *Ann. Math. Stat.*, Vol. 27 (1956), pp. 797–805 and Vol. 29 (1958), p. 936.

for $y_1 = \pm 1$, $-\infty < y_2, \cdots, y_n < \infty$. Putting $w = v\sqrt{\Sigma_{i=1}^n y_i^2}/\sigma$ and $z_n = \Sigma_{i=1}^n y_i/\sqrt{\Sigma_{j=1}^n y_j^2}$, this becomes

$$h_\delta(y_1, \cdots, y_n) = \frac{\exp[-(\delta^2/2)(n - z_n^2)]}{(\sqrt{2\pi})^n(\Sigma y_j^2)^{n/2}} \int_0^\infty w^{n-1} \exp\left[-\frac{1}{2}(w - \delta z_n)^2\right] dw$$

and since z_n is a function of t_n, this proves that t_n is sufficient for δ on the basis of $t_1, \cdots, t_n$.*

(iii) Make the transformation $v' = v/\sigma$ and compare with the ratio of the h_δ's.]

10. *Two-sided t-test.* (i) Let $X_1, \cdots, X_n$ be a sample from $N(\xi, \sigma^2)$. For testing $\xi = 0$ against $\xi \neq 0$, there exists a UMP invariant test with respect to the group $X_i' = cX_i$, $c \neq 0$, given by the two-sided t-test (17) of Chapter 5.

(ii) Let $X_1, \cdots, X_m$ and $Y_1, \cdots, Y_n$ be samples from $N(\xi, \sigma^2)$ and $N(\eta, \sigma^2)$ respectively. For testing $\eta = \xi$ against $\eta \neq \xi$ there exists a UMP invariant test with respect to the group $X_i' = aX_i + b$, $Y_j' = aY_j + b$, $a \neq 0$, given by the two-sided t-test (30) of Chapter 5.

[(i) Sufficiency and invariance reduce the problem to $|t|$, which in the notation of Section 4 has the probability density $p_\delta(t) + p_\delta(-t)$ for $t > 0$. The ratio of this density for $\delta = \delta_1$ to its value for $\delta = 0$ is proportional to $\int_0^\infty (e^{\delta_1 v} + e^{-\delta_1 v})g_{t^2}(v)\,dv$, which is an increasing function of t^2 and hence of $|t|$.]

11. *Testing a correlation coefficient.* Let $(X_1, Y_1), \cdots, (X_n, Y_n)$ be a sample from a bivariate normal distribution.

(i) For testing $\rho \leq \rho_0$ against $\rho > \rho_0$ there exists a UMP invariant test with respect to the group of all transformations $X_i' = aX_i + b$, $Y_i' = cY_i + d$ for which a and c are >0. This test rejects when the sample correlation coefficient R is too large.

(ii) The problem of testing $\rho = 0$ against $\rho \neq 0$ remains invariant in addition under the transformation $Y_i' = -Y_i$, $X_i' = X_i$. With respect to the group generated by this transformation and those of (i) there exists a UMP invariant test, with rejection region $|R| \geqq C$.

[(i) To show that the probability density $p_\rho(r)$ of R has monotone likelihood ratio, apply the condition of Chapter 3, Problem 6(i), to the expression (85) given for this density in Chapter 5. Putting $t = \rho r + 1$, the second derivative $\partial^2 \log p_\rho(r)/\partial\rho\,\partial r$ up to a positive factor is

$$\sum_{i,j=0}^\infty c_i c_j t^{i+j-2}[(j - i)^2(t - 1) + (i + j)]/2\left[\sum_{i=0}^\infty c_i t^i\right]^2.$$

To see that the numerator is positive for all $t > 0$, note that it is greater than

$$2\sum_{i=0}^\infty c_i t^{i-2} \sum_{j=i+1}^\infty c_j t^j[(j - i)^2(t - 1) + (i + j)].$$

* An alternative proof of (ii) based on the facts that $\Sigma x_i/n$ and s_n are sufficient for (ξ, σ) on the basis of the original observations $x_1, \cdots, x_n$; that t_n is a maximal invariant function of these sufficient statistics under changes of scale; and that $t_1, \cdots, t_n$ are invariant under these transformations, is given by Cox, "Sequential tests for composite hypotheses," *Proc. Camb. Phil. Soc.*, Vol. 48 (1952), pp. 290–299, where a number of other examples are treated by the same method.

Holding i fixed and using the inequality $c_{j+1} < \frac{1}{2}c_j$, the coefficient of t^j in the interior sum is $\geqq 0$.]

12. For testing the hypothesis that the correlation coefficient ρ of a bivariate normal distribution is $\leqq \rho_0$, determine the power against the alternative $\rho = \rho_1$ when the level of significance α is .05, $\rho_0 = .3$, $\rho_1 = .5$ and the sample size n is 50, 100, 200.

Section 5

13. Almost invariance of a test ϕ with respect to the group G of either Problem 6(i) or of Example 6 implies that ϕ is equivalent to an invariant test.

Section 6

14. Consider a testing problem which is invariant under a group G of transformations of the sample space, and let $\mathscr{C}$ be a class of tests which is closed under G so that $\phi \in \mathscr{C}$ implies $\phi g \in \mathscr{C}$ where ϕg is the test defined by $\phi g(x) = \phi(gx)$. If there exists an a.e. unique UMP member ϕ_0 of $\mathscr{C}$, then ϕ_0 is almost invariant.

15. *Envelope power function.* Let $S(\alpha)$ be the class of all level α tests of a hypothesis H, and let $\beta_\alpha^*(\theta)$ be the *envelope power function*, defined by

$$\beta_\alpha^*(\theta) = \sup_{\theta \in S(\alpha)} \beta_\phi(\theta)$$

where β_ϕ denotes the power function of ϕ. If the problem of testing H is invariant under a group G, then $\beta_\alpha^*(\theta)$ is invariant under the induced group $\bar{G}$.

16. (i) A generalization of equation (1) is

$$\int_A f(x)\, dP_\theta(x) = \int_{gA} f(g^{-1}x)\, dP_{\bar{g}\theta}(x).$$

(ii) If P_{θ_1} is absolutely continuous with respect to P_{θ_0}, then $P_{\bar{g}\theta_1}$ is absolutely continuous with respect to $P_{\bar{g}\theta_0}$ and

$$\frac{dP_{\theta_1}}{dP_{\theta_0}}(x) = \frac{dP_{\bar{g}\theta_1}}{dP_{\bar{g}\theta_0}}(gx) \qquad (\text{a.e. } P_{\theta_0}).$$

(iii) The distribution of $dP_{\theta_1}/dP_{\theta_0}(X)$ when X is distributed as P_{θ_0} is the same as that of $dP_{\bar{g}\theta_1}/dP_{\bar{g}\theta_0}(X')$ when X' is distributed as $P_{\bar{g}\theta_0}$.

17. *Invariance of likelihood ratio.* Let the family of distributions $\mathscr{P} = \{P_\theta, \theta \in \Omega\}$ be dominated by μ, let $p_\theta = dP_\theta/d\mu$, let μg^{-1} be the measure defined by $\mu g^{-1}(A) = \mu[g^{-1}(A)]$, and suppose that μ is absolutely continuous with respect to μg^{-1} for all $g \in G$.

(i) Then

$$p_\theta(x) = p_{\bar{g}\theta}(gx)\frac{d\mu}{d\mu g^{-1}}(gx) \qquad (\text{a.e. } \mu).$$

(ii) Let Ω and ω be invariant under $\bar{G}$, and countable. Then the likelihood ratio $\sup_\Omega p_\theta(x)/\sup_\omega p_\theta(x)$ is almost invariant under G.

(iii) Suppose that $p_\theta(x)$ is continuous in θ for all x, that Ω is separable, and that Ω and ω are invariant. Then the likelihood ratio is almost invariant under G.

18. *Inadmissible likelihood ratio test.* In many applications in which a UMP invariant test exists, it coincides with the likelihood ratio test. That this is, however, not always the case is seen from the following example. Let $P_1, \cdots, P_n$ be n equidistant points on the circle $x^2 + y^2 = 4$, and $Q_1, \cdots, Q_n$ on the

circle $x^2 + y^2 = 1$. Denote the origin in the x, y-plane by O, let $0 < \alpha \leqq \frac{1}{2}$ be fixed, and let (X, Y) be distributed over the $2n + 1$ points $P_1, \cdots, P_n$, $Q_1, \cdots, Q_n$, O with probabilities given by the following table:

	P_i	Q_i	O
H	α/n	$(1 - 2\alpha)/n$	α
K	p_i/n	0	$(n - 1)/n$

where $\Sigma p_i = 1$. The problem remains invariant under rotations of the plane by the angles $2k\pi/n$ $(k = 0, 1, \cdots, n - 1)$. The rejection region of the likelihood ratio test consists of the points $P_1, \cdots, P_n$, and its power is $1/n$. On the other hand, the UMP invariant test rejects when $X = Y = 0$, and has power $(n - 1)/n$.

19. Let G be a group of transformations of $\mathscr{X}$, and let $\mathscr{A}$ be a σ-field of subsets of $\mathscr{X}$ and μ a measure over $(\mathscr{X}, \mathscr{A})$. Then a set $A \in \mathscr{A}$ is said to be almost invariant if its indicator function is almost invariant.

(i) The totality of almost invariant sets forms a σ-field $\mathscr{A}_0$, and a critical function is almost invariant if and only if it is $\mathscr{A}_0$-measurable.

(ii) Let $\mathscr{P} = \{P_\theta, \theta \in \Omega\}$ be a dominated family of probability distributions over $(\mathscr{X}, \mathscr{A})$, and suppose that $\bar{g}\theta = \theta$ for all $\bar{g} \in \bar{G}$, $\theta \in \Omega$. Then the σ-field $\mathscr{A}_0$ of almost invariant sets is sufficient for $\mathscr{P}$.

[Let $\lambda = \Sigma c_i P_{\theta_i}$ be equivalent to $\mathscr{P}$. Then

$$\frac{dP_\theta}{d\lambda}(gx) = \frac{dP_{g^{-1}\theta}}{\Sigma c_i\, dP_{g^{-1}\theta_i}}(x) = \frac{dP_\theta}{d\lambda}(x) \qquad \text{(a.e. } \lambda)$$

so that $dP_\theta/d\lambda$ is almost invariant and hence $\mathscr{A}_0$-measurable.]

Section 8

20. *Wilcoxon two-sample test.* Let $U_{ij} = 1$ or 0 as $X_i < Y_j$ or $X_i > Y_j$, and let $U = \Sigma\Sigma U_{ij}$ be the number of pairs X_i, Y_j with $X_i < Y_j$.

(i) Then $U = \Sigma S_i - \frac{1}{2}n(n + 1)$ where $S_1 < \cdots < S_n$ are the ranks of the Y's, so that the test with rejection region $U > C$ is equivalent to the Wilcoxon test.

(ii) Any given arrangement of x's and y's can be transformed into the arrangement $x \cdots xy \cdots y$ through a number of interchanges of neighboring elements. The smallest number of steps in which this can be done for the observed arrangement is $m + n - U$.

21. *Expectation and variance of Wilcoxon statistic.* If the X's and Y's are samples from continuous distributions F and G respectively, the expectation and variance of the Wilcoxon statistic U defined in the preceding problem are given by

$$E(U/mn) = P\{X < Y\} = \int F\, dG \tag{29}$$

and

$$mn \operatorname{Var}(U/mn) = \int F\, dG + (n - 1)\int (1 - G)^2\, dF + (m - 1)\int F^2\, dG - (m + n - 1)\left(\int F\, dG\right)^2. \tag{30}$$

Under the hypothesis $G = F$, these reduce to

(31) $$E(U/mn) = \tfrac{1}{2}, \qquad \operatorname{Var}(U/mn) = (m + n + 1)/12mn.$$

22. (i) Let $Z_1, \cdots, Z_N$ be independently distributed with densities $f_1, \cdots, f_N$, and let the rank of Z_i be denoted by T_i. If f is any probability density which is positive whenever at least one of the f_i is positive, then

(32) $$P\{T_1 = t_1, \cdots, T_N = t_N\} = \frac{1}{N!} E\left[\frac{f_1(V^{(t_1)})}{f(V^{(t_1)})} \cdots \frac{f_N(V^{(t_N)})}{f(V^{(t_N)})}\right]$$

where $V^{(1)} < \cdots < V^{(N)}$ is an ordered sample from a distribution with density f.

(ii) If $N = m + n$, $f_1 = \cdots = f_m = f$, $f_{m+1} = \cdots = f_{m+n} = g$, and $S_1 < \cdots < S_n$ denote the ordered ranks of $Z_{m+1}, \cdots, Z_{m+n}$ among all the Z's, the probability distribution of $S_1, \cdots, S_n$ is given by (18).

[(i) The probability in question is $\int \cdots \int f_1(z_1) \cdots f_N(z_N)\, dz_1 \cdots dz_N$ integrated over the set in which z_i is the t_ith smallest of the z's for $i = 1, \cdots, N$. Under the transformation $w_{t_i} = z_i$ the integral becomes $\int \cdots \int f_1(w_{t_1}) \cdots f_N(w_{t_N})\, dw_1 \cdots dw_N$, integrated over the set $w_1 < \cdots < w_N$. The desired result now follows from the fact that the probability density of the order statistics $V^{(1)} < \cdots < V^{(N)}$ is $N! f(w_1) \cdots f(w_N)$ for $w_1 < \cdots < w_N$.]

23. (i) For any continuous cumulative distribution function F, define $F^{-1}(0) = -\infty$, $F^{-1}(y) = \inf\{x\colon\ F(x) = y\}$ for $0 < y < 1$, $F^{-1}(1) = \infty$ if $F(x) < 1$ for all finite x and otherwise as $\inf\{x\colon F(x) = 1\}$. Then $F[F^{-1}(y)] = y$ for all $0 \leqq y \leqq 1$, but $F^{-1}[F(y)]$ may be $<y$.

(ii) Let Z have a cumulative distribution function $G(z) = h[F(z)]$ where F and h are continuous cumulative distribution functions, the latter defined over $(0, 1)$. If $Y = F(Z)$, then $P\{Y < y\} = h(y)$ for all $0 \leqq y \leqq 1$.

(iii) If Z has the continuous cumulative distribution function F, then $F(Z)$ is uniformly distributed over $(0, 1)$.

[(ii) $P\{F(Z) < y\} = P\{Z < F^{-1}(y)\} = F[F^{-1}(y)] = y$.]

24. Let Z_i have a continuous cumulative distribution function F_i $(i = 1, \cdots, N)$, and let G be the group of all transformations $Z_i' = f(Z_i)$ such that f is continuous and strictly increasing.

(i) The transformation induced by f in the space of distributions is $F_i' = F_i(f^{-1})$.

(ii) Two N-tuples of distributions $(F_1, \cdots, F_N)$ and $(F_1', \cdots, F_N')$ belong to the same orbit with respect to $\bar{G}$ if and only if there exist continuous distribution functions $h_1, \cdots, h_N$ defined on $(0, 1)$ and strictly increasing continuous distribution functions F and F' such that $F_i = h_i(F)$ and $F_i' = h_i(F')$.

[(i) $P\{f(Z_i) \leqq y\} = P\{Z_i \leqq f^{-1}(y)\} = F_i[f^{-1}(y)]$.

(ii) If $F_i = h_i(F)$ and the F_i' are on the same orbit so that $F_i' = F_i(f^{-1})$, then $F_i' = h_i(F')$ with $F' = F(f^{-1})$. Conversely, if $F_i = h_i(F)$, $F_i' = h_i(F')$, then $F_i' = F_i(f^{-1})$ with $f = F'^{-1}(F)$.]

25. Under the assumptions of the preceding problem, if $F_i = h_i(F)$, the distribution of the ranks $T_1, \cdots, T_N$ of $Z_1, \cdots, Z_N$ depends only on the h_i, not on F. If the h_i are differentiable, the distribution of the T_i is given by

(33) $$P\{T_1 = t_1, \cdots, T_N = t_N\} = E[h_1'(U^{(t_1)}) \cdots h_N'(U^{(t_N)})]/N!$$

where $U^{(1)} < \cdots < U^{(N)}$ is an ordered sample of size N from the uniform distribution $R(0, 1)$ over $(0, 1)$.

[The left-hand side of (33) is the probability that of the quantities $F(Z_1), \cdots, F(Z_N)$, the ith one is the t_ith smallest for $i = 1, \cdots, N$. This is given by $\int \cdots \int h_1'(y_1) \cdots h_N'(y_N)\, dy$ integrated over the region in which y_i is the t_ith smallest of the y's for $i = 1, \cdots, N$. The proof is completed as in Problem 22.

26. *Distribution of order statistics.* (i) If $Z_1, \cdots, Z_N$ is a sample from a cumulative distribution function F with density f, the joint density of $Y_i = Z^{(s_i)}$, $i = 1, \cdots, n$, is

$$\text{(34)} \quad \frac{N! f(y_1) \cdots f(y_n)}{(s_1 - 1)!(s_2 - s_1 - 1)! \cdots (N - s_n)!} [F(y_1)]^{s_1 - 1}[F(y_2) - F(y_1)]^{s_2 - s_1 - 1} \cdots [1 - F(y_n)]^{N - s_n}$$

for $y_1 < \cdots < y_n$.

(ii) For the particular case that the Z's are a sample from the uniform distribution on $(0, 1)$, this reduces to

$$\text{(35)} \quad \frac{N!}{(s_1 - 1)!(s_2 - s_1 - 1)! \cdots (N - s_n)!} y_1^{s_1 - 1}(y_2 - y_1)^{s_2 - s_1 - 1} \cdots (1 - y_n)^{N - s_n}.$$

For $n = 1$, (35) is the density of the beta-distribution $B_{s, N-s+1}$, which therefore is the distribution of the single order statistic $Z^{(s)}$ from $R(0, 1)$.

(iii) Let the distribution of $Y_1, \cdots, Y_n$ be given by (35), and let V_i be defined by $Y_i = V_i V_{i+1} \cdots V_n$ for $i = 1, \cdots, n$. Then the joint distribution of the V_i is

$$\frac{N!}{(s_1 - 1)! \cdots (N - s_n)!} \prod_{i=1}^{n} v_i^{s_i - 1}(1 - v_i)^{s_{i+1} - s_i - 1} \qquad (s_{n+1} = N + 1)$$

so that the V_i are independently distributed according to the beta-distribution $B_{s_i, s_{i+1} - s_i}$.

[(i) If $Y_1 = Z^{(s_1)}, \cdots, Y_n = Z^{(s_n)}$ and $Y_{n+1}, \cdots, Y_N$ are the remaining Z's in the original order of their subscripts, the joint density of $Y_1, \cdots, Y_n$ is $N(N - 1) \cdots (N - n + 1) \int \cdots \int f(y_{n+1}) \cdots f(y_N)\, dy_{n+1} \cdots dy_N$ integrated over the region in which $s_1 - 1$ of the y's are $< y_1$, $s_2 - s_1 - 1$ between y_1 and $y_2, \cdots$, and $N - s_n > y_n$. Consider any set where a particular $s_1 - 1$ of the y's is $< y_1$, a particular $s_2 - s_1 - 1$ of them is between y_1 and y_2, etc. There are $N!/(s_1 - 1)! \cdots (N - s_n)!$ of these regions, and the integral has the same value over each of them, namely $[F(y_1)]^{s_1 - 1}[F(y_2) - F(y_1)]^{s_2 - s_1 - 1} \cdots [1 - F(y_n)]^{N - s_n}$.]

27. (i) If $X_1, \cdots, X_m$ and $Y_1, \cdots, Y_n$ are samples with continuous cumulative distribution functions F and $G = h(F)$ respectively, and if h is differentiable, the distribution of the ranks $S_1 < \cdots < S_n$ of the Y's is given by

$$\text{(36)} \qquad P\{S_1 = s_1, \cdots, S_n = s_n\} = \frac{E[h'(U^{(s_1)}) \cdots h'(U^{(s_n)})]}{\binom{m + n}{m}}$$

where $U^{(1)} < \cdots < U^{(m+n)}$ is an ordered sample from the uniform distribution $R(0, 1)$.

(ii) If in particular $G = F^k$ where k is a positive integer, (36) reduces to

(37) $P\{S_1 = s_1, \cdots, S_n = s_n\} =$

$$\frac{k^n}{\binom{m+n}{m}} \prod_{j=1}^{n} \frac{\Gamma(s_j + jk - j)}{\Gamma(s_j)} \cdot \frac{\Gamma(s_{j+1})}{\Gamma(s_{j+1} + jk - j)}$$

28. For sufficiently small $\theta > 0$, the Wilcoxon test maximizes the power (among rank tests) against the alternatives (F, G) with $G = (1 - \theta)F + \theta F^2$.

29. An alternative proof of the optimum property of the Wilcoxon test for detecting a shift in the logistic distribution is obtained from the preceding problem by equating $F(x - \theta)$ with $(1 - \theta)F(x) + \theta F^2(x)$, neglecting powers of θ higher than the first. This leads to the differential equation $F - \theta F' = (1 - \theta)F + \theta F^2$, the solution of which is the logistic distribution.

30. Let $\mathscr{F}_0$ be a family of probability measures over $(\mathscr{X}, \mathscr{A})$ and let $\mathscr{C}$ be a class of transformations of the space $\mathscr{X}$. Define a class $\mathscr{F}_1$ of distributions by: $F_1 \in \mathscr{F}_1$ if there exists $F_0 \in \mathscr{F}_0$ and $f \in \mathscr{C}$ such that the distribution of $f(X)$ is F_1 when that of X is F_0. If ϕ is any test satisfying (a) $E_{F_0}\phi(X) = \alpha$ for all $F_0 \in \mathscr{F}_0$, and (b) $\phi(x) \leq \phi[f(x)]$ for all x and all $f \in \mathscr{C}$, then ϕ is unbiased for testing $\mathscr{F}_0$ against $\mathscr{F}_1$.

31. Let $X_1, \cdots, X_m$; $Y_1, \cdots, Y_n$ be samples from a common continuous distribution F. Then the Wilcoxon statistic U defined in Problem 20 is distributed symmetrically about $\frac{1}{2}mn$ even when $m \neq n$.

32. *Confidence intervals for a shift.* Let $X_1, \cdots, X_m$; $Y_1, \cdots, Y_n$ be samples from distributions $F(x)$ and $G(y) = F(y - \Delta)$ respectively. The hypothesis $\Delta = \Delta_0$ can be tested by applying the two-sided Wilcoxon test to the observations X_i and $Y_j - \Delta_0$.

(i) The resulting confidence intervals for Δ are

$$Y^{(1)} - X^{(m)} < \Delta < Y^{(n)} - X^{(1)}$$

when the confidence coefficient is $1 - 2\Big/\binom{m+n}{m}$, and

$$\min(Y^{(1)} - X^{(m-1)}, Y^{(2)} - X^{(m)}) < \Delta < \max(Y^{(n)} - X^{(2)}, Y^{(n-1)} - X^{(1)})$$

when the confidence coefficient is $1 - 4\Big/\binom{m+n}{m}$.

(ii) Determine the confidence interval for Δ when the confidence coefficient is 20/21, $m = n = 6$, and the observations are: x: .113, .212, .249, .522, .709, .788; y: .221, .433, .724, .913, .917, 1.58.

33. (i) Let X, X' and Y, Y' be independent samples of size 2 from continuous distributions F and G respectively. Then

$$p = P\{\max(X, X') < \min(Y, Y')\} + P\{\max(Y, Y') < \min(X, X')\} = \tfrac{1}{3} + 2\Delta$$

where $\Delta = \int (F - G)^2 \, d[(F + G)/2]$.

(ii) $\Delta = 0$ if and only if $F = G$.

[(i) $p = \int (1 - F)^2 \, dG^2 + \int (1 - G)^2 \, dF^2$, which after some computation reduces to the stated form.

(ii) $\Delta = 0$ implies $F(x) = G(x)$ except on a set N which has measure zero both under F and G. Suppose that $G(x_1) - F(x_1) = \eta > 0$. Then there exists x_0 such that $G(x_0) = F(x_0) + \frac{1}{2}\eta$ and $F(x) < G(x)$ for $x_0 \leq x \leq x_1$. Since $G(x_1) - G(x_0) > 0$, it follows that $\Delta > 0$.]

34. *Continuation.* (i) There exists at every significance level α a test of H: $G = F$ which has power $>\alpha$ against all continuous alternatives (F, G) with $F \neq G$.

(ii) There does not exist a nonrandomized unbiased test of H against all $G \neq F$ at level $\alpha = 1\Big/\binom{m+n}{m}$.

[(i) Let X_i, X_i'; Y_i, Y_i' $(i = 1, \cdots, n)$ be independently distributed, the X's with distribution F, the Y's with distribution G, and let $V_i = 1$ if $\max(X_i, X_i') < \min(Y_i, Y_i')$ or $\max(Y_i, Y_i') < \min(X_i, X_i')$, and $V_i = 0$ otherwise. Then ΣV_i has a binomial distribution with the probability p defined in Problem 33, and the problem reduces to that of testing $p = 1/3$ against $p > 1/3$.

(ii) Consider the particular alternatives for which $P\{X < Y\}$ is either 1 or 0.]

Section 9

35. (i) Let m and n be the number of negative and positive observations among $Z_1, \cdots, Z_N$, and let $S_1 < \cdots < S_n$ denote the ranks of the positive Z's among $|Z_1|, \cdots, |Z_N|$. Consider the $N + \frac{1}{2}N(N-1)$ distinct sums $Z_i + Z_j$ with $i = j$ as well as $i \neq j$. The Wilcoxon rank sum ΣS_j is equal to the number of these sums that are positive.

(ii) If the common distribution of the Z's is D, then

$$E(\Sigma S_j) = \tfrac{1}{2}N(N+1) - ND(0) - \tfrac{1}{2}N(N-1)\int D(-z)\,dD(z).$$

[(i) Let K be the required number of positive sums. Since $Z_i + Z_j$ is positive if and only if the larger of $|Z_i|$ and $|Z_j|$ is positive, $K = \Sigma_{i=1}^N \Sigma_{j=1}^N U_{ij}$ where $U_{ij} = 1$ if $Z_j > 0$ and $|Z_i| \leq Z_j$ and $U_{ij} = 0$ otherwise.]

36. Let $Z_1, \cdots, Z_N$ be a sample from a distribution with density $f(z - \theta)$ where $f(z)$ is positive for all z and f is symmetric about 0, and let m, n, and the S_j be defined as in the preceding problem.

(i) The distribution of n and the S_j is given by

$$P\{\text{the number of positive } Z\text{'s is } n \text{ and } S_1 = s_1, \cdots, S_n = s_n\} \tag{38}$$

$$= \frac{1}{2^N} E\left[\frac{f(V^{(r_1)} + \theta) \cdots f(V^{(r_m)} + \theta) f(V^{(s_1)} - \theta) \cdots f(V^{(s_n)} - \theta)}{f(V^{(1)}) \cdots f(V^{(N)})}\right]$$

where $V^{(1)} < \cdots < V^{(N)}$ is an ordered sample from a distribution with density $2f(v)$ for $v > 0$, and 0 otherwise.

(ii) The rank test of the hypothesis of symmetry with respect to the origin, which maximizes the derivative of the power function at $\theta = 0$ and hence maximizes the power for sufficiently small $\theta > 0$, rejects when

$$-E\left[\sum_{j=1}^{n} \frac{f'(V^{(s_j)})}{f(V^{(s_j)})}\right] > C.$$

(iii) In the particular case that $f(z)$ is a normal density with zero mean, the rejection region of (ii) reduces to $\Sigma E(V^{(s_j)}) > C$, where $V^{(1)} < \cdots < V^{(N)}$ is an ordered sample from a χ-distribution with 1 degree of freedom.

(iv) Determine a density f such that the one-sample Wilcoxon test is most powerful against the alternatives $f(z - \theta)$ for sufficiently small positive θ.

[(i) Apply Problem 22(i) to find an expression for $P\{S_1 = s_1, \cdots, S_n = s_n$ given that the number of positive Z's is $n\}$.]

37. An alternative expression for (38) is obtained if the distribution of Z is characterized by (ρ, F, G). If then $G = h(F)$ and h is differentiable, the distribution of n and the S_j is given by

$$\rho^m(1 - \rho)^n E[h'(U^{(s_1)}) \cdots h'(U^{(s_n)})] \tag{39}$$

where $U^{(1)} < \cdots < U^{(N)}$ is an ordered sample from $R(0, 1)$.

38. *Unbiased tests of symmetry.* Let $Z_1, \cdots, Z_N$ be a sample, and let ϕ be any rank test of the hypothesis of symmetry with respect to the origin such that $z_i \leqq z_i'$ for all i implies $\phi(z_1, \cdots, z_N) \leqq \phi(z_1', \cdots, z_N')$. Then ϕ is unbiased against the one-sided alternatives that the Z's are stochastically larger than some random variable that has a symmetric distribution with respect to the origin.

39. *The hypothesis of randomness.* Let $Z_1, \cdots, Z_N$ be independently distributed with distributions $F_1, \cdots, F_N$, and let T_i denote the rank of Z_i among the Z's. For testing the *hypothesis of randomness*: $F_1 = \cdots = F_N$ against the alternatives K of an *upward trend*, namely that Z_i is stochastically increasing with i, consider the rejection regions

$$\Sigma i t_i > C \tag{40}$$

and

$$\Sigma i E(V^{(t_i)}) > C \tag{41}$$

where $V^{(1)} < \cdots < V^{(N)}$ is an ordered sample from a standard normal distribution and where t_i is the value taken on by T_i.

(i) The second of these tests is most powerful among rank tests against the normal alternatives $F = N(\gamma + i\delta, \sigma^2)$ for sufficiently small δ.

(ii) Determine alternatives against which the first test is a most powerful rank test.

(iii) Both tests are unbiased against the alternatives of an upward trend; so is any rank test ϕ satisfying $\phi(z_1, \cdots, z_N) \leqq \phi(z_1', \cdots, z_N')$ for any two points for which $i < j$, $z_i < z_j$ implies $z_i' < z_j'$ for all i and j.

[(iii) Apply Problem 30 with $\mathscr{C}$ the class of transformations $z_1' = z_1$, $z_i' = f_i(z_i)$ for $i > 1$, where $z < f_2(z) < \cdots < f_N(z)$ and each f_i is nondecreasing. If $\mathscr{F}_0$ is the class of N-tuples $(F_1, \cdots, F_N)$ with $F_1 = \cdots = F_N$, then $\mathscr{F}_1$ coincides with the class K of alternatives.]

40. Let $U_{ij} = 1$ if $(j - i)(Z_j - Z_i) > 0$, and $= 0$ otherwise.

(i) The test statistic $\Sigma i T_i$ can be expressed in terms of the U's through the relation

$$\sum_{i=1}^{N} i T_i = \sum_{i<j} (j - i) U_{ij} + \frac{N(N + 1)(N + 2)}{6}.$$

(ii) The smallest number of steps [in the sense of Problem 20(ii)] by which $(Z_1, \cdots, Z_N)$ can be transformed into the ordered sample $(Z^{(1)}, \cdots, Z^{(N)})$ is $[N(N - 1)/2] - U$, where $U = \Sigma_{i<j} U_{ij}$. This suggests $U > C$ as another rejection region for the preceding problem.

[(i) Let $V_{ij} = 1$ or 0 as $Z_i \leqq Z_j$ or $Z_i > Z_j$. Then $T_j = \Sigma_{i=1}^N V_{ij}$ and $V_{ij} = U_{ij}$ or $1 - U_{ij}$ as $i < j$ or $i \geqq j$. Expressing $\Sigma_{j=1}^N jT_j = \Sigma_{j=1}^N j\Sigma_{i=1}^N V_{ij}$ in terms of the U's and using the fact that $U_{ij} = U_{ji}$, the result follows by a simple calculation.]

41. *The hypothesis of independence.* Let $(X_1, Y_1), \cdots, (X_N, Y_N)$ be a sample from a bivariate distribution and $(X^{(1)}, Z_1), \cdots, (X^{(N)}, Z_N)$ be the same sample arranged according to increasing values of the X's so that the Z's are a permutation of the Y's. Let R_i be the rank of X_i among the X's, S_i the rank of Y_i among the Y's, and T_i the rank of Z_i among the Z's, and consider the hypothesis of independence of X and Y against the alternatives of positive dependence.

(i) In terms of the T's this problem is equivalent to testing the hypothesis of randomness of the Z's against the alternatives of an upward trend.

(ii) The test (40) is equivalent to rejecting when the *rank correlation coefficient*

$$\frac{\Sigma(R_i - \bar{R})(S_i - \bar{S})}{\sqrt{\Sigma(R_i - \bar{R})^2\Sigma(S_i - \bar{S})^2}} = \frac{12}{N^3 - N}\Sigma\left(R_i - \frac{N+1}{2}\right)\left(S_i - \frac{N+1}{2}\right)$$

is too large.

(iii) An alternative expression for the rank correlation coefficient* is

$$1 - \frac{6}{N^3 - N}\Sigma(S_i - R_i)^2 = 1 - \frac{6}{N^3 - N}\Sigma(T_i - i)^2.$$

(iv) The test $U > C$ of Problem 40(ii) is equivalent to rejecting when Kendall's t-statistic* $\Sigma_{i<j}V_{ij}/N(N-1)$ is too large where V_{ij} is $+1$ or -1 as $(Y_j - Y_i)(X_j - X_i)$ is positive or negative.

(v) The tests (ii) and (iv) are unbiased against the alternatives of positive dependence.†

Section 10

42. In Example 11, a family of sets $S(x, y)$ is a class of invariant confidence sets if and only if there exists a set $\mathscr{R}$ of real numbers such that

$$S(x, y) = \bigcup_{r\in\mathscr{R}} \{(\xi, \eta): (x - \xi)^2 + (y - \eta)^2 = r^2\}.$$

43. Let $X_1, \cdots, X_n$; $Y_1, \cdots, Y_n$ be samples from $N(\xi, \sigma^2)$ and $N(\eta, \tau^2)$ respectively. Then the confidence intervals (43) of Chapter 5 for τ^2/σ^2, which can be written as

$$\Sigma(Y_j - \bar{Y})^2/k\Sigma(X_i - \bar{X})^2 \leqq \tau^2/\sigma^2 \leqq k\Sigma(Y_j - \bar{Y})^2/\Sigma(X_i - \bar{X})^2,$$

are uniformly most accurate invariant with respect to the smallest group G containing the transformations $X_i' = aX + b$, $Y_i' = aY + c$ for all $a \neq 0$, b, c and the transformation $X_i' = dY_i$, $Y_i' = X_i/d$ for all $d \neq 0$.

[Cf. Problem 6.]

* For further material on these statistics see Hoeffding, "A class of statistics with asymptotically normal distributions," *Ann. Math. Stat.*, Vol. 19 (1948), particularly section 9, and Kendall, *Rank Correlation Methods*, London, Charles Griffin and Co., 2nd ed., 1955.

† A different type of test, which pays particular attention to the extreme observations, has been proposed by Olmstead and Tukey, "A corner test for association," *Ann. Math. Stat.*, Vol. 18 (1947), pp. 495–513.

44. *One-sided invariant confidence limits.* Let θ be real-valued and suppose that for each θ_0, the problem of testing $\theta \leqq \theta_0$ against $\theta > \theta_0$ (in the presence of nuisance parameters ϑ) remains invariant under a group G_{θ_0} and that $A(\theta_0)$ is a UMP invariant acceptance region for this hypothesis at level α. Let the associated confidence sets $S(x) = \{\theta: x \in A(\theta)\}$ be one-sided intervals $S(x) = \{\theta: \underline{\theta}(x) \leqq \theta\}$, and suppose they are invariant under all G_θ and hence under the group G generated by these. Then the lower confidence limits $\underline{\theta}(X)$ are uniformly most accurate invariant at confidence level $1 - \alpha$ in the sense of minimizing $P_{\theta,\vartheta}\{\underline{\theta}(X) \leqq \theta'\}$ for all $\theta' < \theta$.

45. Let $X_1, \cdots, X_n$ be independently distributed as $N(\xi, \sigma^2)$. The upper confidence limits $\sigma^2 \leqq \Sigma(X_i - \bar{X})^2/C_0$ of Example 5, Chapter 5, are uniformly most accurate invariant under the group $X_i' = X_i + c$, $-\infty < c < \infty$. They are also invariant (and hence uniformly most accurate invariant) under the larger group $X_i' = aX_i + c$, $-\infty < a, c < \infty$.

46. (i) Let $X_1, \cdots, X_n$ be independently distributed as $N(\xi, \sigma^2)$ and let $\theta = \xi/\sigma$. The lower confidence bounds $\underline{\theta}$ for θ, which at confidence level $1 - \alpha$ are uniformly most accurate invariant under the transformations $X_i' = aX_i$, are

$$\underline{\theta} = C^{-1}[\sqrt{n}\bar{X}/\sqrt{\Sigma(X_i - \bar{X})^2/(n-1)}]$$

where the function $C(\theta)$ is determined from a table of noncentral t so that

$$P_\theta\{\sqrt{n}\bar{X}/\sqrt{\Sigma(X_i - \bar{X})^2/(n-1)} \leqq C(\theta)\} = 1 - \alpha.$$

(ii) Determine $\underline{\theta}$ when the x's are 7.6, 21.2, 15.1, 32.0, 19.7, 25.3, 29.1, 18.4 and the confidence level is $1 - \alpha = .95$.

47. (i) Let $(X_1, Y_1), \cdots, (X_n, Y_n)$ be a sample from a bivariate normal distribution and let

$$\underline{\rho} = C^{-1}\left\{\frac{\Sigma(X_i - \bar{X})(Y_i - \bar{Y})}{\sqrt{\Sigma(X_i - \bar{X})^2 \Sigma(Y_i - \bar{Y})^2}}\right\},$$

where $C(\rho)$ is determined such that

$$P_\rho\left\{\frac{\Sigma(X_i - \bar{X})(Y_i - \bar{Y})}{\sqrt{\Sigma(X_i - \bar{X})^2 \Sigma(Y_i - \bar{Y})^2}} \leqq C(\rho)\right\} = 1 - \alpha.$$

Then $\underline{\rho}$ is a lower confidence limit for the population correlation coefficient ρ at confidence level $1 - \alpha$; it is uniformly most accurate invariant with respect to the group of transformations $X_i' = aX_i + b$, $Y_i' = cY_i + d$, with $ac > 0$, $-\infty < b, d < \infty$.

(ii) Determine $\underline{\rho}$ at level $1 - \alpha = .95$ when the observations are (12.9, .56), (9.8, .92), (13.1, .42), (12.5, 1.01), (8.7, .63), (10.7, .58), (9.3, .72), (11.4, .64).

Section 11

48. If the confidence sets $S(x)$ are invariant under the group G, then the probability $P_\theta\{\theta \in S(X)\}$ of their covering the true value is invariant under the induced group $\bar{G}$.

49. Consider the problem of obtaining a (two-sided) confidence band for an unknown continuous cumulative distribution function F.

(i) Show that this problem is invariant both under strictly increasing and strictly decreasing continuous transformations $X'_i = f(X_i)$, $i = 1, \cdots, n$, and determine a maximal invariant with respect to this group.

(ii) Show that the problem is not invariant under the transformation

$$X'_i = \begin{cases} X_i & \text{if } |X_i| \geqslant 1 \\ X_i - 1 & \text{if } 0 < X_i < 1 \\ X_i + 1 & \text{if } -1 < X_i < 1. \end{cases}$$

[(ii) For this transformation g, the set $g^*S(x)$ is no longer a band.]

13. REFERENCES

Invariance considerations were introduced for particular classes of problems by Hotelling and Pitman. (See the references to Chapter 1.) The general theory of invariant and almost invariant tests, together with its principal parametric applications, was developed by Hunt and Stein (1946) in an unpublished paper. In their paper, invariance was not proposed as a desirable property in itself but as a tool for deriving most stringent tests (cf. Chapter 8). Apart from this difference in point of view, the present account is based on the ideas of Hunt and Stein, about which I learned through conversations with Charles Stein during the years 1947–1950.

The field of nonparametric statistics, in which many of the basic problems are still unsolved, is at present in a period of rapid development. It has been possible here to give an indication of only some of the work done, particularly since the principal results so far have been in the area of large-sample theory. More detailed accounts of some aspects of nonparametric statistics are given in the books by Fraser (1957) and Kendall (1955); and in the survey papers by Scheffé (1943), Wilks (1948), Wolfowitz (1949), Moran, Whitfield, and Daniels (1950), Kendall and Sundrum (1953), and van Dantzig and Hemelrijk (1954). An extensive bibliography is given by Savage (1953).

Anderson, T. W.
(1958) *An Introduction to Multivariate Statistical Analysis*, New York, John Wiley & Sons (p. 99).
[Problem 11.]

Arnold, Kenneth J.
(1951) "Tables to facilitate sequential t-tests," *Appl. Math. Ser. Nat. Bur. Standards (U.S.)*, Vol. 7, pp. v–viii.
[Problem 9(ii).]

Barnard, G.
(1950) "The Behrens-Fisher test," *Biometrika*, Vol. 37, pp. 203–207.

Deuchler, Gustav

(1914) "Ueber die Methoden der Korrelationsrechnung in der Paedagogik und Psychologie," *Z. pädag. Psychol.*, Vol. 15, pp. 114–131, 145–159, 229–242.

[Appears to contain the first proposal of the two-sample procedure known as the Wilcoxon test, which was later discovered independently by many different authors. A history of this test is given by W. H. Kruskal, "Historical notes on the Wilcoxon unpaired two-sample test," *J. Am. Stat. Assoc.*, Vol. 52 (1957), pp. 356–360.]

Epstein, Benjamin, and Chia Kuci Tsao

(1953) "Some tests based on ordered observations from two exponential populations," *Ann. Math. Stat.*, Vol. 24, pp. 458–466.

Fisher, R. A.

(1956) *Statistical Methods and Scientific Inference*, Edinburgh and London, Oliver and Boyd.

[In Chapter IV the author gives his views on hypothesis testing and in particular also discusses his ideas on the Behrens-Fisher problem.]

Fisher, R. A., and Frank Yates

(1948) *Statistical Tables for Biological, Agricultural and Medical Research*, London, Oliver and Boyd, 3rd ed.

[Implicit in the introduction to tables XX and XXI is a consideration of rank order tests such as (19).]

Fraser, D. A. S.

(1957) *Nonparametric Methods in Statistics*, New York, John Wiley & Sons.

Hemelrijk, J.

(1950) "A family of parameter-free tests for symmetry with respect to a given point," *Proc. Koninkl. Ned. Akad. Wetenschap.*, Vol. 53, pp. 945–955 and 1186–1198.

[Discusses the relationship of the hypothesis of symmetry with the two-sample problem.]

Hoeffding, Wassily

(1951) " 'Optimum' nonparametric tests," *Proc. 2nd Berkeley Symposium on Mathematical Statistics and Probability*, Berkeley, Univ. Calif. Press, pp. 83–92.

[Derives a basic rank distribution of which (18) is a special case, and from it obtains locally optimum tests of the type (19). His results are specialized to the two-sample problem by Milton E. Terry, "Some rank order tests which are most powerful against specific parametric alternatives," *Ann. Math. Stat.*, Vol. 23 (1952), pp. 346–366.]

Hopf, Eberhard

(1937) "Ergodentheorie," *Ergeb. Math.*, Vol. 5, No. 2.

[Proves a result very similar to Theorem 4 (pp. 9/10).]

Hsu, P. L.

(1938) "Contributions to the theory of Student's t-test as applied to the problem of two samples," *Stat. Res. Mem.*, Vol. II, pp. 1–24.

[Shows that the two-sample t-test, in the case of equal and not very small sample sizes, is approximately unbiased even when the variances are unequal, and that for this case the t-test therefore constitutes an approximate solution to the Behrens-Fisher problem.]

Hunt, G., and C. Stein

(1946) "Most stringent tests of statistical hypotheses," unpublished.

Kendall, M. G.

(1955) *Rank Correlation Methods*, London, Charles Griffin and Co., 2nd ed.

Kendall, M. G., and R. M. Sundrum

(1953) "Distribution-free methods and order properties," *Rev. Int. Stat. Inst.*, Vol. 23, pp. 124–134.

Kruskal, William

(1954) "The monotonicity of the ratio of two non-central t density functions," *Ann. Math. Stat.*, Vol. 25, pp. 162–165.

Lehmann, E. L.

(1950) "Some principles of the theory of testing hypotheses," *Ann. Math. Stat.*, Vol. 21, pp. 1–26.

[Lemma 2; Theorem 6; presents an example of Stein on which Problem 18 is patterned.]

(1951) "Consistency and unbiasedness of certain nonparametric tests," *Ann. Math. Stat.*, Vol. 22, pp. 165–179.

[Problems 33, 34.]

(1953) "The power of rank tests," *Ann. Math. Stat.*, Vol. 24, pp. 28–43.

[Applies invariance considerations to nonparametric problems.]

Moran, P. A. P., J. W. Whitfield, and H. E. Daniels

(1950) "Symposium on ranking methods," *J. Roy. Stat. Soc., Ser. B.*, Vol. 12, pp. 153–191.

Pitman, E. J. G.

(1939) "Tests of hypotheses concerning location and scale parameters," *Biometrika*, Vol. 31, pp. 200–215.

[Invariance considerations are introduced, and are applied to problems similar to that treated in Example 4.]

(1949) "Lecture notes on nonparametric statistical inference," unpublished.

[Develops the concept of relative asymptotic efficiency and applies it to several examples including the Wilcoxon test.]

Rushton, S.

(1950) "On a sequential t-test," *Biometrika*, Vol. 37, pp. 326–333.

(1952) "On a two-sided sequential t-test," *Biometrika*, Vol. 39, pp. 302–308.

[Proposes the sequential t-test of Problem 9 and discusses some of its properties. There is also a reference to related unpublished work by Barnard, Goldberg, and Stein.]

Savage, I. R.

(1953) "Bibliography of nonparametric statistics and related topics," *J. Am. Stat. Assoc.*, Vol. 48, pp. 844–906.

Scheffé, H.

(1943) "Statistical inference in the non-parametric case," *Ann. Math. Stat.*, Vol. 14, pp. 305–332.

(1943) "On solutions of the Behrens-Fisher problem, based on the t-distribution," *Ann. Math. Stat.*, Vol. 14, pp. 35–44.

Sukhatme, P. V.

(1936) "On the analysis of k samples from exponential distributions with especial reference to the problem of random intervals," *Stat. Res. Mem.*, Vol. 1, pp. 94–112.

van Dantzig, D., and J. Hemelrijk

(1954) "Statistical methods based on few assumptions," *Bull. Int. Stat. Inst.*, Vol. 34, 2nd part, pp. 3–31.

van der Waerden, B. L.

(1952, 1953) "Order tests for the two-sample problem and their power," *Proc. Koninkl. Ned. Akad. Wetenschap.*, Vol. 55, pp. 453–458, and Vol. 56, pp. 303–316.

[Proposes the two-sample test based on the inverse normal cumulative distribution function.]

Wald, Abraham

(1955) "Testing the difference between the means of two normal populations with unknown standard deviations," published posthumously in *Selected Papers in Statistics and Probability by Abraham Wald*, Stanford, Stanford Univ. Press.

[Considers invariant tests for the Behrens-Fisher problem in the case of equal sample sizes.]

Walsh, John E.

(1949) "Some significance tests for the median which are valid under very general conditions," *Ann. Math. Stat.*, Vol. 20, pp. 64–81.

[Lemma 3; proposes the Wilcoxon one-sample test in the form given in Problem 35. The equivalence of the two tests was shown by Tukey in an unpublished mimeographed report dated 1949.]

Wilcoxon, Frank

(1945) "Individual comparisons by ranking methods," *Biometrics*, Vol. 1, pp. 80–83.

[Proposes the two tests bearing his name. (See also Deuchler.)]

Wilks, S. S.

(1948) "Order statistics," *Bull. Am. Math. Soc.*, Vol. 54, pp. 6–50.

Wolfowitz, J.

(1949) "Non-parametric statistical inference," *Proc. Berkeley Symposium on Mathematical Statistics and Probability*, Berkeley, Univ. Calif. Press, pp. 93–113.

(1949) "The power of the classical tests associated with the normal distribution," *Ann. Math. Stat.*, Vol. 20, pp. 540–551.

[Proves Lemma 2 for a number of special cases.]

CHAPTER 7

Linear Hypotheses

1. A CANONICAL FORM

Many testing problems concern the means of normal distributions and are special cases of the following *general univariate linear hypothesis*. Let $X_1, \cdots, X_n$ be independently normally distributed with means $\xi_1, \cdots, \xi_n$ and common variance σ^2. The vector of means* $\underline{\xi}$ is known to lie in a given s-dimensional linear subspace Π_Ω ($s < n$), and the hypothesis H is to be tested that $\underline{\xi}$ lies in a given $(s - r)$-dimensional subspace Π_ω of Π_Ω ($r \leqq s$).

Example 1. In the two-sample problem of testing equality of two normal means (considered with a different notation in Chapter 5, Section 3), it is given that $\xi_i = \xi$ for $i = 1, \cdots, n_1$ and $\xi_i = \eta$ for $i = n_1 + 1, \cdots, n_1 + n_2$, and the hypothesis to be tested is $\eta = \xi$. The space Π_Ω is then the space of vectors

$$(\xi, \cdots, \xi, \eta, \cdots, \eta) = \xi(1, \cdots, 1, 0, \cdots, 0) + \eta(0, \cdots, 0, 1, \cdots, 1)$$

spanned by $(1, \cdots, 1, 0, \cdots, 0)$ and $(0, \cdots, 0, 1, \cdots, 1)$, so that $s = 2$. Similarly, Π_ω is the set of all vectors $(\xi, \cdots, \xi) = \xi(1, \cdots, 1)$, and hence $r = 1$.

Another hypothesis that can be tested in this situation is $\eta = \xi = 0$. The space Π_ω is then the origin, $s - r = 0$ and hence $r = 2$. The more general hypothesis $\xi = \xi_0$, $\eta = \eta_0$ is not a linear hypothesis since Π_ω does not contain the origin. However, it reduces to the previous case through the transformation $X_i' = X_i - \xi_0$ $(i = 1, \cdots, n_1)$, $X_i' = X_i - \eta_0$ $(i = n_1 + 1, \cdots, n_1 + n_2)$.

Example 2. The regression problem of Chapter 5, Section 6, is essentially a linear hypothesis. Changing the notation to make it conform with that of the present section, let $\xi_i = \alpha + \beta t_i$, where α, β are unknown, and the t_i known and not all equal. Since Π_Ω is the space of all vectors $\alpha(1, \cdots, 1) + \beta(t_1, \cdots, t_n)$, it has dimension $s = 2$. The hypothesis to be tested may be $\alpha = \beta = 0$ $(r = 2)$ or it may only specify that one of the parameters is zero $(r = 1)$. The more general hypotheses $\alpha = \alpha_0$, $\beta = \beta_0$ can be reduced to the previous case by letting $X_i' = X_i - \alpha_0 - \beta_0 t_i$ since then $E(X_i') = \alpha' + \beta' t_i$ with $\alpha' = \alpha - \alpha_0$, $\beta' = \beta - \beta_0$.

* Throughout this chapter, a fixed coordinate system is assumed given in n-space A vector with components $\xi_1, \cdots, \xi_n$ is denoted by $\underline{\xi}$, and an $n \times 1$ column matrix with elements $\xi_1, \cdots, \xi_n$ by ξ.

Higher polynomial regression and regression in several variables also fall under the linear hypothesis scheme. Thus if $\xi_i = \alpha + \beta t_i + \gamma t_i^2$ or more generally $\xi_i = \alpha + \beta t_i + \gamma u_i$ where the t_i and u_i are known, it can be tested whether one or more of the regression coefficients α, β, γ are zero, and by transforming to the variables $X_i' - \alpha_0 - \beta_0 t_i - \gamma_0 u_i$ also whether these coefficients have specified values other than zero.

In the general case, the hypothesis can be given a simple form by making an orthogonal transformation to variables $Y_1, \cdots, Y_n$

$$Y = CX, \qquad C = (c_{ij}); \qquad i, j = 1, \cdots, n, \tag{1}$$

such that the first s row vectors $\underline{c}_1, \cdots, \underline{c}_s$ of the matrix C span Π_Ω, with $\underline{c}_{r+1}, \cdots, \underline{c}_s$ spanning Π_ω. Then $Y_{s+1} = \cdots Y_n = 0$ if and only if $\underline{X}$ is in Π_Ω, and $Y_1 = \cdots = Y_r = Y_{s+1} = \cdots = Y_n = 0$ if and only if $\underline{X}$ is in Π_ω. Let $\eta_i = E(Y_i)$ so that $\eta = C\xi$. Then since $\underline{\xi}$ lies in Π_Ω a priori and in Π_ω under H, it follows that $\eta_i = 0$ for $i = s + 1, \cdots, n$ in both cases, and $\eta_i = 0$ for $i = 1, \cdots, r$ when H is true. Finally, since the transformation is orthogonal, the variables $Y_1, \cdots, Y_n$ are again independently normally distributed with common variance σ^2, and the problem reduces to the following canonical form.

The variables $Y_1, \cdots, Y_n$ are independently, normally distributed with common variance σ^2 and means $E(Y_i) = \eta_i$ for $i = 1, \cdots, s$ and $E(Y_i) = 0$ for $i = s + 1, \cdots, n$, so that their joint density is

$$\frac{1}{(\sqrt{2\pi}\sigma)^n} \exp\left[-\frac{1}{2\sigma^2}\left(\sum_{i=1}^{s}(y_i - \eta_i)^2 + \sum_{i=s+1}^{n} y_i^2\right)\right]. \tag{2}$$

The η's and σ^2 are unknown, and the hypothesis to be tested is

$$H: \eta_1 = \cdots = \eta_r = 0 \qquad (r \leqq s < n). \tag{3}$$

Example 3. To illustrate the determination of the transformation (1), consider once more the regression model $\xi_i = \alpha + \beta t_i$ of Example 2. It was seen there that Π_Ω is spanned by $(1, \cdots, 1)$ and $(t_1, \cdots, t_n)$. If the hypothesis being tested is $\beta = 0$, Π_ω is the one-dimensional space spanned by the first of these vectors. The row vector $\underline{c}_2$ is in Π_ω and of length 1, and hence $\underline{c}_2 = (1/\sqrt{n}, \cdots, 1/\sqrt{n})$. Since $\underline{c}_1$ is in Π_Ω, of length 1, and orthogonal to $\underline{c}_2$, its coordinates are of the form $a + bt_i$, $i = 1, \cdots, n$, where a and b are determined by the conditions $\Sigma(a + bt_i) = 0$ and $\Sigma(a + bt_i)^2 = 1$. The solutions of these equations are $a = -b\bar{t}$, $b = 1/\sqrt{\Sigma(t_j - \bar{t})^2}$, and therefore $a + bt_i = (t_i - \bar{t})/\sqrt{\Sigma(t_j - \bar{t})^2}$, and

$$Y_1 = \frac{\Sigma X_i(t_i - \bar{t})}{\sqrt{\Sigma(t_j - \bar{t})^2}} = \frac{\Sigma(X_i - \bar{X})(t_i - \bar{t})}{\sqrt{\Sigma(t_j - \bar{t})^2}}.$$

The remaining row vectors of C can be taken to be any set of orthogonal unit vectors which are orthogonal to Π_Ω; it turns out not to be necessary to determine them explicitly.

If the hypothesis to be tested is $\alpha = 0$, Π_ω is spanned by $(t_1, \cdots, t_n)$ so that the ith coordinate of $\underset{\sim}{c}_2$ is $t_i/\sqrt{\Sigma t_j^2}$. The coordinates of $\underset{\sim}{c}_1$ are again of the form $a + bt_i$ with a and b now determined by the equations $\Sigma(a + bt_i)t_i = 0$ and $\Sigma(a + bt_i)^2 = 1$. The solutions are $b = -an\bar{t}/\Sigma t_j^2$, $a = \sqrt{\Sigma t_j^2/n\Sigma(t_j - \bar{t})^2}$ and therefore

$$Y_1 = \sqrt{\frac{n\Sigma t_j^2}{\Sigma(t_j - \bar{t})^2}}\left(\bar{X} - \frac{\bar{t}}{\Sigma t_j^2}\Sigma t_i X_i\right).$$

In the case of the hypothesis $\alpha = \beta = 0$, Π_ω is the origin and $\underset{\sim}{c}_1, \underset{\sim}{c}_2$ can be taken as any two orthogonal unit vectors in Π_Ω. One possible choice is that appropriate to the hypothesis $\beta = 0$, in which case Y_1 is the linear function given there and $Y_2 = \sqrt{n}\bar{X}$.

The general linear hypothesis problem in terms of the Y's remains invariant under the group G_1 of transformations $Y_i' = Y_i + c_i$ for $i = r + 1, \cdots, s$; $Y_i' = Y_i$ for $i = 1, \cdots, r$; $s + 1, \cdots, n$. This leaves $Y_1, \cdots, Y_r$ and $Y_{s+1}, \cdots, Y_n$ as maximal invariants. Another group of transformations leaving the problem invariant is the group G_2 of all orthogonal transformations of $Y_1, \cdots, Y_r$. The middle set of variables having been eliminated, it follows from Chapter 6, Example 1(iii), that a maximal invariant under G_2 is $U = \sum_{i=1}^r Y_i^2$, $Y_{s+1}, \cdots, Y_n$. This can be reduced to U and $V = \sum_{i=s+1}^n Y_i^2$ by sufficiency. Finally, the problem also remains invariant under the group G_3 of scale changes $Y_i' = cY_i$, $c \neq 0$, for $i = 1, \cdots, n$. In the space of U and V this induces the transformation $U^* = c^2U$, $V^* = c^2V$, under which $W = U/V$ is maximal invariant. Thus the principle of invariance reduces the data to the single statistic

$$W = \frac{\sum_{i=1}^{r} Y_i^2}{\sum_{i=s+1}^{n} Y_i^2}. \tag{4}$$

Each of the three transformation groups G_i $(i = 1, 2, 3)$ which lead to the above reduction induces a corresponding group $\bar{G}_i$ in the parameter space. The group $\bar{G}_1$ consists of the translations $\eta_i' = \eta_i + c_i$ $(i = r + 1, \cdots, s)$, $\eta_i' = \eta_i$ $(i = 1, \cdots, r)$, $\sigma' = \sigma$; which leaves $(\eta_1, \cdots, \eta_r, \sigma)$ as maximal invariants. Since any orthogonal transformation of $Y_1, \cdots, Y_r$ induces the same transformation on $\eta_1, \cdots, \eta_r$ and leaves σ^2 unchanged, a maximal invariant under $\bar{G}_2$ is $(\sum_{i=1}^r \eta_i^2, \sigma^2)$. Finally the elements of $\bar{G}_3$ are the transformations $\eta_i' = c\eta_i$, $\sigma' = c\sigma$, and hence a maximal invariant with respect to the totality of these transformations is

$$\psi^2 = \frac{\sum_{i=1}^{r} \eta_i^2}{\sigma^2}. \tag{5}$$

It follows from Theorem 3 of Chapter 6 that the distribution of W depends only on ψ^2, so that the principle of invariance reduces the problem to that of testing the simple hypothesis $H: \psi = 0$. More precisely, the probability density of W is (cf. Problems 2 and 3)

$$p_\psi(w) = e^{-\frac{1}{2}\psi^2} \sum_{k=0}^{\infty} c_k \frac{(\frac{1}{2}\psi^2)^k}{k!} \frac{w^{\frac{1}{2}r-1+k}}{(1+w)^{\frac{1}{2}(r+n-s)+k}} \tag{6}$$

where

$$c_k = \frac{\Gamma[\frac{1}{2}(r+n-s)+k]}{\Gamma(\frac{1}{2}r+k)\Gamma[\frac{1}{2}(n-s)]}.$$

For any ψ_1 the ratio $p_{\psi_1}(w)/p_0(w)$ is an increasing function of w, and it follows from the Neyman–Pearson fundamental lemma that the most powerful invariant test for testing $\psi = 0$ against $\psi = \psi_1$ rejects when W is too large or equivalently when

$$W^* = \frac{\sum_{i=1}^{r} Y_i^2/r}{\sum_{i=s+1}^{n} Y_i^2/(n-s)} > C. \tag{7}$$

The cutoff point C is determined so that the probability of rejection is α when $\psi = 0$. Since in this case W^* is the ratio of two independent χ^2 variables, each divided by the number of its degrees of freedom, the distribution of W^* is the F-distribution with r and $n - s$ degrees of freedom and hence C is determined by

$$\int_C^\infty F_{r,n-s}(y)\,dy = \alpha. \tag{8}$$

The test is independent of ψ_1, and hence is UMP among all invariant tests. By Theorem 5 of Chapter 6, it is also UMP among all tests whose power function depends only on ψ^2.

The rejection region (7) can also be expressed in the form

$$\frac{\sum_{i=1}^{r} Y_i^2}{\sum_{i=1}^{r} Y_i^2 + \sum_{i=s+1}^{n} Y_i^2} > C'. \tag{9}$$

When $\psi = 0$, the left-hand side is distributed according to the beta-distribution with r and $n - s$ degrees of freedom [defined through (24) of Chapter 5], so that C' is determined by

$$\int_{C'}^1 B_{\frac{1}{2}r,\frac{1}{2}(n-s)}(y)\,dy = \alpha. \tag{10}$$

For an alternative value of ψ, the left-hand side of (9) is distributed according to the *noncentral beta*-distribution with noncentrality parameter ψ, the density of which is (Problem 3)

$$g_\psi(y) = e^{-\frac{1}{2}\psi^2} \sum_{k=0}^{\infty} \frac{(\frac{1}{2}\psi^2)^k}{k!} B_{\frac{1}{2}r+k,\frac{1}{2}(n-s)}(y). \tag{11}$$

The power of the test against an alternative ψ is therefore*

$$\beta(\psi) = \int_{C'}^{1} g_\psi(y)\, dy.$$

In the particular case $r = 1$, the rejection region (7) reduces to

$$\frac{|Y_1|}{\sqrt{\sum_{i=s+1}^{n} Y_i^2/(n-s)}} > C_0. \tag{12}$$

This is a two-sided t-test, which by the theory of Chapter 5 (see for example Problem 5 of that chapter) is UMP unbiased. On the other hand, no UMP unbiased test exists for $r > 1$.

2. LINEAR HYPOTHESES AND LEAST SQUARES

In applications to specific problems it is usually not convenient to carry out the reduction to canonical form explicitly. The test statistic W can be expressed in terms of the original variables by noting that $\sum_{i=s+1}^{n} Y_i^2$ is the minimum value of

$$\sum_{i=1}^{s} (Y_i - \eta_i)^2 + \sum_{i=s+1}^{n} Y_i^2 = \sum_{i=1}^{n} [Y_i - E(Y_i)]^2$$

under unrestricted variation of the η's. Also, since the transformation $Y = CX$ is orthogonal and orthogonal transformations leave distances unchanged,

$$\sum_{i=1}^{n} [Y_i - E(Y_i)]^2 = \sum_{i=1}^{n} (X_i - \xi_i)^2.$$

Furthermore, there is a 1 : 1 correspondence between the totality of s-tuples $(\eta_1, \cdots, \eta_s)$ and the totality of vectors $\underline{\xi}$ in Π_Ω. Hence

$$\sum_{i=s+1}^{n} Y_i^2 = \sum_{i=1}^{n} (X_i - \hat{\xi}_i)^2 \tag{13}$$

* A set of charts for the power is given by Pearson and Hartley, "Charts of the power function for analysis of variance tests, derived from the noncentral F distribution," *Biometrika*, Vol. 38 (1951), pp. 112–130, and by Fox, "Charts of the power of the F-test," *Ann. Math. Stat.*, Vol. 27 (1956), pp. 484–497. A computing formula for the noncentral beta-distribution is discussed by Hodges, "On the noncentral beta distribution," *Ann. Math. Stat.*, Vol. 26 (1955), pp. 648–653.

where the $\hat{\xi}$'s are the least squares estimates of the ξ's under Ω, that is, the values that minimize $\sum_{i=1}^n (X_i - \xi_i)^2$ subject to $\underline{\xi}$ in Π_Ω.

In the same way it is seen that

$$\sum_{i=1}^{r} Y_i^2 + \sum_{i=s+1}^{n} Y_i^2 = \sum_{i=1}^{n} (X_i - \hat{\hat{\xi}}_i)^2$$

where the $\hat{\hat{\xi}}$'s are the values that minimize $\Sigma(X_i - \xi_i)^2$ subject to $\underline{\xi}$ in Π_ω. The test (7) therefore becomes

$$W^* = \frac{\left[\sum_{i=1}^{n}(X_i - \hat{\hat{\xi}}_i)^2 - \sum_{i=1}^{n}(X_i - \hat{\xi}_i)^2\right]\Big/ r}{\sum_{i=1}^{n}(X_i - \hat{\xi}_i)^2/(n-s)} > C, \tag{14}$$

where C is determined by (8). Geometrically the vectors $\underline{\hat{\xi}}$ and $\underline{\hat{\hat{\xi}}}$ are the projections of $\underline{X}$ on Π_Ω and Π_ω, so that the triangle formed by $\underline{X}$, $\underline{\hat{\xi}}$, and

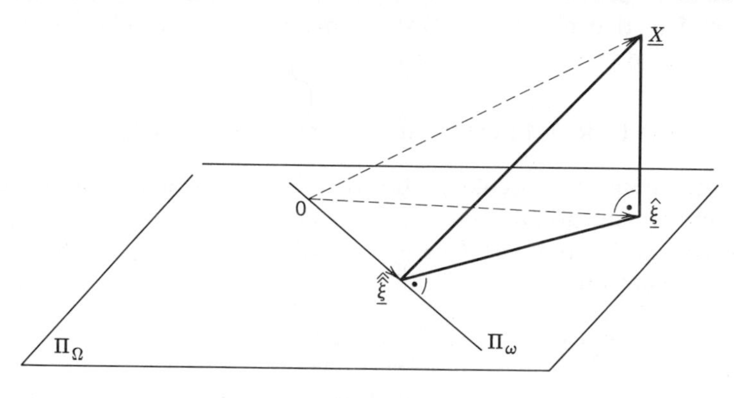

Figure 1.

$\underline{\hat{\hat{\xi}}}$ has a right angle at $\underline{\hat{\xi}}$. (Figure 1.) Thus the denominator and numerator of W^*, except for the factors $1/(n-s)$ and $1/r$, are the squares of the distances between $\underline{X}$ and $\underline{\hat{\xi}}$ and between $\underline{\hat{\xi}}$ and $\underline{\hat{\hat{\xi}}}$ respectively. An alternative expression for W^* is therefore

$$W^* = \frac{\sum_{i=1}^{n}(\hat{\xi}_i - \hat{\hat{\xi}}_i)^2/r}{\sum_{i=1}^{n}(X_i - \hat{\xi}_i)^2/(n-s)}. \tag{15}$$

It is desirable to express also the *noncentrality parameter* $\psi^2 = \sum_{i=1}^r \eta_i^2/\sigma^2$ in terms of the ξ's. Now $X = C^{-1}Y$, $\xi = C^{-1}\eta$, and

$$\sum_{i=1}^{r} Y_i^2 = \sum_{i=1}^{n} (X_i - \hat{\hat{\xi}}_i)^2 - \sum_{i=1}^{n} (X_i - \hat{\xi}_i)^2. \tag{16}$$

If the right-hand side of (16) is denoted by $f(X)$, it follows that $\sum_{i=1}^r \eta_i^2 = f(\xi)$

A slight generalization of a linear hypothesis is the inhomogeneous hypothesis which specifies for the vector of means $\underline{\xi}$ a subhyperplane Π'_ω of Π_Ω not passing through the origin. Let Π_ω denote the subspace of Π_Ω which passes through the origin and is parallel to Π'_ω. If $\underline{\xi}^0$ is any point of Π'_ω, the set Π'_ω consists of the totality of points $\underline{\xi} = \underline{\xi}^* + \underline{\xi}^0$ as $\underline{\xi}^*$ ranges over Π_ω. Applying the transformation (1) with respect to Π_ω, the vector of means $\underline{\eta}$ for $\underline{\xi} \in \Pi'_\omega$ is then given by $\eta = C\xi = C\xi^* + C\xi^0$ in the canonical form (2), and the totality of these vectors is therefore characterized by the equations $\eta_1 = \eta_1^0, \cdots, \eta_r = \eta_r^0, \eta_{s+1} = \cdots = \eta_n = 0$ where η_i^0 is the ith coordinate of $C\xi^0$. In the canonical form, the inhomogeneous hypothesis $\underline{\xi} \in \Pi'_\omega$ therefore becomes $\eta_i = \eta_i^0$ $(i = 1, \cdots, r)$. This reduces to the homogeneous case by replacing Y_i by $Y_i - \eta_i^0$, and it follows from (7) that the UMP invariant test has the rejection region

$$\frac{\sum_{i=1}^r (Y_i - \eta_i^0)^2/r}{\sum_{i=s+1}^n Y_i^2/(n-s)} > C, \tag{17}$$

and that the noncentrality parameter is $\psi^2 = \sum_{i=1}^r (\eta_i - \eta_i^0)^2/\sigma^2$.

In applications it is usually most convenient to apply the transformation $X_i - \xi_i^0$ directly to (14) or (15). It follows from (17) that such a transformation always leaves the denominator unchanged. This can also be seen geometrically since the transformation is a translation of n-space parallel to Π_Ω and therefore leaves the distance $\Sigma(X_i - \hat{\xi}_i)^2$ from $\underline{X}$ to Π_Ω unchanged. The noncentrality parameter can be computed as before by replacing X by ξ in the transformed numerator (16).

Some examples of linear hypotheses, all with $r = 1$, were already discussed in Chapter 5. The following treats two of these from the present point of view.

Example 4. Let $X_1, \cdots, X_n$ be independently, normally distributed with common mean μ and variance σ^2, and consider the hypothesis $H: \mu = 0$. Here Π_Ω is the line $\xi_1 = \cdots = \xi_n$, Π_ω is the origin, and s and r are both equal to 1. From the identity

$$\Sigma(X_i - \mu)^2 = \Sigma(X_i - \bar{X})^2 + n(\bar{X} - \mu)^2, \qquad (\bar{X} = \Sigma X_i/n)$$

it is seen that $\hat{\xi}_i = \bar{X}$, while $\hat{\hat{\xi}}_i = 0$. The test statistic and ψ^2 are therefore given by

$$W = n\bar{X}^2/\Sigma(X_i - \bar{X})^2 \qquad \text{and} \qquad \psi^2 = n\mu^2/\sigma^2.$$

Under the hypothesis, the distribution of $(n - 1)W$ is that of the square of a variable having Student's t-distribution with $n - 1$ degrees of freedom.

Example 5. In the two-sample problem considered in Example 1, the sum of squares

$$\sum_{i=1}^{n_1}(X_i - \xi)^2 + \sum_{i=n_1+1}^{n}(X_i - \eta)^2$$

is minimized by

$$\hat{\xi} = X_{\cdot}^{(1)} = \sum_{i=1}^{n_1} X_i/n_1, \qquad \hat{\eta} = X_{\cdot}^{(2)} = \sum_{i=n_1+1}^{n} X_i/n_2,$$

while under the hypothesis $\eta - \xi = 0$

$$\hat{\hat{\xi}} = \hat{\hat{\eta}} = \bar{X} = [n_1 X_{\cdot}^{(1)} + n_2 X_{\cdot}^{(2)}]/n.$$

The numerator of the test statistic (15) is therefore

$$n_1(X_{\cdot}^{(1)} - \bar{X})^2 + n_2(X_{\cdot}^{(2)} - \bar{X})^2 = \frac{n_1 n_2}{n_1 + n_2}[X_{\cdot}^{(2)} - X_{\cdot}^{(1)}]^2.$$

The more general hypothesis $\eta - \xi = \theta_0$ reduces to the previous case by replacing X_i by $X_i - \theta_0$ for $i = n_1 + 1, \cdots, n$, and is therefore rejected when

$$\frac{(X_{\cdot}^{(2)} - X_{\cdot}^{(1)} - \theta_0)^2 \Big/ \left(\dfrac{1}{n_1} + \dfrac{1}{n_2}\right)}{\left[\sum_{i=1}^{n_1}(X_i - X_{\cdot}^{(1)})^2 + \sum_{i=n_1+1}^{n}(X_i - X_{\cdot}^{(2)})^2\right] \Big/ (n_1 + n_2 - 2)} > C.$$

The noncentrality parameter is $\psi^2 = (\eta - \xi - \theta_0)^2 \Big/ \left(\dfrac{1}{n_1} + \dfrac{1}{n_2}\right)\sigma^2$. Under the hypothesis, the square root of the test statistic has the t-distribution with $n_1 + n_2 - 2$ degrees of freedom.

3. TESTS OF HOMOGENEITY

The UMP invariant test obtained in the preceding section for testing the equality of the means of two normal distributions with common variance is also UMP unbiased (Section 3 of Chapter 5). However, when a number of populations greater than 2 is to be tested for homogeneity of means, a UMP unbiased test no longer exists so that invariance considerations lead to a new result. Let X_{ij} $(j = 1, \cdots, n_i;\ i = 1, \cdots, s)$ be independently distributed as $N(\mu_i, \sigma^2)$ and consider the hypothesis

$$H: \mu_1 = \cdots = \mu_s.$$

This arises, for example, in the comparison of a number of different treatments, processes, varieties, locations, etc., when one wishes to test whether these differences have any effect on the outcome X. It may arise more generally in any situation involving a *one-way classification* of the outcomes, that is, in which the outcomes are classified according to a single factor.

The hypothesis H is a linear hypothesis with $r = s - 1$, with Π_Ω given by the equations $\xi_{ij} = \xi_{ik}$ for $j, k = 1, \cdots, n$; $i = 1, \cdots, s$ and with Π_ω the line on which all $n = \Sigma n_i$ coordinates ξ_{ij} are equal. We have

$$\Sigma\Sigma(X_{ij} - \mu_i)^2 = \Sigma\Sigma(X_{ij} - X_{i\cdot})^2 + \Sigma n_i(X_{i\cdot} - \mu_i)^2$$

with $X_{i\cdot} = \sum_{j=1}^{n_i} X_{ij}/n_i$, and hence $\hat{\xi}_{ij} = X_{i\cdot}$. Also,

$$\Sigma\Sigma(X_{ij} - \mu)^2 = \Sigma\Sigma(X_{ij} - X_{\cdot\cdot})^2 + n(X_{\cdot\cdot} - \mu)^2$$

with $X_{\cdot\cdot} = \Sigma\Sigma X_{ij}/n$ so that $\hat{\hat{\xi}}_{ij} = X_{\cdot\cdot}$. Using the form (15) of W^*, the test therefore becomes

$$W^* = \frac{\Sigma n_i(X_{i\cdot} - X_{\cdot\cdot})^2/(s-1)}{\Sigma\Sigma(X_{ij} - X_{i\cdot})^2/(n-s)} > C. \tag{18}$$

The noncentrality parameter is

$$\psi^2 = \Sigma n_i(\mu_i - \mu_\cdot)^2/\sigma^2$$

with

$$\mu_\cdot = \Sigma n_i \mu_i/n.$$

The sum of squares in both numerator and denominator of (18) admits three interpretations, which are closely related: (i) as the two components in the decomposition of the total variation

$$\Sigma\Sigma(X_{ij} - X_{\cdot\cdot})^2 = \Sigma\Sigma(X_{ij} - X_{i\cdot})^2 + \Sigma n_i(X_{i\cdot} - X_{\cdot\cdot})^2,$$

of which the first represents the variation within, and the second the variation between populations; (ii) as a basis, through the test (18), for comparing these two sources of variation; (iii) as estimates of their expected values, $(n - s)\sigma^2$ and $(s - 1)\sigma^2 + \Sigma n_i(\mu_i - \mu_\cdot)^2$ (Problem 9). This breakdown of the total variation, together with the various interpretations of the components, is an example of an *analysis of variance*, which will be applied to more complex problems in the succeeding sections.

We shall now digress for a moment from the linear hypothesis scheme to consider the hypothesis of equality of variances when the variables X_{ij} are distributed as $N(\mu_i, \sigma_i^2)$, $i = 1, \cdots, s$. A UMP unbiased test of this hypothesis was obtained in Chapter 5, Section 3, for the case $s = 2$, but does not exist for $s > 2$ (see, for example, Problem 6 of Chapter 4). Unfortunately, neither is there available for this problem a group for which there exists a UMP invariant test. To obtain a test, we shall now give a large-sample approximation, which for sufficiently large n essentially reduces the problem to that of testing the equality of s means.

It is convenient first to reduce the observations to the set of sufficient statistics $X_{i\cdot} = \sum_j X_{ij}/n_i$ and $S_i^2 = \sum_j (X_{ij} - X_{i\cdot})^2$, $i = 1, \cdots, s$. The hypothesis

$$H: \sigma_1 = \cdots = \sigma_s$$

remains invariant under the transformations $X'_{ij} = X_{ij} + c_i$, which in the space of sufficient statistics induce the transformations $S_i'^2 = S_i^2$, $X_{i\cdot}' = X_{i\cdot} + c_i\pi$. A set of maximal invariants under this group are $S_1^2, \cdots, S_s^2$. Each statistic S_i^2 is the sum of squares of $n_i - 1$ independent normal variables with zero mean and variance σ_i^2, and it follows from the central limit theorem that for large n_i

$$\sqrt{n_i - 1}\left(\frac{S_i^2}{n_i - 1} - \sigma_i^2\right)$$

is approximately distributed as $N(0, 2\sigma_i^4)$. This approximation is inconvenient for the present purpose since the unknown parameters σ_i enter not only into the mean but also the variance of the limiting distribution.

The difficulty can be avoided through the use of a suitable *variance stabilizing* transformation. Such transformations can be obtained by the following observation.* *If T_n is a sequence of real-valued statistics such that $\sqrt{n}(T_n - \theta)$ has the limiting distribution $N(0, \tau^2)$, then for any continuously differentiable function f, the limiting distribution of $\sqrt{n}[f(T_n) - f(\theta)]$ is normal with zero mean and variance $\tau^2(df/d\theta)^2$.* The variance of this limiting distribution is therefore independent of θ provided the derivative of $f(\theta)$ is proportional to $1/\tau(\theta)$.

This applies to the present case with $n = n_i - 1$, $T_n = S_i^2/(n_i - 1)$, $\theta = \sigma_i^2$, and $\tau^2 = 2\theta^2$, and leads to the transformation $f(\theta) = \log \theta$ for which the derivative is proportional to $1/\theta$. The limiting distribution of $\sqrt{n_i - 1}\,\{\log [S_i^2/(n_i - 1)] - \log \sigma_i^2\}$ is the normal distribution with zero mean and variance 2, so that for large n_i the variable $Z_i = \log [S_i^2/(n_i - 1)]$ has the approximate distribution $N(\zeta_i, a_i^2)$ with $\zeta_i = \log \sigma_i^2$, $a_i^2 = 2/(n_i - 1)$.

The problem is now reduced to that of testing the equality of means of s independent variables Z_i distributed as $N(\zeta_i, a_i^2)$ where the a_i are known. In the particular case that the n_i are equal, the variances a_i^2 are equal and the asymptotic problem is a simpler version (in that the variance is known) of the problem considered at the beginning of the section. The hypothesis $\zeta_1 = \cdots = \zeta_s$ is invariant under addition of a common constant to each of the Z's and under orthogonal transformations of the hyperplanes which are perpendicular to the line $Z_1 = \cdots = Z_s$. The UMP invariant rejection region is then

$$\Sigma(Z_i - \bar{Z})^2/a^2 > C \tag{19}$$

where a^2 is the common variance of the Z_i and where C is determined by

$$\int_C^\infty \chi_{s-1}^2(y)\,dy = \alpha. \tag{20}$$

* For a proof see for example Rao, *Advanced Statistical Methods in Biometric Research*, New York, John Wiley & Sons, 1952, Section 5e.

In the more general case of unequal a_i, the problem reduces to a linear hypothesis with known variance through the transformation $Z_i' = Z_i/a_i$, and the UMP invariant test under a suitable group of linear transformations rejects when

$$\Sigma\frac{1}{a_i^2}\left(Z_i - \frac{\Sigma Z_j/a_j^2}{\Sigma 1/a_j^2}\right)^2 = \Sigma\left(\frac{Z_i}{a_i}\right)^2 - \frac{(\Sigma Z_j/a_j^2)^2}{\Sigma(1/a_j^2)} > C \tag{21}$$

(see Problem 10) where C is again determined by (20). This rejection region, which is UMP invariant for testing $\zeta_1 = \cdots = \zeta_s$ in the limiting distribution, can then be said to have this property asymptotically for testing the original hypothesis H: $\sigma_1 = \cdots = \sigma_s$.*

The same method can be used to test the homogeneity of a number of binomial or Poisson distributions. The details are indicated in Problem 11.

When applying the principle of invariance, it is important to make sure that the underlying symmetry assumptions really are satisfied. In the problem of testing the equality of a number of normal means $\mu_1, \cdots, \mu_s$ for example, all parameter points, which have the same value of $\psi^2 = \Sigma n_i(\mu_i - \mu.)^2/\sigma^2$, are identified under the principle of invariance. This is appropriate only when these alternatives can be considered as being equidistant from the hypothesis. In particular, it should then be immaterial whether the given value of ψ^2 is built up by a number of small contributions or a single large one. Situations where instead the main emphasis is on the detection of large individual deviations do not possess the required symmetry, and the test based on (18) need no longer be optimum.

Usually in such situations a more complex procedure is called for than the testing of a single hypothesis. When comparing a number of varieties or treatments for example, one would typically wish to decide not only whether they are equal but in case this hypothesis is rejected would like to rank or group them or at least pick out those that are best. Suppose for simplicity that the sample sizes are equal, $n_i = n$ for $i = 1, \cdots, s$. A natural procedure which leads to a grouping of the values μ_i consists in claiming μ_i and μ_j to be different if $|X_{j\cdot} - X_{i\cdot}| > CS/\sqrt{sn(n-1)}$ where $S^2 = \Sigma\Sigma(X_{kl} - X_{k\cdot})^2$.† The over-all hypothesis H of equality

* A more commonly used asymptotic test of H is Bartlett's test (see for example Section 6a of Rao, *op. cit.*), which is essentially the likelihood ratio test.

† Other types of multiple decision procedures for this and more general linear hypothesis situations have been proposed by, among others, Duncan, "Multiple range and multiple F tests," *Biometrics*, Vol. 11 (1955), pp. 1–42; Scheffé, "A method for judging all contrasts in the analysis of variance," *Biometrika*, Vol. 40 (1953), pp. 87–104; and Tukey, in "Comparing individual means in the analysis of variance," *Biometrics*, Vol. 5 (1949), pp. 99–114, and in an unpublished work.

of all means is then accepted if

$$(22)\qquad \frac{\max |X_{j\cdot} - X_{i\cdot}|}{S/\sqrt{sn(n-1)}} \leqq C.*$$

When H is rejected, it is asserted that $\mu_j > \mu_i$ for all pairs (i, j) for which $X_{j\cdot} - X_{i\cdot} > CS/\sqrt{sn(n-1)}$. The significance level at which H is tested here is the probability of declaring any of the differences $\mu_j - \mu_i$ significant when actually the μ's are all equal. The left-hand side of (22) is the *studentized range* of the sample means.

An analogous approach is possible in the comparison of several variances. Suppose again that the sample sizes are equal and let the variances σ_i^2 and σ_j^2 be classified as $\sigma_i^2 < \sigma_j^2$ or $\sigma_i^2 > \sigma_j^2$ if S_j^2/S_i^2 is $> C$ or $< 1/C$ $(C > 1)$ and as being equal if neither of these inequalities holds. The over-all hypothesis $\sigma_1 = \cdots = \sigma_s$ is then accepted if

$$(23)\qquad \max_{i,j} (S_j^2/S_i^2) = \max_k S_k^2/\min_k S_k^2 \leqq C.\dagger$$

The studentized range and maximum F-ratio tests do not appear to possess any optimum properties when viewed as tests of the hypotheses $\mu_1 = \cdots = \mu_s$ and $\sigma_1 = \cdots = \sigma_s$ respectively. However, they do possess such properties when considered as solutions to the problem of ranking the means or variances (ties being permitted).‡

4. TWO-WAY CLASSIFICATION: ONE OBSERVATION PER CELL

The hypothesis of equality of several means arises when a number of different treatments, procedures, varieties, or manifestations of some other factors are to be compared. Frequently one is interested in studying the effects of more than one factor, or the effects of one factor as certain other conditions of the experiment vary, which then play the role of additional factors. In the present section we shall consider the case that the number of factors affecting the outcomes of the experiment is two.

Suppose that one observation is obtained at each of a number of levels of these factors, and denote by X_{ij} $(i = 1, \cdots, a; j = 1, \cdots, b)$ the value observed when the first factor is at the ith and the second at the jth level. It is assumed that the X_{ij} are independently normally distributed with

* Tables of C are given in *Biometrika Tables*, Vol. 1, Cambridge Univ. Press, 1954, Table 29, and by May, "Extended and corrected tables of the upper percentage points of the studentized range," *Biometrika*, Vol. 39 (1952), pp. 192–193.

† Tables of C are given in *Biometrika Tables*, *op. cit.*, Table 31.

‡ Lehmann, "A theory of some multiple decision problems," *Ann. Math. Stat.*, Vol. 28 (1957), pp. 1–25 and 547–572.

constant variance σ^2, and for the moment also that the two factors act independently (they are then said to be *additive*) so that ξ_{ij} is of the form $\alpha'_i + \beta'_j$. Putting $\mu = \alpha'_. + \beta'_.$ and $\alpha_i = \alpha'_i - \alpha'_.$, $\beta_j = \beta'_j - \beta'_.$, this can be written as

$$\xi_{ij} = \mu + \alpha_i + \beta_j, \qquad \Sigma\alpha_i = \Sigma\beta_j = 0, \tag{24}$$

where the α's, β's, and μ are uniquely determined by (24) as

$$\alpha_i = \xi_{i.} - \xi_{..}, \qquad \beta_j = \xi_{.j} - \xi_{..}, \qquad \mu = \xi_{..}.^* \tag{25}$$

Consider the hypothesis

$$H: \alpha_1 = \cdots = \alpha_a = 0$$

that the first factor has no effect on the outcome being observed. This arises in two quite different contexts. The factor of interest, corresponding say to a number of treatments, may be β while α corresponds to a classification according to, for example, the site on which the observations are obtained (farm, laboratory, city, etc.). The hypothesis then represents the possibility that this subsidiary classification has no effect on the experiment so that it need not be controlled. Alternatively, α may be the (or a) factor of primary interest. In this case, the formulation of the problem as one of hypothesis testing would usually be an oversimplification since in case of rejection of H, one would require estimates of the α's or at least a grouping according to high and low values.

The hypothesis H is a linear hypothesis with $r = a - 1$, $s = 1 + (a - 1) + (b - 1) = a + b - 1$ and $n - s = (a - 1)(b - 1)$. The least squares estimates of the parameters under Ω can be obtained from the identity

$$\begin{aligned}\Sigma\Sigma(X_{ij} - \xi_{ij})^2 &= \Sigma\Sigma(X_{ij} - \mu - \alpha_i - \beta_j)^2 \\ &= \Sigma\Sigma[(X_{ij} - X_{i.} - X_{.j} + X_{..}) + (X_{i.} - X_{..} - \alpha_i) \\ &\quad + (X_{.j} - X_{..} - \beta_j) + (X_{..} - \mu)]^2 \\ &= \Sigma\Sigma(X_{ij} - X_{i.} - X_{.j} + X_{..})^2 + b\Sigma(X_{i.} - X_{..} - \alpha_i)^2 \\ &\quad + a\Sigma(X_{.j} - X_{..} - \beta_j)^2 + ab(X_{..} - \mu)^2,\end{aligned}$$

which is valid since in the expansion of the third sum of squares the cross-product terms vanish. It follows that

$$\hat{\alpha}_i = X_{i.} - X_{..}, \qquad \hat{\beta}_j = X_{.j} - X_{..}, \qquad \hat{\mu} = X_{..},$$

and that

$$\Sigma\Sigma(X_{ij} - \hat{\xi}_{ij})^2 = \Sigma\Sigma(X_{ij} - X_{i.} - X_{.j} + X_{..})^2.$$

* The replacing of a subscript by a dot indicates that the variable has been averaged with respect to that subscript.

Under the hypothesis H we still have $\hat{\hat{\beta}}_j = X_{\cdot j} - X_{\cdot\cdot}$ and $\hat{\hat{\mu}} = X_{\cdot\cdot}$, and hence $\hat{\xi}_{ij} - \hat{\hat{\xi}}_{ij} = X_{i\cdot} - X_{\cdot\cdot}$. The best invariant test therefore rejects when

$$(26) \qquad W^* = \frac{b\Sigma(X_{i\cdot} - X_{\cdot\cdot})^2/(a-1)}{\Sigma\Sigma(X_{ij} - X_{i\cdot} - X_{\cdot j} + X_{\cdot\cdot})^2/(a-1)(b-1)} > C.$$

The noncentrality parameter, on which the power of the test depends, is given by

$$(27) \qquad \psi^2 = b\Sigma(\xi_{i\cdot} - \xi_{\cdot\cdot})^2/\sigma^2 = b\Sigma\alpha_i^2/\sigma^2.$$

This problem provides another example of an analysis of variance. The total variation can be broken into three components,

$$\Sigma\Sigma(X_{ij} - X_{\cdot\cdot})^2 = b\Sigma(X_{i\cdot} - X_{\cdot\cdot})^2 + a\Sigma(X_{\cdot j} - X_{\cdot\cdot})^2 + \Sigma\Sigma(X_{ij} - X_{i\cdot} - X_{\cdot j} + X_{\cdot\cdot})^2.$$

Of these, the first contains the variation due to the α's, the second that due to the β's. The last component, in the canonical form of Section 1, is equal to $\sum_{i=s+1}^{n} Y_i^2$. It is therefore the sum of squares of those variables whose means are zero even under Ω. Since this residual part of the variation, which on division by $n - s$ is an estimate of σ^2, cannot be put down to any effects such as the α's or β's, it is frequently labeled "error," as an indication that it is due solely to the randomness of the observations, not to any differences of the means. Actually, the breakdown is not quite as sharp as is suggested by the above description. Any component such as that attributed to the α's always also contains some "error," as is seen for example from its expectation, which is

$$E\Sigma(X_{i\cdot} - X_{\cdot\cdot})^2 = (a-1)\sigma^2 + b\Sigma\alpha_i^2.$$

Instead of testing whether a certain factor has any effect, one may wish to estimate the size of the effect at the various levels of the factor. Other parameters, which it is sometimes interesting to estimate, are the average outcomes (for example yields) $\xi_{1\cdot}, \cdots, \xi_{a\cdot}$ when the factor is at the various levels. If $\theta_i = \mu + \alpha_i = \xi_{i\cdot}$, confidence sets for $(\theta_1, \cdots, \theta_a)$ are obtained by considering the hypotheses $H(\theta^0)$: $\theta_i = \theta_i^0$ $(i = 1, \cdots, a)$. For testing $\theta_1 = \cdots = \theta_a = 0$, the least squares estimates of the ξ_{ij} are $\hat{\xi}_{ij} = X_{i\cdot} + X_{\cdot j} - X_{\cdot\cdot}$ and $\hat{\hat{\xi}}_{ij} = X_{\cdot j} - X_{\cdot\cdot}$. The denominator sum of squares is therefore $\Sigma\Sigma(X_{ij} - X_{i\cdot} - X_{\cdot j} + X_{\cdot\cdot})^2$ as before, while the numerator sum of squares is

$$\Sigma\Sigma(\hat{\xi}_{ij} - \hat{\hat{\xi}}_{ij})^2 = b\Sigma X_{i\cdot}^2.$$

The general hypothesis reduces to this special case by replacing X_{ij} by the variable $X_{ij} - \theta_i^0$. Since $s = a + b - 1$ and $r = a$, the hypothesis $H(\theta^0)$ is rejected when

$$\frac{b\Sigma(X_{i\cdot} - \theta_i^0)^2/a}{\Sigma\Sigma(X_{ij} - X_{i\cdot} - X_{\cdot j} + X_{\cdot\cdot})^2/(a-1)(b-1)} > C.$$

The associated confidence sets for $(\theta_1, \cdots, \theta_a)$ are the spheres

$$\Sigma(\theta_i - X_{i\cdot})^2 \leqq aC\Sigma\Sigma(X_{ij} - X_{i\cdot} - X_{\cdot j} + X_{\cdot\cdot})^2/(a-1)(b-1)b.$$

When considering confidence sets for the effects $\alpha_1, \cdots, \alpha_a$ one must take account of the fact that the α's are not independent. Since they add up to zero, it would be enough to restrict attention to $\alpha_1, \cdots, \alpha_{a-1}$. However, an easier and more symmetric solution is found by retaining all the α's. The rejection region of $H: \alpha_i = \alpha_i^0$ for $i = 1, \cdots, a$ (with $\Sigma\alpha_i^0 = 0$) is obtained from (26) by letting $X'_{ij} = X_{ij} - \alpha_i^0$, and hence is given by

$$b\Sigma(X_{i\cdot} - X_{\cdot\cdot} - \alpha_i^0)^2 > C\Sigma\Sigma(X_{ij} - X_{i\cdot} - X_{\cdot j} + X_{\cdot\cdot})^2/(b-1).$$

The associated confidence set consists of the totality of points $(\alpha_1, \cdots, \alpha_a)$ satisfying $\Sigma\alpha_i = 0$ and

$$\Sigma[\alpha_i - (X_{i\cdot} - X_{\cdot\cdot})]^2 \leqq C\Sigma\Sigma(X_{ij} - X_{i\cdot} - X_{\cdot j} + X_{\cdot\cdot})^2/b(b-1).$$

In the space of $(\alpha_1, \cdots, \alpha_a)$, this inequality defines a sphere whose center $(X_{1\cdot} - X_{\cdot\cdot}, \cdots, X_{a\cdot} - X_{\cdot\cdot})$ lies on the hyperplane $\Sigma\alpha_i = 0$. The confidence sets for the α's therefore consist of the interior and surface of the great hyperspheres obtained by cutting the a-dimensional spheres with the hyperplane $\Sigma\alpha_i = 0$.

In both this and the previous case, the usual method shows the class of confidence sets to be invariant under the appropriate group of linear transformations, and the sets are therefore uniformly most accurate invariant.

5. TWO-WAY CLASSIFICATION: m OBSERVATIONS PER CELL

In the preceding section it was assumed that the effects of the two factors α and β are independent and hence additive. The factors may, however, interact in the sense that the effect of one depends on the level of the other. Thus the effectiveness of a teacher depends for example on the quality or the age of the students, and the benefit derived by a crop from various amounts of irrigation depends on the type of soil as well as on the variety being planted. If the additivity assumption is dropped, the means ξ_{ij} of X_{ij} are no longer given by (24) under Ω but are completely arbitrary. More than ab observations, one for each combination of levels, are then required since otherwise $s = n$. We shall here consider only the simple case in which the number of observations is the same at each combination of levels.

Let X_{ijk} $(i = 1, \cdots, a;\ j = 1, \cdots, b;\ k = 1, \cdots, m)$ be independent normal with common variance σ^2 and mean $E(X_{ijk}) = \xi_{ij}$. In analogy with the previous notation we write

$$\xi_{ij} = \xi_{..} + (\xi_{i.} - \xi_{..}) + (\xi_{.j} - \xi_{..}) + (\xi_{ij} - \xi_{i.} - \xi_{.j} + \xi_{..})$$
$$= \mu + \alpha_i + \beta_j + \gamma_{ij}$$

with $\Sigma_i \alpha_i = \Sigma_j \beta_j = \Sigma_i \gamma_{ij} = \Sigma_j \gamma_{ij} = 0$. Then α_i is the average effect of factor 1 at level i, averaged over the b levels of factor 2, and a similar interpretation holds for the β's. The γ's are called *interactions*, since γ_{ij} measures the extent to which the joint effect $\xi_{ij} - \xi_{..}$ of factors 1 and 2 at levels i and j exceeds the sum $(\xi_{i.} - \xi_{..}) + (\xi_{.j} - \xi_{..})$ of the individual effects. Consider again the hypothesis that the α's are zero. Then $r = a - 1$, $s = ab$, and $n - s = (m - 1)ab$. From the decomposition

$$\Sigma\Sigma\Sigma(X_{ijk} - \xi_{ij})^2 = \Sigma\Sigma\Sigma(X_{ijk} - X_{ij.})^2 + m\Sigma\Sigma(X_{ij.} - \xi_{ij})^2$$

and

$$\Sigma\Sigma(X_{ij.} - \xi_{ij})^2 = \Sigma\Sigma(X_{ij.} - X_{i..} - X_{.j.} + X_{...} - \gamma_{ij})^2$$
$$+ b\Sigma(X_{i..} - X_{...} - \alpha_i)^2 + a\Sigma(X_{.j.} - X_{...} - \beta_j)^2 + ab(X_{...} - \mu)^2$$

it follows that

$$\hat{\mu} = \hat{\hat{\mu}} = \hat{\xi}_{..} = X_{...}, \qquad \hat{\alpha}_i = \hat{\xi}_{i.} - \hat{\xi}_{..} = X_{i..} - X_{...},$$

$$\hat{\beta}_j = \hat{\hat{\beta}}_j = \hat{\xi}_{.j} - \hat{\xi}_{..} = X_{.j.} - X_{...}, \quad \hat{\gamma}_{ij} = \hat{\hat{\gamma}}_{ij} = X_{ij.} - X_{i..} - X_{.j.} + X_{...},$$

and hence that

$$\Sigma\Sigma\Sigma(X_{ijk} - \hat{\xi}_{ij})^2 = \Sigma\Sigma\Sigma(X_{ijk} - X_{ij.})^2,$$
$$\Sigma\Sigma\Sigma(\hat{\xi}_{ij} - \hat{\hat{\xi}}_{ij})^2 = mb\Sigma(X_{i..} - X_{...})^2.$$

The most powerful invariant test therefore rejects when

$$W^* = \frac{mb\Sigma(X_{i..} - X_{...})^2/(a - 1)}{\Sigma\Sigma\Sigma(X_{ijk} - X_{ij.})^2/(m - 1)ab} > C, \tag{28}$$

and the noncentrality parameter in the distribution of W^* is

$$mb\Sigma(\xi_{i.} - \xi_{..})^2/\sigma^2 = mb\Sigma\alpha_i^2/\sigma^2. \tag{29}$$

Another hypothesis of interest is the hypothesis H' that the two factors are independent,†

$$H': \gamma_{ij} = 0 \quad \text{for all} \quad i, j.$$

† A test of H' against certain restricted alternatives has been proposed for the case of one observation per cell by Tukey, "One degree of freedom for non-additivity," *Biometrics*, Vol. 5 (1949), pp. 232–242.

The least squares estimates of the parameters are easily derived as before, and the UMP invariant test is seen to have the rejection region (Problem 12)

$$W^* = \frac{m\Sigma\Sigma(X_{ij\cdot} - X_{i\cdot\cdot} - X_{\cdot j\cdot} + X_{\cdots})^2/(a-1)(b-1)}{\Sigma\Sigma\Sigma(X_{ijk} - X_{ij\cdot})^2/(m-1)ab} > C. \tag{30}$$

Under H', the statistic W^* has the F-distribution with $(a-1)(b-1)$ and $(m-1)ab$ degrees of freedom; the noncentrality parameter for any alternative set of γ's is

$$\psi^2 = m\Sigma\Sigma\gamma_{ij}^2/\sigma^2. \tag{31}$$

The decomposition of the total variation into its various components, in the present case is given by

$$\Sigma\Sigma\Sigma(X_{ijk} - X_{\cdots})^2 = mb\Sigma(X_{i\cdot\cdot} - X_{\cdots})^2 + ma\Sigma(X_{\cdot j\cdot} - X_{\cdots})^2$$
$$+ m\Sigma\Sigma(X_{ij\cdot} - X_{i\cdot\cdot} - X_{\cdot j\cdot} + X_{\cdots})^2 + \Sigma\Sigma\Sigma(X_{ijk} - X_{ij\cdot})^2.$$

Here the first three terms contain the variation due to the α's, β's, and γ's respectively, and the last component corresponds to error. The tests for the hypotheses that the α's, β's, or γ's are zero, the first and third of which have the rejection regions (28) and (30), are then obtained by comparing the α, β, or γ sum of squares with that for error.

An analogous decomposition is possible when the γ's are assumed a priori to be equal to zero. In that case, the third component which previously was associated with γ represents an additional contribution to error, and the breakdown becomes

$$\Sigma\Sigma\Sigma(X_{ijk} - X_{\cdots})^2 = mb\Sigma(X_{i\cdot\cdot} - X_{\cdots})^2 + ma\Sigma(X_{\cdot j\cdot} - X_{\cdots})^2$$
$$+ \Sigma\Sigma\Sigma(X_{ijk} - X_{i\cdot\cdot} - X_{\cdot j\cdot} + X_{\cdots})^2,$$

with the last term corresponding to error. The hypothesis $H: \alpha_1 = \cdots = \alpha_a = 0$ is then rejected when

$$\frac{mb\Sigma(X_{i\cdot\cdot} - X_{\cdots})^2/(a-1)}{\Sigma\Sigma\Sigma(X_{ijk} - X_{i\cdot\cdot} - X_{\cdot j\cdot} + X_{\cdots})^2/(abm - a - b + 1)} > C.$$

Suppose now that the assumption of no interaction, under which this test was derived, is not justified. The denominator sum of squares then has a noncentral χ^2-distribution instead of a central one, and is therefore stochastically larger than was assumed (Problem 13). It follows that the actual rejection probability is less than it would be for $\Sigma\Sigma\gamma_{ij}^2 = 0$. This shows that the probability of an error of the first kind will not exceed the nominal level of significance regardless of the values of the γ's. However, the power also decreases with $\Sigma\Sigma\gamma_{ij}^2/\sigma^2$ and tends to zero as this ratio tends to infinity.

The analysis of variance and the associated tests derived in this section for two factors extend in a straightforward manner to a larger number of factors (see for example Problem 14). On the other hand, if the number of observations is not the same for each combination of levels (each *cell*), the problem, while remaining a linear hypothesis, becomes more complex.

Of great importance are arrangements in which only certain combinations of levels occur since they permit reducing the size of the experiment. Thus for example three independent factors, at m levels each, can be analyzed with only m^2 observations, instead of the m^3 required if 1 observation were taken at each combination of levels, by adopting a Latin square design (Problem 15).

The class of problems considered here contains as a special case the two-sample problem treated in Chapter 5, which concerns a single factor with only two levels. The questions discussed in that connection regarding possible inhomogeneities of the experimental material and the randomization required to offset it are of equal importance in the present, more complex situations. If inhomogeneous material is subdivided into more homogeneous groups, this classification can be treated as constituting one or more additional factors. The choice of these groups is an important aspect in the determination of a suitable experimental design.† A very simple example of this is discussed in Problems 26 and 27 of Chapter 5.

To guard against possible inhomogeneities (and other departures from the assumptions made) even in the subgroups, randomization is used in the assignment of treatment factors within the groups. As was the case in the two-sample problem, the process of randomization alone without any assumptions concerning the method of sampling the experimental units, normality, independence, etc., makes it possible to obtain level α tests of the various hypotheses of interest. These permutation tests in the present case consist in computing the appropriate F-statistic W^*, but comparing it only with the values obtained from it by applying to the observations the permutations associated with the randomization procedure.‡ These tests are as before asymptotically equivalent to the corresponding F-tests, by which they can therefore be approximated.

† For a discussion of various designs and the conditions under which they are appropriate see, for example, Kempthorne, *The Design and Analysis of Experiments*, New York, John Wiley & Sons, 1952, and Cochran and Cox, *Experimental Designs*, New York, John Wiley & Sons, 2nd ed., 1957. Optimum properties of certain designs, proved by Wald, Ehrenfeld, Kiefer, and others, are discussed by Kiefer, "On the nonrandomized optimality and randomized nonoptimality of symmetrical designs," *Ann. Math. Stat.*, Vol. 29 (1958), pp. 675–699.

‡ For details see Kempthorne, *loc. cit.*

6. REGRESSION

Hypotheses specifying one or both of the regression coefficients α, β when $X_1, \cdots, X_n$ are independently normally distributed with common variance σ^2 and means

$$\xi_i = \alpha + \beta t_i \tag{32}$$

are essentially linear hypotheses, as was pointed out in Example 2. The hypotheses $H_1: \alpha = \alpha_0$ and $H_2: \beta = \beta_0$ were treated in Chapter 5, Section 6, where they were shown to possess UMP unbiased tests. We shall now consider H_1 and H_2, as well as the hypothesis H_3: $\alpha = \alpha_0$, $\beta = \beta_0$, from the present point of view. By the general theory of Section 1 the resulting tests will be UMP invariant under suitable groups of linear transformations. For the first two cases, in which $r = 1$, this also provides by the argument of Chapter 6, Section 6, an alternative proof of their being UMP unbiased.

The space Π_Ω is the same for all three hypotheses. It is spanned by the vectors $(1, \cdots, 1)$ and $(t_1, \cdots, t_n)$ and has therefore dimension $s = 2$ unless the t_i are all equal, which we shall assume not to be the case. The least squares estimates α and β under Ω are obtained by minimizing $\Sigma(X_i - \alpha - \beta t_i)^2$. For any fixed value of β, this is achieved by the value $\alpha = \bar{X} - \beta\bar{t}$, for which the sum of squares reduces to $\Sigma[(X_i - \bar{X}) - \beta(t_i - \bar{t})]^2$. By minimizing this with respect to β one finds

$$\hat{\beta} = \frac{\Sigma(X_i - \bar{X})(t_i - \bar{t})}{\Sigma(t_j - \bar{t})^2}, \qquad \hat{\alpha} = \bar{X} - \hat{\beta}\bar{t}; \tag{33}$$

and

$$\Sigma(X_i - \hat{\alpha} - \hat{\beta}t_i)^2 = \Sigma(X_i - \bar{X})^2 - \hat{\beta}^2\Sigma(t_i - \bar{t})^2$$

is the denominator sum of squares for all three hypotheses. The numerator of the test statistic (7) for testing the two hypotheses $\alpha = 0$ and $\beta = 0$ is Y_1^2 and for testing $\alpha = \beta = 0$ is $Y_1^2 + Y_2^2$.

For the hypothesis $\alpha = 0$, the statistic Y_1 was shown in Example 3 to be equal to

$$(\bar{X} - \bar{t}\Sigma t_i X_i/\Sigma t_j^2)\sqrt{n\Sigma t_j^2/\Sigma(t_j - \bar{t})^2} = \hat{\alpha}\sqrt{n\Sigma(t_j - \bar{t})^2/\Sigma t_j^2}.$$

Since then

$$E(Y_1) = \alpha\sqrt{n\Sigma(t_j - \bar{t})^2/\Sigma t_j^2},$$

the hypothesis $\alpha = \alpha_0$ is equivalent to the hypothesis $E(Y_1) = \eta_1^0 = \alpha_0\sqrt{n\Sigma(t_j - \bar{t})^2/\Sigma t_j^2}$, for which the rejection region (17) is $(n - s)(Y_1 - \eta_1^0)^2/\sum_{i=s+1}^n Y_i^2 > C_0$ and hence

$$\frac{|\hat{\alpha} - \alpha_0|\sqrt{n\Sigma(t_j - \bar{t})^2/\Sigma t_j^2}}{\sqrt{\Sigma(X_i - \hat{\alpha} - \hat{\beta}t_i)^2/(n-2)}} > C_0. \tag{34}$$

For the hypothesis $\beta = 0$, Y_1 was shown to be equal to

$$\Sigma(X_i - \bar{X})(t_i - \bar{t})/\sqrt{\Sigma(t_j - \bar{t})^2} = \hat{\beta}\sqrt{\Sigma(t_j - \bar{t})^2}.$$

Since then $E(Y_1) = \beta\sqrt{\Sigma(t_j - \bar{t})^2}$, the hypothesis $\beta = \beta_0$ is equivalent to $E(Y_1) = \eta_1^0 = \beta_0\sqrt{\Sigma(t_j - \bar{t})^2}$ and the rejection region is

$$\frac{|\hat{\beta} - \beta_0|\sqrt{\Sigma(t_j - \bar{t})^2}}{\sqrt{\Sigma(X_i - \hat{\alpha} - \hat{\beta}t_i)^2/(n-2)}} > C_0. \tag{35}$$

For testing $\alpha = \beta = 0$, it was shown in Example 3 that

$$Y_1 = \hat{\beta}\sqrt{\Sigma(t_j - \bar{t})^2}, \qquad Y_2 = \sqrt{n}\bar{X} = \sqrt{n}(\hat{\alpha} + \hat{\beta}\bar{t});$$

and the numerator of (7) is therefore

$$(Y_1^2 + Y_2^2)/2 = [n(\hat{\alpha} + \hat{\beta}\bar{t})^2 + \hat{\beta}^2\Sigma(t_j - \bar{t})^2]/2.$$

The more general hypothesis $\alpha = \alpha_0$, $\beta = \beta_0$ is equivalent to $E(Y_1) = \eta_1^0$, $E(Y_2) = \eta_2^0$ where $\eta_1^0 = \beta_0\sqrt{\Sigma(t_j - \bar{t})^2}$, $\eta_2^0 = \sqrt{n}(\alpha_0 + \beta_0\bar{t})$, and the rejection region (17) can therefore be written as

$$\frac{[n(\hat{\alpha} - \alpha_0)^2 + 2n\bar{t}(\hat{\alpha} - \alpha_0)(\hat{\beta} - \beta_0) + \Sigma t_i^2(\hat{\beta} - \beta_0)^2]/2}{\Sigma(X_i - \hat{\alpha} - \hat{\beta}t_i)^2/(n-2)} > C. \tag{36}$$

The associated confidence sets for (α, β) are obtained by reversing this inequality and replacing α_0 and β_0 by α and β. The resulting sets are ellipses centered at $(\hat{\alpha}, \hat{\beta})$.

The simple regression model (32) can be generalized in many directions; the means ξ_i may for example be polynomials in t_i of higher than the first degree (see Problem 18), or more complex functions such as trigonometric polynomials; or they may be functions of several variables, t_i, u_i, v_i. Some further extensions will now be illustrated by a number of examples.

Example 6. A variety of problems arise when there is more than one regression line. Suppose that the variables X_{ij} are independently normally distributed with common variance and means

$$\xi_{ij} = \alpha_i + \beta_i t_{ij} \qquad (j = 1, \cdots, n_i;\ i = 1, \cdots, b). \tag{37}$$

The hypothesis that these regression lines have equal slopes

$$H: \beta_1 = \cdots = \beta_b$$

may occur for example when the equality of a number of growth rates is to be tested. The parameter space Π_Ω has dimension $s = 2b$ provided none of the sums $\Sigma_j(t_{ij} - t_{i\cdot})^2$ is zero; the number of constraints imposed by the hypothesis is $r = b - 1$. The minimum value of $\Sigma\Sigma(X_{ij} - \xi_{ij})^2$ under Ω is obtained by minimizing $\Sigma_j(X_{ij} - \alpha_i - \beta_i t_{ij})^2$ for each i, so that by (33),

$$\hat{\beta}_i = \frac{\sum_j (X_{ij} - X_{i\cdot})(t_{ij} - t_{i\cdot})}{\sum_j (t_{ij} - t_{i\cdot})^2} \qquad \hat{\alpha}_i = X_{i\cdot} - \hat{\beta}_i t_{i\cdot} \quad .$$

Under H, one must minimize $\Sigma\Sigma(X_{ij} - \alpha_i - \beta t_{ij})^2$, which for any fixed β leads to $\alpha_i = X_{i\cdot} - \beta t_{i\cdot}$ and reduces the sum of squares to $\Sigma\Sigma[(X_{ij} - X_{i\cdot}) - \beta(t_{ij} - t_{i\cdot})]^2$. Minimizing this with respect to β one finds

$$\hat{\hat{\beta}} = \frac{\Sigma\Sigma(X_{ij} - X_{i\cdot})(t_{ij} - t_{i\cdot})}{\Sigma\Sigma(t_{ij} - t_{i\cdot})^2}; \quad \hat{\hat{\alpha}}_i = X_{i\cdot} - \hat{\hat{\beta}} t_{i\cdot} \quad .$$

Since

$$X_{ij} - \hat{\xi}_{ij} = X_{ij} - \hat{\alpha}_i - \hat{\beta}_i t_{ij} = (X_{ij} - X_{i\cdot}) - \hat{\beta}_i(t_{ij} - t_{i\cdot})$$

and

$$\hat{\xi}_{ij} - \hat{\hat{\xi}}_{ij} = (\hat{\alpha}_i - \hat{\hat{\alpha}}_i) + t_{ij}(\hat{\beta}_i - \hat{\hat{\beta}}) = (\hat{\beta}_i - \hat{\hat{\beta}})(t_{ij} - t_{i\cdot}),$$

the rejection region (15) is

$$\text{(38)} \qquad \frac{\Sigma_i(\hat{\beta}_i - \hat{\hat{\beta}})^2\Sigma_j(t_{ij} - t_{i\cdot})^2/(b - 1)}{\Sigma\Sigma[(X_{ij} - X_{i\cdot}) - \hat{\beta}_i(t_{ij} - t_{i\cdot})]^2/(n - 2b)} > C$$

where the left-hand side under H has the F-distribution with $b - 1$ and $n - 2b$ degrees of freedom.

Since

$$E(\hat{\beta}_i) = \beta_i \qquad \text{and} \qquad E(\hat{\hat{\beta}}) = \Sigma_i\beta_i\Sigma_j(t_{ij} - t_{i\cdot})^2/\Sigma\Sigma(t_{ij} - t_{i\cdot})^2,$$

the noncentrality parameter of the distribution for an alternative set of β's is $\psi^2 = \Sigma_i(\beta_i - \tilde{\beta})^2\Sigma_j(t_{ij} - t_{i\cdot})^2/\sigma^2$, where $\tilde{\beta} = E(\hat{\hat{\beta}})$. In the particular case that the n_i and the t_{ij} are independent of i, $\tilde{\beta}$ reduces to $\bar{\beta} = \Sigma\beta_j/b$.

Example 7. The regression model (37) arises in the comparison of a number of treatments when the experimental units are treated as fixed and the unit effects u_{ij} (defined in Chapter 5, Section 10) are proportional to known constants t_{ij}. Here t_{ij} might for example be a measure of the fertility of the i, jth piece of land or the weight of the i, jth experimental animal prior to the experiment. It is then frequently possible to assume that the proportionality factor β_i does not depend on the treatment, in which case (37) reduces to

$$\text{(39)} \qquad \xi_{ij} = \alpha_i + \beta t_{ij}$$

and the hypothesis of no treatment effect becomes

$$H: \alpha_1 = \cdots = \alpha_b.$$

The space Π_Ω coincides with Π_ω of the previous example, so that $s = b + 1$ and

$$\hat{\beta} = \frac{\Sigma\Sigma(X_{ij} - X_{i\cdot})(t_{ij} - t_{i\cdot})}{\Sigma\Sigma(t_{ij} - t_{i\cdot})^2}; \qquad \hat{\alpha}_i = X_{i\cdot} - \hat{\beta} t_{i\cdot}$$

Minimization of $\Sigma\Sigma(X_{ij} - \alpha - \beta t_{ij})^2$ gives

$$\hat{\hat{\beta}} = \frac{\Sigma\Sigma(X_{ij} - X_{\cdot\cdot})(t_{ij} - t_{\cdot\cdot})}{\Sigma\Sigma(t_{ij} - t_{\cdot\cdot})^2}, \qquad \hat{\hat{\alpha}} = X_{\cdot\cdot} - \hat{\hat{\beta}} t_{\cdot\cdot} \quad ,$$

where $X.. = \Sigma\Sigma X_{ij}/n$, $t.. = \Sigma\Sigma t_{ij}/n$, $n = \Sigma n_i$. The sum of squares in the numerator of W^* in (15) is thus

$$\Sigma\Sigma(\hat{\xi}_{ij} - \hat{\hat{\xi}}_{ij})^2 = \Sigma\Sigma[(X_{i}. - X..) + \hat{\beta}(t_{ij} - t_{i}.) - \hat{\hat{\beta}}(t_{ij} - t..)]^2.$$

The hypothesis H is therefore rejected when

$$\frac{\Sigma\Sigma[(X_{i}. - X..) + \hat{\beta}(t_{ij} - t_{i}.) - \hat{\hat{\beta}}(t_{ij} - t..)]^2/(b - 1)}{\Sigma\Sigma[(X_{ij} - X_{i}.) - \hat{\beta}(t_{ij} - t_{i}.)]^2/(n - b - 1)} > C, \tag{40}$$

where under H the left-hand side has the F-distribution with $b - 1$ and $n - b - 1$ degrees of freedom.

The hypothesis H can be tested without first ascertaining the values of the t_{ij}; it is then the hypothesis of no effect in a one-way classification considered in Section 3, and the test is given by (18). Actually, since the unit effects u_{ij} are assumed to be constants, which are now completely unknown, the treatments are assigned to the units either completely at random or at random within subgroups. The appropriate test is then a randomization test for which (18) is an approximation.

Example 7 illustrates the important class of situations in which an analysis of variance (in the present case concerning a one-way classification) is combined with a regression problem (in the present case linear regression on the single "concomitant variable" t). Both parts of the problem may of course be considerably more complex than was assumed here. Quite generally, in such combined problems one can test (or estimate) the treatment effects as was done above, and a similar analysis can be given for the regression coefficients. The breakdown of the variation into its various treatment and regression components is the so-called *analysis of covariance*.

7. MODEL II: ONE-WAY CLASSIFICATION

The analysis of the effect of one or more factors has been seen to depend on whether the experimental units are fixed or constitute a random sample from a population of such units. The same distinction also arises with respect to the factor effects themselves, which in some applications are constants and in others unobservable random variables. If all these effects are constant or all random one speaks of *model I* or *model II* respectively, and the term *mixed model* refers to situations in which both types occur. Of course, only the model I case constitutes a linear hypothesis according to the definition given at the beginning of the chapter. In the present section we shall treat as model II the case of a single factor (one-way classification), which was analyzed under the model I assumption in Section 3.

As an illustration of this problem, consider a material such as steel, which is manufactured or processed in batches. Suppose that a sample of size n is taken from each of s batches and that the resulting measurements X_{ij} $(j = 1, \cdots, n; \quad i = 1, \cdots, s)$ are independently normally distributed with variance σ^2 and mean ξ_i. If the factor corresponding to i were constant, with the same effect α_i in each replication of the experiment, we would have

$$\xi_i = \mu + \alpha_i \qquad (\Sigma\alpha_i = 0)$$

and

$$X_{ij} = \mu + \alpha_i + U_{ij}$$

where the U_{ij} are independently distributed as $N(0, \sigma^2)$. The hypothesis of no effect is $\xi_1 = \cdots = \xi_s$ or equivalently $\alpha_1 = \cdots = \alpha_s = 0$. However, the effect is associated with the batches, of which a new set will be involved in each replication of the experiment; and the effect therefore does not remain constant. Instead, we shall suppose that the batch effects constitute a sample from a normal distribution, and to indicate their random nature we shall write A_i for α_i so that

$$X_{ij} = \mu + A_i + U_{ij}. \tag{41}$$

The assumption of additivity (lack of interaction) of batch and unit effect, in the present model, implies that the A's and U's are independent. If the expectation of A_i is absorbed into μ, it follows that the A's and U's are independently normally distributed with zero means and variances σ_A^2 and σ^2 respectively. The X's of course are no longer independent.

The hypothesis of no batch effect, that the A's are zero and hence constant, takes the form

$$H: \sigma_A^2 = 0.$$

This is not realistic in the present situation, but is the limiting case of the hypothesis

$$H(\Delta_0): \sigma_A^2/\sigma^2 \leqq \Delta_0$$

that the batch effect is small relative to the variation of the material within a batch. These two hypotheses correspond respectively to the model I hypotheses $\Sigma\alpha_i^2 = 0$ and $\Sigma\alpha_i^2/\sigma^2 \leqq \Delta_0$.

To obtain a test of $H(\Delta_0)$ it is convenient to begin with the same transformation of variables that reduced the corresponding model I problem to canonical form. Each set $(X_{i1}, \cdots, X_{in})$ is subjected to an orthogonal transformation $Y_{ij} = \sum_{k=1}^n c_{jk} X_{ik}$ such that $Y_{i1} = \sqrt{n} X_{i\cdot}$. Since $c_{1k} = 1/\sqrt{n}$ for $k = 1, \cdots, n$ (see Example 3), it follows from the assumption of orthogonality that $\sum_{k=1}^n c_{jk} = 0$ for $j = 2, \cdots, n$ and hence that $Y_{ij} = \sum_{k=1}^n c_{jk} U_{ik}$ for $j > 1$. The Y_{ij} with $j > 1$ are therefore independently normally distributed with zero mean and variance σ^2. They are also independent of $U_{i\cdot}$ since $(\sqrt{n} U_{i\cdot} Y_{i2} \cdots Y_{in})' = C(U_{i1} U_{i2} \cdots U_{in})'$, (a prime indicates the transpose of a matrix). On the other hand, the variables $Y_{i1} = \sqrt{n}X_{i\cdot} = \sqrt{n}(\mu + A_i + U_{i\cdot})$ are also independently normally distributed but with mean $\sqrt{n}\mu$ and variance $\sigma^2 + n\sigma_A^2$. If an additional orthogonal transformation is made from $(Y_{11}, \cdots, Y_{s1})$ to $(Z_{11}, \cdots, Z_{s1})$ such that $Z_{11} = \sqrt{s}Y_{\cdot 1}$, the Z's are independently normally distributed with common variance $\sigma^2 + n\sigma_A^2$ and means $E(Z_{11}) = \sqrt{sn}\mu$ and $E(Z_{i1}) = 0$ for $i > 1$. Putting $Z_{ij} = Y_{ij}$ for $j > 1$ for the sake of conformity, the joint density of the Z's is then

$$(2\pi)^{-ns/2}\sigma^{-(n-1)s}(\sigma^2 + n\sigma_A^2)^{-s/2} \exp\left[-\frac{1}{2(\sigma^2 + n\sigma_A^2)}\left((z_{11} - \sqrt{sn}\,\mu)^2 + \sum_{i=2}^{s} z_{i1}^2\right) - \frac{1}{2\sigma^2}\sum_{i=1}^{s}\sum_{j=2}^{n} z_{ij}^2\right]. \tag{42}$$

The problem of testing $H(\Delta_0)$ is invariant under addition of an arbitrary constant to Z_{11}, which leaves the remaining Z's as a maximal set of invariants. These constitute samples of size $s(n - 1)$ and $s - 1$ from two normal distributions with means zero and variances σ^2 and $\tau^2 = \sigma^2 + n\sigma_A^2$. The hypothesis $H(\Delta_0)$ is equivalent to $\tau^2/\sigma^2 \leq 1 + \Delta_0 n$, and the problem reduces to that of comparing two normal variances, which was considered in Example 6 of Chapter 6 without the restriction to zero means. The UMP invariant test, under multiplication of all Z_{ij} by a common positive constant, has the rejection region

$$W^* = \frac{1}{(1 + \Delta_0 n)} \cdot \frac{S_A^2/(s-1)}{S^2/(n-1)s} > C \tag{43}$$

where

$$S_A^2 = \sum_{i=2}^{n} Z_{i1}^2 \quad \text{and} \quad S^2 = \sum_{i=1}^{s}\sum_{j=2}^{n} Z_{ij}^2 = \sum_{i=1}^{s}\sum_{j=2}^{n} Y_{ij}^2.$$

The constant C is determined by

$$\int_C^\infty F_{s-1,(n-1)s}(y)\, dy = \alpha.$$

Since

$$\sum_{j=1}^{n} Y_{ij}^2 - Y_{i1}^2 = \sum_{j=1}^{n} U_{ij}^2 - nU_{i\cdot}^2$$

and

$$\sum_{i=1}^{s} Z_{i1}^2 - Z_{11}^2 = \sum_{i=1}^{s} Y_{i1}^2 - sY_{\cdot 1}^2,$$

the numerator and denominator sum of squares of W^*, expressed in terms of the X's, become

$$S_A^2 = n\sum_{i=1}^{s}(X_{i\cdot} - X_{\cdot\cdot})^2 \quad \text{and} \quad S^2 = \sum_{i=1}^{s}\sum_{j=1}^{n}(X_{ij} - X_{i\cdot})^2.$$

In the particular case $\Delta_0 = 0$, the test (43) is equivalent to the corresponding model I test (18) but they are of course solutions of different problems, and also have different power functions. Instead of being distributed according to a noncentral χ^2-distribution as in model I, the numerator sum of squares of W^* is proportional to a central χ^2 variable even when the hypothesis is false, and the power of the test (43) against an alternative value of Δ is obtained from the F-distribution through

$$\beta(\Delta) = P_\Delta\{W^* > C\} = \int_{\frac{1+\Delta_0 n}{1+\Delta n}C}^{\infty} F_{s-1,(n-1)s}(y)\, dy.$$

The family of tests (43) for varying Δ_0 is equivalent to the confidence statements

$$\underline{\Delta} = \frac{1}{n}\left[\frac{S_A^2/(s-1)}{CS^2/(n-1)s} - 1\right] \leqq \Delta. \tag{44}$$

The corresponding upper confidence bounds for Δ are obtained from the tests of the hypotheses $\Delta \geqq \Delta_0$. These have the acceptance regions $W^* \geqq C'$, where W^* is given by (43) and C' is determined by $\int_{C'}^{\infty} F_{s-1,(n-1)s} = 1 - \alpha$, and the resulting confidence bounds are

$$\Delta \leqq \frac{1}{n}\left[\frac{S_A^2/(s-1)}{C'S^2/(n-1)s} - 1\right] = \bar{\Delta}. \tag{45}$$

Both the confidence sets (44) and (45) are invariant with respect to the group of transformations generated by those considered for the testing problems, and hence are uniformly most accurate invariant.

When $\underline{\Delta}$ is negative, the confidence set $(\underline{\Delta}, \infty)$ contains all possible values of the parameter Δ. For small Δ, this will happen with high probability ($1 - \alpha$ for $\Delta = 0$), as must be the case since $\underline{\Delta}$ is then required to be a safe lower bound for a quantity which is equal to or near zero. More awkward is the possibility that $\bar{\Delta}$ is negative, so that the confidence set $(-\infty, \bar{\Delta})$ is empty. An interpretation is suggested by the fact that this occurs if and only if the hypothesis $\Delta \geqq \Delta_0$ is rejected for all positive values of Δ_0. This may be taken as an indication that the assumed model is not appropriate, although it must be realized that for small Δ the probability of the event $\bar{\Delta} < 0$ is near α even when the assumptions are satisfied, so that this outcome will occasionally be observed.

The tests of $\Delta \leqq \Delta_0$ and $\Delta \geqq \Delta_0$ are not only UMP invariant but also UMP unbiased, and UMP unbiased tests also exist for testing $\Delta = \Delta_0$ against the two-sided alternatives $\Delta \neq \Delta_0$. This follows from the fact that the joint density of the Z's constitutes an exponential family. The confidence sets associated with these three families of tests are then uniformly most accurate unbiased (Problem 19). That optimum unbiased procedures exist in the model II case but not in the corresponding model I problem is explained by the different structure of the two hypotheses. The model II hypothesis $\sigma_A^2 = 0$ imposes one constraint since it concerns the single parameter σ_A^2. On the other hand, the corresponding model I hypothesis $\sum_{i=1}^s \alpha_i^2 = 0$ specifies the values of the s parameters $\alpha_1, \cdots, \alpha_s$, and since $s - 1$ of these are independent, imposes $s - 1$ constraints.

8. NESTED CLASSIFICATIONS

The theory of the preceding section does not carry over even to so simple a situation as the general one-way classification with unequal numbers in the different classes (Problem 22). However, the unbiasedness approach does extend to the important case of a *nested* (hierarchical) classification with equal numbers in each class. This extension is sufficiently well indicated by carrying it through for the case of two factors; it follows for the general case by induction with respect to the number of factors.

Returning to the illustration of a batch process, suppose that a single batch of raw material suffices for several batches of the finished product. Let the experimental material consist of ab batches, b coming from each of a batches of raw material, and let a sample of size n be taken from each. Then (41) becomes

$$X_{ijk} = \mu + A_i + B_{ij} + U_{ijk} \qquad (i = 1, \cdots, a; \; j = 1, \cdots, b; \; k = 1, \cdots, n) \tag{46}$$

where A_i denotes the effect of the ith batch of raw material, B_{ij} that of the jth batch of finished product obtained from this material, and U_{ijk} the effect of the kth unit taken from this batch. All these variables are assumed to be independently normally distributed with zero means and with variances σ_A^2, σ_B^2, and σ^2 respectively. The main part of the induction argument consists in proving the existence of an orthogonal transformation to variables Z_{ijk} the joint density of which, except for a constant, is

$$\exp\left[-\frac{1}{2(\sigma^2 + n\sigma_B^2 + bn\sigma_A^2)}\left((z_{111} - \sqrt{abn}\,\mu)^2 + \sum_{i=2}^{a} z_{i11}^2\right) - \frac{1}{2(\sigma^2 + n\sigma_B^2)}\sum_{i=1}^{a}\sum_{j=2}^{b} z_{ij1}^2 - \frac{1}{2\sigma^2}\sum_{i=1}^{a}\sum_{j=1}^{b}\sum_{k=2}^{n} z_{ijk}^2\right]. \tag{47}$$

As a first step, there exists for each fixed i, j an orthogonal transformation from $(X_{ij1}, \cdots, X_{ijn})$ to $(Y_{ij1}, \cdots, Y_{ijn})$ such that

$$Y_{ij1} = \sqrt{n}\, X_{ij\cdot} = \sqrt{n}\,\mu + \sqrt{n}\,(A_i + B_{ij} + U_{ij\cdot}).$$

As in the case of a single classification, the variables Y_{ijk} with $k > 1$ depend only on the U's, are independently normally distributed with zero mean and variance σ^2, and are independent of the $U_{ij\cdot}$. On the other hand, the variables Y_{ij1} have exactly the structure of the Y_{ij} in the one-way classification,

$$Y_{ij1} = \mu' + A'_i + U'_{ij},$$

where $\mu' = \sqrt{n}\,\mu$, $A'_i = \sqrt{n}\,A_i$, $U'_{ij} = \sqrt{n}\,(B_{ij} + U_{ij\cdot})$, and where the variances of A'_i and U'_{ij} are $\sigma'^2_A = n\sigma^2_A$ and $\sigma'^2 = \sigma^2 + n\sigma^2_B$ respectively. These variables can therefore be transformed to variables Z_{ij1} whose density is given by (42) with Z_{ij1} in place of Z_{ij}. Putting $Z_{ijk} = Y_{ijk}$ for $k > 1$, the joint density of all Z_{ijk} is then given by (47).

Two hypotheses of interest can be tested on the basis of (47): H_1: $\sigma^2_A/(\sigma^2 + n\sigma^2_B) \leq \Delta_0$ and H_2: $\sigma^2_B/\sigma^2 \leq \Delta_0$, which state that one or the other of the classifications has little effect on the outcome. Let

$$S^2_A = \sum_{i=2}^{a} Z^2_{i11}, \qquad S^2_B = \sum_{i=1}^{a}\sum_{j=2}^{b} Z^2_{ij1}, \qquad S^2 = \sum_{i=1}^{a}\sum_{j=1}^{b}\sum_{k=2}^{n} Z^2_{ijk}.$$

To obtain a test of H_1, one is tempted to eliminate S^2 through invariance under multiplication of Z_{ijk} for $k > 1$ by an arbitrary constant. However these transformations do not leave (47) invariant since they do not always preserve the fact that σ^2 is the smallest of the three variances σ^2, $\sigma^2 + n\sigma^2_B$, and $\sigma^2 + n\sigma^2_B + bn\sigma^2_A$. We shall instead consider the problem from the point of view of unbiasedness. For any unbiased test of H_1, the probability of rejection is α whenever $\sigma^2_A/(\sigma^2 + n\sigma^2_B) = \Delta_0$, and hence in particular when the three variances are σ^2, τ^2_0, and $(1 + bn\Delta_0)\tau^2_0$ for any fixed τ^2_0 and all $\sigma^2 < \tau^2_0$. It follows by the techniques of Chapter 4 that the conditional probability of rejection given $S^2 = s^2$ must be equal to α for almost all values of s^2. With S^2 fixed, the joint distribution of the remaining variables is of the same type as (42) after the elimination of Z_{111}, and a UMP unbiased conditional test given $S^2 = s^2$ has the rejection region

$$W^*_1 = \frac{1}{1 + bn\Delta_0} \cdot \frac{S^2_A/(a - 1)}{S^2_B/(b - 1)a} \geq C_1. \tag{48}$$

Since S^2_A and S^2_B are independent of S^2, the constant C_1 is determined by the fact that when $\sigma^2_A/(\sigma^2 + n\sigma^2_B) = \Delta_0$, the statistic W^*_1 is distributed as $F_{a-1,(b-1)a}$ and hence in particular does not depend on s. The test (48) is clearly unbiased and hence UMP unbiased.

The argument with respect to H_2 is completely analogous and shows the UMP unbiased test to have the rejection region

$$W^*_2 = \frac{1}{1 + n\Delta_0} \cdot \frac{S^2_B/(b - 1)a}{S^2/(n - 1)ab} \geq C_2, \tag{49}$$

where C_2 is determined by the fact that for $\sigma_B^2/\sigma^2 = \Delta_0$, the statistic W_2^* is distributed as $F_{(b-1)a,(n-1)ab}$.

It remains to express the statistics S_A^2, S_B^2 and S^2 in terms of the X's. From the corresponding expressions in the one-way classification, it follows that

$$S_A^2 = \sum_{i=1}^{a} Z_{i11}^2 - Z_{111}^2 = b\Sigma(Y_{i\cdot 1} - Y_{\cdot\cdot 1})^2,$$

$$S_B^2 = \sum_{i=1}^{a}\left[\sum_{j=1}^{b} Z_{ij1}^2 - Z_{i11}^2\right] = \Sigma\Sigma(Y_{ij1} - Y_{i.1})^2$$

and

$$S^2 = \sum_{i=1}^{a}\sum_{j=1}^{b}\left[\sum_{k=1}^{n} Y_{ijk}^2 - Y_{ij1}^2\right] = \sum_i\sum_j\left[\sum_{k=1}^{n} U_{ijk}^2 - nU_{ij\cdot}^2\right]$$

$$= \sum_i\sum_j\sum_k (U_{ijk} - U_{ij\cdot})^2.$$

Hence

$$(50) \quad S_A^2 = bn\Sigma(X_{i\cdot\cdot} - X_{\cdots})^2, \qquad S_B^2 = n\Sigma\Sigma(X_{ij\cdot} - X_{i\cdot\cdot})^2,$$
$$S^2 = \Sigma\Sigma\Sigma(X_{ijk} - X_{ij\cdot})^2.$$

It is seen from the expression of the statistics in terms of the Z's that their expectations are $E[S_A^2/(a-1)] = \sigma^2 + n\sigma_B^2 + bn\sigma_A^2$, $E[S_B^2/(b-1)a] = \sigma^2 + n\sigma_B^2$, and $E[S^2/(n-1)ab] = \sigma^2$. The decomposition

$$\Sigma\Sigma\Sigma(X_{ijk} - X_{\cdots})^2 = S_A^2 + S_B^2 + S^2$$

therefore forms a basis for the analysis of the variance of X_{ijk},

$$\text{Var}(X_{ijk}) = \sigma_A^2 + \sigma_B^2 + \sigma^2$$

by providing estimates of the *components of variance* σ_A^2, σ_B^2, and σ^2, and tests of certain ratios of these components.

Nested two-way classifications also occur as mixed models. Suppose for example that a firm produces the material of the previous illustrations in different plants. If α_i denotes the effect of the ith plant (which is fixed since the plants do not change in a replication of the experiment), B_{ij} the batch effect, and U_{ijk} the unit effect, the observations have the structure

$$(51) \qquad X_{ijk} = \mu + \alpha_i + B_{ij} + U_{ijk}.$$

Instead of reducing the X's to the fully canonical form in terms of the Z's as before, it is convenient to carry out only the reduction to the Y's (such that $Y_{ij1} = \sqrt{n}\,X_{ij\cdot}$) and the first of the two transformations which take the Y's into the Z's. If the resulting variables are denoted by W_{ijk}, they satisfy $W_{i11} = \sqrt{b}\, Y_{i\cdot 1}$, $W_{ijk} = Y_{ijk}$ for $k > 1$ and

$$\sum_{i=1}^{a} (W_{i11} - W_{\cdot 11})^2 = S_A^2, \qquad \sum_{i=1}^{a}\sum_{j=2}^{b} W_{ij1}^2 = S_B^2, \qquad \sum_{i=1}^{a}\sum_{j=1}^{b}\sum_{k=2}^{n} W_{ijk}^2 = S^2$$

where S_A^2, S_B^2, and S^2 are given by (50). The joint density of the W's is, except for a constant,

$$\exp\left[-\frac{1}{2(\sigma^2 + n\sigma_B^2)}\left(\sum_{i=1}^{a}(w_{i11} - \mu - \alpha_i)^2 + \sum_{i=1}^{a}\sum_{j=2}^{b} w_{ij1}^2\right) - \frac{1}{2\sigma^2}\sum_{i=1}^{a}\sum_{j=1}^{b}\sum_{k=2}^{n} w_{ijk}^2\right]. \tag{52}$$

This shows clearly the different nature of the problem of testing that the plant effect is small,

$$H\colon \alpha_1 = \cdots = \alpha_a = 0 \qquad \text{or} \qquad H'\colon \Sigma\alpha_i^2/(\sigma^2 + n\sigma_B^2) \leqq \Delta_0$$

and testing the corresponding hypothesis for the batch effect: $\sigma_B^2/\sigma^2 \leqq \Delta_0$. The first of these is essentially a model I problem (linear hypothesis). As before, unbiasedness implies that the conditional rejection probability given $S^2 = s^2$ is equal to α a.e. With S^2 fixed, the problem of testing H is a linear hypothesis, and the rejection region of the UMP invariant conditional test given $S^2 = s^2$ has the rejection region (48) with $\Delta_0 = 0$. The constant C_1 is again independent of S^2 and the test is UMP among all tests that are both unbiased and invariant. A test with the same property also exists for testing H'. Its rejection region is

$$\frac{S_A^2/(a-1)}{S_B^2/(b-1)a} \geqq C'$$

where C' is determined from the noncentral F-distribution instead of as before, the (central) F-distribution (see Problem 5).

On the other hand, the hypothesis $\sigma_B^2/\sigma^2 \leqq \Delta_0$ is essentially model II. It is invariant under addition of an arbitrary constant to each of the variables W_{i11}, which leaves $\sum_{i=1}^{a}\sum_{j=2}^{b} W_{ij1}^2$ and $\sum_{i=1}^{a}\sum_{j=1}^{b}\sum_{k=2}^{n} W_{ijk}^2$ as maximal invariants, and hence reduces the structure to pure model II with one classification. The test is then given by (49) as before. It is both UMP invariant and UMP unbiased.

A two-factor mixed model in which there is interaction between the two factors will be considered in Example 11 below.

9. THE MULTIVARIATE LINEAR HYPOTHESIS

The univariate linear models of Section 1 arise in the study of the effects of various experimental conditions (factors) on a single characteristic such as yield, weight, length of life, blood pressure, etc. This characteristic is assumed to be normally distributed with a mean which depends on the various factors under investigation, and a variance which is independent of these factors. We shall now consider the multivariate analogue of this model, which is appropriate when one is concerned with the effect of one or more factors simultaneously on several characteristics, for example the effect of a change in the diet of dairy cows on both fat content and quantity of milk.

The multivariate generalization of a real-valued normally distributed random variable is a random vector $(X_1, \cdots, X_p)$ with the *multivariate normal probability density*

$$\frac{\sqrt{|A|}}{(2\pi)^{\frac{1}{2}p}} \exp\left[-\tfrac{1}{2}\Sigma\Sigma a_{ij}(x_i - \xi_i)(x_j - \xi_j)\right] \tag{53}$$

where the matrix $A = (a_{ij})$ is positive definite, and $|A|$ denotes its determinant. The means and covariance matrix of the X's are given by

$$E(X_i) = \xi_i, \qquad E(X_i - \xi_i)(X_j - \xi_j) = \sigma_{ij}, \qquad (\sigma_{ij}) = A^{-1}. \tag{54}$$

Consider now n independent multivariate normal vectors $X_\alpha = (X_{\alpha 1}, \cdots, X_{\alpha p})$, $\alpha = 1, \cdots, n$ with means $E(X_{\alpha i}) = \xi_{\alpha i}$ and common covariance matrix A^{-1}. As in the univariate case, a *multivariate linear hypothesis* is defined in terms of two linear subspaces Π_Ω and Π_ω of n-dimensional space having dimensions $s < n$ and $0 \leq s - r < s$. It is assumed known that for all $i = 1, \cdots, p$, the vectors $(\xi_{1i}, \cdots, \xi_{ni})$ lie in Π_Ω; the hypothesis to be tested specifies that they lie in Π_ω. This problem is reduced to canonical form by applying to each of the p vectors $(X_{1i}, \cdots, X_{ni})$ the orthogonal transformation (1). If

$$X = \begin{pmatrix} X_{11} \cdots X_{1p} \\ \cdot \quad\quad \cdot \\ \cdot \quad\quad \cdot \\ \cdot \quad\quad \cdot \\ X_{n1} \cdots X_{np} \end{pmatrix}$$

and the transformed variables are denoted by $X_{\alpha i}^*$, the transformation may be written in matrix form as

$$X^* = CX,$$

where $C = (c_{\alpha\beta})$ is an orthogonal matrix.

To obtain the joint distribution of the $X_{\alpha i}^*$ consider first the covariance of any two of them, say $X_{\alpha i}^* = \sum_{\gamma=1}^n c_{\alpha\gamma} X_{\gamma i}$ and $X_{\beta j}^* = \sum_{\delta=1}^n c_{\beta\delta} X_{\delta j}$.

Using the fact that the covariance of $X_{\gamma i}$ and $X_{\delta j}$ is zero when $\gamma \neq \delta$ and σ_{ij} when $\gamma = \delta$, we have

$$\text{Cov}\,(X^*_{\alpha i}, X^*_{\beta j}) = \sum_{\gamma=1}^{n}\sum_{\delta=1}^{n} c_{\alpha\gamma}c_{\beta\delta}\,\text{Cov}\,(X_{\gamma i}, X_{\delta j})$$

$$= \sigma_{ij}\sum_{\gamma=1}^{n} c_{\alpha\gamma}c_{\beta\gamma} = \begin{cases}\sigma_{ij} & \text{when} \quad \alpha = \beta \\ 0 & \text{when} \quad \alpha \neq \beta.\end{cases}$$

The rows of X^* are therefore again independent multivariate normal vectors with common covariance matrix A^{-1}. It follows as in the univariate case that the vectors of means satisfy

$$\xi^*_{s+1,i} = \cdots = \xi^*_{ni} = 0 \qquad (i = 1, \cdots, p)$$

under Ω, and that the hypothesis becomes

$$H: \xi^*_{1i} = \cdots = \xi^*_{ri} = 0 \qquad (i = 1, \cdots, p).$$

Changing notation so that Y's, U's, and Z's denote the first r, the next $s - r$ and the last $m = n - s$ sample vectors, we therefore arrive at the following *canonical form*. The vectors Y_α, U_β, Z_γ $(\alpha = 1, \cdots, r; \beta = 1, \cdots, s - r; \gamma = 1, \cdots, m)$ are independently distributed according to p-variate normal distributions with common covariance matrix A^{-1}. The means of the Z's are given to be zero, and the hypothesis H is to be tested that the means of the Y's are zero. If

$$Y = \begin{pmatrix} Y_{11} & \cdots & Y_{1p} \\ \cdot & & \cdot \\ \cdot & & \cdot \\ \cdot & & \cdot \\ Y_{r1} & \cdots & Y_{rp} \end{pmatrix} \quad \text{and} \quad Z = \begin{pmatrix} Z_{11} & \cdots & Z_{1p} \\ \cdot & & \cdot \\ \cdot & & \cdot \\ \cdot & & \cdot \\ Z_{m1} & \cdots & Z_{mp} \end{pmatrix},$$

invariance and sufficiency will be shown below to reduce the observations to the $p \times p$ matrices $Y'Y$ and $Z'Z$. It will then be convenient to have an expression of these statistics in terms of the original observations.

As in the univariate case, let $(\hat{\xi}_{1i}, \cdots, \hat{\xi}_{ni})$ and $(\hat{\hat{\xi}}_{1i}, \cdots, \hat{\hat{\xi}}_{ni})$ denote the projection of the vector $(X_{1i}, \cdots, X_{ni})$ on Π_Ω and Π_ω. Then

$$\sum_{\alpha=1}^{n}(X_{\alpha i} - \hat{\xi}_{\alpha i})(X_{\alpha j} - \hat{\xi}_{\alpha j})$$

is the inner product of two vectors, each of which is the difference between a given vector and its projection on Π_Ω. It follows that this quantity is unchanged under orthogonal transformations of the coordinate system in which the variables are expressed. Now the transformation

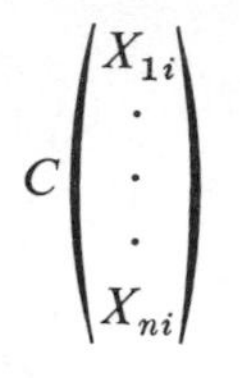

may be interpreted as expressing the vector $(X_{1i}, \cdots, X_{ni})$ in a new coordinate system, the first s coordinate axes of which lie in Π_Ω. The projection on Π_Ω of the transformed vector $(Y_{1i}, \cdots, Y_{ri}, U_{1i}, \cdots, U_{s-r,i}, Z_{1i}, \cdots, Z_{mi})$ is $(Y_{1i}, \cdots, Y_{ri}, U_{1i}, \cdots, U_{s-r,i}, 0, \cdots, 0)$, so that the difference between the vector and its projection is $(0, \cdots, 0, Z_{1i}, \cdots, Z_{mi})$. The ijth element of $Z'Z$ is therefore given by

$$\sum_{\gamma=1}^{m} Z_{\gamma i} Z_{\gamma j} = \sum_{\alpha=1}^{n} (X_{\alpha i} - \hat{\xi}_{\alpha i})(X_{\alpha j} - \hat{\xi}_{\alpha j}). \tag{55}$$

Analogously, the projection of the transformed vector $(Y_{1i}, \cdots, Y_{ri}, U_{1i}, \cdots, U_{s-r,i}, 0, \cdots, 0)$ on Π_ω is $(0, \cdots, 0, U_{1i}, \cdots, U_{s-r,i}, 0, \cdots, 0)$ and the difference between the projections on Π_Ω and Π_ω is therefore $(Y_{1i}, \cdots, Y_{ri}, 0, \cdots, 0, \cdots, 0)$. It follows that the sum $\sum_{\beta=1}^{r} Y_{\beta i} Y_{\beta j}$ is equal to the inner product (for the ith and jth vector) of the difference of these projections. On comparing this sum with the expression of the same inner product in the original coordinate system, it is seen that the i, jth element of $Y'Y$ is given by

$$\sum_{\beta=1}^{r} Y_{\beta i} Y_{\beta j} = \sum_{\alpha=1}^{n} (\hat{\xi}_{\alpha i} - \hat{\hat{\xi}}_{\alpha i})(\hat{\xi}_{\alpha j} - \hat{\hat{\xi}}_{\alpha j}). \tag{56}$$

10. REDUCTION BY INVARIANCE

The multivariate linear hypothesis, described in the preceding section in canonical form, remains invariant under certain groups of transformations. To obtain maximal invariants under these groups we require, in addition to some of the standard theorems concerning quadratic forms, the following lemma.

Lemma 1. *If M is any $m \times p$ matrix, then*

(i) *$M'M$ is positive semidefinite,*

(ii) *the rank of $M'M$ equals the rank of M, so that in particular $M'M$ is nonsingular if and only if $m \geq p$ and M is of rank p.*

Proof. (i) Consider the quadratic form $Q = u'(M'M)u$. If $w = Mu$, then

$$Q = w'w \geqq 0.$$

(ii) The sum of squares $w'w$ is zero if and only if the vector w is zero, and the result follows from the fact that the solutions u of the system of equations $Mu = 0$ form a linear space of dimension $p - \rho$ where ρ is the rank of M.

We shall now consider three groups under which the problem remains invariant.

G_1. Addition of an arbitrary constant $d_{\beta i}$ to each of the variables $U_{\beta i}$ leaves the problem invariant, and this eliminates the U's since the Y's and Z's are maximal invariant under this group.

G_2. In the process of reducing the problem to canonical form it was seen that an orthogonal transformation

$$Y^* = CY$$

affects neither the independence of the row vectors of Y nor the covariance matrix of these vectors. The means of the Y^*'s are zero if and only if those of the Y's are, and hence the problem remains invariant under these transformations.

The matrix $Y'Y$ of inner products of the column vectors of Y is invariant under G_2 since $Y^{*\prime}Y^* = Y'C'CY = Y'Y$. The matrix $Y'Y$ will be proved to be maximal invariant by showing that $Y'Y = Y^{*\prime}Y^*$ implies the existence of an orthogonal matrix C such that $Y^* = CY$. Consider first the case $r = p$. Without loss of generality the p column vectors of Y can be assumed to be linearly independent since the exceptional set of Y's for which this does not hold has measure zero. The equality $Y'Y = Y^{*\prime}Y^*$ implies that $C = Y^*Y^{-1}$ is orthogonal and that $Y^* = CY$, as was to be proved. Suppose next that $r > p$. There is again no loss of generality in assuming the p column vectors of Y to be linearly independent. Since for any two p-dimensional subspaces of r-space there exists an orthogonal transformation taking one into the other, it can be assumed that (after a suitable orthogonal transformation) the p column vectors of Y and Y^* lie in the same p-space, and the problem is therefore reduced to the case $r = p$. If finally $r < p$, the first r column vectors of Y can be assumed to be linearly independent. Denoting the matrices formed by the first r and last $p - r$ columns of Y by Y_1 and Y_2 so that

$$Y = (Y_1 \quad Y_2),$$

one has $Y_1^{*\prime}Y_1^* = Y_1'Y_1$ and by the previous argument there exists an orthogonal matrix B such that $Y_1^* = BY_1$. From the relation $Y_1^{*\prime} Y_2^* = Y_1' Y_2$ it now follows that $Y_2^* = (Y_1^{*\prime})^{-1} Y_1' Y_2 = BY_2$, and this completes the proof.

Similarly the problem remains invariant under the orthogonal transformations

$$Z^* = DZ,$$

which leave $Z'Z$ as maximal invariant. Alternatively the reduction to $Z'Z$ can be argued from the fact that $Z'Z$ together with the Y's and U's form a set of sufficient statistics. In either case the problem under the groups G_1 and G_2 reduces to the two matrices $V = Y'Y$ and $S = Z'Z$.

G_3. We now impose the restriction $m \geqq p$ (see Problem 24), which assures that there are enough degrees of freedom to provide a reasonable estimate of the covariance matrix, and consider the transformations

$$Y^* = YB, \qquad Z^* = ZB,$$

where B is any nonsingular $p \times p$ matrix. These transformations act separately on each of the independent multivariate normal vectors $(Y_{\beta 1}, \cdots, Y_{\beta p})$, $(Z_{\gamma 1}, \cdots, Z_{\gamma p})$, and clearly leave the problem invariant. The induced transformation in the space of $V = Y'Y$ and $S = Z'Z$ is

$$V^* = B'VB, \qquad S^* = B'SB.$$

Since $|B'(V - \lambda S)B| = |B|^2 |V - \lambda S|$, the roots of the determinantal equation

$$|V - \lambda S| = 0 \tag{57}$$

are invariant under this group. To see that they are maximal invariant, suppose that the equations $|V - \lambda S| = 0$ and $|V^* - \lambda S^*| = 0$ have the same roots. One may again without loss of generality restrict attention to the case that p of the row vectors of Z are linearly independent, so that the matrix Z has rank p, and that the same is true of Z^*. The matrix S is then positive definite by Lemma 1 and it follows from the theory of the simultaneous reduction to diagonal form of two quadratic forms† that there exists a nonsingular matrix B_1 such that

$$B_1' V B_1 = \Lambda, \qquad B_1' S B_1 = I$$

where Λ is a diagonal matrix whose elements are the roots of (57) and I is the identity matrix. There also exists B_2 such that

$$B_2' V^* B_2 = \Lambda, \qquad B_2' S^* B_2 = I$$

and thus $B = B_1 B_2^{-1}$ transforms V into V^* and S into S^*.

† See for example Anderson, *An Introduction to Multivariate Statistical Analysis*, New York, John Wiley & Sons, 1958, Theorem 3 of Appendix 1.

Of the roots of (57), which constitute a maximal set of invariants, some may be zero. In fact, since these roots are the diagonal elements of Λ, the number of nonzero roots is equal to the rank of Λ and hence to the rank of $V = B_1'^{-1}\Lambda B_1^{-1}$, which by Lemma 1 is min (p, r). When this number is > 1, a UMP invariant test does not exist. The case $p = 1$ is that of a univariate linear hypothesis treated in Section 1. We shall now consider the remaining possibility that $r = 1$.

When $r = 1$, the equation (57), and hence the equivalent equation

$$|VS^{-1} - \lambda I| = 0,$$

has only one nonzero root. All coefficients of powers of λ of degree $< p - 1$ therefore vanish in the expression of the determinant as a polynomial in λ, and the equation becomes

$$(-\lambda)^p + W(-\lambda)^{p-1} = 0$$

where W is the sum of the diagonal elements (trace) of VS^{-1}. If S^{ij} denotes the i, jth element of S^{-1} and the single Y-vector is $(Y_1, \cdots, Y_p)$, an easy computation shows that

$$W = \sum_{i=1}^{p} \sum_{j=1}^{p} S^{ij} Y_i Y_j. \tag{58}$$

A necessary and sufficient condition for a test to be invariant under G_1, G_2, and G_3 is therefore that it depends only on W.

The distribution of W depends only on the maximal invariant in the parameter space; this is found to be

$$\psi^2 = \sum_{i=1}^{p} \sum_{j=1}^{p} a_{ij}\eta_i\eta_j \tag{59}$$

where $\eta_i = E(Y_i)$, and the probability density of W is given by (Problems 28–30)

$$p_\psi(w) = e^{-\frac{1}{2}\psi^2} \sum_{k=0}^{\infty} \frac{(\frac{1}{2}\psi^2)^k}{k!} c_k \frac{w^{\frac{1}{2}p-1+k}}{(1 + w)^{\frac{1}{2}(m+1)+k}}. \tag{60}$$

This is the same as the density (6) of the test statistic in the univariate case with $r = p$ and $n - s = m + 1 - p$. For any $\psi_0 < \psi_1$ the ratio $p_{\psi_1}(w)/p_{\psi_0}(w)$ is an increasing function of w and it follows from the Neyman-Pearson lemma that the most powerful invariant test for testing $H: \eta_1 = \cdots = \eta_p = 0$ rejects when W is too large or equivalently when

$$\frac{m + 1 - p}{p} W > C. \tag{61}$$

The quantity mW, which for $p = 1$ reduces to the square of Student's t, is essentially Hotelling's T^2-statistic to which it specializes in Example 8 below. The constant C is determined from the fact that for $\psi = 0$ the statistic $(m + 1 - p)W/p$ has the F-distribution with p and $m + 1 - p$ degrees of freedom. As in the univariate case, there also exists a UMP invariant test of the more general hypothesis H': $\psi^2 \leqq \psi_0^2$, with rejection region $W > C'$.

Since a UMP invariant test does not exist when $\min(p, r) > 1$, various functions of the roots λ_i of (57) have been proposed as test statistics for this case, among them the sum of the roots, the maximum or minimum root, and the product $\Pi_{i=1}^p (1 + \lambda_i)^{-1}$, which is the likelihood ratio criterion.

11. APPLICATIONS

The various univariate linear hypotheses with $r = 1$ such as that specifying the mean of a normal distribution, the difference of the means of two normal distributions with equal variance, the slope of a regression line, etc., can now be extended to the multivariate case.

Example 8. Let $(X_{\alpha 1}, \cdots, X_{\alpha p})$, $\alpha = 1, \cdots, n$, be a sample from a multivariate normal distribution with mean $(\xi_1, \cdots, \xi_p)$ and covariance matrix A^{-1} both unknown, and consider the problem of testing the hypothesis H: $\xi_1 = \cdots = \xi_p = 0$. It is seen from Example 4 that

$$\hat{\xi}_{\alpha i} = \sum_{\beta=1}^{n} X_{\beta i}/n = X_{\cdot i}; \qquad \hat{\hat{\xi}}_{\alpha i} = 0.$$

By (55), the ijth element S_{ij} of $S = Z'Z$ is therefore

$$S_{ij} = \sum_{\alpha=1}^{n} (X_{\alpha i} - X_{\cdot i})(X_{\alpha j} - X_{\cdot j})$$

and by (56)

$$Y_i Y_j = n X_{\cdot i} X_{\cdot j} \quad .$$

With these expressions the test statistic is the quantity W of (58), and the test is given by (61) with $s = 1$ and hence with $m = n - s = n - 1$. The statistic $T^2 = (n - 1)W$ is known as *Hotelling's* T^2.

Example 9. Let $(X_{\alpha 1}^{(1)}, \cdots, X_{\alpha p}^{(1)})$, $\alpha = 1, \cdots, n_1$, and $(X_{\beta 1}^{(2)}, \cdots, X_{\beta p}^{(2)})$, $\beta = 1, \cdots, n_2$, be independent samples from multivariate normal distributions with common covariance matrix A^{-1} and means $(\xi_1^{(1)}, \cdots, \xi_p^{(1)})$ and $(\xi_1^{(2)}, \cdots, \xi_p^{(2)})$, and consider the hypothesis H: $\xi_i^{(1)} = \xi_i^{(2)}$ for $i = 1, \cdots, p$. Then $s = 2$ and it follows from Example 5 that for all α and β

$$\hat{\xi}_{\alpha i}^{(1)} = X_{\cdot i}^{(1)}, \qquad \hat{\xi}_{\beta i}^{(2)} = X_{\cdot i}^{(2)}$$

and

$$\hat{\hat{\xi}}_{\alpha i}^{(1)} = \hat{\hat{\xi}}_{\beta i}^{(2)} = \left(\sum_{\alpha=1}^{n_1} X_{\alpha i}^{(1)} + \sum_{\beta=1}^{n_2} X_{\beta i}^{(2)} \right) \Big/ (n_1 + n_2) = \bar{X}_i \quad .$$

Hence

$$S_{ij} = \sum_{\alpha=1}^{n_1} (X_{\alpha i}^{(1)} - X_{\cdot i}^{(1)})(X_{\alpha j}^{(1)} - X_{\cdot j}^{(1)}) + \sum_{\beta=1}^{n_2} (X_{\beta i}^{(2)} - X_{\cdot i}^{(2)})(X_{\beta j}^{(2)} - X_{\cdot j}^{(2)}),$$

and the expression for $Y_i Y_j$ can be simplified to

$$Y_i Y_j = n_1(X_{\cdot i}^{(1)} - \bar{X}_i)(X_{\cdot j} - \bar{X}_j) + n_2(X_{\cdot i}^{(2)} - \bar{X}_i)(X_{\cdot j}^{(2)} - \bar{X}_j).*$$

In addition to the above and other similar extensions of univariate hypotheses, the test (61) can also be applied to certain problems which are not themselves linear hypotheses as defined in Section 9, but which reduce to this form through invariance considerations. Let $(X_{\alpha 1}, \cdots, X_{\alpha p})$, $\alpha = 1, \cdots, n$, be a sample from a multivariate normal distribution with mean $(\xi_1, \cdots, \xi_p)$ and covariance matrix A^{-1}, and consider the hypothesis that the vector $(\xi_1, \cdots, \xi_p)$ lies in a $(p - r)$-dimensional subspace of p-space. The observations can be transformed in the usual manner to a set of variables $(Y_{\alpha 1}, \cdots, Y_{\alpha r}, Z_{\alpha 1}, \cdots, Z_{\alpha l})$, $p = r + l$, constituting a sample from a p-variate normal distribution with mean $(\eta_1, \cdots, \eta_r, \zeta_1, \cdots, \zeta_l)$ such that the hypothesis becomes H: $\eta_1 = \cdots = \eta_r = 0$.

This problem remains invariant under a group of linear transformations for which the Y's are a maximal set of invariants, and in terms of the Y's the hypothesis reduces to that treated above in Example 8. There exists therefore a UMP invariant test of H given by (61) with $p = r$ and $m = n - 1$. Before proving that the Z's can be discarded, we shall give two illustrations of this type of problem.

Example 10. Let $(X_{\alpha 1}, \cdots, X_{\alpha q}, X_{\alpha, q+1}, \cdots, X_{\alpha, 2q})$, $\alpha = 1, \cdots, n$, be a sample from a multivariate normal distribution, and consider the problem of testing H: $\xi_{q+i} = \xi_i$ for $i = 1, \cdots, q$. This might arise for example when $X_{\alpha 1}, \cdots, X_{\alpha q}$ and $X_{\alpha, q+1}, \cdots, X_{\alpha, 2q}$ are q measurements taken on the same subject at two different periods after a certain treatment, or taken on the left and right sides of the subject. In terms of the variables

$$Y_{\alpha i} = X_{\alpha, q+i} - X_{\alpha i}; \qquad Z_{\alpha i} = X_{\alpha i} \qquad (\alpha = 1, \cdots, n;\ i = 1, \cdots, q),$$

the hypothesis becomes $\eta_i = E(Y_{\alpha i}) = 0$ for $i = 1, \cdots, q$, and the UMP invariant test consists in applying the test of Example 8 to the Y's with q in place of p.

Example 11. Let $(X_{\alpha 1}, \cdots, X_{\alpha p})$, $\alpha = 1, \cdots, n$, be a sample from a p-variate normal distribution, and consider the problem of testing the hypothesis H: $\xi_1 = \cdots = \xi_p$. In terms of the new variables $Y_{\alpha i} = X_{\alpha i} - X_{\alpha p}$ $(i = 1, \cdots, p - 1)$ and $Z_\alpha = X_{\alpha p}$, the hypothesis again has the canonical form $\eta_1 = \cdots = \eta_{p-1} = 0$, and the problem reduces to that of Example 8 with $p - 1$ in place of p. As an application suppose that a shop has p machines for manufacturing a certain product, the quality of which is measured by a random variable X. In an experiment, n workers are put on each of the machines, with $X_{\alpha i}$ being the result of the αth worker on the ith machine. If the n workers are considered as a

* A test of H for the case that $p > n_1 + n_2 - 2$ is discussed by Dempster, "A high dimensional two-sample significance test," *Ann. Math. Stat.*, Vol. 29 (1958), pp. 995–1010.

random sample from a large population, the vectors $(X_{\alpha 1}, \cdots, X_{\alpha p})$ may be assumed to be a sample from a p-variate normal distribution. Of the two factors involved in this experiment one is fixed (machines) and one random (workers), in the sense that a replication of the experiment would employ the same machines but a new sample of workers. The hypothesis being tested is that the fixed effect is absent. The test in this mixed model is quite different from the corresponding model I test where both effects are fixed, and which was treated in Section 4.

We return now to the general case of a sample $(Y_{\alpha 1}, \cdots, Y_{\alpha r}, Z_{\alpha 1}, \cdots, Z_{\alpha l})$ $\alpha = 1, \cdots, n$, from a p-variate normal distribution with mean $(\eta_1, \cdots, \eta_r, \zeta_1, \cdots, \zeta_l)$ and the hypothesis $\eta_1 = \cdots = \eta_r = 0$ which was illustrated by Examples 10 and 11. Interpreting the set of pn variables for a moment as the set of $p = r + l$ vectors in n-space, $(Y_{1i}, \cdots, Y_{ni})$, $i = 1, \cdots, r$, and $(Z_{1j}, \cdots, Z_{nj})$, $j = 1, \cdots, l$, consider an orthogonal transformation of n-space which transforms $(x_1, \cdots, x_n)$ into $(x_1', \cdots, x_n')$ such that $x_1' = \sqrt{n}\, \bar{x}$. Let this transformation be applied to each of the p observation vectors, and let the transforms of $(Y_{1i}, \cdots, Y_{ni})$ and $(Z_{1j}, \cdots, Z_{nj})$ be denoted by $(U_{1i}, \cdots, U_{ni})$ and $(V_{1j}, \cdots, V_{nj})$ respectively. Then in particular $U_{1i} = \sqrt{n}\, Y_{\cdot i}$, $V_{1j} = \sqrt{n}\, Z_{\cdot j}$, and the sets of variables $(U_{\alpha 1}, \cdots, U_{\alpha r}, V_{\alpha 1}, \cdots, V_{\alpha l})$, $\alpha = 1, \cdots, n$, are independently distributed, each according to an $(r + l)$-variate normal distribution with common covariance matrix, and means $E(U_{1i}) = \sqrt{n}\, \eta_i$, $E(V_{1j}) = \sqrt{n}\, \zeta_j$, and $E(U_{\alpha i}) = E(V_{\alpha j}) = 0$ for $\alpha > 1$.

Letting

$$U = \begin{pmatrix} U_{21} & \cdots & U_{2r} \\ \cdot & & \cdot \\ \cdot & & \cdot \\ \cdot & & \cdot \\ U_{n1} & \cdots & U_{nr} \end{pmatrix} \quad \text{and} \quad V = \begin{pmatrix} V_{21} & \cdots & V_{2l} \\ \cdot & & \cdot \\ \cdot & & \cdot \\ \cdot & & \cdot \\ V_{n1} & \cdots & V_{nl} \end{pmatrix},$$

it is seen that the following two groups leave the problem invariant.

G_1. Addition of an arbitrary constant c_j to each of the variables V_{1j}, $j = 1, \cdots, l$.

G_2. The transformations

$$V^* = UB + VC, \qquad U^* = U$$

where B is any $r \times l$ and C any nonsingular $l \times l$ matrix.

Before applying the principle of invariance, it will be convenient to reduce the problem by sufficiency.

The variables U_{1i}, V_{1j} together with the matrices of inner products $U'U$, $U'V$, and $V'V$ form a set of sufficient statistics for the unknown vector mean and covariance matrix, and by Problem 1 of Chapter 6 the groups G_1 and G_2 also leave the problem invariant if it is first reduced to the sufficient statistics. A maximal set of invariants with respect to G_1 are the U_{1i} and the matrices $U'U$, $U'V$, and $V'V$. We shall now prove that under the group which G_2 induces on this set of statistics, the U_{1i} and $U'U$ are maximal invariant. This will complete the desired elimination of the V's and hence of the Z's.

To prove this, it is necessary to show that for any given $(n - 1) \times l$ matrix V^{**} there exist B and C such that $V^* = UB + VC$ satisfies

$$U'V^* = U'V^{**} \quad \text{and} \quad V^{*\prime}V^* = V^{**\prime}V^{**}.$$

Geometrically, these equations state that there exist vectors $(V^*_{2i}, \cdots, V^*_{ni})$ $i = 1, \cdots, l$, which lie in the space E spanned by the column vectors of U and V, and which have a preassigned set of inner products among each other and with the column vectors of U.

Consider first the case $l = 1$. If $r + 1 \geqq n - 1$, one can assume that V and the columns of U span the $(n - 1)$-dimensional space, and one can then take $V^* = V^{**}$. If $r + 1 < n - 1$, V and the columns of U may be assumed to be linearly independent. There then exists a rotation about the space spanned by the columns of U as axis, which takes V^{**} into a vector lying in E, and this vector has the properties required of V^*.

The proof is now completed by repeated application of the result for this special case. It can be applied first to the vector $(V_{21}, \cdots, V_{n1})$, to determine the first column of B and a number c_{11} to which one may add zeros to construct the first column of C. By adjoining the transformed vector $(V^*_{21}, \cdots, V^*_{n1})$ to the columns of U and applying the result to the vector $(V_{22}, \cdots, V_{n2})$, one obtains a vector $(V^*_{22}, \cdots, V^*_{n2})$ which lies in the space spanned by $(V_{21}, \cdots, V_{n1})$, $(V_{22}, \cdots, V_{n2})$ and the column vectors of U, and which in addition has the preassigned inner products with $(V^*_{21}, \cdots, V^*_{n1})$, with the columns of U and with itself. This second step determines the second column of B and two numbers c_{12}, c_{22} to which zeros can be added to provide the second column of C. Proceeding inductively in this way, one obtains for C a triangular matrix with zeros below the main diagonal, so that C is nonsingular. Since U, V, and V^{**} can be assumed to have maximal rank, it follows from Lemma 1 and the equation $V^{*\prime}V^* = V^{**\prime}V^{**}$ that the rank of V^* is also maximal, and this completes the proof.

12. χ^2 TESTS: SIMPLE HYPOTHESIS AND UNRESTRICTED ALTERNATIVES

UMP invariant tests exist only for rather restricted classes of problems, among which linear hypotheses are perhaps the most important. However, when the number of observations is large, there frequently exist tests which possess this property at least approximately. Although a detailed treatment of large-sample theory is outside the scope of this book, we shall indicate briefly the theory of two types of tests possessing such properties: χ^2 tests and likelihood ratio tests. In both cases the approximate optimum property is a consequence of the asymptotic equivalence of the problem with one of testing a linear hypothesis. This relationship will be sketched in the next section. As preparation we discuss first a special class of χ^2 problems.

It will be convenient to begin by considering the following modification of the linear hypothesis model. Let $Y = (Y_1, \cdots, Y_q)$ have the multivariate normal probability density

$$\frac{\sqrt{|A|}}{(2\pi)^{\frac{1}{2}q}} \exp\left[-\tfrac{1}{2}\sum_{i=1}^{q}\sum_{j=1}^{q} a_{ij}(y_i - \eta_i)(y_j - \eta_j)\right] \tag{62}$$

with known covariance matrix A^{-1}. The point of means $\eta = (\eta_1, \cdots, \eta_q)$ is known to lie in a given s-dimensional linear space Π_Ω with $s \leqq q$; the hypothesis to be tested is that η lies in a given $(s - r)$-dimensional linear subspace Π_ω of Π_Ω $(r \leqq s)$. This problem is invariant under a suitable group G of linear transformations, and there exists a UMP invariant test with respect to G, given by the rejection region

$$\begin{aligned} \Sigma\Sigma a_{ij}(y_i - \hat{\hat{\eta}}_i)(y_j - \hat{\hat{\eta}}_j) - \Sigma\Sigma a_{ij}(y_i - \hat{\eta}_i)(y_j - \hat{\eta}_j) \\ = \Sigma\Sigma a_{ij}(\hat{\eta}_i - \hat{\hat{\eta}}_i)(\hat{\eta}_j - \hat{\hat{\eta}}_j) \geqq C. \end{aligned} \tag{63}$$

Here $\hat{\eta}$ is the point of Π_Ω which is closest to the sample point y in the metric defined by the quadratic form $\Sigma\Sigma a_{ij}x_i x_j$, that is, which minimizes the quantity $\Sigma\Sigma a_{ij}(y_i - \eta_i)(y_j - \eta_j)$ for η in Π_Ω. Similarly $\hat{\hat{\eta}}$ is the point in Π_ω minimizing this quantity.

When the hypothesis is true, the left-hand side of (63) has a χ^2-distribution with r degrees of freedom, so that C is determined by

$$\int_C^\infty \chi_r^2(z)\, dz = \alpha. \tag{64}$$

When η is not in Π_ω, the probability of rejection is*

$$\int_C^\infty p_\lambda(z)\, dz \tag{65}$$

where $p_\lambda(z)$ is the noncentral χ^2 density [(86) of Problem 2] with r degrees

* Tables are given by Patnaik, "The non-central χ^2 and F-distributions and their applications," *Biometrika*, Vol. 36 (1949), pp. 202–232; by Fix, "Tables of noncentral χ^2," *Univ. Calif. Publ. Statistics*, Vol. 1 (1949), pp. 15–19; and by Fix, Hodges, and Lehmann, "The restricted χ^2 test," in *Studies in Probability and Statistics Dedicated to Harald Cramér*, Almquist and Wiksell, Stockholm, 1959.

of freedom and noncentrality parameter λ^2 obtained by replacing $y_i, \hat{\eta}_i, \hat{\hat{\eta}}_i$ in (63) by their expectations, or equivalently, if (63) is considered as a function of y, by replacing y by η throughout. This expression for the power is valid even when the assumed model is not correct so that $E(Y) = \eta$ does not lie in Π_Ω. For the particular case that $\eta \in \Pi_\Omega$, the second term in this expression for λ^2 equals 0. A proof of the above statements is obtained by reducing the problem to a linear hypothesis through a suitable linear transformation. (See Problem 33).

Returning to the theory of χ^2 tests, which deals with hypotheses concerning multinomial distributions, consider n multinomial trials with m possible outcomes. If $p = (p_1, \cdots, p_m)$ denotes the probabilities of these outcomes and X_i the number of trials resulting in the ith outcome, the distribution of $X = (X_1, \cdots, X_m)$ is

$$(66) \qquad P(x_1, \cdots, x_m) = \frac{n!}{x_1! \cdots x_m!} p_1^{x_1} \cdots p_m^{x_m} \qquad (\Sigma x_i = n, \Sigma p_i = 1).$$

The simplest χ^2 problems are those of testing a hypothesis $H: p = \pi$ where $\pi = (\pi_1, \cdots, \pi_m)$ is given, against the unrestricted alternatives $p \neq \pi$. As $n \to \infty$, the power of the tests to be considered will tend to one against any fixed alternative.* In order to study the power function of such tests for large n, it is of interest to consider a sequence of alternatives $p^{(n)}$ tending to π as $n \to \infty$. If the rate of convergence is faster than $1/\sqrt{n}$, the power of even the most powerful test will tend to the level of significance α. The sequences reflecting the aspects of the power that are of greatest interest, and which are most likely to provide a useful approximation to the actual power for large but finite n, are the sequences for which $\sqrt{n}(p^{(n)} - \pi)$ tends to a nonzero limit, so that

$$(67) \qquad p_i^{(n)} = \pi_i + \frac{\Delta_i}{\sqrt{n}} + R_n$$

say, where $\sqrt{n}R_n$ tends to zero as n tends to infinity.

Let

$$(68) \qquad Y_i = (X_i - n\pi_i)/\sqrt{n}.$$

Then $\sum_{i=1}^m Y_i = 0$, and the mean of Y_i is zero under H and tends to Δ_i under the alternatives (67). The covariance matrix of the Y's is

$$(69) \qquad \sigma_{ij} = -\pi_i\pi_j \quad \text{if} \quad i \neq j; \qquad \sigma_{ii} = \pi_i(1 - \pi_i)$$

when H is true, and tends to these values for the alternatives (67). As $n \to \infty$, the distribution of $Y = (Y_1, \cdots, Y_{m-1})$ tends to the multivariate normal distribution with means $E(Y_i) = 0$ under H and $E(Y_i) = \Delta_i$ for

* A sequence of tests with this property is called *consistent*.

the sequence of alternatives (67), and with covariance matrix (69) in both cases.* The density of the limiting distribution is

$$c \exp\left[-\frac{1}{2}\left(\sum_{i=1}^{m-1}\frac{(y_i-\Delta_i)^2}{\pi_i}+\frac{\left(\sum_{j=1}^{m-1}(y_j-\Delta_j)\right)^2}{\pi_m}\right)\right] \tag{70}$$

and the hypothesis to be tested becomes $H: \Delta_1 = \cdots = \Delta_{m-1} = 0$.

According to (63), the UMP invariant test in this asymptotic model rejects when

$$\sum_{i=1}^{m-1}\frac{y_i^2}{\pi_i}+\frac{1}{\pi_m}\left(\sum_{j=1}^{m-1}y_j\right)^2 > C$$

and hence when

$$n\sum_{i=1}^{m}\frac{(\nu_i-\pi_i)^2}{\pi_i} > C \tag{71}$$

where $\nu_i = X_i/n$ and C is determined by (64) with $r = m - 1$. The limiting power of the test against the sequence of alternatives (67) is given by (65) with $\lambda^2 = \sum_{i=1}^{m}\Delta_i^2/\pi_i$. This provides an approximation to the power for fixed n and a particular alternative p if one identifies p with $p^{(n)}$ for this value of n. From (67) one finds approximately $\Delta_i = \sqrt{n}(p_i - \pi_i)$, so that the noncentrality parameter becomes

$$\lambda^2 = n\sum_{i=1}^{m}\frac{(p_i-\pi_i)^2}{\pi_i}. \tag{72}$$

Example 12. Suppose the hypothesis is to be tested that certain events (births, deaths, accidents) occur uniformly over a stated time interval such as a day or a year. If the time interval is divided into m equal parts and p_i denotes the probability of an occurrence in the ith subinterval, the hypothesis becomes H: $p_i = 1/m$ for $i = 1, \cdots, m$. The test statistic is then

$$mn\sum_{i=1}^{m}\left(\nu_i-\frac{1}{m}\right)^2$$

where ν_i is the relative frequency of occurrence in the ith subinterval. The approximate power of the test is given by (65) with $r = m - 1$ and $\lambda^2 = mn\Sigma_{i=1}^{m}[p_i - (1/m)]^2$.

13. χ^2 AND LIKELIHOOD RATIO TESTS

It is both a strength and a weakness of the χ^2 test of the preceding section that its asymptotic power depends only on the weighted sum of squared deviations (72), not on the signs of these deviations and their distribution over the different values of i. This is an advantage if no

* A proof assuming H is given for example by Cramér, *Mathematical Methods of Statistics*, Princeton Univ. Press, 1946, Section 30.1. It carries over with only the obvious changes to the case that H is not true.

knowledge is available concerning the alternatives since the test then provides equal protection against all alternatives that are equally distant from $H: p = \pi$ in the metric (72). However, frequently one does know the type of deviations to be expected if the hypothesis is not true, and in such cases the test can be modified so as to increase its asymptotic power against the alternatives of interest by concentrating it on these alternatives.

To derive the modified test, suppose that a restricted class of alternatives to H has been defined

$$K: p \in \mathscr{S}, \qquad p \neq \pi.$$

Let the surface $\mathscr{S}$ have a parametric representation

$$p_i = f_i(\theta_1, \cdots, \theta_s), \qquad i = 1, \cdots, m$$

and let

$$\pi_i = f_i(\theta_1^0, \cdots, \theta_s^0).$$

Suppose that the θ_j are real-valued, that the derivatives $\partial f_i/\partial\theta_j$ exist and are continuous at θ^0, and that the Jacobian matrix $(\partial f_i/\partial\theta_j)$ has rank s at θ^0. If $\theta^{(n)}$ is any sequence such that

$$\sqrt{n}(\theta_j^{(n)} - \theta_j^0) \to \delta_j, \tag{73}$$

the limiting distribution of the variables $(Y_1, \cdots, Y_{m-1})$ of the preceding section is normal with mean

$$E(Y_i) = \Delta_i = \sum_{j=1}^{s} \delta_j \left.\frac{\partial f_i}{\partial\theta_j}\right|_{\theta^0} \tag{74}$$

and covariance matrix (69). This is seen by expanding f_i about the point θ^0 and applying the limiting distribution (70). The problem of testing H against all sequences of alternatives in K satisfying (73) is therefore asymptotically equivalent to testing the hypothesis

$$\Delta_1 = \cdots = \Delta_{m-1} = 0$$

in the family (70) against the alternatives $\bar{K}$: $(\Delta_1, \cdots, \Delta_{m-1}) \in \Pi_\Omega$ where Π_Ω is the linear space formed by the totality of points with coordinates

$$\Delta_i = \sum_{j=1}^{s} \delta_j \left.\frac{\partial f_i}{\partial\theta_j}\right|_{\theta^0}. \tag{75}$$

We note for later use that for any fixed n, the totality of points

$$p_i = \pi_i + \frac{\Delta_i}{\sqrt{n}}, \qquad i = 1, \cdots, m$$

with the Δ_i satisfying (75), constitute the tangent plane to $\mathscr{S}$ at π, which will be denoted by $\bar{\mathscr{S}}$.

Let $(\hat{\Delta}_1, \cdots, \hat{\Delta}_m)$ be the values minimizing $\sum_{i=1}^m (y_i - \Delta_i)^2/\pi_i$ subject to the conditions $(\Delta_1, \cdots, \Delta_{m-1}) \in \Pi_\Omega$ and $\Delta_m = -(\Delta_1 + \cdots + \Delta_{m-1})$.

Then by (63), the asymptotically UMP invariant test rejects H in favor of $\bar{K}$ if

$$\frac{\sum_{i=1}^{m} y_i^2}{\pi_i} - \frac{\sum_{i=1}^{m} (y_i - \hat{\Delta}_i)^2}{\pi_i} = \frac{\sum_{i=1}^{m} \hat{\Delta}_i^2}{\pi_i} > C,$$

or equivalently if

$$\text{(76)} \qquad \frac{n\sum_{i=1}^{m} (\nu_i - \pi_i)^2}{\pi_i} - \frac{n\sum_{i=1}^{m} (\nu_i - \hat{p}_i)^2}{\pi_i} = \frac{n\sum_{i=1}^{m} (\hat{p}_i - \pi_i)^2}{\pi_i} > C,$$

where the $\hat{p}_i$ minimize $\Sigma(\nu_i - p_i)^2/\pi_i$ subject to $p \in \bar{\mathscr{S}}$. The constant C is determined by (64) with $r = s$. An asymptotically equivalent test, which, however, frequently is more difficult to compute explicitly, is obtained by letting the $\hat{p}_i$ be the minimizing values subject to $p \in \mathscr{S}$ instead of $p \in \bar{\mathscr{S}}$. An approximate expression for the power of the test against an alternative p is given by (65) with λ^2 obtained from (76) by substituting p_i for ν_i when $\hat{p}_i$ are considered as functions of the ν_i.

Example 13. Suppose that in Example 12, where the hypothesis of a uniform distribution is being tested, the alternatives of interest are those of a cyclic movement, which may be represented at least approximately by a sine wave

$$p_i = \frac{1}{m} + \rho \int_{(i-1)\frac{2\pi}{m}}^{i\frac{2\pi}{m}} \sin(u - \theta)\, du, \qquad i = 1, \cdots, m.$$

Here ρ is the amplitude and θ the phasing of the cyclic disturbance. Putting $\xi = \rho\cos\theta$, $\eta = \rho\sin\theta$, we get

$$p_i = \frac{1}{m}(1 + a_i\xi + b_i\eta)$$

where

$$a_i = 2m \sin\frac{\pi}{m}\sin(2i-1)\frac{\pi}{m}, \qquad b_i = -2m\sin\frac{\pi}{m}\cos(2i-1)\frac{\pi}{m}.$$

The equations for p_i define the surface $\mathscr{S}$, which in the present case is a plane so that it coincides with $\bar{\mathscr{S}}$.

The quantities $\hat{\xi}$, $\hat{\eta}$ minimizing $\Sigma(\nu_i - p_i)^2/\pi_i$ subject to $p \in \mathscr{S}$ are

$$\hat{\xi} = \Sigma a_i\nu_i/\Sigma a_i^2\pi_i, \qquad \hat{\eta} = \Sigma b_i\nu_i/\Sigma b_i^2\pi_i$$

with $\pi_i = 1/m$. Let $m > 2$. Using the fact that $\Sigma a_i = \Sigma b_i = \Sigma a_i b_i = 0$ and that

$$\sum_{i=1}^{m} \sin^2(2i-1)\frac{\pi}{m} = \sum_{i=1}^{m}\cos^2(2i-1)\frac{\pi}{m} = \frac{m}{2},$$

the test becomes after some simplification

$$2n\left[\sum_{i=1}^{m} \nu_i \sin(2i-1)\frac{\pi}{m}\right]^2 + 2n\left[\sum_{i=1}^{m}\nu_i\cos(2i-1)\frac{\pi}{m}\right]^2 > C$$

where the number of degrees of freedom of the left-hand side is $s = 2$. The noncentrality parameter determining the approximate power is

$$\lambda^2 = n\left(\xi m \sin\frac{\pi}{m}\right)^2 + n\left(\eta m \sin\frac{\pi}{m}\right)^2 = n\rho^2 m^2 \sin^2\frac{\pi}{m}.$$

The χ^2 tests discussed so far were for simple hypotheses. Consider now the more general problem of testing $H: p \in \mathscr{T}$ against the alternatives $K: p \in \mathscr{S}, p \notin \mathscr{T}$ where $\mathscr{T} \subset \mathscr{S}$ and where $\mathscr{S}$ and $\mathscr{T}$ have parametric representations

$$\mathscr{S}: p_i = f_i(\theta_1, \cdots, \theta_s); \qquad \mathscr{T}: p_i = f_i(\theta_1^0, \cdots, \theta_r^0, \theta_{r+1}, \cdots, \theta_s).$$

The basis for a large-sample analysis of this problem is the fact that for large n a sphere of radius $\rho/\sqrt{n}$ can be located which for sufficiently large ρ contains the true point p with arbitrarily high probability. Attention can therefore be restricted to sequences of points $p^{(n)} \in \mathscr{S}$ which tend to some fixed point $\pi \in \mathscr{T}$ at the rate of $1/\sqrt{n}$. More specifically, let $\pi_i = f_i(\theta_1^0, \cdots, \theta_s^0)$ and let $\theta^{(n)}$ be a sequence satisfying (73). Then the variables $(Y_1, \cdots, Y_{m-1})$ have a normal limiting distribution with covariance matrix (69) and a vector of means given by (74). Let Π_Ω be defined as before, let Π_ω be the linear space

$$\Pi_\omega: \Delta_i = \sum_{j=r+1}^{s} \delta_j \left.\frac{\partial p_i}{\partial \theta_j}\right|_{\theta^0},$$

and consider the problem of testing that $p^{(n)}$ is a sequence in H for which $\theta^{(n)}$ satisfies (73) against all sequences in K satisfying this condition. This is asymptotically equivalent to the problem, discussed at the beginning of Section 12, of testing $(\Delta_1, \cdots, \Delta_{m-1}) \in \Pi_\omega$ in the family (70) when it is given that $(\Delta_1, \cdots, \Delta_{m-1}) \in \Pi_\Omega$. By (63), the rejection region for this problem is

$$\Sigma(y_i - \hat{\hat{\Delta}}_i)^2/\pi_i - \Sigma(y_i - \hat{\Delta}_i)^2/\pi_i > C$$

where the $\hat{\Delta}_i$ and $\hat{\hat{\Delta}}_i$ minimize $\Sigma(y_i - \Delta_i)^2/\pi_i$ subject to $\Delta_m = -(\Delta_1 + \cdots + \Delta_{m-1})$ and $(\Delta_1, \cdots, \Delta_{m-1})$ in Π_Ω and Π_ω respectively. In terms of the original variables, the rejection region becomes

$$n\Sigma(\nu_i - \hat{\hat{p}}_i)^2/\pi_i - n\Sigma(\nu_i - \hat{p}_i)^2/\pi_i > C. \tag{77}$$

Here the $\hat{p}_i$ and $\hat{\hat{p}}_i$ minimize

$$\Sigma(\nu_i - p_i)^2/\pi_i \tag{78}$$

when p is restricted to lie in the tangent plane at π to $\mathscr{S}$ and $\mathscr{T}$ respectively, and the constant C is determined by (64).

The above solution of the problem depends on the point π which is not given. A test which is asymptotically equivalent to (77) and does not depend on π is obtained if $\hat{p}_i$ and $\hat{\hat{p}}_i$ are replaced by p_i^* and p_i^{**} which minimize (78) for p restricted to $\mathscr{S}$ and $\mathscr{T}$ instead of to their tangents, and if further π_i is replaced in (77) and (78) by a suitable estimate, for example by ν_i. This leads to the rejection region

(79) $$n\Sigma(\nu_i - p_i^{**})^2/\nu_i - n\Sigma(\nu_i - p_i^*)^2/\nu_i = n\Sigma(p_i^* - p_i^{**})^2/\nu_i > C$$

where the p_i^{**} and p_i^* minimize

(80) $$\Sigma(\nu_i - p_i)^2/\nu_i$$

subject to $p \in \mathscr{T}$ and $p \in \mathscr{S}$ respectively, and where C is determined by (64) as before. An approximation to the power of the test for fixed n and a particular alternative p is given by (65) with λ^2 obtained from (79) by substituting p_i for ν_i when the p_i^* and p_i^{**} are considered as functions of the ν_i.†

A more general large-sample approach, which unlike χ^2 is not tied to the multinomial distribution, is based on the method of maximum likelihood. We shall here indicate this theory only briefly, and in particular shall state the main facts without the rather complex regularity assumptions required for their validity.‡

Let $p_\theta(x)$, $\theta = (\theta_1, \cdots, \theta_r)$ be a family of univariate probability densities and consider the problem of testing, on the basis of a (large) sample $X_1, \cdots, X_n$, the simple hypothesis $H: \theta_i = \theta_i^0$, $i = 1, \cdots, r$. Let $\hat{\theta} = (\hat{\theta}_1, \cdots, \hat{\theta}_r)$ be the maximum likelihood estimate of θ, that is, the parameter vector maximizing $p_\theta(x_1) \cdots p_\theta(x_n)$. Then asymptotically as $n \to \infty$, attention can be restricted to the $\hat{\theta}_i$ since they are "asymptotically sufficient."§ The power of the tests to be considered will tend to one against any fixed alternative, and the alternatives of interest similarly as in the χ^2 case are sequences $\theta_i^{(n)}$ satisfying

(81) $$\sqrt{n}\,(\theta_i^{(n)} - \theta_i^0) \to \Delta_i.$$

If $Y_i = \sqrt{n}(\hat{\theta}_i - \theta_i^0)$, the limiting distribution of $Y_1, \cdots, Y_r$ is the multivariate normal distribution (62) with

(82) $$a_{ij} = a_{ij}(\theta^0) = -E\left(\frac{\partial^2 \log p_\theta(X)}{\partial\theta_i\, \partial\theta_j}\right)\Bigg|_{\theta=\theta^0}$$

† For a proof of the above statements and a discussion of certain tests which are asymptotically equivalent to (76) and sometimes easier to determine explicitly, see Fix, Hodges, and Lehmann, *loc. cit.*

‡ For a detailed treatment see Wald, "Tests of statistical hypotheses concerning several parameters when the number of observations is large," *Trans. Am. Math. Soc.*, Vol. 54 (1943), pp. 426–483.

§ This was shown by Wald, *loc. cit.*; for a definition of asymptotic sufficiency and further results concerning this concept see LeCam, "On the asymptotic theory of estimation and testing hypotheses," *Proc. Third Berkeley Symposium on Mathematical Statistics and Probability*, Univ. Calif. Press, 1956.

and with $\eta_i = 0$ under H and $\eta_i = \Delta_i$ for the alternatives satisfying (81).

By (63), the UMP invariant test in this asymptotic model rejects when

$$-\sum_{i=1}^{r}\sum_{j=1}^{r} a_{ij}n(\hat{\theta}_i - \theta_i^0)(\hat{\theta}_j - \theta_j^0) > C. \tag{83}$$

Under H, the left-hand side has a limiting χ^2-distribution with r degrees of freedom, while under the alternatives (81) the limiting distribution is noncentral χ^2 with noncentrality parameter

$$\lambda^2 = -\sum_{i=1}^{r}\sum_{j=1}^{r} a_{ij}n(\theta_i^{(n)} - \theta_i^0)(\theta_j^{(n)} - \theta_j^0). \tag{84}$$

The approximate power against a specific alternative θ is therefore given by (65), with λ^2 obtained from (84) by substituting θ for $\theta^{(n)}$.

The test (83) is asymptotically equivalent to the likelihood ratio test, which rejects when

$$\Lambda_n = \frac{p_{\theta^0}(x_1)\cdots p_{\theta^0}(x_n)}{p_{\hat{\theta}}(x_1)\cdots p_{\hat{\theta}}(x_n)} < k. \tag{85}$$

This is seen by expanding $\sum_{\nu=1}^{n} \log p_{\theta^0}(x_\nu)$ about $\sum_{\nu=1}^{n} \log p_{\hat{\theta}}(x_\nu)$ and using the fact that at $\theta = \hat{\theta}$ the derivatives $\partial\Sigma \log p_\theta(x_\nu)/\partial\theta_i$ are zero. Application of the law of large numbers shows that $-2 \log \Lambda_n$ differs from the left-hand side of (83) by a term tending to zero in probability as $n \to \infty$. In particular, the two statistics therefore have the same limiting distribution.

The extension of this method to composite hypotheses is quite analogous to the corresponding extension in the χ^2 case. Let $\theta = (\theta_1, \cdots, \theta_s)$ and $H: \theta_i = \theta_i^0$ for $i = 1, \cdots, r$ $(r < s)$. If attention is restricted to sequences $\theta^{(n)}$ satisfying (81) for $i = 1, \cdots, s$ and some arbitrary $\theta_{r+1}^0, \cdots, \theta_s^0$, the asymptotic problem becomes that of testing $\eta_1 = \cdots = \eta_r = 0$ against unrestricted alternatives $(\eta_1, \cdots, \eta_s)$ for the distributions (62) with $a_{ij} = a_{ij}(\theta^0)$ given by (82). Then $\hat{\eta}_i = Y_i$ for all i, while $\hat{\hat{\eta}}_i = 0$ for $i = 1, \cdots, r$ and $= Y_i$ for $i = r + 1, \cdots, s$, so that the UMP invariant test is given by (83). The coefficients $a_{ij} = a_{ij}(\theta^0)$ depend on $\theta_{r+1}^0, \cdots, \theta_s^0$ but as before an asymptotically equivalent test statistic is obtained by replacing $a_{ij}(\theta^0)$ by $a_{ij}(\hat{\theta})$. Again, the statistic is also asymptotically equivalent to minus twice the logarithm of the likelihood ratio, and the test is therefore asymptotically equivalent to the likelihood ratio test.*

* The asymptotic theory of likelihood ratio tests has been extended to more general types of problems, including in particular the case of restricted classes of alternatives, by Chernoff, "On the distribution of the likelihood ratio," *Ann. Math. Stat.*, Vol. 25 (1954), pp. 573–578.

14. PROBLEMS

Section 1

1. *Expected sums of squares.* The expected value of the numerator and denominator of the statistic W^* defined by (7) is

$$E\left(\sum_{i=1}^{r} Y_i^2/r\right) = \sigma^2 + \frac{1}{r}\sum_{i=1}^{r}\eta_i^2 \quad \text{and} \quad E\left[\sum_{i=s+1}^{n} Y_i^2/(n-s)\right] = \sigma^2.$$

2. *Noncentral χ^2-distribution.* (i) If X is distributed as $N(\psi, 1)$, the probability density of $V = X^2$ is $p_\psi^V(v) = \sum_{k=0}^\infty P_k(\psi)f_{2k+1}(v)$, where $P_k(\psi) = (\psi^2/2)^k e^{-\frac{1}{2}\psi^2}/k!$ and where f_{2k+1} is the probability density of a χ^2 variable with $2k + 1$ degrees of freedom.

(ii) Let $Y_1, \cdots, Y_r$ be independently normally distributed with unit variance and means $\eta_1, \cdots, \eta_r$. Then $U = \sum Y_i^2$ is distributed according to the noncentral χ^2-distribution with r degrees of freedom and noncentrality parameter $\psi^2 = \sum_{i=1}^r \eta_i^2$, which has probability density

$$p_\psi^U(u) = \sum_{k=0}^{\infty} P_k(\psi) f_{r+2k}(u). \tag{86}$$

Here $P_k(\psi)$ and $f_{r+2k}(u)$ have the same meaning as in (i) so that the distribution is a mixture of χ^2-distributions with Poisson weights.

[(i) This is seen from

$$p_\psi^V(v) = e^{-\frac{1}{2}(\psi^2+v)}(e^{\psi\sqrt{v}} + e^{-\psi\sqrt{v}})/2\sqrt{2\pi v}$$

by expanding the expression in parentheses into a power series, and using the fact that $\Gamma(2k) = 2^{2k-1}\Gamma(k)\Gamma(k+\frac{1}{2})/\sqrt{\pi}$.

(ii) Consider an orthogonal transformation to $Z_1, \cdots, Z_r$ such that $Z_1 = \sum \eta_i Y_i/\psi$. Then the Z's are independent normal with unit variance and means $E(Z_1) = \psi$ and $E(Z_i) = 0$ for $i > 1$.]

3. *Noncentral F- and beta-distribution.* Let $Y_1, \cdots, Y_r; Y_{s+1}, \cdots, Y_n$ be independently normally distributed with common variance σ^2 and means $E(Y_i) = \eta_i (i = 1, \cdots, r)$; $E(Y_i) = 0$ $(i = s+1, \cdots, n)$.

(i) The probability density of $W = \sum_{i=1}^r Y_i^2/\sum_{i=s+1}^n Y_i^2$ is given by (6). The distribution of the constant multiple $(n-s)W/r$ of W is the *noncentral F-distribution.*

(ii) The distribution of the statistic $B = \sum_{i=1}^r Y_i^2/(\sum_{i=1}^r Y_i^2 + \sum_{i=s+1}^n Y_i^2)$ is the *noncentral beta-distribution*, which has probability density

$$\sum_{k=0}^{\infty} P_k(\psi) g_{\frac{1}{2}r+k,\frac{1}{2}(n-s)}(b) \tag{87}$$

where

$$g_{p,q}(b) = \frac{\Gamma(p+q)}{\Gamma(p)\Gamma(q)} b^{p-1}(1-b)^{q-1}, \qquad 0 \le b \le 1 \tag{88}$$

is the probability density of the (central) beta-distribution.

4. (i) If $p_\psi(x)$ is the noncentral χ^2 or the noncentral F density, then the ratio $p_{\psi_1}(x)/p_{\psi_0}(x)$ is an increasing function of x for all $\psi_0 < \psi_1$.

(ii) Under the assumptions of Section 1, the hypothesis H': $\psi^2 \leq \psi_0^2$ ($\psi_0 > 0$ given) remains invariant under the transformations $G_i (i = 1, 2, 3)$ that were used to reduce H: $\psi = 0$, and there exists a UMP invariant test with rejection region $W > C'$. The constant C' is determined by $P_{\psi_0}\{W > C'\} = \alpha$, with the density of W given by (6).

[(i) Let $f(z) = \sum_{k=0}^{\infty} b_k z^k / \sum_{k=0}^{\infty} a_k z^k$ where the constants a_k, b_k are >0 and $\Sigma a_k z^k$ and $\Sigma b_k z^k$ converge for all $z > 0$, and suppose that $b_k/a_k < b_{k+1}/a_{k+1}$ for all k. Then

$$f'(z) = \frac{\sum\sum_{k<n} (n-k)(a_k b_n - a_n b_k) z^{k+n-1}}{\left(\sum_{k=0}^{\infty} a_k z^k\right)^2}$$

is positive since $(n-k)(a_k b_n - a_n b_k) > 0$ for $k < n$, and hence f is increasing.]

5. *Best average power.* (i) Consider the general linear hypothesis H in the canonical form given by (2) and (3) of Section 1, and for any $\eta_{r+1}, \cdots, \eta_s, \sigma$, and ρ let $S = S(\eta_{r+1}, \cdots, \eta_s, \sigma; \rho)$ denote the sphere $\{(\eta_1, \cdots, \eta_r): \sum_{i=1}^r \eta_i^2/\sigma^2 = \rho^2\}$. If $\beta_\phi(\eta_1, \cdots, \eta_s, \sigma)$ denotes the power of a test ϕ of H, then the test (9) maximizes the average power

$$\int_S \beta_\phi(\eta_1, \cdots, \eta_s, \sigma)\, dA \Big/ \int_S dA$$

for every $\eta_{r+1}, \cdots, \eta_s, \sigma$, and ρ among all unbiased (or similar) tests. Here dA denotes the differential of area on the surface of the sphere.

(ii) The result (i) provides an alternative proof of the fact that the test (9) is UMP among all tests whose power function depends only on $\sum_{i=1}^r \eta_i^2/\sigma^2$.

[(i) If $U = \sum_{i=1}^r Y_i^2$, $V = \sum_{i=s+1}^n Y_i^2$, unbiasedness (or similarity) implies that the conditional probability of rejection given $Y_{r+1}, \cdots, Y_s$ and $U + V$ equals α a.e. Hence for any given $\eta_{r+1}, \cdots, \eta_s, \sigma$, and ρ, the average power is maximized by rejecting when the ratio of the average density to the density under H is larger than a suitable constant $C(y_{r+1}, \cdots, y_s, u + v)$ and hence when

$$g(y_1, \cdots, y_r; \eta_1, \cdots, \eta_r) = \int_S \exp\left(\sum_{i=1}^r \eta_i y_i/\sigma^2\right) dA > C(y_{r+1}, \cdots, y_s, u + v).$$

As will be indicated below, the function g depends on $y_1, \cdots, y_r$ only through u and is an increasing function of u. Since under the hypothesis $U/(U + V)$ is independent of $Y_{r+1}, \cdots, Y_s$ and $U + V$, it follows that the test is given by (9).

The exponent in the integral defining g can be written as $\sum_{i=1}^r \eta_i y_i/\sigma^2 = \rho\sqrt{u} \cos\beta/\sigma$ where β is the angle $(0 \leq \beta \leq \pi)$ between $(\eta_1, \cdots, \eta_r)$ and $(y_1, \cdots, y_r)$. Because of the symmetry of the sphere, this is unchanged if β is replaced by the angle γ between $(\eta_1, \cdots, \eta_r)$ and an arbitrary fixed vector. This shows that g depends on the y's only through u; for fixed $\eta_1, \cdots, \eta_r, \sigma$ denote it by $h(u)$. Let S' be the subset of S in which $0 \leq \gamma \leq \pi/2$. Then

$$h(u) = \int_{S'} [\exp(\rho\sqrt{u}\cos\gamma/\sigma) + \exp(-\rho\sqrt{u}\cos\gamma/\sigma)]\, dA,$$

which proves the desired result.]

Section 2

6. Under the assumptions of Section 1 suppose that the means ξ_i are given by

$$\xi_i = \sum_{j=1}^{s} a_{ij}\beta_j$$

where the constants a_{ij} are known and the matrix $A = (a_{ij})$ has full rank, and where the β_j are unknown parameters. Let $\theta = \Sigma_{j=1}^{s} e_j\beta_j$ be a given linear combination of the β_j.

(i) If $\hat{\beta}_j$ denotes the values of the β_j minimizing $\Sigma(X_i - \xi_i)^2$ and if $\hat{\theta} = \Sigma_{j=1}^{s} e_j\hat{\beta}_j = \Sigma_{j=1}^{n} d_i X_i$, the rejection region of the hypothesis H: $\theta = \theta_0$ is

$$\frac{|\hat{\theta} - \theta_0|/\sqrt{\Sigma d_i^2}}{\sqrt{\Sigma(X_i - \hat{\xi}_i)^2/(n - s)}} > C_0 \tag{89}$$

where the left-hand side under H has the distribution of the absolute value of Student's t with $n - s$ degrees of freedom.

(ii) The associated confidence intervals for θ are

$$\hat{\theta} - k\sqrt{\Sigma(X_i - \hat{\xi}_i)^2/(n - s)} \leq \theta \leq \hat{\theta} + k\sqrt{\Sigma(X_i - \hat{\xi}_i)^2/(n - s)} \tag{90}$$

with $k = C_0\sqrt{\Sigma d_i^2}$. These intervals are uniformly most accurate invariant under a suitable group of transformations.

[(i) Consider first the hypothesis $\theta = 0$ and suppose without loss of generality that $\theta = \beta_1$; the general case can be reduced to this by making a linear transformation in the space of the β's. If $\underline{a}_1, \cdots, \underline{a}_s$ denote the column vectors of the matrix A which by assumption span Π_Ω, then $\underline{\xi} = \beta_1\underline{a}_1 + \cdots + \beta_s\underline{a}_s$ and since $\hat{\underline{\xi}}$ is in Π_Ω also $\hat{\underline{\xi}} = \hat{\beta}_1\underline{a}_1 + \cdots + \hat{\beta}_s\underline{a}_s$. The space Π_ω defined by the hypothesis $\beta_1 = 0$ is spanned by the vectors $\underline{a}_2, \cdots, \underline{a}_s$ and also by the row vectors $\underline{c}_2, \cdots, \underline{c}_s$ of the matrix C of (1), while $\underline{c}_1$ is orthogonal to Π_ω. By (1), the vector $\underline{X}$ is given by $\underline{X} = \Sigma_{i=1}^{n} Y_i\underline{c}_i$ and its projection $\hat{\underline{\xi}}$ on Π_Ω therefore satisfies $\hat{\underline{\xi}} = \Sigma_{i=1}^{s} Y_i\underline{c}_i$. Equating the two expressions for $\hat{\underline{\xi}}$ and taking the inner product of both sides of this equation with $\underline{c}_1$ gives $Y_1 = \hat{\beta}_1\Sigma_{i=1}^{n} a_{i1}c_{1i}$ since the $\underline{c}$'s are an orthogonal set of unit vectors. This shows that Y_1 is proportional to $\hat{\beta}_1$ and since the variance of Y_1 is the same as that of the X's that $|Y_1| = |\hat{\beta}_1|/\sqrt{\Sigma d_i^2}$. The result for testing $\beta_1 = 0$ now follows from (12) and (13). The test for $\beta_1 = \beta_1^0$ is obtained by making the transformation $X_i^* = X_i - a_{i1}\beta_1^0$.

(ii) The invariance properties of the intervals (90) can again be discussed without loss of generality by letting θ be the parameter β_1. In the canonical form of Section 1, one then has $E(Y_1) = \eta_1 = \lambda\beta_1$ with $|\lambda| = 1/\sqrt{\Sigma d_1^2}$ while $\eta_2, \cdots, \eta_s$ do not involve β_1. The hypothesis $\beta_1 = \beta_1^0$ is therefore equivalent to $\eta_1 = \eta_1^0$ with $\eta_1^0 = \lambda\beta_1^0$. This is invariant (a) under addition of arbitrary constants to $Y_2, \cdots, Y_s$; (b) under the transformations $Y_1^* = -(Y_1 - \eta_1^0) + \eta_1^0$; (c) under the scale changes $Y_i^* = cY_i$ $(i = 2, \cdots, n)$, $Y_1^* - \eta_1^{0*} = c(Y_1 - \eta_1^0)$. The confidence intervals for $\theta = \beta_1$ are then uniformly most accurate invariant under the group obtained from (a), (b), and (c) by varying η_1^0.]

7. Let $X_{ij}(j = 1, \cdots, m_i)$ and $Y_{ik}(k = 1, \cdots, n_i)$ be independently normally distributed with common variance σ^2 and means $E(X_{ij}) = \xi_i$ and $E(Y_{ij}) = \xi_i + \Delta$. Then the UMP invariant test of H: $\Delta = 0$ is given by (89) with $\theta = \Delta$, $\theta_0 = 0$

$$\hat{\theta} = \frac{\sum_i \frac{m_i n_i}{N_i}(Y_{i\cdot} - X_{i\cdot})}{\sum_i \frac{m_i n_i}{N_i}}, \qquad \hat{\xi}_i = \frac{\sum_{j=1}^{m_i} X_{ij} + \sum_{k=1}^{n_i}(Y_{ik} - \hat{\theta})}{N_i}$$

where $N_i = m_i + n_i$.

8. Let $X_1, \cdots, X_n$ be independently normally distributed with known variance σ_0^2 and means $E(X_i) = \xi_i$, and consider any linear hypothesis with $s \leqq n$ (instead of $s < n$ which is required when the variance is unknown). This remains invariant under a subgroup of that employed when the variance was unknown, and the UMP invariant test has rejection region

$$\Sigma(X_i - \hat{\hat{\xi}}_i)^2 - \Sigma(X_i - \hat{\xi}_i)^2 = \Sigma(\hat{\xi}_i - \hat{\hat{\xi}}_i)^2 > C\sigma_0^2 \tag{91}$$

with C determined by

$$\int_C^\infty \chi_r^2(y)\, dy = \alpha. \tag{92}$$

Section 3

9. If the variables $X_{ij}\,(j = 1, \cdots, n_i;\, i = 1, \cdots, s)$ are independently distributed as $N(\mu_i, \sigma^2)$, then

$$E[\Sigma n_i(X_{i\cdot} - X_{\cdot\cdot})^2] = (s-1)\sigma^2 + \Sigma n_i(\mu_i - \mu_\cdot)^2$$
$$E[\Sigma\Sigma(X_{ij} - X_{i\cdot})^2] = (n-s)\sigma^2.$$

10. Let $Z_1, \cdots, Z_s$ be independently distributed as $N(\zeta_i, a_i^2)$, $i = 1, \cdots, s$, where the a_i are known constants.

(i) With respect to a suitable group of linear transformations there exists a UMP invariant test of H: $\zeta_1 = \cdots = \zeta_s$ given by the rejection region (21).

(ii) The power of this test is the integral from C to ∞ of the noncentral χ^2 density with $s - 1$ degrees of freedom and noncentrality parameter λ^2 obtained by substituting ζ_i for Z_i in the left-hand side of (21).

11. (i) If X has a Poisson distribution with mean $E(X) = \lambda$, then for large λ the statistic $\sqrt{X}$ is approximately distributed as $N(\sqrt{\lambda}, \frac{1}{4})$.

(ii) If X has the binomial distribution $b(p, n)$, then for large n the quantity arc sin $\sqrt{X/n}$ is approximately distributed as $N(\text{arc sin } \sqrt{p}, 1/4n)$.*

Section 5

12. The linear hypothesis test of the hypothesis of no interaction in a two-way classification with m observations per cell is given by (30).

* A detailed discussion of these transformations is given by Eisenhart in Chapter 16 of *Selected Techniques of Statistical Analysis*, New York, McGraw-Hill Book Co., 1947. Certain refinements are discussed by Anscombe, "Transformations of Poisson, binomial and negative binomial data," *Biometrika*, Vol. 35 (1948), pp. 246–254, and by Freeman and Tukey, "Transformations related to the angular and the square root," *Ann. Math. Stat.*, Vol. 21 (1950), pp. 607–611.

13. Let X_λ denote a random variable distributed as noncentral χ^2 with f degrees of freedom and noncentrality parameter λ^2. Then $X_{\lambda'}$ is stochastically larger than X_λ if $\lambda < \lambda'$.

[It is enough to show that if Y is distributed as $N(0, 1)$, then $(Y + \lambda')^2$ is stochastically larger than $(Y + \lambda)^2$. The equivalent fact that for any $z > 0$,

$$P\{|Y + \lambda'| \leq z\} \leq P\{|Y + \lambda| \leq z\},$$

is an immediate consequence of the shape of the normal density function. An alternative proof is obtained by combining Problem 4 with Lemma 2 of Chapter 3.

14. Let $X_{ijk}(i = 1, \cdots, a;\ j = 1, \cdots, b;\ k = 1, \cdots, m)$ be independently normally distributed with common variance σ^2 and mean

$$E(X_{ijk}) = \mu + \alpha_i + \beta_j + \gamma_k \qquad (\Sigma\alpha_i = \Sigma\beta_j = \Sigma\gamma_k = 0).$$

Determine the linear hypothesis test for testing H: $\alpha_1 = \cdots = \alpha_a = 0$.

15. In the three-factor situation of the preceding problem, suppose that $a = b = m$. The hypothesis H can then be tested on the basis of m^2 observations as follows. At each pair of levels (i, j) of the first two factors one observation is taken, to which we refer as being in the ith row and the jth column. If the levels of the third factor are chosen in such a way that each of them occurs once and only once in each row and column, the experimental design is a *Latin square*. The m^2 observations are denoted by $X_{ij(k)}$ where the third subscript indicates the level of the third factor when the first two are at levels i and j. It is assumed that $E(X_{ij(k)}) = \xi_{ij(k)} = \mu + \alpha_i + \beta_j + \gamma_k$, with $\Sigma\alpha_i = \Sigma\beta_j = \Sigma\gamma_k = 0$.

(i) The parameters are determined from the ξ's through the equations

$$\xi_{i\cdot(\cdot)} = \mu + \alpha_i, \qquad \xi_{\cdot j(\cdot)} = \mu + \beta_j, \qquad \xi_{\cdot\cdot(k)} = \mu + \gamma_k, \qquad \xi_{\cdot\cdot(\cdot)} = \mu.$$

(Summation over j with i being held fixed automatically causes summation also over k.)

(ii) The least squares estimates of the parameters may be obtained from the identity

$$\begin{aligned}\sum_i\sum_j [x_{ij(k)} - \xi_{ij(k)}]^2 &= m\Sigma[x_{i\cdot(\cdot)} - x_{\cdot\cdot(\cdot)} - \alpha_i]^2 + m\Sigma[x_{\cdot j(\cdot)} - x_{\cdot\cdot(\cdot)} - \beta_j]^2 \\ &\quad + m\Sigma[x_{\cdot\cdot(k)} - x_{\cdot\cdot(\cdot)} - \gamma_k]^2 + m^2[x_{\cdot\cdot(\cdot)} - \mu]^2 \\ &\quad + \sum_i\sum_k [x_{ij(k)} - x_{i\cdot(\cdot)} - x_{\cdot j(\cdot)} - x_{\cdot\cdot(k)} + 2x_{\cdot\cdot(\cdot)}]^2.\end{aligned}$$

(iii) For testing the hypothesis H: $\alpha_1 = \cdots = \alpha_m = 0$, the test statistic W^* of (15) is

$$\frac{m\Sigma[X_{i\cdot(\cdot)} - X_{\cdot\cdot(\cdot)}]^2}{\Sigma\Sigma[X_{ij(k)} - X_{i\cdot(\cdot)} - X_{\cdot j(\cdot)} - X_{\cdot\cdot(k)} + 2X_{\cdot\cdot(\cdot)}]^2/(m-2)}.$$

The degrees of freedom are $m - 1$ for the numerator and $(m - 1)(m - 2)$ for the denominator, and the noncentrality parameter is $\psi^2 = m\Sigma\alpha_i^2/\sigma^2$.

Section 6

16. In a regression situation, suppose that the observed values X_j and Y_j of the independent and dependent variable differ from certain true values X_j' and Y_j' by errors U_j, V_j which are independently normally distributed with zero

means and variances σ_U^2 and σ_V^2. The true values are assumed to satisfy a linear relation: $Y_j' = \alpha + \beta X_j'$. However, the variables which are being controlled, and which are therefore constants, are the X_j rather than the X_j'. Writing x_j for X_j, we have $x_j = X_j' + U_j$, $Y_j = Y_j' + V_j$, and hence $Y_j = \alpha + \beta x_j + W_j$, where $W_j = V_j - \beta U_j$. The results of Section 6 can now be applied to test that β or $\alpha + \beta x_0$ have a specified value.

17. Let $X_1, \cdots, X_m$; $Y_1, \cdots, Y_n$ be independently normally distributed with common variance σ^2 and means $E(X_i) = \alpha + \beta(u_i - \bar{u})$, $E(Y_j) = \gamma + \delta(v_j - \bar{v})$ where the u's and v's are known numbers. Determine the UMP invariant tests of the linear hypotheses H: $\beta = \delta$ and H: $\alpha = \gamma$, $\beta = \delta$.

18. Let $X_1, \cdots, X_n$ be independently normally distributed with common variance σ^2 and means $\xi_i = \alpha + \beta t_i + \gamma t_i^2$ where the t_i are known. If the coefficient vectors $(t_1^k, \cdots, t_n^k)$, $k = 0, 1, 2$, are linearly independent, the parameter space Π_Ω has dimension $s = 3$, and the least squares estimates $\hat{\alpha}$, $\hat{\beta}$, $\hat{\gamma}$ are the unique solutions of the system of equations

$$\alpha \Sigma t_i^k + \beta \Sigma t_i^{k+1} + \gamma \Sigma t_i^{k+2} = \Sigma t_i^k X_i \qquad (k = 0, 1, 2).$$

The solutions are linear functions of the X's and if $\hat{\gamma} = \Sigma c_i X_i$, the hypothesis $\gamma = 0$ is rejected when

$$\frac{|\hat{\gamma}|/\sqrt{\Sigma c_i^2}}{\sqrt{\Sigma(X_i - \hat{\alpha} - \hat{\beta} t_i - \hat{\gamma} t_i^2)^2/(n-3)}} > C_0.$$

Section 7

19. (i) The test (43) of H: $\Delta \leq \Delta_0$ is UMP unbiased.

(ii) Determine the UMP unbiased test of H: $\Delta = \Delta_0$ and the associated uniformly most accurate unbiased confidence sets for Δ.

20. In the model (41), the correlation coefficient ρ between two observations X_{ij}, X_{ik} belonging to the same class, the so-called *intraclass correlation coefficient*, is given by $\rho = \sigma_A^2/(\sigma_A^2 + \sigma^2)$.

Section 8

21. The tests (48) and (49) are UMP unbiased.

22. If X_{ij} is given by (41) but the number n_i of observations per batch is not constant, obtain a canonical form corresponding to (42) by letting $Y_{i1} = \sqrt{n_i} X_{i.}$. Note that the set of sufficient statistics has more components than when n_i is constant.

23. The general nested classification with a constant number of observations per cell, under model II has the structure

$$X_{ijk\cdots} = \mu + A_i + B_{ij} + C_{ijk} + \cdots + U_{ijk\cdots}, \tag{93}$$

$i = 1, \cdots, a$; $j = 1, \cdots, b$; $k = 1, \cdots, c$; $\cdots$.

(i) This can be reduced to a canonical form generalizing (47).

(ii) There exist UMP unbiased tests of the hypotheses

$$H_A\colon \sigma_A^2/(cd \cdots \sigma_B^2 + d \cdots \sigma_C^2 + \cdots + \sigma^2) \leq \Delta_0,$$

$$H_B\colon \sigma_B^2/(d \cdots \sigma_C^2 + \cdots + \sigma^2) \leq \Delta_0.$$

Section 10

24. (i) If $m < p$, the matrix S, and hence the matrix S/m which is an unbiased estimate of the unknown covariance matrix of the underlying p-variate distribution, is singular. If $m \geqq p$, it is nonsingular with probability 1.

(ii) If $r + m \leqq p$, the test $\varphi(y, u, z) \equiv \alpha$ is the only test that is invariant under the groups G_1 and G_3 of Section 10.

[(ii) The U's are eliminated through G_1. Since the $r + m$ row vectors of the matrices Y and Z may be assumed to be linearly independent, any such set of vectors can be transformed into any other through an element of G_3.]

25. (i) If $p < r + m$, and $V = Y'Y$, $S = Z'Z$, the $p \times p$ matrix $V + S$ is nonsingular with probability 1, and the characteristic roots of the equation

$$|V - \lambda(V + S)| = 0 \tag{94}$$

constitute a maximal set of invariants under G_1, G_2, and G_3.

(ii) Of the roots of (94), $p - \min(r, p)$ are zero and $p - \min(m, p)$ are equal to one. There are no other constant roots so that the number of variable roots, which constitute a maximal invariant set, is $\min(r, p) + \min(m, p) - p$.

[The multiplicity of the root $\lambda = 1$ is p minus the rank of S, and hence $p - \min(m, p)$. Equation (94) cannot hold for any constant $\lambda \neq 0, 1$ for almost all V, S since for any $\mu \neq 0$, $V + \mu S$ is nonsingular with probability 1.]

26. (i) If A and B are $k \times m$ and $m \times k$ matrices respectively, then the product matrices AB and BA have the same nonzero characteristic roots.

(ii) This provides an alternative derivation of the fact that W defined by (58) is the only nonzero characteristic root of the determinantal equation (57).

[(i) If x is a nonzero solution of the equation $ABx = \lambda x$ with $\lambda \neq 0$, then $y = Bx$ is a nonzero solution of $BAy = \lambda y$.]

27. In the case $r = 1$, the statistic W given by (58) is maximal invariant under the group induced by G_1 and G_3 on the statistics Y_i, $U_{\alpha i}(i = 1, \cdots, p;\ \alpha = 1, \cdots, s - 1)$ and $S = Z'Z$.

[There exists a nonsingular matrix B such that $B'SB = I$ and such that only the first coordinate of YB is nonzero. This is seen by first finding B_1 such that $B_1'SB_1 = I$ and then an orthogonal Q such that only the first coordinate of YB_1Q is nonzero.]

28. Let $Z_{\alpha i}$ ($\alpha = 1, \cdots, m;\ i = 1, \cdots, p$) be independently distributed as $N(0, 1)$ and let $Q = Q(Y)$ be an orthogonal $m \times m$ matrix depending on a random variable Y that is independent of the Z's. If $Z_{\alpha i}^*$ is defined by

$$(Z_{1i}^* \cdots Z_{mi}^*) = (Z_{1i} \cdots Z_{mi})Q',$$

then the $Z_{\alpha i}^*$ are independently distributed as $N(0, 1)$ and are independent of Y.

[For each y, the conditional distribution of the $(Z_{1i} \cdots Z_{mi})Q'(y)$, given $Y = y$, is as stated.]

29. Let Z be the $m \times p$ matrix $(Z_{\alpha i})$ where $p \leqq m$ and the $Z_{\alpha i}$ are independently distributed as $N(0, 1)$, let $S = Z'Z$, and let S_1 be the matrix obtained by omitting the last row and column of S. Then the ratio of determinants $|S|/|S_1|$ has a χ^2-distribution with $m - p + 1$ degrees of freedom.

[Let Q be an orthogonal matrix (dependent on $Z_{11}, \cdots, Z_{m1}$) such that $(Z_{11} \cdots Z_{m1})Q' = (R0 \cdots 0)$, where $R^2 = \Sigma_{\alpha=1}^m Z_{\alpha 1}^2$. Then

$$S = Z'Q'QZ = \begin{pmatrix} R & 0 \cdots & 0 \\ Z_{12}^* & \cdots & Z_{m2}^* \\ \cdot & & \cdot \\ \cdot & & \cdot \\ Z_{1p}^* & \cdots & Z_{mp}^* \end{pmatrix} \begin{pmatrix} R & Z_{12}^* & \cdots & Z_{1p}^* \\ 0 & \cdot & & \cdot \\ \cdot & \cdot & & \cdot \\ \cdot & \cdot & & \cdot \\ 0 & Z_{m2}^* & \cdots & Z_{mp}^* \end{pmatrix}$$

where the $Z_{\alpha i}^*$ denote the transforms under Q. The first of the matrices on the right-hand side is equal to the product

$$\left(\begin{array}{c|c} R & 0 \\ \hline Z_1^* & I \end{array}\right) \left(\begin{array}{c|c} 1 & 0 \\ \hline 0 & Z^{*\prime} \end{array}\right)$$

where Z^* is the $(m-1) \times (p-1)$ matrix with elements $Z_{\alpha i}^*$ ($\alpha = 2, \cdots, m$; $i = 2, \cdots, p$), I is the $(p-1) \times (p-1)$ identity matrix, Z_1^* is the column vector $(Z_{12}^* \cdots Z_{1p}^*)'$, and 0 indicates a row or column of zeros. It follows that $|S|$ is equal to R^2 multiplied by the determinant of $Z^{*\prime}Z^*$. Since S_1 is the product of the $m \times (p-1)$ matrix obtained by omitting the last column of Z multiplied on the left by the transpose of this $m \times (p-1)$ matrix, $|S_1|$ is equal to R^2 multiplied by the determinant of the matrix obtained by omitting the last row and column of $Z^{*\prime}Z^*$. The ratio $|S|/|S_1|$ has therefore been reduced to the corresponding ratio in terms of the $Z_{\alpha i}^*$ with m and p replaced by $m-1$ and $p-1$, and by induction the problem is seen to be unchanged if m and p are replaced by $m-k$ and $p-k$ for any $k < p$. In particular, $|S|/|S_1|$ can be evaluated under the assumption that m and p have been replaced by $m-(p-1)$ and $p-(p-1) = 1$. In this case, the matrix Z' is a row matrix $(Z_{11} \cdots Z_{m-p+1,1})$; the determinant of S is $|S| = \Sigma_{\alpha=1}^{m-p+1} Z_{\alpha 1}^2$ which has a χ_{m-p+1}^2-distribution; and since S is a 1×1 matrix, $|S_1|$ is replaced by 1.]

30. The statistic $W = YS^{-1}Y'$ defined by (58), where Y is a row vector, has the distribution of a ratio, of which the numerator and denominator are distributed independently, as noncentral χ^2 with noncentrality parameter ψ^2 and p degrees of freedom and as central χ^2 with $m + 1 - p$ degrees of freedom respectively.

[Since the distribution of W is unchanged if the same nonsingular transformation is applied to $(Y_1, \cdots, Y_p)$ and each of the m vectors $(Z_{\alpha 1}, \cdots, Z_{\alpha p})$, the common covariance matrix of these vectors can be assumed to be the identity matrix. Let Q be an orthogonal matrix (depending on the Y's) such that $(Y_1 \cdots Y_p)Q = (00 \cdots T)$ where $T^2 = \Sigma Y_i^2$. Since QQ' is the identity matrix one has

$$W = (YQ)(Q'S^{-1}Q)(Q'Y') = (0 \cdots 0T)(Q'S^{-1}Q)(0 \cdots 0T)'.$$

Hence W is the product of T^2, which has a noncentral χ^2-distribution with p degrees of freedom and noncentrality parameter ψ^2, and the element which lies in the pth row and the pth column of the matrix $Q'S^{-1}Q = (Q'SQ)^{-1} = (Q'Z'ZQ)^{-1}$. By Problems 28 and 29, this matrix is distributed independently of the Y's and the reciprocal of the element in question is distributed as χ_{m-p+1}^2.]

31. Let $(X_{\nu 1}, \cdots, X_{\nu p})$, $\nu = 1, \cdots, N$, be a sample from a p-variate normal distribution with unknown covariance matrix and mean $E(X_{\nu i}) = \alpha_i + \beta_i(u_\nu - u.)$ where the u's are known numbers. The hypothesis H: $\beta_1 = \cdots = \beta_p = 0$ is a multivariate linear hypothesis with $r = 1$, $s = 2$. One has $\hat{\alpha}_i = \hat{\hat{\alpha}}_i = X._i$, $\hat{\beta}_i = \Sigma_{\nu=1}^N (u_\nu - u.)(X_{\nu i} - X._i)/\Sigma_{\nu=1}^N (u_\nu - u.)^2$, and the test statistic W is given by (58) with

$$S_{ij} = \Sigma[X_{\nu i} - \hat{\alpha}_i - \hat{\beta}_i(u_\nu - u.)][X_{\nu j} - \hat{\alpha}_j - \hat{\beta}_j(u_\nu - u.)]$$

and

$$Y_i Y_j = \hat{\beta}_i \hat{\beta}_j \Sigma(u_\nu - u.)^2.$$

32. Let $X = (X_{\alpha i})$, $i = 1, \cdots, p$; $\alpha = 1, \cdots, N$, be a sample from a p-variate normal distribution, let $q < p$, $\max(q, p - q) \leqq N$, and consider the hypothesis H that $(X_{11}, \cdots, X_{1q})$ is independent of $(X_{1q+1}, \cdots, X_{1p})$, that is, that the covariances $\sigma_{ij} = E(X_{\alpha i} - \xi_i)(X_{\alpha j} - \xi_j)$ are zero for all $i \leqq q$, $j > q$. The problem of testing H remains invariant under the transformations $X^*_{\alpha i} = X_{\alpha i} + b_i$ and $X^* = XC$ where C is any nonsingular $p \times p$ matrix of the structure

$$C = \begin{pmatrix} C_{11} & 0 \\ 0 & C_{22} \end{pmatrix}$$

with C_{11} and C_{22} being $q \times q$ and $(p - q) \times (p - q)$ respectively.

(i) A set of maximal invariants under the induced transformations in the space of the sufficient statistics $X._i$ and the matrix S, which we partition as

$$S = \begin{pmatrix} S_{11} & S_{12} \\ S_{21} & S_{22} \end{pmatrix}.$$

are the q roots of the equation

$$|S_{12}S_{22}^{-1}S_{21} - \lambda S_{11}| = 0.$$

(ii) In the case $q = 1$, a maximal invariant is the statistic $R^2 = S_{12}S_{22}^{-1}S_{21}/S_{11}$, which is the square of the *multiple correlation coefficient* between X_{11} and $(X_{12}, \cdots, X_{1p})$. The distribution of R^2 depends only on the square ρ^2 of the population multiple correlation coefficient, which is obtained from R^2 by replacing the elements of S by their expected values σ_{ij}.

(iii) Using the fact that the distribution of R^2 has the density†

$$\frac{(1 - R^2)^{\frac{1}{2}(N-p-2)}(R^2)^{\frac{1}{2}(p-1)-1}(1 - \rho^2)^{\frac{1}{2}(N-1)}}{\Gamma[\frac{1}{2}(N - 1)]\Gamma[\frac{1}{2}(N - p)]} \sum_{h=0}^{\infty} \frac{(\rho^2)^h(R^2)^h\Gamma^2[\frac{1}{2}(N - 1) + h]}{h!\Gamma[\frac{1}{2}(p - 1) + h]}$$

and that the hypothesis H for $q = 1$ is equivalent to $\rho = 0$, show that the UMP invariant test rejects this hypothesis when $R^2 > C_0$.

(iv) When $\rho = 0$, the statistic

$$\frac{R^2}{1 - R^2} \cdot \frac{N - p}{p - 1}$$

has the F-distribution with $p - 1$ and $N - p$ degrees of freedom.

† See for example Anderson, *An Introduction to Multivariate Statistical Analysis* New York, John Wiley & Sons, 1958.

[(i) Suppose that $|S_{12}S_{22}^{-1}S_{21} - \lambda S_{11}| = 0$ and $|S_{12}^*S_{22}^{*-1}S_{21}^* - \lambda S_{11}^*| = 0$ have the same roots. Then there exist matrices B and C such that $BS_{11}B' = I = CS_{11}^*C'$ and $BS_{12}S_{22}^{-1}S_{21}B' = CS_{12}^*S_{22}^{*-1}S_{21}^* = \Lambda$ where Λ is the diagonal matrix whose diagonal elements are the roots λ. Since S_2^{-1} and S_2^{*-1} are positive definite there exist nonsingular matrices E and E^* such that $S_2^{-1} = EE'$ and $S_2^{*-1} = E^*E^{*\prime}$. (This can be seen by reducing E to diagonal form through an orthogonal transformation.) Then

$$(BS_{12}E)(BS_{12}E)' = (CS_{12}^*E^*)(CS_{12}^*E^*)'$$

and it follows from the argument given in Section 10 in connection with G_2 that there exists an orthogonal matrix Q such that $BS_{12}EQ = CS_{12}^*E^*$, so that $C_{11} = C^{-1}B$ and $C_{22}' = EQE^{*-1}$.]

Section 12

33. The problem of testing the hypothesis $\eta \in \Pi_\omega$ when the distribution of Y is given by (62) with $\eta \in \Pi_\Omega$ remains invariant under a suitable group of linear transformations, and with respect to this group the test (63) is UMP invariant. The probability of rejecting with this test is given by (65) and (66) for all points $(\eta_1, \cdots, \eta_q)$.

[There exists a nonsingular linear transformation $Z = CY$ for which $C'A^{-1}C$ is the identity matrix, and in terms of Z the problem reduces to a linear hypothesis with known variance.]

Section 13

34. Let the equation of the tangent $\overline{\mathscr{S}}$ at π be $p_i = \pi_i(1 + a_{i1}\xi_1 + \cdots + a_{is}\xi_s)$ and suppose that the vectors $(a_{i1}, \cdots, a_{is})$ are orthogonal in the sense that $\Sigma a_{ik}a_{il}\pi_i = 0$ for all $k \neq l$.

(i) If $(\hat{\xi}_1, \cdots, \hat{\xi}_s)$ minimizes $\Sigma(\nu_i - p_i)^2/\pi_i$ subject to $p \in \overline{\mathscr{S}}$, then $\hat{\xi}_j = \Sigma_i a_{ij}\nu_i / \Sigma_i a_{ij}^2\pi_i$.

(ii) The test statistic (76) for testing H: $p = \pi$ reduces to

$$n \sum_{j=1}^{s} \left[\left(\sum_{i=1}^{m} a_{ij}\nu_i \right)^2 \Big/ \sum_{i=1}^{m} a_{ij}^2\pi_i \right].$$

35. *Independence in contingency tables.* Consider a twofold classification of n elements into classes $A_1, \cdots, A_a$ and $B_1, \cdots, B_b$ respectively. If n_{ij} is the number of elements belonging to both A_i and B_j, the likelihood ratio test for testing the hypothesis H that the A and B classifications are independent rejects when

$$\Lambda = \frac{\prod_i n_{i\cdot}^{bn_{i\cdot}} \prod_j n_{\cdot j}^{an_{\cdot j}}}{\prod_{i,j} n_{ij}^{n_{ij}}}$$

is too large, where $n_{i\cdot} = \Sigma_j n_{ij}/b$, $n_{\cdot j} = \Sigma_i n_{ij}/a$. For large n, the distribution of $-2 \log \Lambda$ under H is χ^2 with $(r-1)(s-1)$ degrees of freedom.

[The likelihood of a multinomial sample $x_1, \cdots, x_m$ with m classes is proportional to $p_1^{x_1} \cdots p_m^{x_m}$ which has the maximum value $(x_1/n)^{x_1} \cdots (x_m/n)^{x_m}$. This can be seen for example by considering n numbers of which x_i are equal to p_i/x_i for $i = 1, \cdots, m$ and noting that their geometric mean is less than or equal to their arithmetic mean. The result follows by applying this result to the multinomial situations with probabilities $p_{ij}(i = 1, \cdots, a;\ j = 1, \cdots, b)$ which constitute Ω, and with $p_{ij} = p_i p'_j (\Sigma p_i = \Sigma p'_j = 1)$ which constitute ω.]

15. REFERENCES

The present chapter contains brief introductions to three subjects which between them cover a large part of present-day statistical method: the analysis of variance, multivariate analysis, and χ^2 tests.

The analysis of variance has its origin principally in the work of R. A. Fisher, much of it contained in his books (1925, 1935). A comprehensive treatment is given by Kempthorne (1952) and in a forthcoming book by Scheffé (1959). Certain aspects of the field are surveyed in the papers by Scheffé (1956) and by Cochran (1957).

A detailed account of multivariate analysis is found in the books by Anderson (1958), Rao (1952), and Kendall (1957).

A survey of the χ^2 method is given by Cochran in two papers (1952, 1954).

Anderson, T. W.

(1958) *An Introduction to Multivariate Analysis*, New York, John Wiley & Sons.

Bartlett, M. S.

(1947) "The use of transformations," *Biometrics*, Vol. 3, pp. 39–52.

[Discussion of, among others, the logarithmic, square root, and arc sine transformations.]

Box, G. E. P.

(1949) "A general distribution theory for a class of likelihood ratio criteria," *Biometrika*, Vol. 36, pp. 317–346.

Cochran, W. G.

(1952) "The χ^2 test of goodness of fit," *Ann. Math. Stat.*, Vol. 2, pp. 315–345.

(1954) "Some methods for strengthening the common χ^2 tests," *Biometrics*, Vol. 10, pp. 417–451.

(1957) "Analysis of covariance: Its nature and uses," *Biometrics*, Vol. 13, pp. 261–281.

Eisenhart, C.

(1947) "The assumptions underlying the analysis of variance," *Biometrics*, Vol. 3, pp. 1–21.

[Discusses the distinction between model I and model II.]

Fisher, R. A.

(1924a) "The conditions under which chi square measures the discrepancy between observation and hypothesis," *J. Roy. Stat. Soc.*, Vol. 87, pp. 442–450.

[Obtains the limiting distribution (under the hypothesis) of the χ^2 statistic for the case of composite hypotheses and discusses the dependence of this distribution on the method used to estimate the parameters.]

(1924b) "On a distribution yielding the error functions of several well-known statistics," *Proc. Int. Math. Congress*, Toronto, pp. 805–813.

[Discusses the use of the z-distribution (which is equivalent to the F-distribution) in analysis of variance (model I) and regression analysis. Obtains the distribution of the sample multiple correlation coefficient when the population multiple correlation coefficient is zero.]

(1925) *Statistical Methods for Research Workers*, Edinburgh, Oliver and Boyd, 1st ed.

(1928) "The general sampling distribution of the multiple correlation coefficient," *Proc. Roy. Soc., Ser. A*, Vol. 121, pp. 654–673.

[Derives the noncentral χ^2 and noncentral beta-distribution and the distribution of the sample multiple correlation coefficient for arbitrary values of the population multiple correlation coefficient.]

(1935) *The Design of Experiments*, Edinburgh, Oliver and Boyd, 1st ed.

Hartley, H. O.

(1950) "Maximum F ratio as a short-cut test for heterogeneity of variance," *Biometrika*, Vol. 37, pp. 308–312.

[Proposes the test (23).]

Herbach, Leon H.

(1957) "Optimum properties of analysis of variance tests based on model II and some generalizations of model II," Scientific Paper No. 6, *Engin. Stat. Lab.*, New York University.

Hotelling, Harold

(1931) "The generalization of Student's ratio," *Ann. Math. Stat.*, Vol. 2, pp. 360–378.

[Proposes the statistic (58) as a multivariate extension of Student's t, and obtains the distribution of the statistic under the hypothesis.]

Hsu, P. L.

(1938) "Notes on Hotelling's generalized T^2," *Ann. Math. Stat.*, Vol. 9, pp. 231–243.

[Obtains the distribution of T^2 in the noncentral case and applies the statistic to the class of problems described in Section 11 following Example 9. The derivation of the T^2-distribution indicated in Problems 29 and 30 is that of Wijsman, "Random orthogonal transformations and their use in some classical distribution problems in multivariate analysis," *Ann. Math. Stat.*, Vol. 28 (1957), pp. 415–423, which was noted also by Stein (cf. Wijsman, p. 416) and by Bowker (cf. Anderson, *op. cit.*, p. 107).]

(1941) "Analysis of variance from the power function stand-point," *Biometrika*, Vol. 32, pp. 62–69.

[Shows that the test (7) is UMP among all tests whose power function depends only on the noncentrality parameter.]

Hunt, G., and C. M. Stein

(1946) "Most stringent tests of statistical hypotheses," unpublished.

[Proves the tests (7) and (61) to be UMP almost invariant, and the roots of (57) to constitute a maximal set of invariants.]

Kempthorne, O.

(1952) *The Design and Analysis of Experiments*, New York, John Wiley & Sons.

Kendall, M. G.

(1957) *A Course in Multivariate Analysis*, London, Charles Griffin.

Kolodziejczyk, S.

(1935) "An important class of statistical hypotheses," *Biometrika*, Vol. 37, pp. 161–190.

[Discussion of the general linear univariate hypothesis from the likelihood ratio point of view.]

Newman, D.
(1939) "Range in samples from a normal population," *Biometrika*, Vol. 31, pp. 20–30.
[Discusses the test (22), which he attributes to "Student" (W. S. Gossett).]

Neyman, J.
(1949) "Contribution to the theory of the χ^2 test," *Proc. Berkeley Symposium on Mathematical Statistics and Probability*, Berkeley, Univ. Calif. Press, pp. 239–273.
[Gives a theory of χ^2 tests with restricted alternatives.]

Neyman, J., and E. S. Pearson
(1928) "The use and interpretation of certain test criteria for purposes of statistical inference," *Biometrika*, Vol. 20A, pp. 175–240 and 263–294.
[Proposes the likelihood ratio criterion as a method that takes account of both types of error, and applies it to a variety of testing problems.]

Pearson, Karl
(1900) "On the criterion that a given system of deviations from the probable in the case of a correlated system of variables is such that it can be reasonably supposed to have arisen from random sampling," *Phil. Mag., Ser.* 5, Vol. 50, pp. 157–172.
[The χ^2 test (71) is proposed for testing a simple multinomial hypothesis, and the limiting distribution of the test criterion is obtained under the hypothesis. The test is extended to composite hypotheses but contains an error in the degrees of freedom of the limiting distribution; a correct solution for the general case was found by Fisher (1924a). Applications.]

Rao, C. R.
(1952) *Advanced Statistical Methods in Biometric Research*, New York, John Wiley & Sons.

Scheffé, Henry
(1956) "A 'mixed model' for the analysis of variance," *Ann. Math. Stat.*, Vol. 27, pp. 23–36.
[Example 11.]
(1956) "Alternative models for the analysis of variance," *Ann. Math. Stat.*, Vol. 27, pp. 251–271.
(1958) "Fitting straight lines when one variable is controlled," *J. Am. Stat. Assoc.*, Vol. 53, pp. 106–117.
[Problem 16.]
(1959) *Analysis of Variance*, New York, John Wiley & Sons.

Simaika, J. B.
(1941) "An optimum property of two statistical tests," *Biometrika*, Vol. 32, pp. 70–80.
[Shows that the test (58) is UMP among all tests whose power function depends only on the noncentrality parameter (59), and establishes the corresponding property for the test of multiple correlation given in Problem 32(iii).]

Tang, P. C.
(1938) "The power function of the analysis of variance test with tables and illustrations of their use," *Stat. Res. Mem.*, Vol. II, pp. 126–149.

Wald, Abraham
(1942) "On the power function of the analysis of variance test," *Ann. Math. Stat.*, Vol. 13, pp. 434–439.

[Problem 5. This problem is also treated by Hsu, "On the power function of the E^2-test and the T^2-test," *Ann. Math. Stat.*, Vol. 16 (1945), pp. 278–286.]

(1943) "Tests of statistical hypotheses concerning several parameters when the number of observations is large," *Trans. Am. Math. Soc.*, Vol. 54, pp. 426–482.

[General asymptotic distribution and optimum theory of likelihood ratio (and asymptotically equivalent) tests.]

Wilks, S. S.

(1938) "The large-sample distribution of the likelihood ratio for testing composite hypotheses," *Ann. Math. Stat.*, Vol. 9, pp. 60–62.

[Derives the asymptotic distribution of the likelihood ratio when the hypothesis is true.]

CHAPTER 8

The Minimax Principle

1. TESTS WITH GUARANTEED POWER

The criteria discussed so far, unbiasedness and invariance, suffer from the disadvantage of being applicable, or leading to optimum solutions, only in rather restricted classes of problems. We shall therefore turn now to an alternative approach, which potentially is of much wider applicability. Unfortunately, its application to specific problems is in general not easy, and has so far been carried out successfully mainly in cases in which there exists a UMP invariant test.

One of the important considerations in planning an experiment is the number of observations required to insure that the resulting statistical procedure will have the desired precision or sensitivity. For problems of hypothesis testing this means that the probabilities of the two kinds of errors should not exceed certain preassigned bounds, say α and $1 - \beta$, so that the tests must satisfy the conditions

$$
\begin{aligned}
E_\theta \varphi(X) &\leqq \alpha \quad \text{for} \quad \theta \in \Omega_H \\
E_\theta \varphi(X) &\geqq \beta \quad \text{for} \quad \theta \in \Omega_K.
\end{aligned} \tag{1}
$$

If the power function $E_\theta \varphi(X)$ is continuous and if $\alpha < \beta$, (1) cannot hold when the sets Ω_H and Ω_K are contiguous. This mathematical difficulty corresponds in part to the fact that the division of the parameter values θ into the classes Ω_H and Ω_K for which the two different decisions are appropriate is frequently not sharp. Between the values for which one or the other of the decisions is clearly correct there may lie others for which the relative advantages and disadvantages of acceptance and rejection are approximately in balance. Accordingly we shall assume that Ω is partitioned into three sets

$$\Omega = \Omega_H + \Omega_I + \Omega_K,$$

of which Ω_I designates the *indifference zone*, and Ω_K the class of parameter values differing so widely from those postulated by the hypothesis that false acceptance of H is a serious error, which should occur with probability at most $1 - \beta$.

To see how the sample size is determined in this situation, suppose that $X_1, X_2, \cdots$ constitute the sequence of available random variables, and for a moment let n be fixed and let $X = (X_1, \cdots, X_n)$. In the usual applicational situations (for a more precise statement, see Problem 1) there exists a test φ_n which maximizes

$$\inf_{\Omega_K} E_\theta \varphi(X) \tag{2}$$

among all level α tests based on X. Let $\beta_n = \inf_{\Omega_K} E_\theta \varphi(X)$ and suppose that for sufficiently large n there exists a test satisfying (1).* The desired sample size, which is the smallest value of n for which $\beta_n \geqq \beta$, is then obtained by trial and error. This requires the ability of determining for each fixed n the test that maximizes (2) subject to

$$E_\theta \varphi(X) \leqq \alpha \quad \text{for} \quad \theta \in \Omega_H. \tag{3}$$

A method for determining a test with this *maximin* property (of maximizing the minimum power over Ω_K) is obtained by generalizing Theorem 7 of Chapter 3. It will be convenient in this discussion to make a change of notation, and to denote by ω and ω' the subsets of Ω previously denoted by Ω_H and Ω_K. Let $\mathscr{P} = \{P_\theta, \theta \in \omega \cup \omega'\}$ be a family of probability distributions over a sample space $(\mathscr{X}, \mathscr{A})$ with densities $p_\theta = dP_\theta/d\mu$ with respect to a σ-finite measure μ, and suppose that the densities $p_\theta(x)$ considered as functions of the two variables (x, θ) are measurable $(\mathscr{A} \times \mathscr{B})$ and $(\mathscr{A} \times \mathscr{B}')$ where $\mathscr{B}$ and $\mathscr{B}'$ are given σ-fields over ω and ω'. Under these assumptions, the following theorem gives conditions under which a solution of a suitable Bayes problem provides a test with the required properties.

Theorem 1. *For any distributions λ and λ' over $\mathscr{B}$ and $\mathscr{B}'$, let $\varphi_{\lambda,\lambda'}$ be the most powerful test for testing*

$$h(x) = \int_\omega p_\theta(x)\, d\lambda(\theta)$$

at level α against

$$h'(x) = \int_{\omega'} p_\theta(x)\, d\lambda'(\theta)$$

and let $\beta_{\lambda,\lambda'}$ be its power against the alternative h'. If there exist λ and λ' such that

$$\begin{gathered} \sup_\omega E_\theta \varphi_{\lambda,\lambda'}(X) \leqq \alpha \\ \inf_{\omega'} E_\theta \varphi_{\lambda,\lambda'}(X) = \beta_{\lambda,\lambda'}, \end{gathered} \tag{4}$$

* Conditions under which this is the case are given by Berger, "On uniformly consistent tests," *Ann. Math. Stat.*, Vol. 22 (1951), pp. 289–293.

then

(i) $\varphi_{\lambda,\lambda'}$ *maximizes* $\inf_{\omega'} E_\theta\varphi(X)$ *among all level* α *tests of the hypothesis* $H: \theta \in \omega$ *and is the unique test with this property if it is the unique most powerful level* α *test for testing h against* h'.

(ii) *The pair of distributions* λ, λ' *is least favorable in the sense that for any other pair* ν, ν' *we have*

$$\beta_{\lambda,\lambda'} \leqq \beta_{\nu,\nu'}.$$

Proof. (i) If φ^* is any other level α test of H, it is also of level α for testing the simple hypothesis that the density of X is h, and the power of φ^* against h' therefore cannot exceed $\beta_{\lambda,\lambda'}$. It follows that

$$\inf_{\omega'} E_\theta\varphi^*(X) \leqq \int_{\omega'} E_\theta\varphi^*(X)\, d\lambda'(\theta) \leqq \beta_{\lambda,\lambda'} = \inf_{\omega'} E_\theta\varphi_{\lambda,\lambda'}(X),$$

and the second inequality is strict if $\varphi_{\lambda,\lambda'}$ is unique.

(ii) Let ν, ν' be any other distributions over $(\omega, \mathscr{B})$ and $(\omega', \mathscr{B}')$, and let

$$g(x) = \int_\omega p_\theta(x)\, d\nu(\theta); \qquad g'(x) = \int_{\omega'} p_\theta(x)\, d\nu'(\theta).$$

Since both $\varphi_{\lambda,\lambda'}$ and $\varphi_{\nu,\nu'}$ are level α tests of the hypothesis that $g(x)$ is the density of X, it follows that

$$\beta_{\nu,\nu'} \geqq \int \varphi_{\lambda,\lambda'}(x) g'(x)\, d\mu(x) \geqq \inf_{\omega'} E_\theta\varphi_{\lambda,\lambda'}(X) = \beta_{\lambda,\lambda'}.$$

Corollary 1. *Let* λ, λ' *be two probability distributions and C a constant such that*

$$\text{(5)} \qquad \varphi_{\lambda,\lambda'}(x) = \begin{cases} 1 & \text{if } \int_{\omega'} p_\theta(x)\, d\lambda'(\theta) > C \int_\omega p_\theta(x)\, d\lambda(\theta) \\ \gamma & \text{if } \int_{\omega'} p_\theta(x)\, d\lambda'(\theta) = C \int_\omega p_\theta(x)\, d\lambda(\theta) \\ 0 & \text{if } \int_{\omega'} p_\theta(x)\, d\lambda'(\theta) < C \int_\omega p_\theta(x)\, d\lambda(\theta) \end{cases}$$

is a size α *test for testing that the density of X is* $\int_\omega p_\theta(x)\, d\lambda(\theta)$ *and such that*

$$\text{(6)} \qquad \lambda(\omega_0) = \lambda'(\omega_0') = 1$$

where

$$\omega_0 = \{\theta: \theta \in \omega \text{ and } E_\theta\varphi_{\lambda,\lambda'}(X) = \sup_{\theta' \in \omega} E_{\theta'}\varphi_{\lambda,\lambda'}(X)\}$$

$$\omega_0' = \{\theta: \theta \in \omega' \text{ and } E_\theta\varphi_{\lambda,\lambda'}(X) = \inf_{\theta' \in \omega'} E_{\theta'}\varphi_{\lambda,\lambda'}(X)\}.$$

Then the conclusions of Theorem 1 *hold.*

Proof. If h, h', and $\beta_{\lambda,\lambda'}$ are defined as in Theorem 1, the assumptions imply that $\varphi_{\lambda,\lambda'}$ is a most powerful level α test for testing h against h', that

$$\sup_{\omega} E_\theta \varphi_{\lambda,\lambda'}(X) = \int_{\omega} E_\theta \varphi_{\lambda,\lambda'}(X)\, d\lambda(\theta) = \alpha,$$

and that

$$\inf_{\omega'} E_\theta \varphi_{\lambda,\lambda'}(X) = \int_{\omega'} E_\theta \varphi_{\lambda,\lambda'}(X)\, d\lambda'(\theta) = \beta_{\lambda,\lambda'}.$$

Condition (4) is thus satisfied and Theorem 1 applies.

Suppose that the sets Ω_H, Ω_I, and Ω_K are defined in terms of a nonnegative function d, which is a measure of the distance of θ from H, by

$$\Omega_H = \{\theta: d(\theta) = 0\}; \quad \Omega_I = \{\theta: 0 < d(\theta) < \Delta\}; \quad \Omega_K = \{\theta: d(\theta) \geqq \Delta\}.$$

Suppose also that the power function of any test is continuous in θ. In the limit as $\Delta = 0$, there is no indifference zone. Then Ω_K becomes the set $\{\theta: d(\theta) > 0\}$ and the infimum of $\beta(\theta)$ over Ω_K is $\leqq \alpha$ for any level α test. This infimum is therefore maximized by any test satisfying $\beta(\theta) \geqq \alpha$ for all $\theta \in \Omega_K$, that is, by any unbiased test, so that unbiasedness is seen to be a limiting form of the maximin criterion. A more useful limiting form, since it will typically lead to a unique test, is given by the following definition. A test φ_0 is said to *maximize the minimum power locally** if, given any other test φ, there exists Δ_0 such that

$$\inf_{\omega_\Delta} \beta_{\varphi_0}(\theta) \geqq \inf_{\omega_\Delta} \beta_{\varphi}(\theta) \quad \text{for all} \quad 0 < \Delta < \Delta_0 \tag{7}$$

where ω_Δ is the set of θ's for which $d(\theta) \geqq \Delta$.

2. EXAMPLES

In Chapter 3 it was shown for a family of probability densities depending on a real parameter θ that a UMP test exists for testing $H: \theta \leqq \theta_0$ against $\theta > \theta_0$ provided for all $\theta < \theta'$ the ratio $p_{\theta'}(x)/p_\theta(x)$ is a monotone function of some real-valued statistic. This assumption, although satisfied for a one-parameter exponential family, is quite restrictive, and a UMP test of H will in fact exist only rarely. A more general approach is furnished by the formulation of the preceding section. If the indifference zone is the set of θ's with $\theta_0 < \theta < \theta_1$, the problem becomes that of maximizing the minimum power over the class of alternatives $\omega': \theta \geqq \theta_1$. Under appropriate assumptions, one would expect the least favorable distributions λ and λ' of Theorem 1 to assign probability 1 to the points θ_0 and θ_1, and hence the maximin test to be given by the rejection region

* For a local optimum property not involving the choice of a distance function d see Problem 4.

$p_{\theta_1}(x)/p_{\theta_0}(x) > C$. The following lemma gives sufficient conditions for this to be the case.

Lemma 1. *Let $X_1, \cdots, X_n$ be identically and independently distributed with probability density $f_\theta(x)$ where θ and x are real-valued, and suppose that for any $\theta < \theta'$ the ratio $f_{\theta'}(x)/f_\theta(x)$ is a nondecreasing function of x. Then the level α test φ of H which maximizes the minimum power over ω' is given by*

$$\varphi(x_1, \cdots, x_n) = \begin{cases} 1 & \text{if } r(x_1, \cdots, x_n) > C \\ \gamma & \text{if } r(x_1, \cdots, x_n) = C \\ 0 & \text{if } r(x_1, \cdots, x_n) < C \end{cases} \tag{8}$$

where $r(x_1, \cdots, x_n) = f_{\theta_1}(x_1) \cdots f_{\theta_1}(x_n)/f_{\theta_0}(x_1) \cdots f_{\theta_0}(x_n)$ and where C and γ are determined by

$$E_{\theta_0}\varphi(X_1, \cdots, X_n) = \alpha. \tag{9}$$

Proof. The function $\varphi(x_1, \cdots, x_n)$ is nondecreasing in each of its arguments, so that by Lemma 2 of Chapter 3

$$E_\theta\varphi(X_1, \cdots, X_n) \leqq E_{\theta'}\varphi(X_1, \cdots, X_n)$$

when $\theta < \theta'$. Hence the power function of φ is monotone and φ is a level α test. Since $\varphi = \varphi_{\lambda,\lambda'}$ where λ and λ' are the distributions assigning probability 1 to the points θ_0 and θ_1, condition (4) is satisfied, which proves the desired result as well as the fact that the pair of distributions (λ, λ') is least favorable.

Example 1. Let θ be a location parameter so that $f_\theta(x) = g(x - \theta)$, and suppose for simplicity that $g(x) > 0$ for all x. We will show that a necessary and sufficient condition for $f_\theta(x)$ to have monotone likelihood ratio in x is that $-\log g$ is convex. The condition of monotone likelihood ratio in x,

$$\frac{g(x - \theta')}{g(x - \theta)} \leqq \frac{g(x' - \theta')}{g(x' - \theta)} \quad \text{for all} \quad x < x', \theta < \theta',$$

is equivalent to

$$\log g(x' - \theta) + \log g(x - \theta') \leqq \log g(x - \theta) + \log g(x' - \theta').$$

Since $x - \theta = t(x - \theta') + (1 - t)(x' - \theta)$ and $x' - \theta' = (1 - t)(x - \theta') + t(x' - \theta)$ where $t = (x' - x)/(x' - x + \theta' - \theta)$, a sufficient condition for this to hold is that the function $-\log g$ be convex. To see that this condition is also necessary, let $a < b$ be any real numbers and let $x - \theta' = a$, $x' - \theta = b$, and $x' - \theta' = x - \theta$. Then $x - \theta = \frac{1}{2}(x' - \theta + x - \theta') = \frac{1}{2}(a + b)$, and the condition of monotone likelihood ratio implies

$$\tfrac{1}{2}[\log g(a) + \log g(b)] \leqq \log g[\tfrac{1}{2}(a + b)].$$

Since $\log g$ is measurable, this in turn implies that $-\log g$ is convex.*

* See Sierpinski, "Sur les fonctions convexes mesurables," *Fundamenta Mathematicae*, Vol. 1 (1920), pp. 125–129.

Two distributions which satisfy the above condition [besides the normal distribution for which the resulting densities $p_\theta(x_1, \cdots, x_n)$ form an exponential family] are the *double exponential distribution* with

$$g(x) = \tfrac{1}{2}e^{-|x|}$$

and the *logistic distribution* whose cumulative distribution function is

$$G(x) = \frac{1}{1 + e^{-x}}$$

so that the density is $g(x) = e^{-x}/(1 + e^{-x})^2$.

Example 2. To consider the corresponding problem for a scale parameter, let $f_\theta(x) = \theta^{-1}h(x/\theta)$ where h is an even function. Without loss of generality one may then restrict x to be nonnegative, since the absolute values $|X_1|, \cdots, |X_n|$ form a set of sufficient statistics for θ. If $Y_i = \log X_i$ and $\eta = \log \theta$, the density of Y_i is

$$h(e^{y-\eta})e^{y-\eta}.$$

By Example 1, if $h(x) > 0$ for all $x \geqq 0$, a necessary and sufficient condition for $f_{\theta'}(x)/f_\theta(x)$ to be a nondecreasing function of x for all $\theta < \theta'$ is that $-\log [e^y h(e^y)]$ or equivalently $-\log h(e^y)$ is a convex function of y. An example in which this holds, in addition to the normal and double exponential distributions where the resulting densities form an exponential family, is the *Cauchy distribution* with

$$h(x) = \frac{1}{\pi}\frac{1}{1 + x^2}.$$

Since the convexity of $-\log h(y)$ implies that of $-\log h(e^y)$, it follows that if h is an even function and $h(x - \theta)$ has monotone likelihood ratio, so does $h(x/\theta)$. When h is the normal or double exponential distribution, this property of $h(x/\theta)$ follows therefore also from Example 1. That monotone likelihood ratio for the scale parameter family does not conversely imply the same property for the associated location parameter family is illustrated by the Cauchy distribution. The condition is therefore more restrictive for a location than for a scale parameter.

The chief difficulty in the application of Theorem 1 to specific problems is the necessity of knowing, or at least being able to guess correctly, a pair of least favorable distributions (λ, λ'). Guidance for obtaining these distributions is sometimes provided by invariance considerations. If there exists a group G of transformations of X such that the induced group $\bar{G}$ leaves both ω and ω' invariant, the problem is symmetric in the various θ's that can be transformed into each other under $\bar{G}$. It then seems plausible that unless λ and λ' exhibit the same symmetries, they will make the statistician's task easier, and hence will not be least favorable.

Example 3. In the problem of paired comparisons considered in Example 7 of Chapter 6, the observations X_i $(i = 1, \cdots, n)$ are independent variables taking on the values 1 and 0 with probabilities p_i and $q_i = 1 - p_i$. The hypothesis H to be tested specifies the set ω: $\max p_i \leqq \frac{1}{2}$. Only alternatives with $p_i \geqq \frac{1}{2}$ for all i are considered, and as ω' we take the subset of those alternatives for which $\max p_i \geqq \frac{1}{2} + \delta$. One would expect λ to assign probability 1 to the point $p_1 = \cdots = p_n = \frac{1}{2}$, and λ' to assign positive probability only to the n points $(p_1, \cdots, p_n)$ which have $n - 1$ coordinates equal to $\frac{1}{2}$ and the remaining coordinate equal to $\frac{1}{2} + \delta$. Because of the symmetry with regard to the n variables it seems plausible that λ' should assign equal probability $1/n$ to each of these n points. With these choices, the test $\varphi_{\lambda,\lambda'}$ rejects when

$$\sum_{i=1}^{n} \left(\frac{\frac{1}{2} + \delta}{\frac{1}{2} - \delta}\right)^{x_i} > C.$$

This is equivalent to

$$\sum_{i=1}^{n} x_i > C,$$

which had previously been seen to be UMP invariant for this problem. Since the critical function $\varphi_{\lambda,\lambda'}(x_1, \cdots, x_n)$ is nondecreasing in each of its arguments, it follows from Lemma 2 of Chapter 3 that $p_i \leqq p_i'$ for $i = 1, \cdots, n$ implies

$$E_{p_1,\ldots,p_n}\varphi_{\lambda,\lambda'}(X_1, \cdots, X_n) \leqq E_{p_1',\ldots,p_n'}\varphi_{\lambda,\lambda'}(X_1, \cdots, X_n)$$

and hence the conditions of Theorem 1 are satisfied.

Example 4. Let $X = (X_1, \cdots, X_n)$ be a sample from $N(\xi, \sigma^2)$, and consider the problem of testing H: $\sigma = \sigma_0$ against the set of alternatives ω': $\sigma \leqq \sigma_1$ or $\sigma \geqq \sigma_2$ $(\sigma_1 < \sigma_0 < \sigma_2)$. This problem remains invariant under the transformations $X_i' = X_i + c$ which in the parameter space induce the group $\bar{G}$ of transformations $\xi' = \xi + c$, $\sigma' = \sigma$. One would therefore expect the least favorable distribution λ over the line ω: $-\infty < \xi < \infty$, $\sigma = \sigma_0$, to be invariant under $\bar{G}$. Such invariance implies that λ assigns to any interval a measure proportional to the length of the interval. Hence λ cannot be a probability measure and Theorem 1 is not directly applicable. The difficulty can be avoided by approximating λ by a sequence of probability distributions, in the present case for example by the sequence of normal distributions $N(0, k)$, $k = 1, 2, \cdots$.

In the particular problem under consideration, it happens that there also exist least favorable distributions λ and λ', which are true probability distributions and therefore not invariant. These distributions can be obtained by an examination of the corresponding one-sided problem in Chapter 3, Section 9, as follows. On ω, where the only variable is ξ, the distribution λ of ξ is taken as the normal distribution with an arbitrary mean ξ_1 and with variance $(\sigma_2^2 - \sigma_0^2)/n$. Under λ' all probability should be concentrated on the two lines $\sigma = \sigma_1$ and $\sigma = \sigma_2$ in the (ξ, σ)-plane, and we put $\lambda' = p\lambda_1' + q\lambda_2'$ where λ_1' is the normal distribution with mean ξ_1 and variance $(\sigma_2^2 - \sigma_1^2)/n$ while λ_2' assigns probability 1 to the point (ξ_1, σ_2). A computation analogous to that carried out in Chapter 3, Section 9, then shows the acceptance region to be given by

$$\frac{\dfrac{p}{\sigma_1^{n-1}\sigma_2} \exp\left[\dfrac{-1}{2\sigma_1^2}\Sigma(x_i - \bar{x})^2 - \dfrac{n}{2\sigma_2^2}(\bar{x} - \xi_1)^2\right] + \dfrac{q}{\sigma_2^n} \exp\left[\dfrac{-1}{2\sigma_2^2}\left[\Sigma(x_i - \bar{x})^2 + n(\bar{x} - \xi_1)^2\right]\right]}{\dfrac{1}{\sigma_0^{n-1}\sigma_2} \exp\left[\dfrac{-1}{2\sigma_0^2}\Sigma(x_i - \bar{x})^2 - \dfrac{n}{2\sigma_2^2}(\bar{x} - \xi_1)^2\right]} < C$$

which is equivalent to

$$C_1 \leqq \Sigma(x_i - \bar{x})^2 \leqq C_2.$$

The probability of this inequality is independent of ξ, and hence C_1 and C_2 can be determined so that the probability of acceptance is $1 - \alpha$ when $\sigma = \sigma_0$, and is equal for the two values $\sigma = \sigma_1$ and $\sigma = \sigma_2$.

It follows from Section 7 of Chapter 3 that there exist p and C which lead to these values of C_1 and C_2 and that the above test satisfies the conditions of Corollary 1 with $\omega_0 = \omega$, and with ω_0' consisting of the two lines $\sigma = \sigma_1$ and $\sigma = \sigma_2$.

3. MAXIMIN TESTS AND INVARIANCE

When the problem of testing Ω_H against Ω_K remains invariant under a certain group of transformations, it seems reasonable to expect the existence of an invariant pair of least favorable distributions (or at least of sequences of distributions which in some sense are least favorable and invariant in the limit), and hence also of a maximin test which is invariant. This suggests the possibility of bypassing the somewhat cumbersome approach of the preceding sections. If it could be proved that for an invariant problem there always exists an invariant test that maximizes the minimum power over Ω_K, attention could be restricted to invariant tests; in particular, a UMP invariant test would then automatically have the desired maximin property. These speculations turn out to be correct for an important class of problems, although unfortunately not in general. To find out under what conditions they hold, it is convenient first to separate out the statistical aspects of the problem from the group theoretic ones by means of the following lemma.

Lemma 2. *Let $\mathcal{P} = \{P_\theta, \theta \in \Omega\}$ be a dominated family of distributions on $(\mathcal{X}, \mathcal{A})$, and let G be a group of transformations of $(\mathcal{X}, \mathcal{A})$, such that the induced group $\bar{G}$ leaves the two subsets Ω_H and Ω_K of Ω invariant. Suppose that for any critical function φ there exists an (almost) invariant critical function ψ satisfying*

$$\inf_{\bar{G}} E_{\bar{g}\theta}\varphi(X) \leqq E_\theta\psi(X) \leqq \sup_{\bar{G}} E_{\bar{g}\theta}\varphi(X) \tag{10}$$

for all $\theta \in \Omega$. Then if there exists a level α test φ_0 maximizing $\inf_{\Omega_K} E_\theta\varphi(X)$, there also exists an (almost) invariant test with this property.

Proof. Let $\inf_{\Omega_K} E_\theta\varphi_0(X) = \beta$, and let ψ_0 be an (almost) invariant test such that (10) holds with $\varphi = \varphi_0$, $\psi = \psi_0$. Then

$$E_\theta\psi_0(X) \leqq \sup_{\bar{G}} E_{\bar{g}\theta}\varphi_0(X) \leqq \alpha \quad \text{for all} \quad \theta \in \Omega_H$$

and

$$E_\theta\psi_0(X) \geqq \inf_{\bar{G}} E_{\bar{g}\theta}\varphi_0(X) \geqq \beta \quad \text{for all} \quad \theta \in \Omega_K$$

as was to be proved.

To determine conditions under which there exists an invariant or almost invariant test ψ satisfying (10), consider first the simplest case that G is a finite group, $G = \{g_1, \cdots, g_N\}$ say. If ψ is then defined by

$$\psi(x) = \frac{1}{N} \sum_{i=1}^{N} \varphi(g_i x), \tag{11}$$

it is clear that ψ is again a critical function, and that it is invariant under G. It also satisfies (10), since $E_\theta \varphi(gX) = E_{\bar{g}\theta} \varphi(X)$ so that $E_\theta \psi(X)$ is the average of a number of terms of which the first and last member of (10) are the minimum and maximum respectively.

An illustration of the finite case is furnished by Example 3. Here the problem remains invariant under the $n!$ permutations of the variables $(X_1, \cdots, X_n)$. Lemma 2 is applicable and shows that there exists an invariant test maximizing $\inf_{\Omega_K} E_\theta \varphi(X)$. Thus in particular the UMP invariant test obtained in Example 7 of Chapter 6 has this maximin property and therefore constitutes a solution of the problem.

The definition (11) suggests the possibility of obtaining $\psi(x)$ also in other cases by averaging the values of $\varphi(gx)$ with respect to a suitable probability distribution over the group G. To see what conditions would be required of this distribution, let $\mathscr{B}$ be a σ-field of subsets of G and ν a probability distribution over $(G, \mathscr{B})$. Disregarding measurability problems for the moment, let ψ be defined by

$$\psi(x) = \int \varphi(gx) \, d\nu(g). \tag{12}$$

Then $0 \leqq \psi \leqq 1$, and (10) is seen to hold by applying Fubini's theorem (Theorem 3 of Chapter 2) to the integral of ψ with respect to the distribution P_θ. For any $g_0 \in G$,

$$\psi(g_0 x) = \int \varphi(g g_0 x) \, d\nu(g) = \int \varphi(hx) \, d\nu^*(h)$$

where $h = gg_0$ and where ν^* is the measure defined by

$$\nu^*(B) = \nu(Bg_0^{-1}) \quad \text{for all} \quad B \in \mathscr{B},$$

into which ν is transformed by the transformation $h = gg_0$. Thus ψ will have the desired invariance property, $\psi(g_0 x) = \psi(x)$ for all $g_0 \in G$, if ν is *right invariant*, that is, if it satisfies

$$\nu(Bg) = \nu(B) \quad \text{for all} \quad B \in \mathscr{B}, g \in G. \tag{13}$$

The measurability assumptions required for the above argument are: (i) For any $A \in \mathscr{A}$, the set of pairs (x, g) with $gx \in A$ is measurable $\mathscr{A} \times \mathscr{B}$. This insures that the function ψ defined by (12) is again measurable. (ii) For any $B \in \mathscr{B}$, $g \in G$, the set Bg belongs to $\mathscr{B}$.

Example 5. If G is a finite group with elements $g_1, \cdots, g_N$, let $\mathscr{B}$ be the class of all subsets of G and ν the probability measure assigning probability $1/N$ to each of the N elements. Condition (13) is then satisfied and the definition (12) of ψ in this case reduces to (11).

Example 6. Consider the group G of orthogonal $n \times n$ matrices Γ, with the group product $\Gamma_1\Gamma_2$ defined as the corresponding matrix product. Each matrix can be interpreted as the point in n^2-dimensional Euclidean space whose coordinates are the n^2 elements of the matrix. The group then defines a subset of this space; the Borel subsets of G will be taken as the σ-field $\mathscr{B}$. To prove the existence of a right invariant probability measure over $(G, \mathscr{B})$,* we shall define a random orthogonal matrix whose probability distribution satisfies (13) and is therefore the required measure. With any nonsingular matrix $x = (x_{ij})$, associate the orthogonal matrix $y = f(x)$ obtained by applying the following Gram-Schmidt orthogonalization process to the n row vectors $x_i = (x_{i1}, \cdots, x_{in})$ of x: y_1 is the unit vector in the direction of x_1; y_2 the unit vector in the plane spanned by x_1 and x_2, which is orthogonal to y_1 and forms an acute angle with x_2; etc. Let $y = (y_{ij})$ be the matrix whose ith row is y_i.
Suppose now that the variables $X_{ij}(i, j = 1, \cdots, n)$ are independently distributed as $N(0, 1)$, let X denote the random matrix (X_{ij}), and let $Y = f(X)$. To show that the distribution of the random orthogonal matrix Y satisfies (13), consider any fixed orthogonal matrix Γ and any fixed set $B \in \mathscr{B}$. Then $P\{Y \in B\Gamma\} = P\{Y\Gamma' \in B\}$ and from the definition of f it is seen that $Y\Gamma' = f(X\Gamma')$. Since the n^2 elements of the matrix $X\Gamma'$ have the same joint distribution as those of the matrix X, the matrices $f(X\Gamma')$ and $f(X)$ also have the same distribution, as was to be proved.

Examples 5 and 6 are sufficient for the applications to be made here. General conditions for the existence of an invariant probability measure, of which these examples are simple special cases, are given in the theory of Haar measure.†

4. THE HUNT-STEIN THEOREM

Invariant measures exist (and are essentially unique) for a large class of groups, but unfortunately they are frequently not finite and hence cannot be taken to be probability measures. The situation is similar and related to that of the nonexistence of a least favorable pair of distributions in Theorem 1. There it is usually possible to overcome the difficulty by considering instead a sequence of distributions, which has the desired

* A more detailed discussion of this invariant measure is given by James, "Normal multivariate analysis and the orthogonal group," *Ann. Math. Stat.*, Vol. 25 (1954), pp. 40–75.

† This is treated for example in the books by Montgomery and Zippin, *Topological Transformation Groups*, New York, Interscience Publishers, 1955, Chapters I, II, and by Halmos, *Measure Theory*, New York, D. Van Nostrand Co., 1950, Chapters XI, XII.

property in the limit. Analogously we shall now generalize the construction of ψ as an average with respect to a right invariant probability distribution, by considering a sequence of distributions over G, which are approximately right invariant for n sufficiently large.

Let $\mathscr{P} = \{P_\theta, \theta \in \Omega\}$ be a family of distributions over a Euclidean space $(\mathscr{X}, \mathscr{A})$ dominated by a σ-finite measure μ, and let G be a group of transformations of $(\mathscr{X}, \mathscr{A})$ such that the induced group $\bar{G}$ leaves Ω invariant.

Theorem 2. (*Hunt–Stein.*) *Let $\mathscr{B}$ be a σ-field of subsets of G such that for any $A \in \mathscr{A}$ the set of pairs (x, g) with $gx \in A$ is in $\mathscr{A} \times \mathscr{B}$ and for any $B \in \mathscr{B}$ and $g \in G$ the set Bg is in $\mathscr{B}$. Suppose that there exists a sequence of probability distributions ν_n over $(G, \mathscr{B})$ which is asymptotically right invariant in the sense that for any $g \in G$, $B \in \mathscr{B}$*

$$\lim_{n\to\infty} |\nu_n(Bg) - \nu_n(B)| = 0. \tag{14}$$

Then given any critical function φ, there exists a critical function ψ which is almost invariant and satisfies (10).

Proof. Let

$$\psi_n(x) = \int \varphi(gx)\, d\nu_n(g),$$

which as before is measurable and between 0 and 1. By the weak compactness theorem (Theorem 3 of the Appendix) there exists a subsequence $\{\psi_{n_i}\}$ and a measurable function ψ between 0 and 1 satisfying

$$\lim_{i\to\infty} \int \psi_{n_i} p\, d\mu = \int \psi p\, d\mu$$

for all μ-integrable functions p, so that in particular

$$\lim_{i\to\infty} E_\theta \psi_{n_i}(X) = E_\theta \psi(X)$$

for all $\theta \in \Omega$. By Fubini's theorem

$$E_\theta \psi_{n_i}(X) = \int [E_\theta \varphi(gX)]\, d\nu_{n_i}(g) = \int E_{\bar{g}\theta} \varphi(X)\, d\nu_{n_i}(g)$$

so that

$$\inf_{\bar{G}} E_{\bar{g}\theta}\varphi(X) \leqq E_\theta \psi_{n_i}(X) \leqq \sup_{\bar{G}} E_{\bar{g}\theta}\varphi(X)$$

and ψ satisfies (10).

In order to prove that ψ is almost invariant we shall now show that for all x and g,

$$\psi_{n_i}(gx) - \psi_{n_i}(x) \to 0. \tag{15}$$

By the bounded convergence theorem [Theorem 1(ii) of Chapter 2] this will imply that

$$\int_A [\psi_{n_i}(gx) - \psi_{n_i}(x)]\, dP_\theta(x) \to 0$$

for all $\theta \in \Omega$ and $A \in \mathcal{A}$, and hence that $\psi(gx) = \psi(x)$ (a.e. $\mathcal{P}$) as was to be proved.

For fixed x and any integer m, let G be partitioned into the mutually exclusive sets

$$B_k = \left\{h \in G: a_k < \varphi(hx) \leqq a_k + \frac{1}{m}\right\}, \qquad k = 0, \cdots, m$$

where $a_k = (k-1)/m$. In particular, B_0 is the set $\{h \in G: \varphi(hx) = 0\}$. It is seen from the definition of the sets B_k that

$$\sum_{k=0}^{m} a_k \nu_{n_i}(B_k) \leqq \sum_{k=0}^{m} \int_{B_k} \varphi(hx)\, d\nu_{n_i}(h) \leqq \sum_{k=0}^{m} \left(a_k + \frac{1}{m}\right) \nu_{n_i}(B_k)$$
$$\leqq \sum_{k=0}^{m} a_k \nu_{n_i}(B_k) + \frac{1}{m}$$

and analogously that

$$\left|\sum_{k=0}^{m} \int_{B_k g^{-1}} \psi(hgx)\, d\nu_{n_i}(h) - \sum_{k=0}^{m} a_k \nu_{n_i}(B_k g^{-1})\right| \leqq \frac{1}{m}$$

from which it follows that

$$|\psi_{n_i}(gx) - \psi_{n_i}(x)| \leqq \Sigma\, |a_k| \cdot |\nu_{n_i}(B_k g^{-1}) - \nu_{n_i}(B_k)| + \frac{2}{m}.$$

By (14) the first term of the right-hand side tends to zero as i tends to infinity, and this completes the proof.

When there exists a right invariant measure ν over G, and a sequence of subsets G_n of G with $G_n \subseteq G_{n+1}$, $\bigcup G_n = G$, and $\nu(G_n) = c_n < \infty$, it is suggestive to take for the probability measures ν_n of Theorem 2 the measures ν/c_n truncated on G_n. This leads to the desired result in the example below. On the other hand, there are cases in which there exists such a sequence of subsets of G_n but no invariant test satisfying (10) and hence no sequence ν_n satisfying (14).

Example 7. Let $x = (x_1, \cdots, x_n)$, $\mathcal{A}$ be the class of Borel sets in n-space, and G the group of translations $(x_1 + g, \cdots, x_n + g)$, $-\infty < g < \infty$. The elements of G can be represented by the real numbers, and the group product gg' is then the sum $g + g'$. If $\mathcal{B}$ is the class of Borel sets on the real line, the measurability assumptions of Theorem 2 are satisfied. Let ν be Lebesgue measure, which is clearly invariant under G, and define ν_n to be the uniform distribution on the interval $I(-n, n) = \{g: -n \leq g \leq n\}$. Then for all $B \in \mathscr{B}, g \in G$,

$$|\nu_n(B) - \nu_n(Bg)| = \frac{1}{2n} |\nu[B \cap I(-n, n)] - \nu [B \cap I(-n - g, n - g)]| \leq \frac{|g|}{2n}$$

so that (14) is satisfied.

This argument also covers the group of scale transformations $(ax_1, \cdots, ax_n)$, $0 < a < \infty$, which can be transformed into the translation group by taking logarithms.

The reduction of the maximin problem can be carried out in steps under the assumptions of Theorem 2, Chapter 6. Suppose that the problem remains invariant under two groups D and E, and denote by $y = s(x)$ a maximal invariant with respect to D and by E^* the group defined in Theorem 2, Chapter 6, which E induces in y-space. If D and E^* satisfy the conditions of the Hunt-Stein theorem, it follows first that there exists a maximin test depending only on $y = s(x)$, and then that there exists a maximin test depending only on a maximal invariant $z = t(y)$ under E^*.

Example 8. Consider a univariate linear hypothesis in the canonical form in which $Y_1, \cdots, Y_n$ are independently distributed as $N(\eta_i, \sigma^2)$, where it is given that $\eta_{s+1} = \cdots = \eta_n = 0$, and where the hypothesis to be tested is $\eta_1 = \cdots = \eta_r = 0$. It was shown in Section 1 of Chapter 7 that this problem remains invariant under certain groups of transformations and that with respect to these groups there exists a UMP invariant test. The groups involved are the group of orthogonal transformations, translation groups of the kind considered in Example 7, and a group of scale changes. Since each of these satisfies the assumptions of the Hunt-Stein theorem, and since they leave invariant the problem of maximizing the minimum power over the set of alternatives

$$\sum_{i=1}^{r} \eta_i^2/\sigma^2 \geq \psi_1^2 \qquad (\psi_1 > 0), \tag{16}$$

it follows that the UMP invariant test of Chapter 7 is also the solution of this maximin problem. It is also seen slightly more generally that the test which is UMP invariant under the same groups for testing

$$\sum_{i=1}^{r} \eta_i^2/\sigma^2 \leq \psi_0^2$$

(Problem 4 of Chapter 7) maximizes the minimum power over the alternatives (16) for $\psi_0 < \psi_1$.

Example 9. (Stein.) Let G be the group of all nonsingular linear transformations of p-space. That for $p > 1$ this does not satisfy the conditions of Theorem 2 is shown by the following problem, which is invariant under G but for which the UMP invariant test does not maximize the minimum power. Generalizing Example 10 of Chapter 6, let $X = (X_1, \cdots, X_p)$, $Y = (Y_1, \cdots, Y_p)$ be independently distributed according to p-variate normal distributions with zero means and nonsingular covariance matrices $E(X_i X_j) = \sigma_{ij}$ and $E(Y_i Y_j) = \Delta\sigma_{ij}$, and let H: $\Delta \leqq \Delta_0$ be tested against $\Delta \geqq \Delta_1(\Delta_0 < \Delta_1)$, the σ_{ij} being unknown.

This problem remains invariant if the two vectors are subjected to any common nonsingular transformation, and since with probability 1 this group is transitive over the sample space, the UMP invariant test is trivially $\varphi(x, y) \equiv \alpha$. The maximin power against the alternatives $\Delta \geqq \Delta_1$ that can be achieved by invariant tests is therefore α. On the other hand, the test with rejection region $Y_1^2/X_1^2 > C$ has a strictly increasing power function $\beta(\Delta)$, whose minimum over the set of alternatives $\Delta \geqq \Delta_1$ is $\beta(\Delta_1) > \beta(\Delta_0) = \alpha$.

It is a remarkable feature of Theorem 2 that its assumptions concern only the group G and not the distributions P_θ. When these assumptions hold for a certain G, it follows from (10) as in the proof of Lemma 2 that for any testing problem which remains invariant under G and possesses a UMP invariant test, this test maximizes the minimum power over any invariant class of alternatives. Suppose conversely that a UMP invariant test under G has been shown in a particular problem not to maximize the minimum power, as was the case for the group of linear transformations in Example 9. Then the assumptions of Theorem 2 cannot be satisfied. However, this does not rule out the possibility that for another problem remaining invariant under G, the UMP invariant test may maximize the minimum power. Whether or not it does is no longer a property of the group alone but will in general depend also on the particular distributions.

Consider in particular the problem of testing $H: \xi_1 = \cdots = \xi_p = 0$ on the basis of a sample $(X_{\alpha 1}, \cdots, X_{\alpha p})$, $\alpha = 1, \cdots, n$, from a p-variate normal distribution with mean $E(X_{\alpha i}) = \xi_i$ and common covariance matrix $(\sigma_{ij}) = (a_{ij})^{-1}$. This was seen in Section 10 of Chapter 7 to be invariant under a number of groups including that of all nonsingular linear transformations of p-space, and a UMP invariant test was found to exist. An invariant class of alternatives under these groups is

$$\Sigma\Sigma a_{ij}\xi_i\xi_j/\sigma^2 \geqq \psi_1^2. \tag{17}$$

Here Theorem 2 is not applicable, and whether the UMP invariant test maximizes the minimum power against the alternatives (17) is an open question.

5. MOST STRINGENT TESTS

One of the practical difficulties in the consideration of tests that maximize the minimum power over a class Ω_K of alternatives is the determination of an appropriate Ω_K. If no information is available on which to base the choice of this set and if a natural definition is not imposed by invariance arguments, a frequently reasonable definition can be given in terms of the power that can be achieved against the various alternatives. The *envelope power function* β_α^* was defined in Chapter 6, Problem 15, by

$$\beta_\alpha^*(\theta) = \sup \beta_\varphi(\theta), \tag{18}$$

where β_φ denotes the power of a test φ and where the supremum is taken over all level α tests of H. Thus $\beta_\alpha^*(\theta)$ is the maximum power that can be attained at level α against the alternative θ. (That it can be attained follows under mild restrictions from Theorem 3 of the Appendix.) If

$$S_\Delta^* = \{\theta: \beta_\alpha^*(\theta) = \Delta\},$$

then of two alternatives $\theta_1 \in S_{\Delta_1}^*$, $\theta_2 \in S_{\Delta_2}^*$, θ_1 can be considered closer to H, equidistant, or further away than θ_2 as Δ_1 is $<$, $=$, or $>\Delta_2$.

The idea of measuring the distance of an alternative from H in terms of the available information has been encountered before. If for example $X_1, \cdots, X_n$ is a sample from $N(\xi, \sigma^2)$, the problem of testing $H: \xi \leqq 0$ was discussed (Chapter 5, Section 2) both when the alternatives ξ are measured in absolute units and in σ-units. The latter possibility corresponds to the present proposal, since it follows from invariance considerations (Problem 15 of Chapter 6) that $\beta_\alpha^*(\xi, \sigma)$ is constant on the lines $\xi/\sigma =$ constant.

Fixing a value of Δ and taking as Ω_K the class of alternatives θ for which $\beta_\alpha^*(\theta) \geqq \Delta$, one can determine the test that maximizes the minimum power over Ω_K. Another possibility, which eliminates the need of selecting a value of Δ, is to consider for any test φ the difference $\beta_\alpha^*(\theta) - \beta_\varphi(\theta)$. This difference measures the amount by which the actual power $\beta_\varphi(\theta)$ falls short of the maximum power attainable. A test that minimizes

$$\sup_{\Omega - \omega} [\beta_\alpha^*(\theta) - \beta_\varphi(\theta)] \tag{19}$$

is said to be *most stringent*. Thus a test is most stringent if it minimizes its maximum shortcoming.

Let φ_Δ be a test that maximizes the minimum power over S_Δ^*, and hence minimizes the maximum difference between $\beta_\alpha^*(\theta)$ and $\beta_\varphi(\theta)$ over S_Δ^*. If φ_Δ happens to be independent of Δ, it is most stringent. This remark makes it possible to apply the results of the preceding sections to the determination of most stringent tests. Suppose that the problem of testing $H: \theta \in \omega$ against the alternatives $\theta \in \Omega - \omega$ remains invariant under a group G, that there exists a UMP almost invariant test φ_0 with respect to G, and that the assumptions of Theorem 2 hold. Since $\beta_\alpha^*(\theta)$ and hence the set S_Δ^* is invariant under $\bar{G}$ (Problem 15 of Chapter 6), it follows that φ_0 maximizes the minimum power over S_Δ^* for each Δ, and φ_0 is therefore most stringent.

As an example of this method consider the problem of testing H:

$p_1, \cdots, p_n \leqq 1/2$ against the alternative $K: p_i > 1/2$ for all i where p_i is the probability of success in the ith trial of a sequence of n independent trials. If X_i is 1 or 0 as the ith trial is a success or failure, the problem remains invariant under permutations of the X's, and the UMP invariant test rejects (Example 7 of Chapter 6) when $\Sigma X_i > C$. It now follows from the remarks above that this test is also most stringent.

Another illustration is furnished by the general univariate linear hypothesis. Here it follows from the discussion in Example 8 that the standard test for testing $H: \eta_1 = \cdots = \eta_r = 0$ or $H': \sum_{i=1}^r \eta_i^2/\sigma^2 \leqq \psi_0^2$ is most stringent.

The determination of most stringent tests for problems to which the invariance method is not applicable has not yet been carried out for many specific cases. The following is a class of problems for which they are easily obtained by a direct approach. Let the distributions of X constitute a one-parameter exponential family, the density of which is given by (12) of Chapter 3, and consider the hypothesis $H: \theta = \theta_0$. Then according as $\theta > \theta_0$ or $\theta < \theta_0$, the envelope power $\beta_\alpha^*(\theta)$ is the power of the UMP one-sided test for testing H against $\theta > \theta_0$ or $\theta < \theta_0$. Suppose that there exists a two-sided test φ_0 given by (3) of Chapter 4, such that

$$\sup_{\theta<\theta_0} [\beta_\alpha^*(\theta) - \beta_{\varphi_0}(\theta)] = \sup_{\theta>\theta_0} [\beta_\alpha^*(\theta) - \beta_{\varphi_0}(\theta)], \tag{20}$$

and that the supremum is attained on both sides, say at points $\theta_1 < \theta_0 < \theta_2$. If $\beta_{\varphi_0}(\theta_i) = \beta_i$, $i = 1, 2$, an application of the fundamental lemma [Theorem 5(iii) of Chapter 3] to the three points $\theta_1, \theta_2, \theta_0$ shows that among all tests φ with $\beta_\varphi(\theta_1) \geqq \beta_1$ and $\beta_\varphi(\theta_2) \geqq \beta_2$, only φ_0 satisfies $\beta_\varphi(\theta_0) \leqq \alpha$. For any other level α test, therefore, either $\beta_\varphi(\theta_1) < \beta_1$ or $\beta_\varphi(\theta_2) < \beta_2$, and it follows that φ_0 is the unique most stringent test. The existence of a test satisfying (20) can be proved by a continuity consideration [with respect to variation of the constants C_i and γ_i which define the boundary of the test (3) of Chapter 4] from the fact that for the UMP one-sided test against the alternatives $\theta > \theta_0$ the right-hand side of (20) is zero and the left-hand side positive, while the situation is reversed for the other one-sided test.

6. PROBLEMS

Section 1

1. *Existence of maximin tests.* Let $(\mathscr{X}, \mathscr{A})$ be a Euclidean sample space and let the distributions P_θ, $\theta \in \Omega$ be dominated by a σ-finite measure over $(\mathscr{X}, \mathscr{A})$. For any mutually exclusive subsets Ω_H, Ω_K of Ω there exists a level α test maximizing (2).

[Let $\beta = \sup [\inf_{\Omega_K} E_\theta\varphi(X)]$ where the supremum is taken over all level α tests of H: $\theta \in \Omega_H$. Let φ_n be a sequence of level α tests such that $\inf_{\Omega_K} E_\theta\varphi_n(X)$ tends to β. If φ_{n_i} is a subsequence and φ a test (guaranteed by Theorem 3 of the Appendix) such that $E_\theta\varphi_{n_i}(X)$ tends to $E_\theta\varphi(X)$ for all $\theta \in \Omega$, then φ is a level α test and $\inf_{\Omega_K} E_\theta\varphi(X) = \beta$.]

2. *Locally most powerful tests.* Let d be a measure of the distance of an alternative θ from a given hypothesis H. A level α test φ_0 is said to be *locally most powerful* (LMP) if, given any other level α test φ, there exists Δ such that

$$\beta_{\varphi_0}(\theta) \geqq \beta_\varphi(\theta) \quad \text{for all} \quad \theta \quad \text{with} \quad 0 < d(\theta) < \Delta. \tag{21}$$

Suppose that θ is real-valued and that the power function of every test is continuously differentiable at θ_0.

(i) Then a LMP test of H: $\theta = \theta_0$ against $\theta > \theta_0$ exists and is defined by the fact that it maximizes $\beta'(\theta_0)$ among all level α tests of H.

(ii) A LMP test maximizes the minimum power locally provided its power function is bounded away from α for every set of alternatives which is bounded away from H.

(iii) Let $X_1, \cdots, X_n$ be a sample from a Cauchy distribution with unknown location parameter θ so that the joint density of the X's is $\pi^{-n}\prod_{i=1}^n[1 + (x_i - \theta)^2]^{-1}$. The LMP test for testing $\theta = 0$ against $\theta > 0$ at level $\alpha < \frac{1}{2}$ is not unbiased and hence does not maximize the minimum power locally.

[(iii) There exists M so large that any point with $x_i \geqq M$ for all $i = 1, \cdots, n$ lies in the acceptance region of the LMP test. Hence the power of the test tends to zero as θ tends to infinity.]

3. A test φ_0 is *LMP unbiased* if it is unbiased and if, given any other unbiased level α test φ, there exists Δ such that (21) holds. Suppose that θ is real-valued and that the power function of every test is twice continuously differentiable at θ_0. Then a LMP unbiased level α test of H: $\theta = \theta_0$ against $\theta \neq \theta_0$ exists and is defined by the fact that it maximizes $\beta''(\theta_0)$ among all unbiased level α tests of H.

Section 2

4. Let the distribution of X depend on the parameters $(\theta, \vartheta) = (\theta_1, \cdots, \theta_r, \vartheta_1, \cdots, \vartheta_s)$. A test of H: $\theta = \theta^0$ is *locally strictly unbiased* if for each ϑ, (a) $\beta_\varphi(\theta^0, \vartheta) = \alpha$, (b) there exists a θ-neighborhood of θ^0 in which $\beta_\varphi(\theta, \vartheta) > \alpha$ for $\theta \neq \theta^0$.

(i) Suppose that the first and second derivatives

$$\beta_\varphi^i(\vartheta) = \frac{\partial}{\partial\theta_i}\beta_\varphi(\theta, \vartheta)\bigg|_{\theta^0} \quad \text{and} \quad \beta_\varphi^{ij}(\vartheta) = \frac{\partial^2}{\partial\theta_i\,\partial\theta_j}\beta_\varphi(\theta, \vartheta)\bigg|_{\theta^0}$$

exist for all critical functions φ and all ϑ. Then a necessary and sufficient condition for φ to be locally strictly unbiased is that $\beta_\varphi^i(\vartheta) = 0$ for all i and ϑ, and that the matrix $(\beta_\varphi^{ij}(\vartheta))$ is positive definite for all ϑ.

(ii) A test of H is said to be of *type* E (*type* D if $s = 0$ so that there are no nuisance parameters) if it is locally strictly unbiased and among all tests with this property maximizes the determinant $|(\beta_\varphi^{ij})|$. (This determinant under the stated conditions turns out to be equal to the Gaussian curvature of the power surface at θ^0.) Then the test φ_0 given by (7) of Chapter 7 for testing the general linear univariate hypothesis (3) of Chapter 7 is of type E.

[(ii) With $\theta = (\eta_1, \cdots, \eta_r)$ and $\vartheta = (\eta_{r+1}, \cdots, \eta_s, \sigma)$ the test φ_0, by Problem 5 of Chapter 7, has the property of maximizing the surface integral

$$\int_S [\beta_\varphi(\eta, \sigma^2) - \alpha] \, dA$$

among all similar (and hence all locally unbiased) tests where $S = \{(\eta_1, \cdots, \eta_r): \sum_{i=1}^r \eta_i^2 = \rho^2\sigma^2\}$. Letting ρ tend to zero and utilizing the conditions

$$\beta_\varphi^i(\vartheta) = 0, \qquad \int_S \eta_i\eta_j \, dA = 0 \text{ for } i \neq j, \qquad \int_S \eta_i^2 \, dA = k(\rho\sigma),$$

one finds that φ_0 maximizes $\sum_{i=1}^r \beta_\varphi^{ii}(\eta, \sigma^2)$ among all locally unbiased tests. Since for any positive definite matrix, $|(\beta_\varphi^{ij})| \leqq \prod \beta_\varphi^{ii}$, it follows that for any strictly locally unbiased test φ,

$$|(\beta_\varphi^{ij})| \leqq \prod \beta_\varphi^{ii} \leqq [\Sigma \beta_\varphi^{ii}/r]^r \leqq [\Sigma \beta_{\varphi_0}^{ii}/r]^r = [\beta_{\varphi_0}^{11}]^r = |(\beta_{\varphi_0}^{ij})|.]$$

5. Let $Z_1, \cdots, Z_N$ be identically independently distributed according to a continuous distribution D, of which it is assumed only that it is symmetric about some (unknown) point. For testing the hypothesis H: $D(0) = \frac{1}{2}$, the sign test maximizes the minimum power against the alternatives K: $D(0) \leqq q (q < \frac{1}{2})$.

[A pair of least favorable distributions assign probability 1 respectively to the distributions $F \in H$, $G \in K$ with densities

$$f(x) = \frac{1-2q}{2(1-q)}\left(\frac{q}{1-q}\right)^{[|x|]}; \qquad g(x) = (1-2q)\left(\frac{q}{1-q}\right)^{|[x]|}$$

where for all x, positive, negative or zero, $[x]$ denotes the largest integer $\leqq x$.]

6. Let $f_\theta(x) = \theta g(x) + (1-\theta)h(x)$ with $0 \leqq \theta \leqq 1$. Then $f_\theta(x)$ satisfies the assumptions of Lemma 1 provided $g(x)/h(x)$ is a nondecreasing function of x.

7. Let $x = (x_1, \cdots, x_n)$ and let $g_\theta(x, \xi)$ be a family of probability densities depending on $\theta = (\theta_1, \cdots, \theta_r)$ and the real parameter ξ, and jointly measurable in x and ξ. For each θ, let $h_\theta(\xi)$ be a probability density with respect to a σ-finite measure ν such that $p_\theta(x) = \int g_\theta(x, \xi) h_\theta(\xi) \, d\nu(\xi)$ exists. We shall say that a function f of two arguments $u = (u_1, \cdots, u_r)$, $v = (v_1, \cdots, v_s)$ is nondecreasing in (u, v) if $f(u', v)/f(u, v) \leqq f(u', v')/f(u, v')$ for all (u, v) satisfying $u_i \leqq u_i'$; $v_j \leqq v_j' (i = 1, \cdots, r;\ j = 1, \cdots, s)$. Then $p_\theta(x)$ is nondecreasing in (x, θ) provided the product $g_\theta(x, \xi) h_\theta(\xi)$ is (a) nondecreasing in (x, θ) for each fixed ξ; (b) nondecreasing in (θ, ξ) for each fixed x; (c) nondecreasing in (x, ξ) for each fixed θ.

[Interpreting $g_\theta(x, \xi)$ as the conditional density of x given ξ, and $h_\theta(\xi)$ as the a priori density of ξ, let $\rho(\xi)$ denote the a posteriori density of ξ given x and let $\rho'(\xi)$ be defined analogously with θ' in place of θ. That $p_\theta(x)$ is nondecreasing in its two arguments is equivalent to

$$\int \frac{g_\theta(x', \xi)}{g_\theta(x, \xi)} \rho(\xi) \, d\nu(\xi) \leqq \int \frac{g_{\theta'}(x', \xi)}{g_{\theta'}(x, \xi)} \rho'(\xi) \, d\nu(\xi).$$

By (a) it is enough to prove that

$$D = \int [g_\theta(x', \xi)/g_\theta(x, \xi)][\rho'(\xi) - \rho(\xi)]\, d\nu(\xi) \geqq 0.$$

Let $S_- = \{\xi: \rho'(\xi)/\rho(\xi) < 1\}$ and $S_+ = \{\xi: \rho'(\xi)/\rho(\xi) \geqq 1\}$. By (ii) the set S_- lies entirely to the left of S_+. It follows from (iii) that there exists $a \leqq b$ such that

$$D = a\int_{S_-} [\rho'(\xi) - \rho(\xi)]\, d\nu(\xi) + b\int_{S_+} [\rho'(\xi) - \rho(\xi)]\, d\nu(\xi),$$

and hence that $D = (b - a)\int_{S_+} [\rho'(\xi) - \rho(\xi)]\, d\nu(\xi) \geqq 0$.]

8. (i) Let X have binomial distribution $b(p, n)$ and consider testing H: $p = p_0$ at level α against the alternatives $\Omega_K: p/q \leqq \frac{1}{2}p_0/q_0$ or $\geqq 2p_0/q_0$. For $\alpha = .05$ determine the smallest sample size for which there exists a test with power $\geqq .8$ against Ω_K if $p_0 = .1, .2, .3, .4, .5$.

(ii) Let $X_1, \cdots, X_n$ be independently distributed as $N(\xi, \sigma^2)$. For testing $\sigma = 1$ at level $\alpha = .05$, determine the smallest sample size for which there exists a test with power $\geqq .9$ against the alternatives $\sigma^2 \leqq \frac{1}{2}$ and $\sigma^2 \geqq 2$.

[See Problem 5 of Chapter 4.]

9. *Double exponential distribution.* Let $X_1, \cdots, X_n$ be a sample from the double exponential distribution with density $\frac{1}{2}e^{-|x-\theta|}$. The LMP test for testing $\theta \leqq 0$ against $\theta > 0$ is the sign test.

[The following proof is for the case that the level is of the form $\alpha = \sum_{k=0}^{m}\binom{n}{k}/2^n$, so that the level α sign test is nonrandomized. Let $R_k (k = 0, \cdots, n)$ be the subset of the sample space in which k of the X's are positive and $n - k$ are negative. Let $0 \leqq k < l < n$ and let S_k, S_l be subsets of R_k, R_l such that $P_0(S_k) = P_0(S_l) \neq 0$. Then it follows from a consideration of $P_\theta(S_k)$ and $P_\theta(S_l)$ for small θ that there exists Δ such that $P_\theta(S_k) < P_\theta(S_l)$ for $0 < \theta < \Delta$. Suppose now that the rejection region of a nonrandomized test of $\theta = 0$ against $\theta > 0$ does not consist of the upper tail of a sign test. Then it can be converted into a sign test of the same size by a finite number of steps, each of which consists in replacing an S_k by an S_l with $k < l$, and each of which therefore increases the power for θ sufficiently small. For randomized tests the argument is similar, with φ_k, φ_l replacing S_k, S_l.]

Section 4

10. Let $X = (X_1, \cdots, X_p)$ and $Y = (Y_1, \cdots, Y_p)$ be independently distributed according to p-variate normal distributions with zero means and covariance matrices $E(X_iX_j) = \sigma_{ij}$ and $E(Y_iY_j) = \Delta\sigma_{ij}$.

(i) The problem of testing H: $\Delta \leqq \Delta_0$ remains invariant under the group G of transformations $X^* = XA$, $Y^* = YA$ where $A = (a_{ij})$ is any nonsingular $p \times p$ matrix with $a_{ij} = 0$ for $i > j$, and there exists a UMP invariant test under G with rejection region $Y_1^2/X_1^2 > C$.

(ii) The test with rejection region $Y_1^2/X_1^2 > C$ maximizes the minimum power for testing $\Delta \leqq \Delta_0$ against $\Delta \geqq \Delta_1$ $(\Delta_0 < \Delta_1)$.

[(ii) That the Hunt-Stein theorem is applicable to G can be proved in steps by considering the group G_q of transformations $X'_q = \alpha_1 X_1 + \cdots + \alpha_q X_q$, $X'_i = X_i$ for $i = 1, \cdots, q-1, q+1, \cdots, p$, successively for $q = 1, \cdots, p-1$. Here $\alpha_q \neq 0$ since the matrix A is nonsingular if and only if $a_{ii} \neq 0$ for all i. The group product $(\gamma_1, \cdots, \gamma_q)$ of two such transformations $(\alpha_1, \cdots, \alpha_q)$ and $(\beta_1, \cdots, \beta_q)$ is given by $\gamma_1 = \alpha_1\beta_q + \beta_1, \gamma_2 = \alpha_2\beta_q + \beta_2, \cdots, \gamma_{q-1} = \alpha_{q-1}\beta_q + \beta_{q-1}$, $\gamma_q = \alpha_q\beta_q$ which shows G_q to be isomorphic to a group of scale changes (multiplication of all components by β_q) and translations (addition of $(\beta_1, \cdots, \beta_{q-1}, 0)$). The result now follows from the Hunt-Stein theorem and Example 7 since the assumptions of the Hunt-Stein theorem, except for the easily verifiable measurability conditions, concern only the abstract structure $(G, \mathscr{B})$, and not the specific realization of the elements of G as transformations of some space.]

11. Suppose that the problem of testing $\theta \in \Omega_H$ against $\theta \in \Omega_K$ remains invariant under G, that there exists a UMP almost invariant test φ_0 with respect to G, and that the assumptions of Theorem 2 hold. Then φ_0 maximizes $\inf_{\Omega_K}[w(\theta)E_\theta\varphi(X) + u(\theta)]$ for any weight functions $w(\theta) \geq 0$, $u(\theta)$ that are invariant under $\bar{G}$.

Section 5

12. *Existence of most stringent tests.* Under the assumptions of Problem 1 there exists a most stringent test for testing $\theta \in \Omega_H$ against $\theta \in \Omega - \Omega_H$.

13. Let $\{\Omega_\Delta\}$ be a class of mutually exclusive sets of alternatives such that the envelope power function is constant over each Ω_Δ and that $\bigcup \Omega_\Delta = \Omega - \Omega_H$, and let φ_Δ maximize the minimum power over Ω_Δ. If $\varphi_\Delta = \varphi$ is independent of Δ, then φ is most stringent for testing $\theta \in \Omega_H$.

14. Let $(Z_1, \cdots, Z_N) = (X_1, \cdots, X_m, Y_1, \cdots, Y_n)$ be distributed according to the joint density (56) of Chapter 5 and consider the problem of testing H: $\eta = \xi$ against the alternatives that the X's and Y's are independently normally distributed with common variance σ^2 and means $\eta \neq \xi$. Then the permutation test with rejection region $|\bar{Y} - \bar{X}| > C[T(Z)]$, the two-sided version of the test (55) of Chapter 5, is most stringent.

[Apply Problem 13 with each of the sets Ω_Δ consisting of two points (ξ_1, η_1, σ), (ξ_2, η_2, σ) such that

$$\xi_1 = \zeta - \frac{n}{m+n}\delta, \qquad \eta_1 = \zeta + \frac{m}{m+n}\delta; \qquad \xi_2 = \zeta + \frac{n}{m+n}\delta,$$

$$\eta_2 = \zeta - \frac{m}{m+n}\delta$$

for some ζ and δ.]

7. REFERENCES

The concepts and results of Section 1 are essentially contained in the minimax theory developed by Wald for general decision problems. An exposition of this theory and some of its applications is given in Wald's book (1950). The material of Sections 3–5, including in particular Lemma 2, Theorem 2, and Example 8, constitutes the main part of an unpublished paper by Hunt and Stein (1946).

Hunt, G., and C. Stein
(1946) "Most stringent tests of statistical hypotheses," unpublished.

Isaacson, S. L.
(1951) "On the theory of unbiased tests of simple statistical hypotheses specifying the values of two or more parameters," *Ann. Math. Stat.*, Vol. 22, pp. 217–234.
[Introduces type D and E tests.]

Kiefer, J.
(1958) "On the nonrandomized optimality and randomized nonoptimality of symmetrical designs," *Ann. Math. Stat.*, Vol. 29, pp. 675–699.
[Problem 4(ii).]

Lehmann, E. L.
(1947) "On families of admissible tests," *Ann. Math. Stat.*, Vol. 18, pp. 97–104.
[Last example of Section 5.]
(1950) "Some principles of the theory of testing hypotheses," *Ann. Math. Stat.*, Vol. 21, pp. 1–26.
[Theorem 1; Problem 10.]
(1955) "Ordered families of distributions," *Ann. Math. Stat.*, Vol. 26, pp. 399–419.
[Lemma 1; Problems 2, 7,* and 8.]

Lehmann, E. L., and C. Stein
(1949) "On the theory of some nonparametric hypotheses," *Ann. Math. Stat.*, Vol. 20, pp. 28–45.
[Problem 14.]

Neyman, J.
(1935) "Sur la vérification des hypothèses statistiques composées," *Bull. Soc. Math. France*, Vol. 63, pp. 246–266.
[Defines, and shows how to derive, tests of type B, that is, tests which are LMP among locally unbiased tests in the presence of nuisance parameters.]

Neyman, J., and E. S. Pearson
(1936, 1938) "Contributions to the theory of testing statistical hypotheses," *Stat. Res. Mem.*, Vol. I, pp. 1–37; Vol. II, pp. 25–57.
[Discusses tests of type A, that is, tests which are LMP among locally unbiased tests when no nuisance parameters are present.]

Ruist, Erik
(1954) "Comparison of tests for non-parametric hypotheses," *Arkiv Mat.*, Vol. 3, pp. 133–163.
[Problem 5.]

Schoenberg, I. J.
(1951) "On Pólya frequency functions, I." *J. Analyse Math.*, Vol. 1, pp. 331–374.
[Example 1.]

* This problem is a corrected version of Theorem 3 of the paper in question. I am grateful to Mr. R. Blumenthal for pointing out an error in the statement of this theorem in the paper.

Wald, Abraham

(1942) "On the principles of statistical inference," *Notre Dame Math. Lectures No. 1*, Notre Dame, Ind.

[Definition of most stringent tests.]

(1950) *Statistical Decision Functions*, New York, John Wiley & Sons.

Wolfowitz, J.

(1949) "The power of the classical tests associated with the normal distribution," *Ann. Math. Stat.*, Vol. 20, pp. 540–551.

[Proves that the standard tests of the univariate linear hypothesis and for testing the absence of multiple correlation are most stringent among all similar tests and possess certain related optimum properties.]

Appendix

1. EQUIVALENCE RELATIONS; GROUPS

A relation: $x \sim y$ among the points of a space $\mathscr{X}$, is an equivalence relation if it is reflexive, symmetric, and transitive, that is, if

(i) $x \sim x$ for all $x \in \mathscr{X}$;
(ii) $x \sim y$ implies $y \sim x$;
(iii) $x \sim y$, $y \sim z$ implies $x \sim z$.

Example 1. Consider a class of statistical decision procedures as a space, of which the individual procedures are the points. Then the relation: $\delta \sim \delta'$ if the procedures δ and δ' have the same risk function, is an equivalence relation. As another example consider all real-valued functions defined over the real line as points of a space. Then the relation: $f \sim g$ if $f(x) = g(x)$ a.e., is an equivalence relation.

Given an equivalence relation, let D_x denote the set of points of the space that are equivalent to x. Then $D_x = D_y$ if $x \sim y$, and $D_x \cap D_y = 0$ otherwise. Since by (i) each point of the space lies in at least one of the sets D_x, it follows that these sets, the *equivalence classes* defined by the relation $\sim$, constitute a partition of the space.

A set G of elements is called a *group* if it satisfies the following conditions.

(i) There is defined an operation, group multiplication, which with any two elements $a, b \in G$ associates an element c of G. The element c is called the product of a and b and is denoted by ab.

(ii) Group multiplication obeys the associative law

$$(ab)c = a(bc).$$

(iii) There exists an element $e \in G$, called the *identity*, such that

$$ae = ea = a \quad \text{for all} \quad a \in G.$$

(iv) To each element $a \in G$, there exists an element $a^{-1} \in G$, its *inverse*, such that

$$aa^{-1} = a^{-1}a = e.$$

Both the identity element and the inverse a^{-1} of any element a can be shown to be unique.

Example 2. The set of all $n \times n$ orthogonal matrices constitutes a group if matrix multiplication and inverse are taken as group multiplication and inverse respectively, and if the identity matrix is taken as the identity element of the group. With the same specification of the group operations, the class of all nonsingular $n \times n$ matrices also forms a group. On the other hand, the class of all $n \times n$ matrices fails to satisfy condition (iv).

If the elements of G are transformations of some space onto itself, with the group product ba defined as the result of applying first transformation a and following it by b, G is called a *transformation group*. Assumption (ii) is then satisfied automatically. For any transformation group defined over a space $\mathscr{X}$ the relation between points of $\mathscr{X}$:

$$x \sim y \quad \text{if there exists } a \in G \quad \text{such that} \quad y = ax,$$

is an equivalence relation. That it satisfies conditions (i), (ii), and (iii) required of an equivalence follows respectively from the defining properties (iii), (iv), and (i) of a group.

Let $\mathscr{C}$ be any class of 1 : 1 transformations of a space and let G be the class of all finite products $a_1^{\pm 1} a_2^{\pm 1} \cdots a_m^{\pm 1}$, with $a_1, \cdots, a_m \in \mathscr{C}$, $m = 1, 2, \cdots$, where each of the exponents can be $+1$ or -1 and where the elements $a_1, a_2, \cdots$ need not be distinct. Then it is easily checked that G is a group, and is in fact the smallest group containing $\mathscr{C}$.

2. CONVERGENCE OF DISTRIBUTIONS

When studying convergence properties of functions it is frequently convenient to consider a class of functions as a realization of an abstract space $\mathscr{F}$ of points f in which convergence of a sequence f_n to a limit f, denoted by $f_n \to f$, has been defined.

Example 3. Let μ be a measure over a measurable space $(\mathscr{X}, \mathscr{A})$.

(i) Let $\mathscr{F}$ be the class of integrable functions. Then f_n converges to f *in the mean* if*

$$\int |f_n - f| \, d\mu \to 0. \tag{1}$$

(ii) Let $\mathscr{F}$ be a uniformly bounded class of measurable functions. The sequence f_n is said to converge to f *weakly* if

$$\int f_n p \, d\mu \to \int fp \, d\mu \tag{2}$$

for all functions p that are integrable μ.

* Here and in the examples that follow, the limit f is not unique. More specifically, if $f_n \to f$, then $f_n \to g$ if and only if $f = g$ (a.e. μ). Putting $f \sim g$ when $f = g$ (a.e. μ), uniqueness can be obtained by working with the resulting equivalence classes of functions rather than with the functions themselves.

(iii) Let $\mathscr{F}$ be the class of measurable functions. Then f_n converges to f *pointwise* if

$$f_n(x) \to f(x) \quad \text{a.e. } \mu. \tag{3}$$

A subset $\mathscr{F}_0$ of $\mathscr{F}$ is *dense* in $\mathscr{F}$ if, given any $f \in \mathscr{F}$, there exists a sequence in $\mathscr{F}_0$ having f as its limit point. A space $\mathscr{F}$ is *separable* if there exists a countable dense subset of $\mathscr{F}$. A space $\mathscr{F}$ such that every sequence has a convergent subsequence whose limit point is in $\mathscr{F}$ is *compact.** A space $\mathscr{F}$ is a *metric space* if for every pair of points f, g in $\mathscr{F}$ there is defined a distance $d(f, g) \geqq 0$ such that

(i) $d(f, g) = 0$ if and only if $f = g$;

(ii) $d(f, g) = d(g, f)$;

(iii) $d(f, g) + d(g, h) \geqq d(f, h)$ for all f, g, h.

The space is *pseudometric* if (i) is replaced by

(i′) $d(f, f) = 0$ for all $f \in \mathscr{F}$.

A pseudometric space can be converted into a metric space by introducing the equivalence relation $f \sim g$ if $d(f, g) = 0$. The equivalence classes $F, G, \cdots$ then constitute a metric space with respect to the distance $D(F, G) = d(f, g)$ where $f \in F$, $g \in G$.

In any pseudometric space a natural convergence definition is obtained by putting $f_n \to f$ if $d(f_n, f) \to 0$.

Example 4. The space of integrable functions of Example 3(i) becomes a pseudometric space if we put

$$d(f, g) = \int |f - g| \, d\mu$$

and the induced convergence definition is that given by (1).

Example 5. Let $\mathscr{P}$ be a family of probability distributions over $(\mathscr{X}, \mathscr{A})$. Then $\mathscr{P}$ is a metric space with respect to the metric

$$d(P, Q) = \sup_{A \in \mathscr{A}} |P(A) - Q(A)| \tag{4}$$

Lemma 1. *If $\mathscr{F}$ is a separable pseudometric space then every subset of $\mathscr{F}$ is also separable.*

Proof. By assumption there exists a dense countable subset $\{f_n\}$ of $\mathscr{F}$. Let

$$S_{m,n} = \{f : d(f, f_n) < 1/m\}$$

and let A be any subset of $\mathscr{F}$. Select one element from each of the intersections $A \cap S_{m,n}$ that is nonempty, and denote this countable collection of elements by A_0. If a is any element of A and m any positive integer there exists an element f_{n_m} such that $d(a, f_{n_m}) < 1/m$. Therefore

* The term *compactness* is more commonly used for an alternative concept, which coincides with the one given here in metric spaces. The distinguishing term *sequential compactness* is then sometimes given to the notion defined here.

a belongs to S_{m,n_m}, the intersection $A \cap S_{m,n_m}$ is nonempty, and there exists therefore an element of A_0 whose distance to a is $<2/m$. This shows that A_0 is dense in A, and hence that A is separable.

Lemma 2. *A sequence f_n of integrable functions converges to f in the mean if and only if*

$$\int_A f_n\,d\mu \to \int_A f\,d\mu \quad \text{uniformly for} \quad A \in \mathscr{A}. \tag{5}$$

Proof. That (1) implies (5) is obvious since for all $A \in \mathscr{A}$

$$\left|\int_A f_n\,d\mu - \int_A f\,d\mu\right| \leqq \int |f_n - f|\,d\mu.$$

Conversely suppose that (5) holds and denote by A_n and A_n' the set of points x for which $f_n(x) > f(x)$ and $f_n(x) < f(x)$ respectively. Then

$$\int |f_n - f|\,d\mu = \int_{A_n} (f_n - f)\,d\mu - \int_{A_n'} (f_n - f)\,d\mu \to 0.$$

Lemma 3. *A sequence f_n of uniformly bounded functions converges to a bounded function f weakly if and only if*

$$\int_A f_n\,d\mu \to \int_A f\,d\mu \quad \text{for all} \quad A \quad \text{with} \quad \mu(A) < \infty. \tag{6}$$

Proof. That weak convergence implies (6) is seen by taking for p in (2) the indicator function of a set A, which is integrable if $\mu(A) < \infty$. Conversely (6) implies that (2) holds if p is any simple function $s = \Sigma a_i I_{A_i}$ with all the $\mu(A_i) < \infty$. Given any integrable function p there exists, by the definition of the integral, such a simple function s for which $\int |p - s|\,d\mu < \epsilon/3M$ where M is a bound on the $|f|$'s. We then have

$$\left|\int (f_n - f)p\,d\mu\right| \leqq \left|\int f_n(p - s)\,d\mu\right| + \left|\int f(s - p)\,d\mu\right| + \left|\int (f_n - f)s\,d\mu\right|.$$

The first two terms on the right-hand side are $<\epsilon/3$, and the third term tends to zero as n tends to infinity. Thus the left-hand side is $<\epsilon$ for n sufficiently large, as was to be proved.

Lemma 4.* *Let f and f_n, $n = 1, 2, \cdots$, be nonnegative integrable functions with*

$$\int f\,d\mu = \int f_n\,d\mu = 1.$$

Then pointwise convergence of f_n to f implies that $f_n \to f$ in the mean.

* Scheffé, "A useful convergence theorem for probability distributions," *Ann. Math. Stat.*, Vol. 18 (1947), pp. 434–438.

Proof. If $g_n = f_n - f$, then $g_n \geqq -f$, and the negative part $g_n^- = \max(-g_n, 0)$ satisfies $|g_n^-| \leqq f$. Since $g_n(x) \to 0$ (a.e. μ), it follows from Theorem 1(ii) of Chapter 2 that $\int g_n^- \, d\mu \to 0$, and $\int g_n^+ \, d\mu$ then also tends to zero since $\int g_n \, d\mu = 0$. Therefore $\int |g_n| \, d\mu = \int (g_n^+ + g_n^-) \, d\mu \to 0$, as was to be proved.

Let P and P_n, $n = 1, 2, \cdots$, be probability distributions over $(\mathscr{X}, \mathscr{A})$ with densities p_n and p with respect to μ. Consider the convergence definitions

(a) $p_n \to p$ (a.e. μ);

(b) $\int |p_n - p| \, d\mu \to 0$;

(c) $\int g p_n \, d\mu \to \int g p \, d\mu$ for all bounded measurable g;

and

(b′) $P_n(A) \to P(A)$ uniformly for all $A \in \mathscr{A}$;

(c′) $P_n(A) \to P(A)$ for all $A \in \mathscr{A}$.

Then Lemmas 2 and 4 together with a slight modification of Lemma 3 show that (a) implies (b) and (b) implies (c); and that (b) is equivalent to (b′) and (c) to (c′). It can further be shown that neither (a) and (b) nor (b) and (c) are equivalent.*

3. DOMINATED FAMILIES OF DISTRIBUTIONS

Let $\mathscr{M}$ be a family of measures defined over a measurable space $(\mathscr{X}, \mathscr{A})$. Then $\mathscr{M}$ is said to be *dominated* by a σ-finite measure μ defined over $(\mathscr{X}, \mathscr{A})$ if each member of $\mathscr{M}$ is absolutely continuous with respect to μ. The family $\mathscr{M}$ is said to be *dominated* if there exists a σ-finite measure dominating it. Actually, if $\mathscr{M}$ is dominated there always exists a finite dominating measure. For suppose that $\mathscr{M}$ is dominated by μ and that $\mathscr{X} = \bigcup A_i$ with $\mu(A_i)$ finite for all i. If the sets A_i are taken to be mutually exclusive, the measure ν defined by $\nu(A) = \Sigma \mu(A \cap A_i)/2^i \mu(A_i)$ also dominates $\mathscr{M}$ and is finite.

Theorem 1.† *A family $\mathscr{P}$ of probability measures over a Euclidean space $(\mathscr{X}, \mathscr{A})$ is dominated if and only if it is separable with respect to the metric* (4) *or equivalently with respect to the convergence definition*

$$P_n \to P \quad \text{if} \quad P_n(A) \to P(A) \quad \text{uniformly for} \quad A \in \mathscr{A}.$$

Proof. Suppose first that $\mathscr{P}$ is separable and that the sequence $\{P_n\}$

* Robbins, "Convergence of distributions," *Ann. Math. Stat.*, Vol. 19 (1948), pp. 72–76.

† Berger, "Remark on separable spaces of probability measures," *Ann. Math. Stat.*, Vol. 22 (1951), pp. 119–120.

is dense in $\mathscr{P}$, and let $\mu = \Sigma P_n/2^n$. Then $\mu(A) = 0$ implies $P_n(A) = 0$ for all n, and hence $P(A) = 0$ for all $P \in \mathscr{P}$. Conversely suppose that $\mathscr{P}$ is dominated by a measure μ which without loss of generality can be assumed to be finite. Then we must show that the set of integrable functions $dP/d\mu$ is separable with respect to the convergence definition (5) or, because of Lemma 2, with respect to convergence in the mean. It follows from Lemma 1 that it suffices to prove this separability for the class $\mathscr{F}$ of all functions f that are integrable μ. Since by the definition of the integral every integrable function can be approximated in the mean by simple functions, it is enough to prove this for the case that $\mathscr{F}$ is the class of all simple integrable functions. Any simple function can be approximated in the mean by simple functions taking on only rational values, so that it is sufficient to prove separability of the class of functions $\Sigma r_i I_{A_i}$ where the r's are rational and the A's are Borel sets, with finite μ-measure since the f's are integrable. It is therefore finally enough to take for $\mathscr{F}$ the class of functions I_A, which are indicator functions of Borel sets with finite measure. However, any such set can be approximated by finite unions of disjoint rectangles with rational end points. The class of all such unions is denumerable, and the associated indicator functions will therefore serve as the required countable dense subset of $\mathscr{F}$.

An examination of the proof shows that the Euclidean nature of the space $(\mathscr{X}, \mathscr{A})$ was used only to establish the existence of a countable number of sets $A_i \in \mathscr{A}$ such that for any $A \in \mathscr{A}$ with finite measure there exists a subsequence A_{i_j} with $\mu(A_{i_j}) \to \mu(A)$. This property holds quite generally for any σ-field $\mathscr{A}$, which has a *countable number of generators*, that is, for which there exists a countable number of sets B_i such that $\mathscr{A}$ is the smallest σ-field containing the B_i.* It follows that Theorem 1 holds for any σ-field with this property. Statistical applications of such σ-fields occur in sequential analysis, where the sample space $\mathscr{X}$ is the union $\mathscr{X} = \bigcup_i \mathscr{X}_i$ of Borel subsets $\mathscr{X}_i$ of i-dimensional Euclidean space. In these problems, $\mathscr{X}_i$ is the set of points $(x_1, \cdots, x_i)$ for which exactly i observations are taken. If $\mathscr{A}_i$ is the σ-field of Borel subsets of $\mathscr{X}_i$, one can take for $\mathscr{A}$ the σ-field generated by the $\mathscr{A}_i$, and since each $\mathscr{A}_i$ possesses a countable number of generators so does $\mathscr{A}$.

If $\mathscr{A}$ does not possess a countable number of generators, a somewhat weaker conclusion can be asserted. Two families of measures $\mathscr{M}$ and $\mathscr{N}$ are *equivalent* if $\mu(A) = 0$ for all $\mu \in \mathscr{M}$ implies $\nu(A) = 0$ for all $\nu \in \mathscr{N}$ and vice versa.

* A proof of this is given for example by Halmos, *Measure Theory*, New York, D. Van Nostrand Co., 1950. (Theorem B of Section 40.)

Theorem 2.† *A family $\mathcal{P}$ of probability measures is dominated by a σ-finite measure if and only if $\mathcal{P}$ has a countable equivalent subset.*

Proof. Suppose first that $\mathcal{P}$ has a countable equivalent subset $\{P_1, P_2, \cdots\}$. Then $\mathcal{P}$ is dominated by $\mu = \Sigma P_n/2^n$. Conversely let $\mathcal{P}$ be dominated by a σ-finite measure μ, which without loss of generality can be assumed to be finite. Let $\mathcal{Q}$ be the class of all probability measures Q of the form $\Sigma c_i P_i$, where $P_i \in \mathcal{P}$, the c's are positive, and $\Sigma c_i = 1$. The class $\mathcal{Q}$ is also dominated by μ, and we denote by q a fixed version of the density $dQ/d\mu$. We shall prove the fact, equivalent to the theorem, that there exists Q_0 in $\mathcal{Q}$, such that $Q_0(A) = 0$ implies $Q(A) = 0$ for all $Q \in \mathcal{Q}$.

Consider the class $\mathcal{C}$ of sets C in $\mathcal{A}$ for which there exists $Q \in \mathcal{Q}$ such that $q(x) > 0$ a.e. μ on C and $Q(C) > 0$. Let $\mu(C_i)$ tend to $\sup_{\mathcal{C}} \mu(C)$, let $q_i(x) > 0$ a.e. on C_i, and denote the union of the C_i by C_0. Then $q_0^*(x) = \Sigma c_i q_i(x)$ agrees a.e. with the density of $Q_0 = \Sigma c_i Q_i$ and is positive a.e. on C_0, so that $C_0 \in \mathcal{C}$. Suppose now that $Q_0(A) = 0$, let Q be any other member of $\mathcal{Q}$, and let $C = \{x: q(x) > 0\}$. Then $Q_0(A \cap C_0) = 0$, and therefore $\mu(A \cap C_0) = 0$ and $Q(A \cap C_0) = 0$. Also $Q(A \cap \tilde{C}_0 \cap \tilde{C}) = 0$. Finally, $Q(A \cap \tilde{C}_0 \cap C) > 0$ would lead to $\mu(C_0 \cup [A \cap \tilde{C}_0 \cap C]) > \mu(C_0)$ and hence to a contradiction of the relation $\mu(C_0) = \sup_{\mathcal{C}} \mu(C)$ since $A \cap \tilde{C}_0 \cap C$ and therefore $C_0 \cup [A \cap \tilde{C}_0 \cap C]$ belongs to $\mathcal{C}$.

4. THE WEAK COMPACTNESS THEOREM

The following theorem forms the basis for proving the existence of most powerful tests, most stringent tests, etc.

Theorem 3.‡ (*Weak compactness theorem.*) *Let μ be a σ-finite measure over a Euclidean space or more generally over any measurable space $(\mathcal{X}, \mathcal{A})$ for which $\mathcal{A}$ has a countable number of generators. Then the set of measurable functions ϕ with $0 \leqq \phi \leqq 1$ is compact with respect to the weak convergence* (2).

Proof. Given any sequence $\{\phi_n\}$, we must prove the existence of a subsequence $\{\phi_{n_i}\}$ and a function ϕ such that

$$\lim \int \phi_{n_i} p \, d\mu = \int \phi p \, d\mu$$

† Halmos and Savage, "Application of the Radon-Nikodym theorem to the theory of sufficient statistics," *Ann. Math. Stat.*, Vol. 20 (1948), pp. 225–241.

‡ Banach, *Théorie des opérations linéaires*, Warszawa, Fundusg Kultury Narodowej, 1932, p. 131.

for all integrable p. If μ^* is a finite measure equivalent to μ then p^* is integrable μ^* if and only if $p = (d\mu^*/d\mu)p^*$ is integrable μ, and $\int \phi p \, d\mu = \int \phi p^* \, d\mu^*$ for all ϕ. We may therefore assume without loss of generality that μ is finite. Let $\{p_n\}$ be a sequence of p's, which is dense in the p's with respect to convergence in the mean. The existence of such a sequence is guaranteed by Theorem 1 and the remark following it. If

$$\Phi_n(p) = \int \phi_n p \, d\mu$$

the sequence $\Phi_n(p)$ is bounded for each p. A subsequence Φ_{n_k} can be extracted such that $\Phi_{n_k}(p_m)$ converges for each p_m by the following diagonal process. Consider first the sequence of numbers $\{\Phi_n(p_1)\}$ which possesses a convergent subsequence $\Phi_{n_1'}(p_1), \Phi_{n_2'}(p_1), \cdots$. Next the sequence $\Phi_{n_1'}(p_2), \Phi_{n_2'}(p_2), \cdots$ has a convergent subsequence $\Phi_{n_1''}(p_2), \Phi_{n_2''}(p_2), \cdots$. Continuing in this way let $n_1 = n_1', n_2 = n_2'', n_3 = n_3''', \cdots$. Then $n_1 < n_2 < \cdots$, and the sequence $\{\Phi_{n_i}\}$ converges for each p_m. It follows from the inequality

$$\left|\int (\phi_{n_j} - \phi_{n_i}) p \, d\mu\right| \leqq \left|\int (\phi_{n_j} - \phi_{n_i}) p_m \, d\mu\right| + 2\int |p - p_m| \, d\mu$$

that $\Phi_{n_i}(p)$ converges for all p. Denote its limit by $\Phi(p)$, and define a set function Φ^* over $\mathscr{A}$ by putting

$$\Phi^*(A) = \Phi(I_A).$$

Then Φ^* is nonnegative and bounded since for all A, $\Phi^*(A) \leqq \mu(A)$. To see that it is also countably additive let $A = \bigcup A_k$ where the A_k are disjoint. Then $\Phi^*(A) = \lim \Phi^*_{n_i}(\bigcup A_k)$ and

$$\left|\int_{\bigcup A_k} \phi_{n_i} \, d\mu - \Sigma\Phi^*(A_k)\right| \leqq \left|\int_{\bigcup_{k=1}^{m} A_k} \phi_{n_i} \, d\mu - \sum_{k=1}^{m} \Phi^*(A_k)\right|$$

$$+ \left|\int_{\bigcup_{k=m+1}^{\infty} A_k} \phi_{n_i} d\mu - \sum_{k=m+1}^{\infty} \Phi^*(A_k)\right|.$$

Here the second term is to be taken as zero in the case of a finite sum $A = \bigcup_{k=1}^{m} A_k$, and otherwise does not exceed $2\mu(\bigcup_{k=m+1}^{\infty} A_k)$, which can be made arbitrarily small by taking m sufficiently large. For any fixed m the first term tends to zero as i tends to infinity. Thus Φ^* is a finite measure over $(\mathscr{X}, \mathscr{A})$. It is furthermore absolutely continuous with respect to μ, since $\mu(A) = 0$ implies $\Phi_{n_i}(I_A) = 0$ for all i, and therefore $\Phi(I_A) = \Phi^*(A) = 0$. We can now apply the Radon-Nikodym theorem to get

$$\Phi^*(A) = \int_A \phi \, d\mu \quad \text{for all} \quad A,$$

with $0 \leqq \phi \leqq 1$. We then have

$$\int_A \phi_{n_i} \, d\mu \to \int_A \phi \, d\mu \quad \text{for all} \quad A,$$

and weak convergence of the ϕ_{n_i} to ϕ follows from Lemma 3.

Author Index

Subject Index

Applied Probability and Statistics (Continued)

WHITTLE · Optimization under Constraints
WILLIAMS · Regression Analysis
WOLD and JUREEN · Demand Analysis
WONNACOTT and WONNACOTT · Introduction to Econometric Methods
YOUDEN · Statistical Methods for Chemists
ZELLNER · An Introduction to Bayesian Inference in Econometrics

Tracts on Probability and Statistics

BILLINGSLEY · Ergodic Theory and Information
BILLINGSLEY · Convergence of Probability Measures
CRAMÉR and LEADBETTER · Stationary and Related Stochastic Processes
HARDING and KENDALL · Stochastic Geometry
JARDINE and SIBSON · Mathematical Taxonomy
KENDALL and HARDING · Stochastic Analysis
KINGMAN · Regenerative Phenomena
RIORDAN · Combinatorial Identities
TAKÁCS · Combinatorial Methods in the Theory of Stochastic Processes

—